MODERN INDUSTRIAL/ELECTRICAL MOTOR CONTROLS

Operation, Installation, and Troubleshooting

MODERN INDUSTRIAL/ELECTRICAL MOTOR CONTROLS
Operation, Installation, and Troubleshooting

Thomas E. Kissell

Prentice Hall Career & Technology/Englewood Cliffs, NJ 07632

Library of Congress Cataloging-in-Publication Data

Kissell, Thomas E.
    Modern industrial electrical motor controls : operation, installation, and troubleshooting / Thomas E. Kissell.
       p.   cm.
    ISBN 0-13-596164-5
      1. Electric controllers.   2. Electric motors—Electronic control.
    3. Electric controllers—Maintenance and repair.   I. Title.
TK2851.K53   1990
621.46--dc20   89-28244
                                                CIP

Editorial/production supervision and
   interior design: *Eileen Margaret O'Sullivan*
Cover design: *Diane Saxe*
Manufacturing buyer: *Gina Chirco-Brennan*

©1993 Prentice Hall Career & Technology
A Paramount Communications Company
Englewood Cliffs, NJ 07632

All rights reserved. No part of this book may be
reproduced, in any form or by any means,
without permission in writing from the publisher.

Printed in the United States of America

10 9 8 7 6 5 4 3 2

ISBN 0-13-596164-5

Prentice-Hall International (UK) Limited, *London*
Prentice-Hall of Australia Pty. Limited, *Sydney*
Prentice-Hall Canada Inc., *Toronto*
Prentice-Hall Hispanoamericana, S.A., *Mexico*
Prentice-Hall of India Private Limited, *New Delhi*
Prentice-Hall of Japan, Inc., *Tokyo*
Simon & Schuster Asia Pte. Ltd., *Singapore*
Editora Prentice-Hall do Brasil, Ltda., *Rio de Janeiro*

*I would like to dedicate this text book to my
parents and my wife's parents; Jerry and Dorothy Kissell and
Harold and Mary Anne Wilhelm. Little did they know the time
and patience that they provided would lead to this. I would
also like to thank my wife Kathy for typing this manuscript
and helping to keep track of all of the diagrams and art.
Without her valuable help I would never attempt to write.*

Contents

ABOUT THE AUTHOR   xii

PREFACE   xiii

ACKNOWLEDGMENTS   xiv

1 OVERVIEW OF MOTOR CONTROLS   1

The Need for Motor Controls, 2
Control Circuits and Load Circuits, 4
Innovations in Control Systems, 6
Conclusions, 8

2 SAFETY IN WORKING WITH MOTOR CONTROL CIRCUITS   9

Personal Safety, 9
Electrical Safety, 10
Safety Conditions and Hazardous Locations, 13
Working Safely, 20

3 SYMBOLS & DIAGRAMS — 23

Electrical Symbols, 24
Electrical Diagrams, 33
Conclusions, 47

4 POWER DISTRIBUTION — 48

Generating Electricity with an Alternator, 49
Three-Phase Voltage, 50
Meeting Peak Demands, 50
Distributing Generated Power, 51
Transformers, 52
Fuses and Circuit Breakers, 60
Hardware for Power Distribution, 73

5 MANUAL CONTROL DEVICES — 83

Disconnect Switches, 84
Manual Motor Starters, 84
Manual Drum Switches, 103

6 MAGNETICS, SOLENOIDS, AND RELAYS — 108

Magnets, 108
Electromagnets, 110
Solenoids, 112
Relays, 121

7 CONTACTORS AND MOTOR STARTERS — 148

Contactors, 149
Magnetic Motor Starters, 167

8 PILOT DEVICES — 194

Pushbutton Switches, 197
Limit Switches, 201
Selector Switches, 210
Joystick Controls, 211
Foot Switches, 213
Float Switches, 214
Pressure Switches, 217

Flow Switches, 220
Temperature Switches, 222
Indicator Lamps, 225
Connecting Pilot Devices to Robots and
Programmable Controllers, 228

9 TIMERS, COUNTERS, AND SEQUENCERS   234

Timers, 234
Wiring Diagrams, 253
Counters, 256
Sequencers, 263

10 DC MOTORS   266

Magnetic Theory, 267
DC Motor Theory, 268
DC Motor Components, 269
DC Motor Operation, 273
Types of DC Motors, 275
DC Series Motors, 275
DC Shunt Motors, 277
DC Compound Motors, 281

11 AC MOTORS   284

Characteristics of Three-Phase
Voltage, 285
Three-Phase Motor Components, 286
AC Induction Motor Operation, 288
Connecting Motors for Torque, Speed and
Horsepower Conditions, 289
Motor Data Plates, 296
Three-Phase Synchronous Motors, 298
AC Single-Phase Motors, 298
Split-Phase Motors, 303
Capacitor Start, Induction Run
Motors, 304
Capacitor Start, Capacitor Run
Motors, 306
Permanent Split-Capacitor Motors, 308
Shaded-Pole Motors, 309
Repulsion Start Motors, 310
Troubleshooting Three-Phase and
Single-Phase Motors, 310

12 MOTOR CONTROL CIRCUITS — 316

Two-Wire Control Circuits, 316
Three-Wire Control Circuits, 318
Reversing Motor Starters, 321
Jogging Motors, 326
Sequence Controls for Motor Starters, 328
Controlling Circuits with a Programmable Controller, 331

13 PHOTOELECTRIC AND PROXIMITY CONTROLS — 347

Incandescent Photoelectric Controls, 348
Analysis of Light, 349
Modulated Light Source Devices, 350
Through-Beam Scan Photoelectric Switches, 350
Retroreflective Photoelectric Switches, 354
Diffuse Scan Photoelectric Switches, 358
Output Stage for Photoelectric Switches, 359
Selecting a Photoelectric Switch for your Application, 361
Installing the Photoelectric Switch, 363
Proximity Switches, 366
Hall Effect Sensors, 370

14 ADVANCED MOTOR CONTROL CIRCUITS — 372

Acceleration Circuits, 373
Reduced-Voltage Starters, 373
Autotransformer Starters, 376
Wye-Delta Starters, 378
Part-Winding Starters, 380
Secondary Resistor Starters, 382
Troubleshooting Reduced-Voltage Starters, 384
Solid-State Reduced-Voltage Starters, 384
Acceleration Control for DC Motors, 389
Deceleration and Braking Methods, 389
Brakes and Clutches, 395

15 DC AND AC ELECTRONIC MOTOR DRIVE SYSTEMS — 403

DC Motors and Drive Systems, 403
SCR Motor Speed Control, 405
Eddy Current Drives, 411

AC Variable-Speed Drives, 411
Speed and Torque Characteristics of Variable-Frequency Drives, 418
Servo Systems, 420
Resolvers and Encoders, 425
Stepper Motors, 430
Ball Screw Mechanisms, 433

16 PROGRAMMABLE CONTROLLERS 436

Basic Components of the P/C, 437
Program Scan, 440
Addresses, 440
Programmed Functions, 442
Binary and Binary-Coded Decimal Numbers, 450
Analog Inputs and Outputs, 452

17 TROUBLESHOOTING TECHNIQUES AND TEST EQUIPMENT 456

Types of Tests, 456
Voltmeters, 457
Ammeters, 459
Panel Meters, 461
Specialized Test Equipment, 461
Conclusions, 463

INDEX 465

About the author

Thomas Kissell is a technical instructor and consultant in the field of industrial automation. He is currently the Department Head for Automated Technologies at Terra Technical College and spends the majority of his time on the factory floor where he teaches electricians and technicians. This training includes installation, troubleshooting, and repair of major automation and motor control systems. The author has integrated this practical approach to troubleshooting and repair into this text that includes the theory of operation of all traditional motor control circuits as well as state of the art control systems used with programmable controllers and robots.

Thomas Kissell has worked directly with the following companies: Ford Motor Company, General Motors, Texas Instruments, Union Carbide, Libbey Owens Ford, American National Can Company and Kelsey Hayes. He has also presented his unique training and troubleshooting methods to major industries as a key note speaker at seminars across the country sponsored by Texas Instruments. These same proven methods and techniques are now incorporated into an easy-to-understand text that students can use as their primer in formal training and classroom experience, and then carry with them onto the factory floor as they begin their careers in industry.

Thomas has written two other books, *Understanding and Using Programmable Controllers* and *Motors, Controls and Circuits for Air Conditioning and Refrigeration Systems*.

Preface

This book has been written to respond to the need of modern technicians for the enormous amount of information necessary to allow them to work successfully with modern motor controls. In the past 20 years the technology involved with motor controls has changed drastically—to the point of overwhelming some technicians today. Early motor control circuits consisted primarily of a motor starter and several pilot switches. When the pilot switches were closed, the motor would run, and if any of the switches were open, the motor would be deenergized.

Today, the technician is responsible for installing, programming, troubleshooting, and repairing programmable controllers, robots, and other automated systems that are considered part of the motor control system. A technician in any factory today will probably be expected to work on a traditional motor control circuit with motor starters and pilot devices, process control systems for process heating, motion control systems that include servo systems or robotic applications, and sophisticated electronic and microprocessor-controlled devices.

Since these control systems represent a wide variety of technologies and theories, it is important that a book be available to explain all these systems at an introductory level, so that they are easy to understand. Each chapter contains sufficient information to allow an engineering student to design and install complex motor control systems that contain programmable controllers and robots used in automated manufacturing systems. The book also provides extensive reference information that is usable when technicians get jobs and begin to work on the factory floor.

Since numerous motor control systems have been in existence for many years, we have taken great pains to ensure that traditional circuits are included, so that new technicians can be introduced to them and understand their value. It is also important to see that the programmable controller is capable of replacing these traditional circuits and providing many of the same functions.

The book goes into great detail to explain the implementation of old and new controls to robotic and programmable controller–controlled systems. Information is provided that can be obtained only from years of experience on the factory floor, and it is presented in such a way that new technicians can utilize it immediately. The value of hands-on learning is also included, in that each chapter provides detailed diagrams that can be tested to develop wiring skills for installation and troubleshooting skills for making repairs.

Thomas E. Kissell

Acknowledgments

The technical information in this text could not be assembled without the help of the companies who supplied much of it. I would like to personally acknowledge the following people as well as their corporations and companies for all of their help. Without this type of help, it would be impossible to undertake writing a book of this size. Thank you all.

Mr. John Kelly at Cutler Hammer Products Division of Eaton

Mr. Randy Randall at Allen-Bradley in Milwaukee, Wisconsin

Mr. Mike Hughes at Allen-Bradley in Brown Deer, Wisconsin

Mr. Vic Jensen at Square D Company

Ms. Marlyn Smit at Honeywell Micro Switch

Mr. Ray Freiwald at Warner Electric Brake and Clutch

Ms. Kris Smith at Eagle Signal Controls

Mr. Jim Proft at Cincinnati Electro Systems

Mr. Bob Roderique at Bussman Manufacturing Division

Mr. Cy Pfeifer at National Controls Corporation

Mr. Tom Morrow at Omron Electronics Inc.

Mr. Rhea Gustofson at MagnaTek Century Electric

Ms. Diann Helnore at Eaton Corporation

Mr. Jack Saunders and Ms. Joanne Marino at GMF Robotics

Mr. Dennis Berry at National Fire Protection Association (NEC)

Ms. Crystyna Bagnato at National Electric Manufactures Association (NEMA)

Mr. Louis Eruska at Parker Hannifin Corporation

Ms. Mary Jessup and Mr. David Gephart at Electrical Apparatus Service Association Inc.

Mr. Bill Orr at Texas Instruments Inc.

The Joint Industrial Council (JIC)

MODERN INDUSTRIAL/ELECTRICAL MOTOR CONTROLS

Operation, Installation, and Troubleshooting

Overview of motor controls

INTRODUCTION

Modern motor control has become a complex technology that includes simple control of motors and other loads as well as intricate motion control for precise positioning. It includes simple safety controls such as fuses and other controls as complex as programmable controllers. Some of these controls are designed to stand alone while others must be interfaced to robots and automated manufacturing cells. In this book you will learn the basic theories that are used to design and implement modern complex motor controls. You will also experience the changes that have evolved to bring motor control systems into the modern electronic and computer age. You will see how new knowledge of microprocessors, electronics, physics, and mechanics has been blended to provided the most sophisticated controls known to humankind.

Human beings have tried to control their environment and machines ever since the discovery of the wheel. As electricity was developed, the focus changed to controlling the operation of the electric motor. This early control consisted of ways to turn the motor on and off, and methods of protecting the motor from damage. As transducers evolved it became possible to automate the controls so that something other than people could switch the motor on and off.

Some of the early controls included mechanical methods of motor control. For example, the switch that was used to turn the motor on and off could be customized to complement the action the motor provided. If the motor was used to move material like a winch, a limit switch would be used to detect this motion. If the motor was used to pump water, a float device could be mounted on the end of a rod that would move against the switch handle and cause it to open and close to control the motor. An example of this type control is shown in Figure 1-1, where you can see that the controls are related to the function the motor provides. You must remember that these controls began to evolve at a time when human labor was both cheap and plentiful. In fact, the cost of early controls was related directly to the expense of providing people to turn the switches on and off manually.

The motor control circuits that you will encounter on the job will fall into one of several classifications according to the function they are designed to provide. These functions include operation of the equipment, safety of the equipment, and

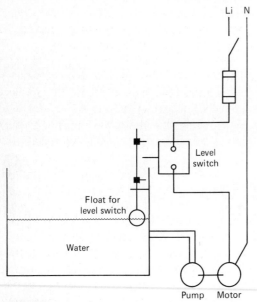

FIGURE 1-1 Liquid-level control used to control the water level in a tank.

safety of personnel who must operate the equipment or work on it.

The circuits that you find in a control system operate in a similar manner regardless of their function. This means that all motor control circuits operate in essentially the same way. They must first sense conditions in their environment. The sensing signal is then sent to a device or circuit that must decide what to do when this condition arises, and the third part of the operation is the action that is prescribed based on the decision. This means that each circuit goes through these three distinct steps to turn a motor on or an output signal on or off. Figure 1-2 shows an example of a simple control application that uses these three steps.

This circuit uses a limit switch to determine if a box is in place on a conveyor line. If the box is in place, a push bar is extended to push the box off the conveyor. The sensor in this circuit is the limit switch, and the decision is determined on the basis of whether a box is or is not in place. The action part of the circuit is the solenoid valve that will cause the push bar to extend. If a box is not in place, the circuit will decide not to energize the solenoid, and if the box is in place, the solenoid will be activated.

A safety circuit could be designed in the same manner. The safety circuit could use a photoelectric device to indicate if anyone was near the operating equipment. If someone entered the danger area around the equipment, all machine operation would be terminated until the area was clear again and the system was restarted. A second type of safety circuit could also be designed to determine if any boxes were jammed in the machinery. If a box was jammed in the machine, it could cause damage to equipment, so a jam detection switch could also be included in the circuit.

Each of these circuits would operate in the same manner, in that they would use sensors to detect the conditions that exist at any time. The decision part of the circuit would be determined by the types of contacts used in the circuit (normally open or normally closed), and the action part of the system would be the solenoid turning on or off, or a relay shutting the equipment down when a dangerous situation was detected. These circuits also show the three main types of circuits used in motor controls: operational control, equipment safety, and personnel safety circuits.

THE NEED FOR MOTOR CONTROLS

From the beginning, motor controls were developed for a variety of reasons. The primary reason was to control an operation or process, such as determining the number of parts a machine would manufacture, or to limit the time of a machine cycle. In process control, the temperature of a material or the level of product in a container was controlled.

Other controls were soon required to protect the motor and other electrical and mechanical

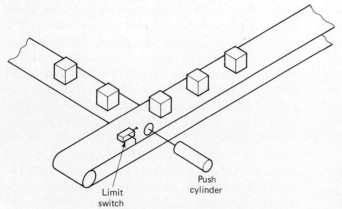

FIGURE 1-2 Automatic control circuit that utilizes a limit switch to control a sorting function.

equipment. These controls included fuses, circuit breakers, and clutches. The main function of the fuse and circuit breaker was to ensure that a motor would not receive too much current or be allowed to become damaged by heat. These devices have been evolved to a point where the newest fuses are programmable in both the amount of current they will allow to flow and the amount of time the condition is allowed to continue. This control is provided by a combination of sensors and microprocessors that are used as stand-alone protection or in combination with motor drives, programmable controllers, and robots.

The final type of control was designed for personal safety. These controls were needed to protect the people who had to work around the equipment and processes. As more knowledge was gained about the safety factors, more controls were necessary to provide adequate protection against them in both the short and the long term.

Machine Operational Controls

As you know, operational controls were first designed to control simple machine operations such as start and stop functions. The simplest of these is a toggle switch, which is similar to the light switch in your home. More complex controls involve other types of switches that are incorporated into controls that can sense movement, level, pressure, flow, temperature, and force.

Some of the controls in use today are still manual or semiautomatic controls. They are used because they are the best type of control for the job. In other cases, control systems installed in the late 1940s are still in operation. Controls that were installed in the 1950s and 1960s may also seem archaic in comparison to today's controls, but as long as they are controlling systems in a satisfactory manner, they will continue to be used.

Some operational controls are used to provide speed control of the motor. These include a variety of belts, clutches, and transmissions to control the speed mechanically, and variable-frequency and variable-voltage controls that control the speed electronically.

Since the advent of the new generation of solid-state devices and microprocessors, they have been actively incorporated to provide more complex control. It is now possible to change ac voltage to dc voltage in one simple circuit and to control the amount of voltage and current that is sent to a dc motor. This provides speed control and torque control that is needed to meet the variety of changing conditions that loads place on the motor. Ac motors can have their speed, torque, and efficiency controlled by the same types of electronic devices that are used to alter the voltage, current, and frequency of the power the motor uses.

These speed controls can be used as stand-alone devices or they can be incorporated into complex control systems such as programmable controllers or robotic work cells. The choice of operational control will depend on the complexity of the system and the amount of money that is to be invested.

The new trend in motor control is to provide data as well as control for a system. It is now as important to know the quality and quantity of the parts being manufactured as it is to produce the parts. In years past, as long as the machines were operational and parts were being produced, the job of the control system was considered complete. Today, it is important to test the parts that are being manufactured, and controls must be provided to make subtle changes in the system to make corrections when defects are found. In some systems the correction is made to machines manually; in more complex systems, the control is closed loop. Controls such as measuring device and vision systems send analog data instead of the more simplified on/off signals to indicate that the product is good or bad. The analog values from the testing sensors can determine not only that the part is bad, but how much correction is needed in the machine to make good parts again.

It is also important to understand where the data that are gathered must be sent. Part of the control system that you will be involved with will be used to send the data that have been gathered to several different offices in the plant. In today's modern control systems the data must be analyzed by production control specialists who determine the number of parts that the system must make, and to quality control specialists, who determine if the parts that are being produced meet the required standards. Data must be sent to product specialists such as engineers, technicians, and machine operators, who understand all facets of the production system. They must be consulted when parts are not meeting standards, so that they may suggest minor changes that can be implemented with a simple modification, or a complex modification that requires intricate changes to the system controls.

If the parts being made are manufactured from metal products, the people receiving the data may include engineers and technicians who understand the chemical and physical makeup of the raw material being used, as well as production specialists who understand molds and dies and metal forming. You will be required to work with all of these people in the plant to interpret their needs and to convert these needs to changes in the electrical control of the system.

Safety for Equipment and Personnel

Most safety devices were first built into equipment to protect the motor and working parts. As designs evolved, more safety features were required to protect the people that must work on or around the equipment. These safety controls included sensors to indicate overtemperature, overpressure, and spillage of dangerous or volatile substances.

Modern control systems provide safety circuits to protect the equipment, as well as protection for people who must work around the equipment. Other safety circuits are provided to protect people from the product that is being produced. In modern technology, many products, such as chemicals, pesticides, and pharmaceuticals, must be controlled during manufacturing and handling to avoid excess contact with human beings, and the production of other products, such as electroplating, that produce dangerous vapors, fumes, and by-products which must be controlled and disposed of safely.

The manufacture of these products requires extensive safety devices and circuits to detect the presence of dangerous conditions and to execute predetermined procedures to control the condition. These procedures can be designed into the control circuitry or programs that are used to provide motor control. The safety components will reside in the same electrical cabinet as the operational components and production controls that are used to make the product. In fact, a particular type of sensor may be used in one part of the control system as an operational control and in another part of the system for a safety circuit. The control will provide a slightly different function and be connected to a different part of the control circuit, but when it comes to installation, calibration, and troubleshooting, you will use the same operational theory to determine if it is operating correctly.

In other systems it is possible to use the same sensor for production and safety control. For instance, a temperature control can be used to control the production temperature of a product and protect against overtemperature. Figure 1-3 shows an example of this type of control circuit. The high safety temperature level is set at 500°F, the high operational temperature is set at 450°F, and the low operational temperature is set for 350°F. In this application, one sensor is used to sense the temperature, but different control strategies are used to determine what should be done when the temperature is in a specific range. If the temperature of the product is between 350 and 450°F, nothing is changed in the system, which means that heat is maintained at the present level. If the product's temperature drops below 350°F, more heat is added to the system. If the temperature increases above 450°F, heat is gradually removed from the product, and if the temperature exceeds 500°F, an unsafe condition is sensed, all heat is removed, and an alarm is sounded. From this circuit you can see that one type of sensor can be used to provide information to be used as operational control and also as safety control.

CONTROL CIRCUITS AND LOAD CIRCUITS

Another way to break down the components to learn their operation and function is to divide the motor control circuit into two distinct parts. One part of the circuit has the function of providing *control* of the circuit. This is where sensing, decision, and action are initiated. The second part of the circuit is called the *load* circuit. This part consists

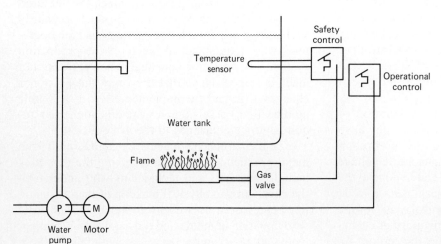

FIGURE 1-3 Example of a sensor being used as an operational control and a safety control.

of the motor and any devices used to provide power to the machinery. The load circuit usually uses higher voltage and more current than the control circuit. Figure 1-4 shows an example of a simple circuit that uses a start switch to energize a motor starter coil. The motor starter contacts will close to provide high voltage and sufficient current to the motor to allow it to turn the load. The load in this case is a conveyor. When the pushbutton is depressed, the motor starter coil is energized, which causes the contacts in the circuit to close. When the contacts close, they allow voltage to flow to the motor. When the stop button is depressed, the motor starter coil is deenergized and the contacts are returned to their open position, which deenergizes the motor.

The start button, stop button, and motor starter coil are all part of the control circuit. This is where the sense, decide, and act functions are executed. The part of the circuit, where the motor starter contacts close and open to energize and deenergize the motor, is part of the load circuit. In this case the motor is the load for the circuit.

In most circuits it is possible to determine which part of the circuit is functioning as the control and which part is operating as the load. In the circuit just described, the start and stop buttons are the control for the circuit, and they are located at some distance from the motor. This type of control is called *remote control* or *automatic control*. If a manual motor starter was used, the start switch would be located very close to the motor and the control circuit would be called a *manual control* circuit. In this book we explain how each component used in these circuits functions and methods for testing them. Other information will be provided for installation. When you understand the function of a component, it will be easier to troubleshoot it since you know what the device is supposed to do.

Motion Control

In some systems a more precise type of control is required. As you know, motor control can turn motors and other loads on and off, depending on the status of the input signal, but in some applications, the speed and position of the load must be controlled. This type of control, called *motion control*, is associated with robotic systems and other automated systems.

Specialized devices are used as controls in these systems to provide the intricate type of positioning. The speed with which a load is moved and the positions to which it is moved can be controlled by servomotors and stepper motors. This type of control can also be provided by hydraulic servo systems. In the past, a robotic technician or a servo specialist was required to install, calibrate, troubleshoot, and repair these systems, but today, every electrical technician or electrician is expected to be able to understand their operation. The information in this book will help you understand the theory of operation used to design and operate motion control systems, and will provide you with diagnostic tests that can be used to determine the status of each component in a system.

Other control systems provide speed and torque control of motors through the use of motor drives. Such drives use variable frequency and variable voltage to change conditions of the motor's speed and load capability. SCR controllers are used to provide voltage and current control for dc motors. A wide variety of magnetic and mechanical clutches are also used to allow the motor to operate

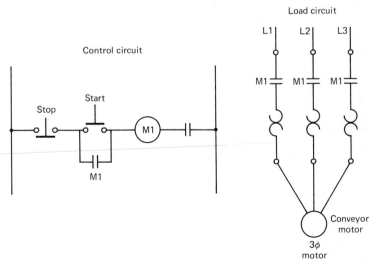

FIGURE 1-4 An electrical diagram showing an example of a control circuit and a load circuit.

at full speed and torque while adjusting the speed and torque characteristics of the load.

Process Control

Another specialized type of control system is known as *process control*. For many years these control systems were installed, calibrated, troubleshooted, and maintained by instrumentation technicians. In fact, most factories had only motor control systems for manufacturing products, or process control systems that were designed for the processing of products in batch or continuous form. This meant that you could study one type of control system, dedicate your career to specialized work on that system, and never have to work on other types of systems.

In modern industry you will find a mixture of motor control and process control built into each machine in both the process industry and the manufacturing industry. In process control the sensors usually transmit analog signals to an amplifier or analog-to-digital converter for use in a computer or controller, and in motor control the signal tends to be on/off in nature. Other differences between the two systems are now blended into complex machine control, and generally no distinction is made between them when you are assigned to work on them.

A Major Change

A major change has occurred in the controls industry that has affected what you will be doing when you work as an electrical technician or as an electrician. This change is the removal of barriers between motor controls, motion controls, and process controls. The advent of the programmable controller has helped to bring about this change.

When you go to work in industry today, you will be expected to work on any of the control systems listed above. Industry can no longer afford to separate the duties and hire three different people to work on these three types of system. You will also find all three systems included in the same piece of machinery because they allow designers to produce machines that perform more functions and maintain more constant quality in the product that is manufactured.

For example, it is common today to work on a system such as a plastic injection molding machine that will have normal motor control circuits to turn it on or off, have process control circuitry to provide closed-loop temperature control, and have a hydraulic servo system to provide intricate control of the pressure, flow, and location of all major parts on the press. With this mixture of components and systems, you can see that you can no longer afford to learn only one of the systems and feel that you are ready to design, install, and service motor control systems. In addition to these controls, many factories are adding programmable controllers to manage the large control circuits, and robots are used to provide material-handling capabilities.

In this text we show you how this blend of controls functions. Even though motor controls, motion controls, and process controls have a slightly different purpose, their operation is similar enough that you will easily understand them all. Some instructors teaching this course will have taught motor control courses for years, and the main topic of those courses will have been limit switches, pushbuttons, timers, and motor starters. If you could take a trip through any factory, whether it is brand new or 30 years old, you would notice that such components and controls are just the basic part of the control system. The controls that are found today include programmable controllers, robots, automated manufacturing, and motion and process control systems. If you cannot go to a factory, open any trade journal and look closely at the machinery advertisements. These ads will show how designers have incorporated the finest advanced controls to make their machines best at what they do. What the ads do not say is that this practice has required a drastic change in the content of what electrical technicians or electricians must study in order to become competent and remain so throughout their careers—even with changes that will not occur until years from now.

INNOVATIONS IN CONTROL SYSTEMS

A variety of innovations have been incorporated in motor control devices over the last few years. They include the use of solid-state devices in all areas of the detection and amplification of signals that were once too small to detect. Another innovation that has been integrated into motor control systems is the *programmable controller* (P/C). The P/C was originally introduced for use in motor control circuits as a relay replacement system. It has since grown into a complex industrial computer-controlled system that can provide motor control, motion control, and process control. In fact, multiple programmable controllers can be connected in a local area network so that data can be transmitted and received between each P/C and the network host. This allows the implementation of complex automated systems that extend beyond the confines of factory walls.

Another innovation that has developed from

solid-state components is the *microprocessor*. The microprocessor is a complete computer on a chip that has become less expensive and more usable, so that it can now be interfaced directly with many motor controls. It is the heart of the P/C system, and it has been customized for use in many other controls. The timing and memory functions that it can provide make it a logical replacement for complex electromechanical devices such as timers, counters, and other sensors. Since microprocessors have been added to motor control devices, the devices now have the ability to be programmed and execute very complex operations.

Solid-State Devices

The advent of the solid-state diode and transistor has brought on another large change in the motor control industry. Solid-state devices allow the control to be much smaller and react much faster than could traditional mechanical and electromechanical devices. The transistor allows very small physical changes to be detected and sensed. For example, transistors added to photoelectric and proximity controls allow them to sense changes in motion as small as 0.01 in. Other control devices, such as resolvers and encoders, have incorporated the transistor together with other solid-state components, such as the operational amplifier, to provide control to within 0.001 in.

The transistor has also been designed to be used in power switching circuits for controlling voltage and current to motors and other devices. They are also quite useful in providing positive and negative bias to controls that must detect bipolar signals. The SCR and triac has also been designed from work on power thyristors. These devices provide control of exceedingly large voltages and currents at speeds that are compatible with microprocessors.

The operational amplifier (op-amp) has been developed from work with traditional transistor amplifiers to provide circuits on a solid-state chip that can detect changes in input sensors that were once thought impossible. These devices are also used to control analog outputs to motor speed controls and control of servo valves.

The discrete (individual) solid-state components that are presently used in motor controls are described throughout this text. The functions they provide are also explained with the theory of their operation. Newer developments have included some of these components in integrated circuits (ICs). The integrated circuit provides complete circuits on a single chip that can easily be tested and quickly removed and replaced when mounted in IC sockets on printed circuit boards. This allows motor control devices to be troubleshooted and maintained with ease. The research and development of these devices has lead to the discovery of new applications for the microprocessor chip used in programmable controllers and other motor control devices.

Programmable Controllers

Programmable controllers have altered the application of motor control as much as any other device or component that has been introduced into motor control circuits. The P/C was introduced in a General Motors Hydramatic Plant in 1969 as a reprogrammable control that could replace the vast network of control relays that were used to provide motor control logic. Since that time the P/C has evolved into a series of controllers that may be as simple as the first sequence controller designed in 1969, or as complex as any mainframe computer with multiple microprocessors that operate asynchronously.

The P/C provides the ability to make changes quickly and easily in motor control systems. The logic control that is executed by the P/C is called a *program* and it can be stored on a variety of memory devices, such as the floppy disk and magnetic tape. This means that the program can be developed in an office and downloaded into the P/C by a technician working near the system, or from the network, which allows the programmer to be hundreds of miles away in another city.

The P/C provides a method of controlling multiple analog and digital circuits that may be some distance from each other. It also provides a method of diagnosing problems that occur in the complex system by using its logic program for troubleshooting as well as control. The P/C also provides a very complex system that can easily be interfaced to a wide variety of electronic sensors and controls. This is important since the variety of sensors and controls covers the complete spectrum of electronic signals from TTL through 220 V ac. This means that modules are available that can be plugged into a P/C's rack that can quickly allow the system to be programmed to control such complex operations as servo positioning, data communications, and report generation, as well as received traditional signals from limit switches and to send voltage to energize motor starter coils.

Microprocessor Control

The microprocessor has also been incorporated into a wide variety of controls used in motor control systems. The microprocessor allows a device to be programmed for operation that may occur many days in advance. The program may be permanently entered into the control, or it may be reprogrammable, such as timers and counters and motor speed drives. The microprocessor has also been added to many devices that are interfaced with P/Cs since they allow the device to execute sensing and control without requiring constant updating.

In some devices the microprocessor allows specific functions to be selected on a "use as needed" basis. This means that a device may be capable of being programmed as a timer or a counter, which allows fewer components to be maintained in inventory. If a timer becomes faulty, the control can be removed from inventory and programmed as a timer, and if the a counter control is needed, the same device could be reprogrammed as a counter.

The microprocessor also provides other selections in motor controls, including choice of voltage, frequency, and range of input and output signals. The maximum value that the signal can reach is also programmable, which allows the microprocessor to provide set-point and dead-band control on process control instruments. All of these choices mean that you must be able to understand the ramifications of all the selections that can or must be made on microprocessor-controlled devices and systems so that the system can be troubleshooted.

CONCLUSIONS

This overview of modern motor controls is intended to provide you with an understanding of where motor controls started many years ago and how quickly they have progressed to the present state. You should also realize that you will need to understand the operation of many electrical, electronic, electromechanical, and mechanical devices in order to install, calibrate, troubleshoot, and maintain modern control systems. You should also understand how fast these controls continue to change. This means that devices you read about today will by next week be installed in your factory on the equipment you are responsible for maintaining. You will be expected to read a small amount of documentation or attend a short school or seminar and be able to maintain the equipment. You will also be expected to provide this information to your fellow workers and to teach them the principles that you have learned. You can also see from past innovations that the state of the art as we know it today will not stand still; rather, you can be prepared to continue to learn about these changes for the rest of your career.

For this reason it is vitally important that you fully understand the concepts and principles of operation for every device that is used in motor control devices and systems. In this way you will be able to understand any changes quickly, as they occur. This also ensures that you will have a job in the years to come. The better you understand these concepts, the more you can contribute to the success of your employer. This will aid in the employer's profitability, which will help to keep the company in business. The knowledge that you gain and possess can never be taken from you. The key to your career success is based on the amount of knowledge that you have. Even if the company that you are working for moves or goes out of business, you will have the necessary skills to be quickly reemployed.

QUESTIONS

1-1. Name three classifications of functions of motor controls.

1-2. Give two examples of motor controls provided to protect personnel who must operate equipment.

1-3. Give two examples of motor controls provided to protect a machine from damaging itself.

1-4. Provide an example of a motor control sensor that could be used for operation and for safety.

1-5. Explain the difference between the control portion of a circuit and the load portion of a circuit.

1-6. Provide an example of a motion control system.

1-7. Provide an example of a process control system.

1-8. Why have programmable controllers brought about so much change in motor controls?

1-9. Explain how electronic components and microprocessors have changed the modern motor controls.

2
Safety in working with motor control circuits

INTRODUCTION

When you are working with motor control circuits, personal safety is the most important part of the job. The motor control system will have specific safety circuits built into it to protect personnel from dangerous conditions, and it will provide protection for all components of the system.

In this chapter we explain the types of unsafe conditions that are present in the factory environment and what precautions should be taken. Types of protective clothing that should be worn on the job are also discussed. Also covered are safety features that are built into all the components in common use in motor control systems. These features are specified by safety regulations provided by associations such as the Occupational Safety and Health Administration (OSHA), the National Electrical Code® (NEC), the National Electrical Manufacturers Association (NEMA), and other state and local codes, including company policies.

PERSONAL SAFETY

When you are working anywhere on the factory floor you must always wear safety glasses. Safety glasses protect you against any flying debris and any material that may become airborne should sparks or arcs be caused by electrical equipment and connections. It is important to understand that electrical equipment may develop a fault at any time and cause a tremendous amount of current to be drawn through switch contacts and other terminals, which could cause metal in these parts to melt and explode. If the proper covers are not in place, the molten metal may spray throughout the area and may cause blindness. When you are working around live motor control and other electrical circuits, even in a laboratory setting, you must be sure that you and all others who are in the area are wearing safety glasses at all times. These rules follow common sense, but they are also recommended and enforced by OSHA and by state and local codes. In fact, all plants will have a safety committee that ensures that this and other regulations are adhered to strictly.

Since you will be working around electrical circuits you should be completely familiar with CPR (cardiopulmonary resuscitation) methods. You should take both the CPR course and a first-aid course so that you will be able to treat properly any emergency that comes up on the job. In many cities the Red Cross will provide training on an annual basis right at your plant or school. There are also scheduled courses presented monthly in nearly every city where you can receive initial training or training update. You need to understand CPR because one effect of electrical shock on the human body is to cause breathing and circulation to stop or become impaired. If you are on the scene of an accident where an electrical shock has occurred, be sure to check the victim for breathing and a pulse. If either of these vital functions has stopped, initiate CPR immediately and send someone for help.

It is also important to be aware of other personal safety conditions to which you are exposed. You must take responsibility for yourself to determine the extent of the exposure. If conditions warrant, you should request proper safety gear, such as particle masks in areas where fibers and dust exist, and request adequate ventilation in areas where fumes may accumulate. Other equipment, such as safety-toe shoes, are required anywhere you work that has conditions that may allow your feet to become exposed to falling material, such as heavy tools or parts. The heavy rubber sole and leather upper of these industrial-grade shoes also provide protection against some minor electrical shocks.

If you are working in or around high-voltage power distribution panels, you should be familiar with the heavy-duty leather and rubber gloves that are available to protect you against electrical shock. If you must work around high-voltage circuits and components, be sure that power is off whenever possible, but you should understand that in some cases you must work on systems that are under full power. In these cases your principal concern is for your safety and those who are working with you. Investigate the type of protective clothing that is needed for each type of job, and be sure to check that all safety systems that you depend on are in working order.

In some cases you will be required to wear a hard hat, especially in areas where construction is in progress. In some factories, the threat of falling material will cause the area to be identified as a hard-hat area. This means that you and any one with you must wear a hard hat at all times. You will also be responsible for supplying hats to anyone who visits the area.

You should also be aware of any loose or baggy clothing such as neckties and shirt tails that might get caught in rotating shafts or other moving machinery parts. You should also be aware of any metal that is worn such as rings, necklaces, and watches that may come in contact with electrical components or that might draw an arc when you are working near live circuits. For this reason, the frames of your safety glasses must be made from a nonconductive material such as plastic, because you may be severely burned on your face if you allow metal frames of glasses to come in contact with an electrical circuit. If you find it uncomfortable to remove a ring, you may apply several coats of electrical tape to prevent it from coming in contact with live circuit parts. You should also be aware that rings pose a threat of becoming snagged or caught on sharp corners. There are many cases on record of workers who have caught a ring on a sharp edge while they were climbing or reaching for something and the quick motion allowed the ring to cut their finger severely or even caused the finger to become detached.

You should also be aware of handling tools around electrical circuits. All tools used in and around these areas should be properly insulated. When the insulation begins to wear, the tools should be rotated to another part of the plant where insulation is not a concern, and the old ones replaced.

ELECTRICAL SAFETY

Electrical circuits present several safety conditions that are dangerous to personnel and to equipment in the system. These conditions include electrical shocks, short circuits, grounded conductors, and ungrounded or improperly grounded circuits. Electrical shocks may occur at any time when a person comes into contact with a live electrical circuit. The amount of electrical shock that person receives will depend on the surface on which he or she is standing, the type of shoes being worn, and the intensity of the voltage. If the surface of the floor is a nonconductor such as rubber or plastic, the shock would be minimal because the floor would not carry current well. A small amount of current (400 mA) can cause death if it is conducted directly into your body. The amount of current that will be conducted into the body will depend on the intensity of the source and the parts of the body that are touching a grounded part of the circuit. The most severe electrical shocks occur when your body comes into contact with two wires that represent different potentials such as L1 and L2 of a three-phase service entrance, or between L1 and neutral of a single-phase circuit.

One way of avoiding severe shock and burns caused by electrical hazards is to work with only one hand in the panel when you must work on a live circuit. The reason for using only one hand is that you will have less of a chance of coming into contact with two different potentials. If you touch two different potentials each with one hand, you will provide a path for current, and that path will be down one arm, through your chest, and out the other arm. This means that your heart will be right in the path of the current, which will cause almost certain heart failure that leads to death. If you are working with only one hand in the area of a live circuit, the shock will tend not to be as severe and usually will not affect the heart directly unless you are well grounded. The most likely way to be grounded is to stand on a wet or damp surface. If you are standing in water, you are well grounded, which represents one potential of the electrical circuit. This means that whenever you come in contact with or are near a live conductor or part of the circuit, you will automatically receive a severe shock that will enter the body where the contact takes place, and will exit the body at your feet, where you are standing in the water.

Electrical Short Circuits

A *short circuit* is caused whenever the amount of resistance in a circuit is lowered to a point that will cause extremely large currents to flow out of control. Figure 2-1 shows two conditions that can cause a short circuit. Figure 2-1a shows a motor connected in a circuit between L1 and L2. The voltage in this circuit is 208 V and the normal current drawn by the motor is 8 A. A short circuit will occur in the motor when the insulation on the conductors in the winding abrate and allow the bare conductors to come into contact with each other. When the two bare wires touch each other, a short circuit occurs, which causes an increase in current. The increase in current causes a hot spot in the motor winding, which further breaks down the remaining insulation, which causes the winding's resistance to be lowered. This process escalates rapidly until the motor is drawing current up to 10 times the amount of normal current. If a fuse, circuit breaker, or overload does not interrupt the circuit, the motor will become hot enough to burn the motor windings completely.

If the location of the short circuit occurs where the L1 and L2 supply voltage can come directly in contact with each other inside the motor, a violent explosion caused by uncontrolled current can occur inside the winding. This means that the closer the short circuit is to the supply voltage lines, the larger the short-circuit current will be.

Another condition that can cause an uncontrolled increase of current is shown in Figure 2-1b. This condition occurs when conductors of the motor winding come into contact with the frame of the motor. This type of condition is called a *grounded winding*. The cause of the ground is similar to the cause of the short circuit, and the result is the same.

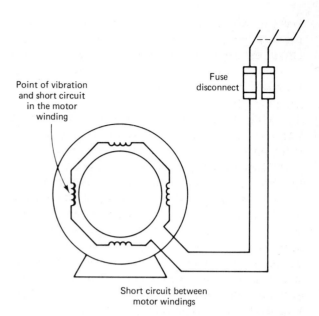

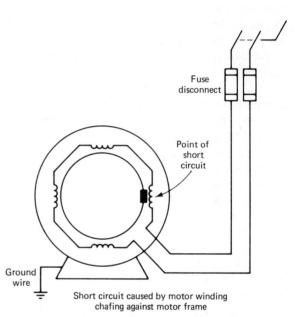

FIGURE 2-1 Conditions that can cause a short circuit.

In fact, if you do not inspect the motor closely, you would not know for sure if the damage to the motor was caused by a short circuit or a grounded winding.

Another cause of these conditions is brought about by too much moisture in the motor windings, which causes deterioration of the insulation. There are several tests that can be performed with a megohm meter (meggar) to determine the serviceability of the motor. A meggar can measure the amount of resistance between the windings and the frame accurately in the range of several million ohms. If the resistance is lower than the rated specification, it indicates that a fault is imminent and the motor can be removed from service and repaired before it is severely damaged by the short-circuit current.

Equipment Safety Ground

The safety ground circuit is used on all equipment in a factory to prevent unsafe voltage potentials from being present in the frame of the equipment. Figure 2-2 shows examples of ways to ground an electrical machine. To fully understand the unsafe condition that can occur when a system is not grounded, you must first understand that the earth is considered to be at a potential difference with all ungrounded conductors in the power distribution system. This means that if you place one lead of a voltmeter on L1 of a grounded system and the other lead on the grounding rod pressed into the earth, a voltage potential will be indicated. The amount of the voltage will depend on the type of transformer connection being used. For example, if the power system was connected as a 230-V delta system, the voltage between L1 and ground would be 115 V.

The other important fact is that a fuse can only detect an increase of current, not the presence of unwanted voltage. If a conductor comes into contact with the frame of the ungrounded motor, the motor frame will have the same voltage potential as the conductor. This means that the frame will be powered with 115 V. If a person is standing on the ground and comes into contact with the frame, a current will be developed through the person's body between the point where they are touching the frame and where they are touching ground, which will cause a deadly shock.

If the frame is grounded when the conductor comes into contact with it, a short circuit will occur, which will cause the fuse that is protecting the motor to open and keep the circuit safe. When the fuse is replaced, it will immediately blow when power is again applied to the circuit. At this point, the technician will know to test the motor for short circuits.

Even though the ungrounded system is protected by a fuse, the fuse will not blow because no current will flow between the shorted winding and the motor frame. In this case the frame is said to be *floating above ground,* which creates a potential electrical shock hazard if anyone touches any part of the metal case of the motor and the machine to which it is bolted. From this example you can see why it is important to ensure that all equipment in the factory is properly grounded. It is also important to ensure that the ground circuit is provided throughout the plant and that it is tested on a regular basis, about every 30 days. On some machines, a lamp is connected between one of the lines and the ground circuit to provide an indicator that the ground circuit is operating correctly. If the lamp is not illuminated, the ground circuit should be tested immediately.

Ground-Fault Detectors

An electronic device has been developed that can be connected to the two power lines of any circuit and accurately measure the amount of current that is present in the supply and return line of the circuit at any instant. If the circuit is operating correctly, the current in the supply line should be exactly the same as the current in the return line. If the amount of current being supplied is larger than the amount of current in the return line, the circuit will assume that some of the current is leaking to ground or to another path, such as through a person, and the device will trip the safety contacts. When the safety contacts are tripped, all power to both sides of the circuit is interrupted and the device must be reset. This device is called a *ground-fault interruptor* (GFI). The GFI can be connected to any circuit

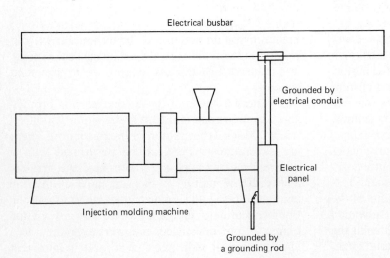

FIGURE 2-2 Example methods of grounding a machine through an electrical metal conduit and by a grounding rod.

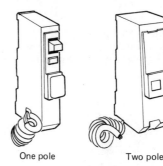

QWIK-GARD©
CIRCUIT BREAKER
WITH GROUND FAULT
CIRCUIT INTERRUPTER

One pole   Two pole

FIGURE 2-3 A ground-fault interruptor is used to detect ground current and protect personnel from unsafe electrical shock. (Courtesy of Square D Company.)

QWIK-GARD circuit breakers provide QO overload and short circuit protection, combined with Class A ground fault protection. Class A denotes a ground fault circuit interrupter that will trip when a fault current to ground is 6 milliamperes or more.

where there is a risk of leak current occurring, such as hand power tools and extension cords. The danger of leak current causing fatal electrical shock is always present when metal case hand power tools are being used and when the work area is exceedingly wet. The National Electrical Code® (NEC) requires the use of GFI devices where the possibility of electrical shock is imminent.

SAFETY CONDITIONS AND HAZARDOUS LOCATIONS

Figure 2-4 shows the three basic classifications of hazardous locations that have been produced by the National Electrical Code®: class I, class II, and class III. A summary of the locations where these hazards are present is provided in the figure. From the list you can see that each class has two divisions.

A list of materials is also presented in this figure, in groups A through G. These hazardous locations and materials determine the types of enclosures that are needed to provide protection against explosive and other unsafe conditions.

Fire Safety

Figure 2-4 includes a classification of dangerous and hazardous environments which contain gases, dust, and fibers that may cause conditions where a fire or explosion could occur. The conditions are broken down into divisions and groups in accordance with the type of hazard that may exist and the potential for danger. In the gas group vapors may accumulate from acetylene, hydrogen, ethylene, and petrochemicals with alcohol.

In the group from sources of dust, hazards include conductive dust and combustible dust. The conductive dust presents a problem in that when it accumulates, it is possible to draw an electric arc to wherever it is located. The dust may come from sources such as machining conductive materials, which may include copper, brass, aluminum, and ferrous metals.

The last group include fibers from production areas and material-handling systems. These are especially dangerous where clothing and other fiber products are manufactured. The main danger from these fibers is that they are easily ignited and pose a threat for fires and small explosions.

These divisions are used to determine classifications of enclosures by the National Electrical Manufacturers Association (NEMA). The National Electrical Code® (NEC) class is also listed, as class I, class II, or class III, for each of these examples.

Types of Fires. It is important to understand that not all fires are the same. In a factory it is possible to have several different types of fires which require different types of fire extinguishers to put them out. The fires are classified by the type of material that is burning. A class A fire consists of normal combustible material such as paper, wood, plastics, and cloth. These fires may be extinguished with water, available in many factories from fire hoses. If a fire extinguisher is available, it must be marked as approved for class A fires. These extinguishers include water, foam, and dry chemical types.

The second type of fire, called a class B fire, includes flammable gases and liquids such as oils, machinery fluids, and grease. You should not try to put water on these types of fires, as it will generally spread the fire around. This type of fire must be smothered to be quickly extinguished. Fire extinguishers rated for class B fires include dry chemical, foam, and carbon dioxide (CO_2) types.

The third type of fire includes fires in which energized electrical equipment is involved. You must be aware of the type of extinguishing material that is used since some of the extinguishers mentioned previously are good conductors of electricity. The approved type of fire extinguishers for class C fires are carbon dioxide and dry chemical types. These materials will quickly extinguish a fire and

Typical Hazardous Locations

Class I
- Petroleum refining facilities
- Petroleum distribution points
- Petrochemical plants
- Dip tanks containing flammable or combustible liquids
- Dry cleaning plants
- Plants manufacturing organic coatings
- Spray finishing areas (residue must be considered)
- Solvent extraction plants
- Locations where inhalation anesthetics are used
- Utility gas plants, operations involving storage and handling of liquified petroleum and natural gas
- Aircraft hangers and fuel servicing areas

Class II
- Grain elevators and bulk handling facilities
- Manufacture and storage of magnesium
- Manufacture and storage of starch
- Fireworks manufacture and storage
- Flour and feed mills
- Areas for packaging and handling of pulverized sugar and cocoa
- Facilities for the manufacture of magnesium and aluminum powder
- Coal preparation plants
- Spice grinding plants
- Confectionary manufacturing plants

Class III
- Rayon, cotton, and other textile mills
- Combustible fiber manufacturing and processing plants
- Cotton gins and cotton-seed mills
- Flax-processing plants
- Clothing manufacturing plants
- Sawmills and other woodworking locations

Materials by Group ① ②

Group	Materials	
A	Acetylene	
B	Acrolein (inhibited) ③ 1, 3-Butadiene ④ Ethylene Oxide ③ Hydrogen	Manufactured Gas (containing more than 30% hydrogen by volume) Propylene Oxide ③ Propyl Nitrate
C	Acetaldehyde Allyl Alcohol n-Butyraldehyde Carbon Monoxide Crotonaldehyde Diethyl Ether Diethylamine Epichlorohydrin Ethylene Ethylenimine Ethyl Mercaptan	Hydrogen Cyanide Hydrogen Sulfide Methylacetylene Methyl Ether Methyl Formal n-Propyl Ether Tetrahydrofuran Triethylamine Unsymetrical Dimethyl Hydrazine (UDMH)

FIGURE 2-4 Summary of National Electrical Code® requirements for hazardous locations. (Courtesy of Allen-Bradley, Inc.)

Materials by Group (continued) ① ②

Group	Materials	
D	Acetone Acrylonitrile Ammonia ⑤ Benzene Butane 1-Butanol (butyl alcohol) 2-Butanol (secondary butyl alcohol) n-Butyl Acetate iso-Butyl Acetate Cyclopropane Di-isobutylene Ethane Ethanol (ethyl alcohol) Ethyl Acetate Ethyl Acrylate (inhibited) Ethylamine Ethylenediamine (anhydrous) Ehtylene Dichloride Gasoline Heptane Hexane Isoprene	Isopropyl Ether Mesityl Oxide Methane (natural gas) Methanol (methyl alcohol) Methyl Ethyl Ketone Methyl Isobutyl Ketone 2-Methyl-1-Propanol (isobutyl alcohol) 2-Methyl-2-Propanol (tertiary butyl alcohol) Naphtha (petroleum) ⑥ Octane Pentane 1-Pentanol (amyl alcohol) Propane 1-Propanol (propyl alcohol) 2-Propanol (isopropyl alcohol) Propylene Pyridine Styrene Toluene Vinyl Acetate Vinyl Chloride Xylenes
E	Atmospheres containing combustible metal dusts regardless of resistivity, or other combustible dusts of similarly hazardous characteristics having resistivity of less than 10^5 ohm-centimeter.	
G	Atmospheres containing combustible dusts having resistivity of 10^5 ohm-centimeter or greater.	

① Certain chemicals may have characteristics that require safeguards beyond those required for any of the following groups. Carbon disulfide is one of these chemicals because of its low autoignition temperature and the small joint clearance to arrest its flame propagation.

② Certain metal dusts may have characteristics that require safeguards beyond those required for atmospheres containing electrically conductive combustible dusts. For example, zirconium, thorium, and uranium dusts have extremely low ignition temperatures (as low as 20°C) and minimum ignition energies lower than any material classified in any of the Class I or Class II groups.

③ Group C equipment shall be permitted for this atmosphere if such equipment is isolated in accordance with Section 501-5(a) of NFPA 70-1984, National Electrical Code, by sealing all conduit 1/2-inch size or larger.

④ Group D equipment shall be permitted for this atmosphere if such equipment is isolated in accordance with Section 501-5(a) of NFPA 70-1984, National Electrical Code, by sealing all conduit 1/2-inch size or larger.

⑤ For classification of areas involving ammonia atmosphere, see Safety Code for Mechanical Refrigeration (ANSI/ASHRAE 15-1978) and Safety Requirements for the Storage and Handling of Anhydrous Ammonia (ANSI/CGA G2.1-1972).

⑥ A saturated hydrocarbon mixture boiling in the range 20-135°C (68-275°F). Also known by the synonyms benzine, ligroin, petroleum ether, or petroleum naphtha.

FIGURE 2-4 Continued

not conduct electricity to the person using the fire extinguisher.

The fourth type of fire is the class D fire. These involve metals such as magnesium, sodium, titanium, and potassium, which burn very hot. These materials are very difficult to contain once they have been ignited because they burn at a very high temperature. The dry powder type of extinguisher is used to extinguish this type of fire.

You should also be aware that most of these fires produce heavy smoke, and in some cases they will produce toxic fumes. This means that it is important to call the fire department as soon as a fire is spotted and to remove all personnel from the building. In some factories, selected employees are trained to fight fires as a first response. They are also trained in proper utilization of firefighting equipment such a respirators and fire extinguishers available in the plant. This provides an extra margin of safety that can be used to control a fire until the fire department arrives.

Safety and Testing Associations

Since the first electrical components and controls were manufactured, it has been important to ensure that specific standards were prepared and utilized. As the number of companies making components increased, they could see the advantage of associations that would meet as often as needed to update old standards and to write new ones. The need for independent testing associations also became apparent to ensure compliance with the standards.

The first organization to form was called the International Electro-Technical Commission (IEC). This commission has grown to become the leader in writing recommendations for electrical equipment performance. Its standards are known as European Standards, and they are commonly used for the design of components sold in world markets. The IEC was conceived in 1904 with 42 member nations, which continue to remain active today. Each member has an equal voice in determining the standards that are included in their recommendations. This organization does not issue tough standards for compliance; rather they provide recommendations for testing electrical components. These recommendations for testing are designed to be used as a basis of comparison between similar products made by different manufacturers. This provides a base that design engineers can use to determine if one product is better than another.

Since IEC was based in Europe, many of the component designs reflect the conditions of the times rather than being written specifically for any reason. This means that components such as motor and switchgear were manufactured to be as small as possible and with the minimum of material since

FIGURE 2-5 Examples of NEMA-type enclosures with test specifications. (Courtesy of Allen-Bradley, Inc., and the National Electrical Manufacturers Association.)

ENCLOSURES FOR NON-HAZARDOUS LOCATIONS

For a degree of protection against:	Designed to Meet Tests No. [1]	For Indoor Use			Outdoor Use		Indoor or Outdoor		
		1	12	13	3R	3	4	4X	6P
Physical contact with internal equipment	6.2	✓	✓	✓	✓	✓	✓	✓	✓
Falling dirt	6.2	✓	✓	✓	✓	✓	✓	✓	✓
Rust	6.8		✓	✓	✓	✓	✓	✓	✓
Dust (windblown for outdoor types)	6.5	✓	✓	✓		✓	✓	✓	✓
Dripping oil or non-corrosive liquids	6.3(2.2)		✓	✓					
Spraying oil or non-corrosive liquids	6.12			✓					
Falling rain	6.4(2.1)				✓	✓	✓	✓	✓
Windblown rain	6.4(2.2)					✓	✓	✓	✓
Hose-directed water	6.7						✓	✓	✓
Submersion at limited depth	6.11								✓
Sleet and external icing	6.6(2.2)				✓	✓	✓	✓	✓
Corrosion	6.9							✓	✓

[1] See below for abridged description of NEMA enclosure test requirements. Refer to NEMA Standards Publication No. 250 for complete test specifications.

ABRIDGED DESCRIPTION OF NEMA ENCLOSURE TEST REQUIREMENTS

6.2 **Rod Entry Test** — A 1/8" diameter rod must not be able to enter enclosure except at locations where nearest live part is more than 4" from an opening — such opening shall not permit a 1/2" diameter rod to enter.

6.3 **Drip Test** — Water is dripped onto enclosure for 30 minutes from an overhead pan having uniformly spaced spouts, one every 20 sq. inches of pan area, each spout having a drip rate of 20 drops per minute.
Evaluation (2.2): No water shall enter enclosure.

6.4 **Rain Test** — Entire top and all exposed sides are sprayed with water at a pressure of 5 psi from nozzles for one hour at a rate to cause water to rise 18 inches in a straight-sided pan beneath the enclosure.
Evaluation (2.1): No water shall reach live parts.
Evaluation (2.2): No water shall enter enclosure.

6.5 **Dust Test (Alternate Method)** — Atomized water at a pressure of 30 psi is sprayed on all seams, joints and external operating mechanisms from a distance of 12 to 15 inches at a rate of three gallons per hour. No less than five ounces of water per linear foot of test length (height plus length plus depth of enclosure) is applied. No water shall enter enclosure.

6.6 **External Icing Test** — Water is sprayed on enclosure for one hour in a cold room (2°C): then room temperature is lowered to approximately −5°C and water spray is controlled so as to cause ice to build up at a rate of 1/4" per hour until 3/4" thick ice has formed on top surface of a metal test bar.
Evaluation (2.2): Equipment shall be undamaged after ice has melted (not required to be operable while ice-laden).

6.7 **Hosedown Test** — Enclosure and external mechanisms are subjected to a stream of water at 65 gallons per minute from a 1" diameter nozzle, directed at all joints from all angles from a distance of 10 to 12 feet. Test time is 48 seconds times the test length (height + width + depth of enclosure in feet). No water shall enter enclosure.

6.8 **Rust Resistance Test** — Enclosure is subjected to a salt spray (fog) for 24 hours, using water with five parts by weight of salt (NaCl), at 35°C., then rinsed and dried. There shall be no rust except where protection is impractical (e.g. machined mating surfaces, sliding surfaces of hinges, shafts, etc.)

6.9 **Corrosion-Resistance Test** — Same test conditions as 6.8 except exposure time 200 hours. There shall be no pitting, cracking or other deterioration more severe than that resulting from a similar 200 hr. test on passivated Type 304 stainless steel.

6.11 **Air Pressure Test (Alternate Method)** — Enclosure is submerged in water at a pressure equal to water depth of six feet, for 24 hours. No water shall enter enclosure.

6.12 **Oil Exclusion Test** — Enclosure is subjected to a stream of test liquid for 30 minutes from a 3/8" diameter nozzle at two gallons a minute. Water with 0.1% wetting agent is directed from all angles from a distance of 12 to 18 inches, while any externally operated device is operated at 30 operations per minute. No test liquid shall enter the enclosure.

FIGURE 2-5(b)

Europe was the location of a major war (World War II) that severely limited raw material and the large amounts of capital required to manufacture electrical equipment. In the 20 years immediately following the war, these conditions created a trend to make electrical controls as small and as rugged as possible. This is a case of external conditions shaping product design rather than compliance with a body of laws or recommendations.

In comparison, in the United States and North America, World War II caused the industrial base to become dependent on electrical products designed to be as rugged and as reliable as possible. Since heat is the main factor that causes motors and controls to deteriorate, they were designed slightly larger than was really necessary to ensure that they would not overheat. These larger sizes lead to some standardization that was not specified by any organization. For example, since the components were slightly larger, the enclosures in which they were housed were also designed to be slightly larger.

Today, control systems tend to be very large, so space becomes limited when the components must be mounted in electrical cabinets. This has forced manufacturers to downsize their components as much as possible. This has also led to a trend to compare the size of components to the smaller sizes found in European markets, and to a tendency to compare the size of components with the products listed in IEC test recommendations. In this way some size specifications for equipment are listed as complying with IEC standards even though no IEC standards specifically address component size.

National Electrical Manufacturers Association (NEMA). NEMA was conceived in 1926 in Washington, D.C., with 550 members. These members represented manufacturers of electrical components from all areas of North America. The manufacturers provide recommendations for the operation of components, such as the minimum number of cycles that a component should be able to provide. They also provide recommendations for types of enclosures and methods that should be employed to protect components from hazardous environments such as explosive vapors.

Even though these associations do not perform testing, the standards that they provide are used by testing associations to ensure that products indicated as meeting NEMA standards actually comply with all tests.

Joint Industrial Council (JIC). The Joint Industrial Council has evolved out of a combination of standards written by two groups, the National Machine Tool Builders Association (NMTBA) and the Joint Industrial Conference. In 1941 the NMTBA issued its first set of standards and revised them periodically through 1960. The Joint Industrial Conference, begun in 1948, produced a set of standards that were updated periodically through 1957. In 1963 these two groups combined efforts to produce a new set of standards and a new organization, called the Joint Industrial Council (JIC). One goal of the JIC was to write standards for electrical equipment that supported the development and advancement of the American National Standards Institute (ANSI).

Since the JIC standards evolved from two organizations with slightly different objectives, the new standards are written in two documents that follow the original intent of these groups of manufacturers. One of the documents is called *JIC Electrical Standards for General Purpose Machine Tools,* and the other is called *Standards for Mass Production Equipment.* Many of the standards are similar, but differences are provided where necessary to be applicable to the type of equipment being produced.

Underwriters' Laboratory (UL). The Underwriters' Laboratory is one of many organizations that exist solely to perform rigorous testing that is specified in many association standards. UL also provides a set of test guidelines with which all equipment should comply to provide safe operational service. UL is different from organizations that write specifications, in that their main objective is to perform tests and report their results. This organization originated in 1869 to provide independent testing of equipment for insurance companies. The insurance companies were asked to write policies that would cover damage from fires. Since electrical components were responsible for many fires during the late nineteenth century, insurance companies recognized the need for independent testing results that could be used as a basis of underwriting fire damage. Since this was one of the earliest organizations that provide testing of electrical components, it has become one of the more common standards used. Today there are many organizations that provide similar testing and they are now as common as UL.

Types of Enclosures and Their Ratings

Figure 2-5 shows a list of enclosures that must be used in hazardous locations. The list provides standards for manufacturers to meet in the design of the enclosures they produce to protect electrical devices. The standards are produced by NEMA and NEC and cover all conditions to which electrical components may be subjected.

Figure 2-5 also shows examples of the types of NEMA enclosures and the conditions they are rated for. From these examples you can see that enclosures are rated for the protection they provide against dust, dirt, moisture, and oils that may penetrate the working parts or contacts of electrical equipment. A table is provided in this figure to help you see the types of conditions to which electrical equipment will be subjected. It is important to understand that the proper enclosure is required by local and national codes. You must know what enclosures are available and ensure that they are specified when a control system is designed and installed. You should inspect enclosures that are installed in existing systems periodically to ensure that the gaskets and protective covers are still in good condition. More information about the application of these enclosures is provided in the chapters that present information about equipment installed in these enclosures.

PERIODIC MAINTENANCE INSPECTION FORM

MACHINE NAME _____ MACHINE # _____ DATE _____

INSPECTION POINT	CONDITION	DATE ACTION TAKEN
1. Inspect electrical panel and ensure it is clean and free of debris.		
2. Check the condition of all wires and electrical terminals (look for loose or damaged wires and terminals)		

3. Check the condition of all bolts, screws and other fasteners at the following locations:
   a. motor mounts
   b. fasteners on moving parts
   c. doors and cabinets
4. Check for missing nuts, bolts and screws.		

5. Check the condition of all belts. Tighten to torque specifications where applicable.		
6. Check the condition of all clutches, brakes and transmissions. (Make sure all guards are in place and operating correctly.)		
7. Check the condition and operation of all safety devices on the machine including emergency stop buttons and lock out devices.		
8. Lubricate all points indicated on the machine lubrication chart. Be sure to check the level of all lubricant reservoirs.		

FIGURE 2-6 Periodic inspection list that indicates points that should be inspected on a regular basis.

NEMA enclosures for starters

Type	Enclosure
1	General purpose—indoor
2	Drip-proof—indoor
3	Dust-tight, raintight, sleet-tight—outdoor
3R	rainproof, sleet resistant—outdoor
3S	Dust-tight, raintight, sleetproof—outdoor
4	Watertight, dust-tight, sleet resistant—indoor
4X	Watertight, dust-tight, corrosion resistant—indoor/outdoor
5	Dust-tight—indoor
6	Submersible, watertight, dust-tight, sleet resistant—indoor/outdoor
7	Class 1, group A, B, C, or D hazardous locations, air-break—indoor
8	Class 1, group A, B, C, or D hazardous locations, oil-immersed—indoor
9	Class II, group E, F, or G hazardous locations, air-break—indoor
10	Bureau of Mines
11	Corrosion-resistant and drip-proof, oil immersed—indoor
12	Industrial use, dust-tight, and driptight—indoor
13	Oiltight and dust-tight—indoor

Test specifications for enclosures

Safety Inspections

It is important to check all parts of working equipment periodically for conditions that may be unsafe and cause damage to personnel, equipment, or to the product being produced. These inspections are generally placed under the heading of periodic inspection or preventive maintenance. Most adverse conditions can be corrected rather easily if they are found when they are beginning. But if any of these are left unattended, they may deteriorate to the point of severely damaging equipment or hurting personnel.

Another point that should be understood is the relationship between the person making the inspection and the person making the repairs. In some plants maintenance personnel are expected to complete the inspection and then come back and make any repairs. In some cases, few conditions are actually detected because the person making the inspection knows that he or she will be responsible for making repairs for any poor conditions that are found. This is clearly not the best way to handle safety inspections. A better method utilizes an independent inspector who is knowledgeable in all aspects of equipment operation and safety procedures. Since the inspector is not responsible for making repairs, a more thorough inspection is apt to be made, and follow-up checks can be instituted to ensure that the repairs are actually accomplished.

Since many motor control systems are controlled by programmable controllers, it is possible to program a clock and counter that will provide an alarm or printed report indicating that an inspection is required because a machine has operated for the allotted time or has operated a predetermined number of cycles. Figure 2-6 shows a typical list of inspection points that should be checked periodically.

WORKING SAFELY

Now that you understand some of the safety conditions that may exist at a work site, including the laboratory at your school, it is important to take the responsibility for working safely. You must look out for your own welfare and the welfare of people with whom you are working. This includes locking out the electrical disconnect of systems on which you are working. Figure 2-7 shows a diagram of a lockout and the method used to ensure that all personnel are aware when power is being returned to the system. As you can see, the lockout mechanism has room for up to six padlocks to be connected. Each person who is working on the equipment must place a padlock in one of the holes. This means that the lockout device cannot be removed from the disconnect until all six workers have removed their padlocks. This ensures that all personnel who are working on the project are aware that power is being returned to the system.

Other safety conditions should be closely monitored when you must work above the ground on ladders or scaffolding or on beams. Be sure to wear all safety equipment and be aware where you walk or place your weight, so that you do not fall. Also be aware of exposed conductors in some bus systems in factory ceilings. If you are working with a ladder or a high lift, be sure that they not come into contact with live wires. A good habit to get into when you start a job is to stand back and survey the situation and identify the possible safety conditions that exist that could turn into an accident. Determine a strategy that will include being aware of safety conditions as you go about your work. In this way you become more aware of the conditions that could turn into accidents and learn to work safely around them. Equipment manufactures also try to identify potential safety hazards by marking them with highly visible signs. Figure 2-8 shows an example of a machine displaying several safety signs.

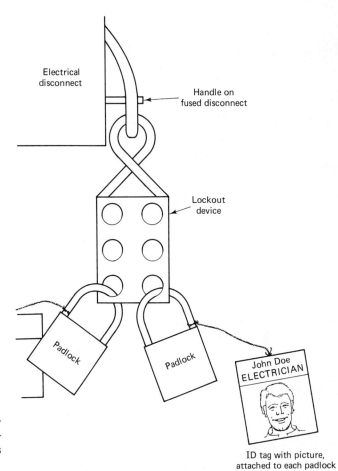

FIGURE 2-7 Typical lockout device used to ensure that all personnel are aware when power is returned to a system.

FIGURE 2-8 Example of typical safety warning signs displayed on equipment. (Courtesy of NATCO, Inc.)

Working Safely 21

QUESTIONS

2-1. Explain why you should learn CPR if your job causes you to work around electrical circuits.

2-2. Explain why you should wear safety glasses at all times when working with electrical circuits. List several hazards that may allow injury to your eyes when you are working with electrical circuits.

2-3. Explain two different ways that you may receive an electrical shock when working around equipment.

2-4. Explain what an electrical short circuit is and what hazard it presents.

2-5. Explain the purpose of a safety ground and how it works to prevent electrical shock.

2-6. Explain the operation of a ground-fault interrupter (GFI).

2-7. Give an example of each of the three classes of hazardous locations indicated by the National Electrical Code®.

2-8. Give an example of a class A, a class B, and a class C fire. Explain why each of type requires a special fire extinguisher.

2-9. Explain what the IEC is and what it is responsible for.

2-10. Explain what NEMA is and what it is responsible for.

2-11. State two basic differences between components that meet IEC standards and components that meet NEMA standards.

2-12. Explain what JIC is and what it is responsible for.

2-13. Explain what UL is and how it differs from JIC, IEC, and NEMA.

Symbols and diagrams

3

INTRODUCTION

Electrical diagrams and other drawings are used to communicate a large amount of information about the location and operation of equipment on which you will be working. This information includes the sequence of operation, operating instructions, and the location of electrical components that control a system. In order to include as much information as possible on a diagram, symbols will be used to represent all the electrical components.

The diagrams that you will use to install, troubleshoot, and repair electrical equipment are drawn on vellum paper and converted to blueprints which are used on the factory floor. The complete set of blueprints is required to understand fully how the machine has been designed to operate. They will include a plant layout diagram that shows the location of the equipment in the factory and its relationship to other equipment, such as conveyors and material-handling equipment. A machine drawing will be included to show the location of all electrical cabinets on the equipment.

The wiring diagram will be included to show the location of all electrical components in the cabinet and the route of all wires between them. Wire numbers and cable numbers will also be included in this print. An elementary diagram will be included in the set of prints to show the sequence of operation. This diagram is also called a *ladder diagram* because it shows the sequence of operation in separate lines that look like the rungs of a wooden ladder. This diagram is invaluable in troubleshooting a system when it becomes inoperative.

If the machinery is controlled by a programmable controller, its program will be shown on a print and it will look very similar to the ladder diagram. A diagram that shows the type and number of input and output modules will be included in this set of prints, which will help you troubleshoot the system.

At the end of each set of blueprints will be a parts list. The parts list will detail all components used in the equipment and provide part numbers and other important information, such as voltage and current specifications. When you determine that a part is inoperative and needs to be replaced, you will use the parts list to find the manufacturer's part number so that you can locate a new part in the parts crib, or provide information that will be needed to order a new one.

In this chapter we provide examples of all of these types of blueprints and explain how to read and use them in installing, troubleshooting, and repairing equipment. Additional information will provide pictures of some of the typical devices that are used in electrical systems and show their electrical symbols. This will help you relate the symbol to the actual component that you will be working with when testing and troubleshooting.

You will also understand how the motor control process is displayed on paper so that electrical technicians will have available the vast amount of information that was used to design the machinery and motor control circuit. You will learn that the motor control process is a three-step process where inputs are used to sense the conditions in the system. This information is presented to another part of the control system, which must decide what action is required to be taken in terms of a signal sent to an output device such as a motor starter or solenoid. The diagrams and controls presented throughout this book will be related to the sequence of events (sense, decide, act) that are used to provide motor control.

In the final part of this chapter we provide vital information about interfaces between electrical control systems and other important systems in the factory. These include electrical mechanical systems, such as robots, material-handling equipment, and hydraulic systems that are used in virtually all types of manufacturing and process equipment. Other interfaces between the electrical systems and computer-controlled devices are also explained. The most common of these interfaces is between the programmable controller and the electrical mechanical controls in the motor control system. When you have completed the chapter you will be able, properly and completely, to use all information that is provided in the blueprints and other diagrams that you will use when working with motor controls.

ELECTRICAL SYMBOLS

Electrical components are identified by symbols on electrical wiring diagrams and ladder diagrams. These symbols have been standardized by the Joint Industrial Council (JIC).

JIC Symbols

Figure 3-1 shows several typical electrical symbols that are used in various electrical diagrams. A picture of the component that is being represented is also shown with these symbols to give you an idea of what each symbol is trying to represent. The most difficult part of understanding symbols is to get a visual image of the component that is being represented by the symbol. This visual image is called a *referent*. After you have the image in your memory, you will be able to look at a symbol on a diagram and instantly visualize what component that symbol is representing and what it looks like when it is installed on the machinery. The link between the symbol and picture of the component is vital if you are to receive the full amount of information that is provided on any electrical diagram. If you find a symbol in an electrical diagram, you should try to find a picture of the component or an actual component to look at so that you will have an image of the component in your memory for future use.

The symbols that are shown in this figure are used in wiring diagrams and elementary (ladder) diagrams. They will show the number of terminals the component has and the identification that is used to mark each terminal. The diagram will also show the number of electrical parts that the component has, such as the number of contacts or poles.

A complete set of JIC electrical symbols is provided in Figure 3-2. From this set you can see that all the electrical components that are commonly used in motor control circuits are included in this set. You should understand that these symbols include several standard methods of representing the operation of components. For example, a dashed line between sets of contacts indicates that each of the sets of contacts is operated by the same actuator. This is used to indicate that multiple-pole switches such as disconnects and circuit breakers will have all of their sets of contacts activated by the same condition. Another method of indication operation is shown in the symbol of plugging switches. A curved line with an arrowhead on each end indicates that this switch moves in a rotary motion when it is in operation.

You should also understand that each switch has a different method of activating its contacts. The symbol of the operator normally looks like the device it is representing. For example, the symbol of the float switch (liquid-level switch) uses a ball to represent the float, and the symbol of the pressure switch uses a pressure element to represent the pressure operator of that switch. The location of the switch contact in the symbol of a switch is also used to indicate whether the contacts will open or close when power or the means of activation is removed. For example, the symbol for the normally open pressure switch shows the switch arm opening below the contact. This indicates that when pressure increases (goes up) the switch will move up

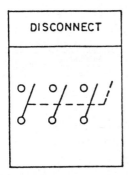

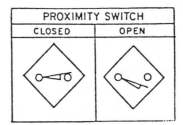

a b c d

FIGURE 3-1 Electrical symbol and picture of electrical components. (Courtesy of Eaton Corp./Cutler-Hammer Division.)

Electrical Symbols 25

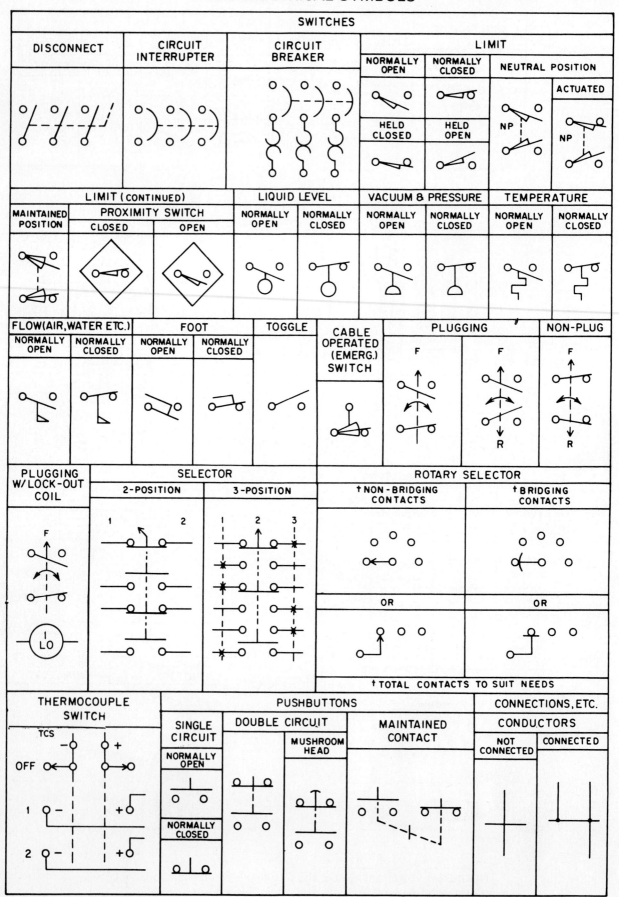

FIGURE 3-2 Complete set of JIC electrical symbols. (Courtesy of Joint Industrial Council.)

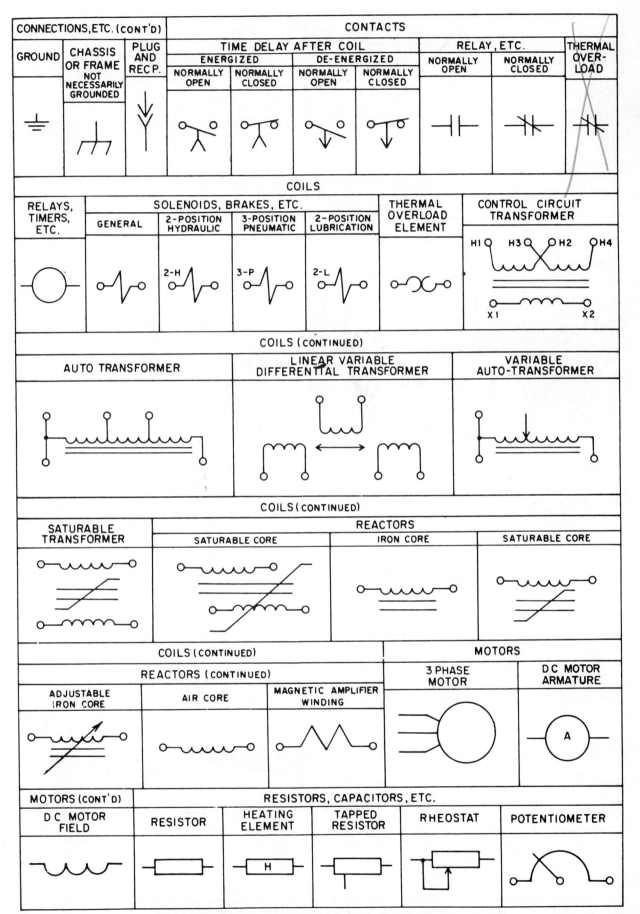

FIGURE 3-2 JIC Electrical Symbols (Continued)

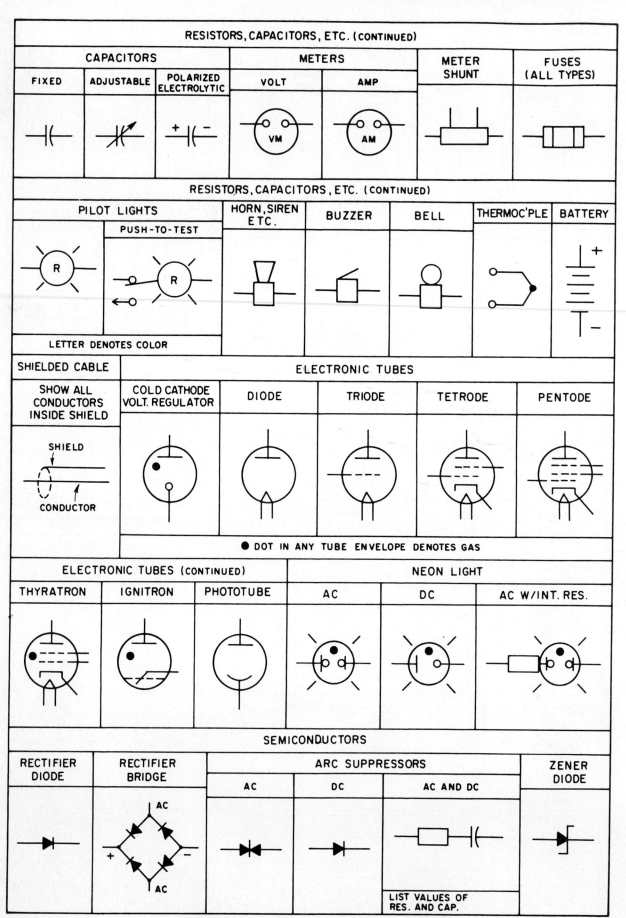

FIGURE 3-2 JIC Electrical Symbols (Continued)

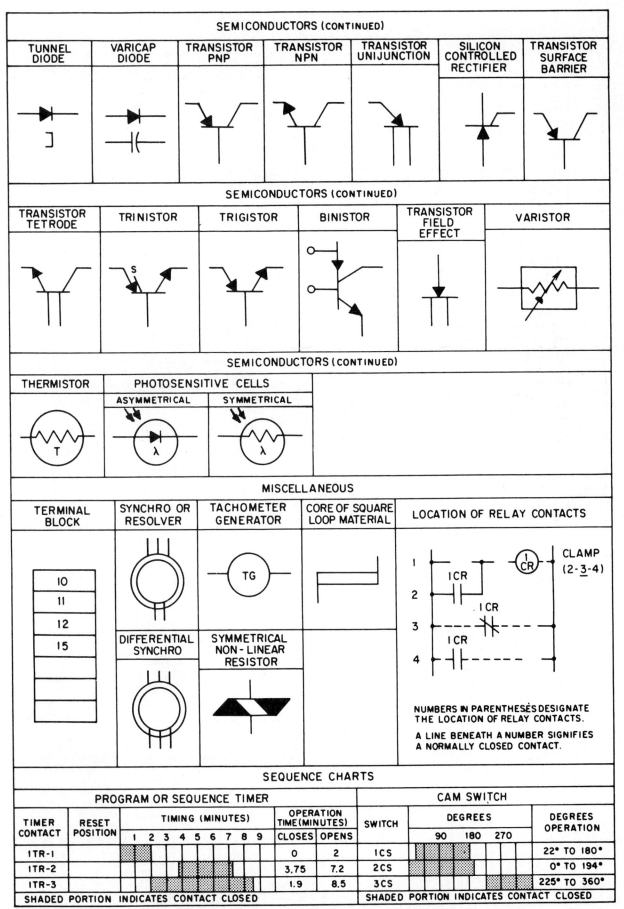

FIGURE 3-2 JIC Electrical Symbols (Continued)

STANDARD ELEMENTARY DIAGRAM SYMBOLS

The diagram symbols shown below have been adopted by the Square D Company and conform where applicable to standards established by the National Electrical Manufacturers Association (NEMA).

SUPPLEMENTARY CONTACT SYMBOLS

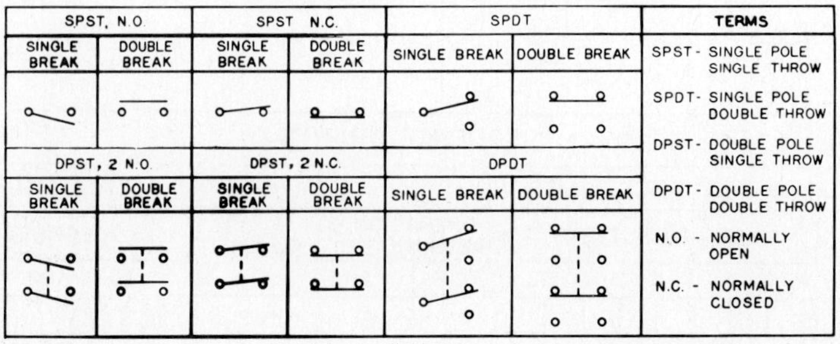

FIGURE 3-3 (a) Electrical symbols used by the Square D Company adopted from the National Electrical Manufacturers Association; (b) fluid power symbols. [(a) Courtesy of Square D Company; (b) courtesy of Parker Fluid Power.]

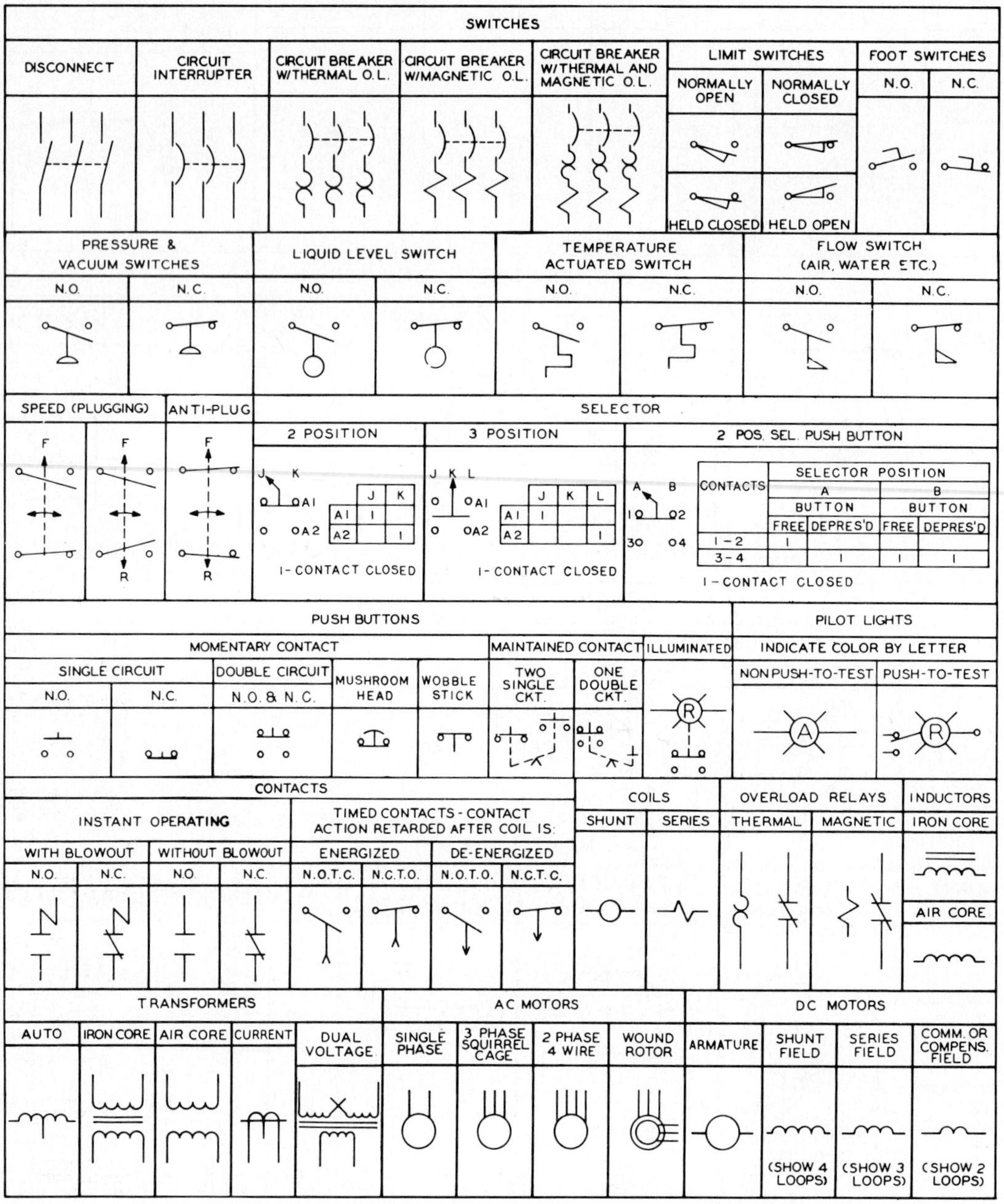

FLUID POWER GRAPHIC SYMBOLS

Parker FRL Series Air Preparation Units

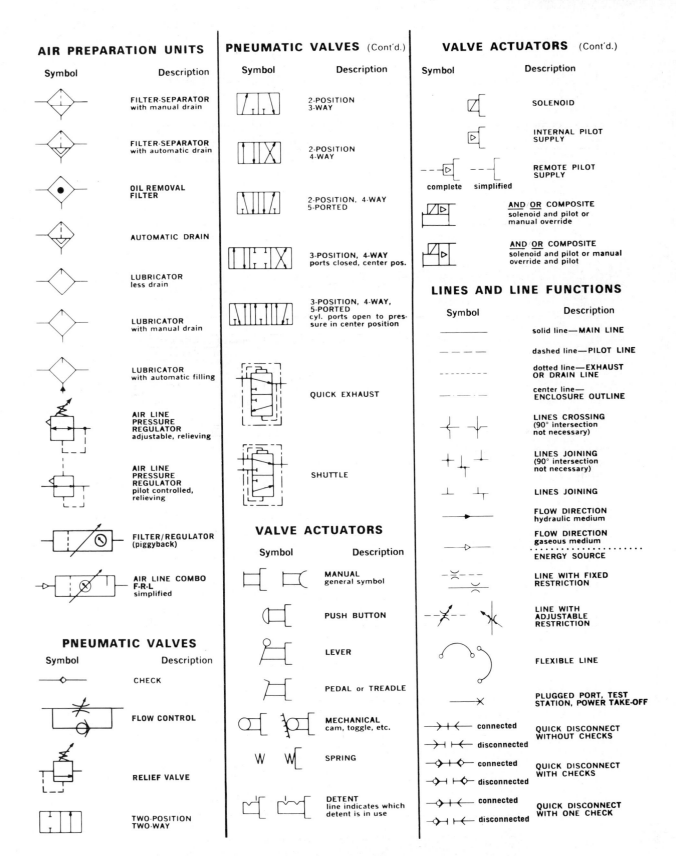

FIGURE 3-3 (Continued)

and close. In the instance of the normally closed pressure switch, the switch arm is shown above the contact, which means that the switch will open when pressure increases. You can assume that when the force (pressure, temperature, flow, etc.) increases, the switch arm shown in the symbol is moved upward, and when it decreases, it moves downward.

Another important feature of the symbols is the symbol for conductors. In a diagram, conduc-

COMMON JIC ELECTRICAL ABBREVIATIONS

Designation	Device	Designation	Device
A	Accelerating Contactor or Relay	MF	Motor Starter – Forward
ABE	Alarm or Annunciator Bell	MR	Motor Starter – Reverse
ABU	Alarm or Annunciator Buzzer	MSH	Meter Shunt
AH	Alarm or Annunciator Horn	MTR	Motor
AM	Ammeter	NLT	Neon Light
AT	Autotransformer	OL	Overload Relay
B	Brake Relay	PB	Pushbutton
CAP	Capacitor	PC	Printed Circuit
CB	Circuit Breaker	PL	Plug
CH	Chassis or Frame (not necessarily grounded)	PLS	Plugging Switch
CI	Circuit Interrupter	POT	Potentiometer
CON	Contactor	PRS	Proximity Switch
COS	Cable Operated (Emergency) Switch	PS	Pressure Switch
CR	Control Relay	PSC	Photosensitive Cell
CRA	Control Relay, Automatic	Q	Transistor
CRE	Control Relay, Electronically-Energized	QBN	Binistor
CRH	Control Relay, Manual	QFE	Transistor, Field-Effect
CRL	Control Relay, Latch	QSB	Transistor, Surface-Barrier
CRM	Control Relay, Master	QT	Transistor, Tetrode
CRU	Control Relay, Unlatch	QTG	Trigistor
CS	Cam Switch	QTM	Thermistor
CT	Current Transformer	QTN	Trinistor
CTR	Counter	QU	Transistor, Unijunction
D	Diode	QVR	Varistor
DAS	Diode Arc Suppressor	R	Reverse
DB	Dynamic Braking Contactor or Relay	REC	Rectifier
DISC	Disconnect Switch	RECP	Receptacle
DT	Tunnel Diode	RES	Resistor
DVC	Varicap Diode	RH	Rheostat
DZ	Zener Diode	RSS	Rotary Selector Switch
F	Forward	S	Switch
FA	Field Accelerating Contactor or Relay	SCR	Silicon Controlled Rectifier
FB	Fuse Block	SOC	Socket
FD	Field Decelerating Contactor or Relay	SOL	Solenoid
FF	Full Field Contactor or Relay	SS	Selector Switch
FL	Field Loss Contactor or Relay	ST	Saturable Transformer
FLD	Field	SX	Saturable Core & Reactor
FLS	Flow Switch	SYN	Synchro or Resolver
FS	Float Switch	T	Transformer
FTB	Fusible Terminal Block	TACH	Tachometer Generator
FTS	Foot Switch	TAS	Temperature-Actuated Switch
FU	Fuse	TB	Terminal Block
FW	Field Weakening	T/C	Thermocouple
GRD	Ground	TCS	Thermocouple Switch
HTR	Heating Element	TGS	Toggle Switch
INST	Instrument	TR	Time Delay Relay
IOL	Instantaneous Overload	TRE	Timer Relay, Electronically-Energized
LO	Lock-Out Coil (located in plugging switch)	TVM	Tachometer Indicator
LS	Limit Switch	V	Electronic Tube
LT	Pilot Light	VAT	Variable Autotransformer
LVT	Linear Variable Differential Transformer	VM	Voltmeter
M	Motor Starter	VS	Vacuum Switch
MAX	Magnetic Amplifier Winding	WLT	Work Light
MB	Magnetic Brake	WM	Wattmeter
MC	Magnetic Clutch	X	Reactor
MCS	Motor Circuit Switch		

FIGURE 3-4 Common abbreviations for JIC electrical symbols. (Courtesy of Joint Industrial Council.)

tors are represented by a line. If the conductor carries high voltage and high current, the line will be heavier than the lines for conductors that carry lower voltage and lower current. This means that the wires used to represent load circuit conductors will be darker and heavier than lines used to represent conductors in the control circuit.

The symbol for conductors will also indicate the presence of a junction of two or more conductors. If conductors cross each other and form an electrical junction, this point will be indicated by a dot or small circle that is filled in. If the wires merely cross each other but are not connected electrically, no terminal point will be shown. As diagrams become increasingly complex, many wires must be shown crossing other wires to make connections at the proper terminal on each component. If a connection is being indicated, the terminal point will be shown.

Other Electrical Symbol Standards

Other electrical associations provide sets of standardized electrical symbols that are commonly used. Some of these standards have been adopted by equipment manufacturers and incorporated into the symbols that they use in their electrical equipment catalogs. Figure 3-3a shows an example of a set of symbols used by the Square D Company that have been adopted from the standards presented by the National Electrical Manufacturers Association (NEMA). You can see that these symbols are very similar to the ones adopted by JIC, and some companies will use a set of standards that have been adopted from both associations.

Other symbols will be used to represent the electrical part of other components, such as pneumatic devices, and they are shown in Figure 3-3b. The electrical part of these components will be connected to other parts of the electrical circuits on which you will be working.

Abbreviations Used with Electrical Symbols

Another important part of the electrical symbols that are used to identify each component in the diagram is the component abbreviation. The abbreviation provides a method of identifying multiple components that are used in the system. For example, every electrical technician can easily recognize the symbol for a limit switch. If more than one limit switch were used in a circuit, it would be difficult to tell them apart if they were not identified by an abbreviation that included a number. For instance, the first limit switch would carry the abbreviation LS1 and the second would be identified as LS2. The abbreviation LS helps identify the component symbol as a limit switch, and the numbers help identify which limit switch is being indicated. Abbreviations are also used to identify components that do not have a distinct symbol. Several electrical components use the symbol of a coil of wire. This symbol could indicate a motor winding, a transformer coil, a relay coil, or a solenoid. The abbreviation helps to identify these components positively.

Another common symbol used for electrical diagrams is a square or rectangular box. An abbreviation is used in conjunction with the box to indicate what type of component or circuitry is housed in the box. Each association uses its own standards for abbreviations. It is acceptable practice to indicate the full name with abbreviations that are not easily identified. A set of abbreviations used by JIC is provided in Figure 3-4.

ELECTRICAL DIAGRAMS

A large variety of electrical diagrams are required to present all the information about the location, installation, operation, and replacement of components that are connected together to provide a motor control system. These diagrams will be used to manufacture the machinery before it is brought into the plant. They will also be used to install the machinery correctly and to start it up.

After the machinery has been installed and is operating, the diagrams will be used to troubleshoot faults that occur and to identify parts that must be replaced. If the machine is controlled by a programmable controller (P/C), a diagram will be used to show the interface between the P/C and the hardware components used as inputs and outputs. A copy of the P/C program will also be provided on a set of diagrams. This allows you to use the P/C to help locate the problem.

When a faulty part is located, another diagram will be used to determine wire numbers and terminal locations. These diagrams will also be used to determine which terminals should be used during troubleshooting tests, and the part number for the replacement part. After the replacement part has been located, the diagrams will be used to remove and replace the part and to make any preliminary setup or calibration adjustments. In this section of the chapter we present several different types of diagrams and explain how to utilize all the information that is provided on them.

Two of the most useful diagrams that are used in installation, troubleshooting, and repair are the wiring diagram and the elementary diagram. Example wiring diagrams are provided in Figure 3-5 and an elementary diagram is provided in Figure 3-6. These two diagrams represent the same set of components in two distinct different ways. The *wiring diagram* shows the location of the components in relation to each other in the electrical cabinet and also show the exact location of terminals on the components and the location of wire bundles within the cabinet. The *elementary diagram,* which is also called a ladder diagram, is used to show the sequence of operation. This means that the diagram is trying to show what must happen in the proper order for the equipment to operate correctly.

General Wiring Diagram

An electrical wiring diagram is shown in Figure 3-5a. From this diagram you can see that the each component is represented by a figure that looks like the outline of the part. This means that the component will generally be represented by a box with terminal locations indicated in the proper locations. The main function of the wiring diagram is to show the location of components in relation to each other, and the location of specific terminals on each component. Wire numbers and terminal numbers are used to indentify all conductors and terminals within the diagram.

The wiring numbers and terminal identification can also be used in conjunction with the ladder diagram shown in Figure 3-6. The ladder diagram will show the sequence of operation, and the wiring diagram provides the location of components and their specific terminals that are used when voltage and current tests are made. Generally, the ladder diagram is used to locate possible faults in the system, and the wiring diagram is used to indicate where the meter leads should be placed for the test.

The wiring diagram is provided by the machine builder since it also used when the machine is being built. If any wiring changes are made, or if any new component is added after the machine is

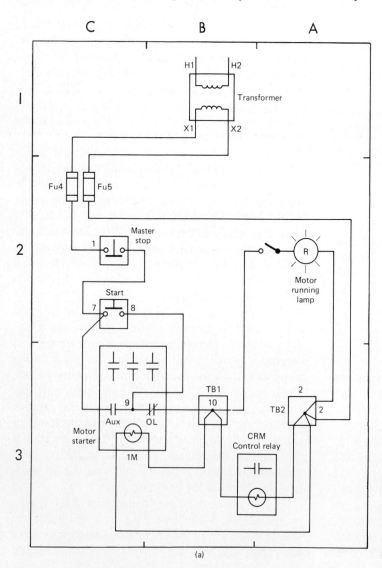

FIGURE 3-5 (a) Electrical wiring diagram of a metal-forming machine; (b) electrical wiring diagram and elementary diagram of a motor starter. [(b) Courtesy of Square D Company.]

WIRING DIAGRAM OF A MOTOR STARTER

WIRING DIAGRAM

A WIRING DIAGRAM shows, as closely as possible, the actual location of all of the component parts of the device. The open terminals (marked by an open circle) and arrows represent connections made by the user.

Since wiring connections and terminal markings are shown, this type of diagram is helpful when wiring the device, or tracing wires when troubleshooting. Note that bold lines denote the power circuit, and thin lines are used to show the control circuit. Conventionally, in ac magnetic equipment, black wires are used in power circuits and red wiring is used for control circuits.

A wiring diagram, however, is limited in its ability to convey a clear picture of the sequence of operation of a controller. Where an illustration of the circuit in its simplest form is desired, the elementary diagram is used.

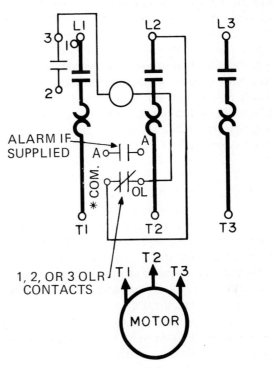

* Marked as "OL" if alarm contact is supplied.

ELEMENTARY DIAGRAM OF A MOTOR STARTER

ELEMENTARY DIAGRAM

The elementary diagram gives a fast, easily understood picture of the circuit. The devices and components are not shown in their actual positions. All the control circuit components are shown as directly as possible, between a pair of vertical lines, representing the control power supply. The arrangement of the components is designed to show the sequence of operation of the devices, and helps in understanding how the circuit operates. The effect of operating various auxiliary contacts, control devices etc. can be readily seen — this helps in trouble shooting, particularly with the more complex controllers. This form of electrical diagram is sometimes referred to as a "schematic" or "line" diagram.

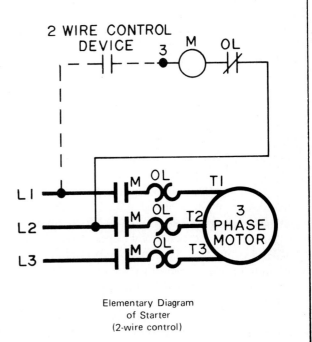

Elementary Diagram
of Starter
(2-wire control)

FIGURE 3-5 (Continued)

installed, this diagram will be used to locate terminal points and spaces in the panel where the component can be located. Some wiring diagrams also provide grid numbers that are used to locate components or terminals in a large diagram. A cross-reference sheet is provided with the grid that indicates the location of each component or conductor on the diagram. For example, in the diagram shown in Figure 3-5a you can see that motor starter 1 (MS1) is located at grid location 3,C. This means that you can move to those grid locations and find the device marked MS1. The grid is necessary when a very large diagram is shown on a blueprint, and in cases where the diagram requires two or three pages of blueprints to display the entire system.

You should also notice that the electrical symbols shown in Figure 3-3 are not used in the wiring diagram. Some parts of the symbol may be used, but generally the wiring diagram uses the outline of the component rather than its symbol. The internal contacts of the device are generally shown using the electrical symbol for switches or contacts. The terminals of each component are shown in close proximity to where they will be found on the actual component. These terminals are generally shown as a small circle on the diagram. Other crucial information about the direction in which a component should be mounted is also listed on the wiring diagram. You can assume that the top of the diagram is the reference for the diagram, and you can determine the direction of mounting from the location of terminals or other identifying marks on the component.

Wiring Diagrams of Components

Wiring diagrams of individual components are also provided for some components, such as motor starters. Figure 3-5b shows the wiring diagram of a full-voltage motor starter. From this diagram you can see that the auxiliary contacts for the contactor are shown on the left side of the diagram and other auxiliary contacts are shown on the right side of the diagram. You should also notice that the coil for each component is shown as a circle. The terminals of all the contacts are identified with letters and numbers, such as T1 or L1. You should also notice that many of the conductors cross each other but are not connected. Terminal points are indicated by a black dot, which indicates that two conductors are connected at that point. The actual location of the connection will be at the terminal connection on the motor starter contacts rather than as a splice in the middle of the conductor.

Some of the components in the diagram are represented by their electrical symbol. These include the pushbutton switch shown on the left side of the motor starter, indicator lamps, and the heater and overload contacts in the overload device. You will also notice in the diagrams provided by some manufacturers that symbols are used that do not belong to any standards that have been listed. These symbols may have originated when new electronic products were developed from traditional products or as a combination of several technologies, such as hydraulics and pneumatics with electronics.

The top of the wiring diagram shows part of an elementary diagram. The elementary diagram (ladder diagram) is used to show the sequence of operation in the control circuit. Notice that the lines that represent the conductors in the control circuit are smaller than the lines used to represent the load circuit conductors. The smaller lines indicate that smaller-size wire is used in the control circuit. In fact, in this diagram the wire size is listed with the lines that are used in the diagram to represent them. You can see that the large dark lines in the load circuit indicate that 8-gage wire is used, the smaller wire in the load circuit should be 14 gage, and the wire in the control circuit is 16 gage. This will help you determine what size wire should be used for replacement, since wire size is not stamped on wire used in original equipment manufacturing (OEM) installations.

Ladder Diagrams

The ladder diagram shown in Figure 3-6 is the most widely used electrical diagram because it provides the sequence of operation for a system, which explains what should happen when the machine is started. The diagram is divided into two distinct sections. The top part of the diagram will show all components and conductors of the load circuit, and the lower part of the diagram shows the control part of the circuit.

The load circuit in this diagram is powered by 460-V three-phase 60-Hz ac, while the control circuit uses 115 V. The three-phase voltage lines in the load circuit are identified as L1, L2, and L3, which indicates where these lines should be connected during installation. The small circles on the left side of the disconnect switch indicate that the three high-voltage lines should be connected in the disconnect switch. Since the symbol shows a set of three fuses with the disconnect symbol, it indicates that the disconnect is a fusible type. The size of the fuses (15 A) is listed directly below the bottom fuse, which will help you check the actual size of the fuse in the panel. If any of the fuses should fail, they would be replaced with a 15-A fuse. The parts list

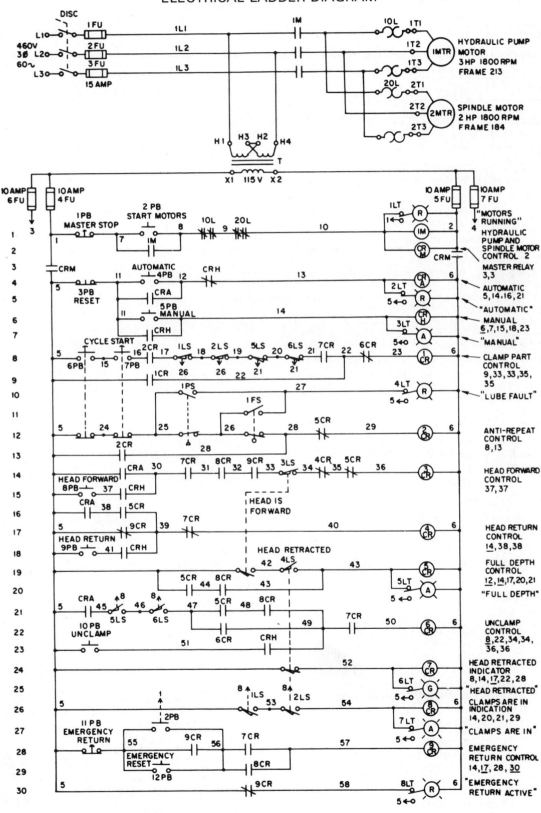

FIGURE 3-6 Ladder diagram of a metal-forming machine shown in Figure 3-5a. (Courtesy of Joint Industrial Council.)

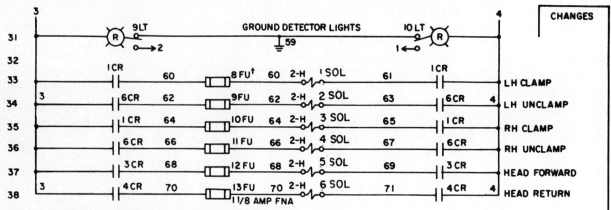

† WHERE SOLENOID FUSE IS PART OF TERMINAL STRIP
ASSEMBLY, CONDUCTOR NUMBER DOES NOT CHANGE

SEQUENCE OF OPERATION

A. PRESS "START MOTORS" PUSHBUTTON "2PB". MOTORS START. "MOTORS RUNNING" LAMP "1LT" AND "CRM" ARE ENERGIZED.

B. PRESS EITHER AUTOMATIC "4PB" OR MANUAL "5PB" PUSHBUTTON. CORRESPONDING RELAY AND LAMP ARE ENERGIZED.
NOTE: TO SWITCH FROM MANUAL TO AUTOMATIC, OPERATOR MUST PRESS "RESET" PUSHBUTTON "3PB" BEFORE PRESSING AUTOMATIC PUSHBUTTON "4PB".

C. AUTOMATIC CYCLE: WITH MOTORS RUNNING AND "CRA" ENERGIZED, MACHINE IS SET FOR AUTOMATIC CYCLE. HEAD MUST BE RETRACTED AND PART UNCLAMPED TO START CYCLE.

  1. OPERATOR LOADS PART IN FIXTURE AND PRESSES BOTH "CYCLE START" PUSHBUTTONS "6PB" AND "7PB", ENERGIZING "1CR" (SOL A AND SOL C) TO CLAMP PART.
  2. CLAMPED PART TRIPS "1LS" AND "2LS", ENERGIZING "8CR". RELAY "3CR" (SOL E) IS ENERGIZED MOMENTARILY, STARTING HEAD FORWARD IN RAPID ADVANCE. HEAD CAMS VALVE INTO FEED.
  3. WHEN HEAD IS IN FORWARD POSITION, "3LS" IS TRIPPED, ENERGIZING RELAY "5CR".
  4. "5CR" CONTACT ENERGIZES RELAY "4CR" (SOL F) AND HEAD RETURNS.
  5. WHEN HEAD IS FULLY RETRACTED, "4LS" IS TRIPPED, DE-ENERGIZING "4CR" AND ENERGIZING "7CR" WHICH ENERGIZES "6CR" (SOL B AND SOL D) UNCLAMPING PART.
  6. WHEN PART IS UNCLAMPED, "5LS" AND "6LS" ARE TRIPPED, DE-ENERGIZING RELAY "6CR".
  7. "2CR" RELAY PREVENTS MACHINE RE-CYCLING IF BOTH "CYCLE START" PUSHBUTTONS ARE NOT RELEASED.

D. MANUAL CYCLE:

  1. WITH HEAD RETRACTED AND PART UNCLAMPED, PRESS "CYCLE START" PUSHBUTTONS "6PB" AND "7PB", ENERGIZING RELAY "1CR" (SOL A AND SOL C) TO CLAMP PART.
  2. PRESS "HEAD FORWARD" PUSHBUTTON "8PB", ENERGIZING "3CR" (SOL E) TO START HEAD FORWARD.
  3. TO RETURN HEAD TO RETRACTED POSITION, PRESS "HEAD RETURN" PUSHBUTTON "9PB", ENERGIZING "4CR" (SOL F).
  4. WITH HEAD RETRACTED, PRESS "UNCLAMP" PUSHBUTTON "10PB", ENERGIZING "6CR" (SOL B AND SOL D) TO UNCLAMP PART.

E. EMERGENCY RETURN: IF "EMERGENCY RETURN" PUSHBUTTON "11PB" IS PRESSED ON EITHER "AUTOMATIC" OR "MANUAL" CYCLE, THE HEAD WILL RETURN AND REMAIN IN THE RETRACTED POSITION. IN ORDER TO START CYCLE, THE "EMERGENCY RESET" PUSHBUTTON "12PB" OR "START MOTORS" PUSHBUTTON "2PB" MUST BE PRESSED.

F. LUBRICATION FAULT: IF OVER-PRESSURE OR INADEQUATE SUPPLY IS INDICATED BY THE OPERATION OF PRESSURE SWITCH "1PS" OR FLOAT SWITCH "1FS", "LUBE FAULT" LAMP "4LT" IS LIGHTED AND RELAY "2CR" WILL REMAIN DE-ENERGIZED AT THE END OF THE MACHINE CYCLE. WHEN THE FAULT IS CORRECTED, "2CR" WILL BE ENERGIZED WHICH ALLOWS THE OPERATOR TO RESUME NORMAL OPERATION OF THE MACHINE.

LIMIT SWITCHES

1LS (8,26) TRIPPED WHEN PART IS CLAMPED
2LS (8,26) TRIPPED WHEN PART IS CLAMPED
3LS (14,19) TRIPPED WHEN HEAD IS FORWARD
4LS (19,24) TRIPPED WHEN HEAD IS RETRACTED
5LS (8,21) TRIPPED WHEN PART IS UNCLAMPED
6LS (8,21) TRIPPED WHEN PART IS UNCLAMPED
1PS (10,12) OPERATED BY LUBE SYSTEM OVER-PRESSURE
1FS (11,12) OPERATED BY ADEQUATE LUBE SUPPLY

FOR PANELS AND CONTROL STATION LAYOUT SEE ____ SHEET 2
FOR HYDRAULIC DIAGRAM SEE _____
FOR LUBRICATION DIAGRAM SEE _____

LAST WIRE NUMBER USED   71
LAST RELAY NUMBER USED 9CR
SUPPLIER'S DWG. NO. _____
SUPPLIER'S NAME _____
PURCHASE ORDER NO.   P.O. 1234
SERIAL NO. OF MACHINE TYP 5678
THESE DIAGRAMS USED FOR
    MACHINE NO. _____

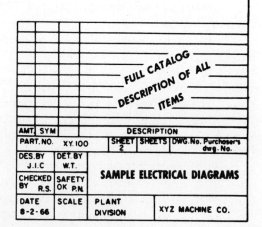

FIGURE 3-6 Electrical Ladder Diagram (Continued)

(Figure 3-8) indicates the type of fuse by brand name and part number, which will help you determine if the fuse should be a time-delay or a fast-blow fuse.

Figure 3-6 shows that the three conductors should be connected between the fused disconnect and the incoming terminals of the motor starter contacts. The contacts are identified as "1M." You will also notice that the primary side of the control transformer is connected to L1 and L2. This is indicated by the small dark dot placed on the point where the connection is located. This diagram does not indicate whether the transformer should be connected at the fuses in the disconnect or at the contacts of the motor starter. That information is shown on the wiring diagram (Figure 3-5).

Notice that two motors are connected to the motor starter contacts in this diagrams. The motors are identified by name, horsepower rating, value, and frame type. They also have an abbreviation (1MTR or 2MTR) that will be used with other components associated with the motor. Each motor has its own set of overload heaters, marked 1OL and 2OL. The contacts that the heaters activate will be shown where they provide the safety function, in line 1 of the control circuit diagram. This is a good example of the organization of the ladder diagram compared to the wiring diagram. In the wiring diagram, the heaters and overload contacts are shown beside each other because that is where they are physically located, and the wiring diagram is trying to show the physical location of all parts. In the ladder diagram, the overload heaters are shown in the load diagram near the motor because that is where they are functionally located, and the overload contacts are shown some distance away in the control diagram, where they will provide the function of opening the circuit to the motor starter coil if the overloads sense that the motor is receiving too much current. These two components are shown in separated circuits because they provide two separate functions, even though they are physically located near each other.

The control portion of this diagram receives its power from the secondary of the control transformer. The lines are connected to terminals X1 and X2 of the transformer, and they are fused with 10-A fuses. You would need to look at the parts list for fuses 4FU and 5FU to determine that these fuses are a FRN-type fuse. Since the primary side of the control transformer is connected to the fuse disconnect, the control circuit will be powered with 115 V any time the disconnect is closed. The vertical line on each side of the control circuit that represents the conductors connected to the secondary of the transformer is called the *power rail* since it supplies power to the control circuit.

As you know, the ladder diagram has been designed to provide you with information concerning the sequence of operation for the system. This means that you should begin reading the diagram from top to bottom, and each line is read from left to right. Each line of the diagram has a line number. In this diagram the line numbers on the first page of the print run from 1 to 30, and continue with numbers 31 to 38 on the second page. Other methods of indicating line numbers of larger diagrams include listing the page number in conjunction with the line number. For example, the fourth line on page 1 of the diagram would be numbered 1/4 and the fifth line on page 8 would be numbered 8/5. Similar line numbering schemes may indicate the fifth line on page 8 in this manner: $\langle \frac{5}{8} \rangle$. You should look for line numbers in the diagram because they will be used to indicate the source of coils for contacts and interconnecting wires from other pages in the print.

The numbers shown beside each wire in the diagram indicate a wire number. The same wire number will be attached to the end of the actual wire that is used to make the connection shown in the diagram. For example, in the diagram the wire that connects the start button to the overload contacts is identified with the number 8. The wire that makes that physical connection will also have the number 8 marked on it with wire number tapes. The wire numbers will help you locate the proper wire when you are troubleshooting the system. They will also be useful if any alterations must be made to the circuit.

You should also notice that each switch in the circuit is identified in several ways. First each switch will be identified by its type, such as PB for pushbutton and LS for limit switch. Second, each switch will be identified by its function in the circuit, such as start, stop, or reset. A third way that each switch is identified is by its electrical symbol. The start pushbutton is shown as a normally open pushbutton, and the stop pushbutton is shown as a normal closed pushbutton. Several sets of contacts are also shown on line 14 and other lines. The contact symbol indicates that their operation is controlled by a relay coil rather than by some other physical operation, such as a limit switch or pushbutton.

Indicating Coils and Contacts

Contacts are controlled by the coils of relays and motor starters. The coils are shown connected against the right power rail. The conductor on this side of the circuit is identified by the number 6, and it is used as the return side of the control circuit. Each coil listed in this diagram has control over one or more sets of contacts. The exact number of contacts and their location in the program are shown at the right of the power rail on the right side of the diagram. For example, the coil for 1CR is shown at the right end of line 8. The numbers at the right side of the coil indicate that it has control of 1CR contacts in diagram lines 9, 33, 33, 35, and 35. Since lines 33 and 35 are listed twice, this means that those lines contain two sets of contacts. In some diagrams the type of contact (normally open or normally closed) is also indicated. If the contact is normally closed, the number is listed with a back slash; the normally closed contacts for 5CR shown in lines 12 and 14 would be shown as /12 and /14. The normally open contacts would be shown as just a number. These numbers help you identify the lines where all the contacts that each coil activates are located.

In some diagrams the contacts will have a number above or below in addition to the coil identification. This number indicates on which line the coil that controls these contacts is located. For instance, the number 24 would be added to the normally open 7CR contacts that are shown in line 14 of the diagram. The number 24 would indicate that the coil that controls 7CR is located on line 24 in this diagram. This number is called a *source number* and helps you identify the location of the coil.

You should also notice the presence of relay contacts identified as CRM. One set of each of these contacts are located in each of the power rails of the diagram. The coil that controls these contacts is located in line 2 of the diagram and is called the *master control relay*. If these contacts open, all power to lines 4 through 38 will be disabled.

In several lines of the diagram, such as lines 20 and 27, you should notice a switch with an arrow that is near the lamps. The arrow also has a number near it. The arrow indicates that the line should be connected to another conductor, as indicated by the number near the arrow. In this case the number 5 is shown near the arrow, which indicates that the wire with the arrow on it is actually connected to the left power rail, which is identified with the number 5. The arrow is used so that multiple lines are not shown crossing all over the diagram.

Indicating Jacks and Pins

Some automated systems found in factories today utilize multiconductor cables that have jack and pin connections. These cables plug into a socket in the control panel and at the equipment, to make connections simple. The connection symbol for the jack and pin ($\underrightarrow{\text{J-P4}}\ \underline{12}\ \underline{12}$) will also indicate the cable number and the pin number so that you can test individual conductors within the cable. In this example, J-P4 indicates that this is cable 4 and the symbol shows the connection is for pin 12. The symbol will also be used to indicate which side of the connection has the jack and which side has the pins. In the ladder diagram, the jack and pins will be listed individually throughout the diagram on the lines where they occur, and in the wiring diagram the pin outline of the cable head will be used to show the location of the pins within the cable head and its physical shape.

Sequence of Operation

The actual sequence of operations is listed at the end of the diagram in English words. The main steps to the sequence of operation are listed by the letters A through F. Substeps within each main step are listed by numbers as required. The sequence of operation also lists the function and operation of each limit switch. Notice that the location of other information within the set of diagrams is listed in this area of the diagram. It lists the last wire number used in the system and the highest number used to identify the control relays.

A title block is provided in the diagram to indicate the machine for which the diagram is prepared and the company and plant where the machine is installed. This block also indicates the sheet or page number of the diagram. This will be useful when a line shows a connection or a set of contacts that are referenced to a line on another sheet in the diagram. You can quickly locate it by looking for the sheet number in the lower right corner of the sheets.

Electrical Panel and Control Station Layout Diagrams

The third diagram in this series of electrical diagrams is the panel layout diagram (Figure 3-7). It is very similar to the wiring diagram in that it shows the location of the components and terminal boards within the panel and the location of all switches on the operator's console. The major difference between the layout diagram and the wiring diagram is that the layout diagram does not show any terminal connections on the components and it does not show the location of any wires or cables.

In some cases the layout diagram is substituted for the wiring diagram, especially if sufficient detail is provided on the ladder diagram. In either case, the information provided in this type of dia-

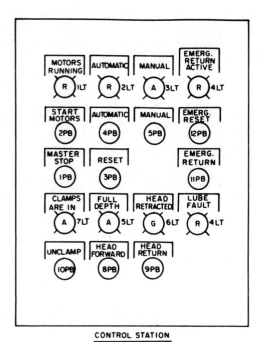

FIGURE 3-7 Panel layout diagram of metal-forming machine shown in Figure 3-5a. (Courtesy of Joint Industrial Council.)

gram is used to locate components in the panel. This may not sound too significant when you are reading it, but when you are troubleshooting, this diagram is invaluable. For instance, you may read the ladder diagram and determine that there is a good possibility that fuse 4FU is open. If you did not have a wiring diagram or a panel layout diagram, you would have to look all over the machine for the location of that fuse, and when you did locate any fuses, you would have to check wire numbers to determine if this is the correct fuse. In some units, you may find up to five separate electrical panels, and the components in each one will look similar to the others. This presents a real problem in locating the correct component and conductors to make the appropriate troubleshooting tests. In this diagram you can also see that all pushbutton switches are identified by a circle that has the switch name above it, and all lamps are identified by a traditional lamp symbol, with the color of lens being indicated by a letter in the center of the symbol. All electrical components such as control relays, motor starters, and transformers are identified by a rectangular box with the abbreviation of the component name and number indicated in the box.

Terminal boards are identified by name, and each terminal on the strip is identified by number. The amount of voltage that is used on each terminal strip is also identified beside each strip. This is quite useful when you must make repeated voltage readings within the panel, because it will indicate at what range to set the voltmeter. You will notice that this diagram may list dimensions of blank space within the panel, or overall dimensions of the panel. These dimensions are provided in case you need space in the cabinet to mount additional components, such as motor speed drives or other solid-state add-on equipment. The panel diagram may be used in conjunction with the ladder diagram during troubleshooting and installation procedures, or on its own to make modifications to the system.

Parts Lists

The parts lists for this system is shown in Figure 3-8. This diagram shows a partial list of the components used to make the motor control system operate. The entire list of components may require several sheets of the diagram. You should notice on the parts list that the parts are grouped by com-

Parts List

Part Name	Part No.	Company	Number Used
Fused disconnect, three-phase, 30 A 60 Hz	DG22INGB	Cutler-Hammer	1
15-A fuse	FRN 15	Bussman Fuse	3
3-hp size 1 motor starter	SCG-1	Square D Company	2
10-A fuse	FRN 10	Bussman Fuse	4
Pushbutton, 110 V	E22MPBCO	Cutler-Hammer	11
Indicator lamp 110 V	1025OT -34R—red -34A—amber	Cutler-Hammer	8

FIGURE 3-8 Partial parts list for the electrical system shown in Figure 3-5a.

ponent type. The manufacturer's part number is used in conjunction with other product specifications, such as voltage and current. If this system is used in a large factory, the inventory number may also be provided so that the availability of parts may be checked immediately by typing the inventory number into the computer terminal. The computer will indicate the number of parts with that number that are on hand and the location of the bin in which they are stored. Some systems also provide a purchase order that will be used to order the part from the storeroom or parts crib. This saves time for the technician, who must ordinarily fill out these forms by hand before the storeroom attendant will be able to locate the part for them. This form is also used to order new parts when the inventory becomes too low.

The parts list diagram is also useful for checking the size of components such as fuses or motor starter heaters. In most factories, when a fuse is blown, it will be replaced with the same type of fuse as that found in the fuse socket. If someone used the wrong-size fuse at some time while troubleshooting, it is likely that the same size of fuse will be used to replace it when it goes bad again. This may cause problems if the fuse is of the wrong type (slow blow instead of fast blow, for example) or is the wrong size.

From the information provided so far from the various types of diagrams, you can see that it is vitally important that all these diagrams be maintained as changes are made. It is also important that all changes to the diagrams be made by the engineering staff, and then all copies of the diagram that are located with the equipment and in files should be updated.

Diagrams of Systems That Use Programmable Controllers

In some systems a programmable controller (P/C) is used for control. In this case several additional diagrams will be provided with the system. These include the program, which will look like a ladder diagram, and a wiring diagram that shows the location and address numbers of all input and output modules. Other pages will be provided that list the cross-reference name and number of all switches and outputs in the system and all variables that are used in the program.

Figure 3-9 shows a P/C program that would be used to provide the same operational sequence as that of the ladder diagram shown in Figure 3-6. In this diagram you will notice that all switches are identified with the contact symbol. The P/C does not provide a set of symbols for each type of switch,

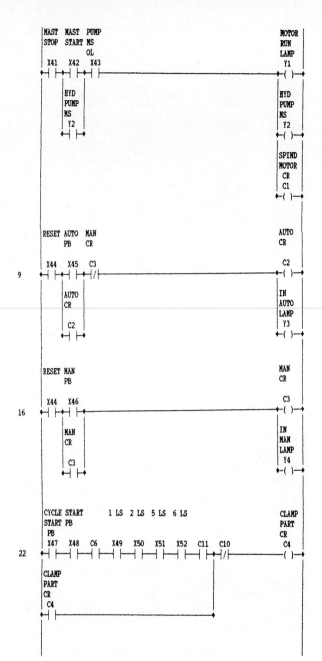

FIGURE 3-9 Programmable controller program for motor control system shown in Figure 3-5a. This program is written specifically for the Texas Instruments 530 and 560 programmable controllers.

such as limit switch or pressure switch, because it would require too much memory. Since each switch is shown in the program as a set of contacts, a comment and an address will be used to identify different switches. This means that you will need the P/C input/output (I/O) diagram and the cross-reference sheet when you are troubleshooting the system.

The programmable controller I/O diagram is shown in Figure 3-10 and the cross-reference sheet

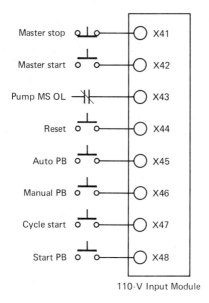
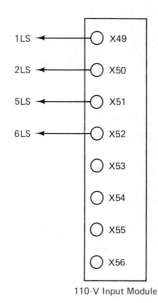
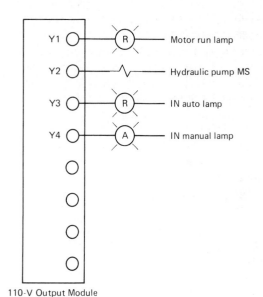

FIGURE 3-10 Partial input and output wiring diagrams for the programmable controller program in Figure 3-9.

is shown in Figure 3-11. This P/C program and the addresses used on it are for the Texas Instruments 530 and 560 programmable controllers. The diagram for Allen-Bradley, General Electric, Modicon, and other systems will be very similar, with only the address numbering systems being substantially different.

The P/C program is often called *relay ladder logic* (RLL) because it looks like a typical ladder diagram and the microprocessor executes the instructions as a logic program. In a logic program, the series and parallel contacts in a diagram are given names that refer to their operation. Figure 3-12 shows an example of each of the common logic circuits, with a truth table to indicate the condition of the output as the switches are opened and closed through all possible variations. In the truth table, a 1 indicates that a switch or output is on, and a 0 indicates it is off. From the diagram you can see that when two switches are in series with each other and one output, the logic circuit is called an AND gate. In this case the word "gate" is used to indicate that components have been put together in a circuit for a specific set of output conditions. The gate is called an AND gate because the only time the output will be on (1) will be when switch A "and" switch B are on (1). You can see that the output will be off (0) for all other conditions of the switches. The AND gate for computer logic uses a bullet-shaped symbol with the inputs connected to the back edge and the output connected to the nose. In relay ladder logic programs found in the P/C, the AND function is represented by contacts connected in series.

Electrical Diagrams **43**

I/O	SYNONYM	I/O	SYNONYM
0001	MOTOR RUN LAMP	0019	
0002	HYD PUMP MS	0020	
0003	IN AUTO LAMP	0021	
0004	IN MAN LAMP	0022	
0005		0023	
0006		0024	
0007		0025	
0008		0026	
0009		0027	
0010		0028	

I/O	SYNONYM	I/O	SYNONYM
0041	MAST STOP	0059	
0042	MAST START	0060	
0043	PUMP MS OL	0061	
0044	RESET	0062	
0045	AUTO PB	0063	
0046	MAN PB	0064	
0047	CYCLE START PB	0065	
0048	START PB	0066	

CONTROL	SYNONYM	CONTROL	SYNONYM
C0001	SPIND MOTOR CR	C0019	
C0002	AUTO CR	C0020	
C0003	MAN CR	C0021	
C0004	CLAMP PART CR	C0022	
C0005	HEAD FWD CR	C0023	
C0006		C0024	
C0007		C0025	
C0008		C0026	
C0009		C0027	
C0010		C0028	

FIGURE 3-11 Partial cross-reference sheet that indicates switch and output names and indicates the lines of the program where these components are located. This diagram also shows a cross-reference list for control relays used in this program.

The OR gate is also shown in this figure, and you can see that in this circuit the two contacts are in parallel with each other. In this condition the output will be on (1) when either switch A "or" switch B is closed. This means that the output will be on for three of the switch conditions and off only when both of the switches are open. The computer logic symbol for the OR gate looks similar to the AND gate except that the back is concave rather than flat. The OR function is identified by parallel contacts in the RLL program for P/Cs.

The NOT gate is similar to a normally closed set of contacts on a switch. You can refer this operation to a switch that has two sets of contacts, one set that is normally open and one set that is normally closed. The NOT gate is concerned only with the output from the normally closed set of contacts. This means that when the switch is "not" on, power will flow through the normally closed set of contacts to the output, and when the switch is on, the normally set of contacts will be open and no power will flow to the output. The symbol that is used in computer logic programs to represent the NOT function is a triangle on its side with a small circle drawn at the output. In the relay ladder logic diagram in a P/C the not function is identified as a normally closed set of contacts.

The NOT function can be used alone or in conjunction with other logic functions. Two other symbols and truth tables are provided in this figure to represent the combination of the AND function and the NOT function to make a NAND gate, and the combination of the OR function and the NOT function to make a NOR gate. Notice that only the small circle at the output end of the gate is used to represent the NOT function. More information about the operation of these logic functions is provided in Chapter 16 with in depth information about programmable controllers.

If the logic functions are used in a computer, the logic gate symbols are used, and when the logic is preformed in the programmable controller, the ladder diagram symbols (coils and contacts) are used. The reason coils and contacts are used for symbols in the ladder diagram is because that is the language that electricians, technicians, and electrical engineers are used to. This means that the P/C will present a program that is readable by the personnel who must work with them daily.

The P/C program used for troubleshooting the control system is like the ladder diagram. Since the P/C program is in memory, and executes control of the circuit from a microprocessor, it is called a *soft-wired system;* the traditional motor control circuit is called a *hard-wired system*. The P/C system also provides a variety of status indicators that allow the technician to diagnose the condition of input switches and outputs such as solenoids and relays. The status indicators on the input modules will be shown on their diagram located on the module. There will be one indicator for each input, which will glow when voltage is received at that input. The status indicator for each output is shown located on the output module in the wiring diagram. Each indicator will glow when its output circuit in energized. These indicators allow the technician to look for the ones that are glowing to indicate which switches or outputs are energized, without testing each terminal for voltage.

The input and output module wiring diagram will also provide information that is necessary for locating the modules in the P/C rack and for identifying wire numbers and voltages. This information

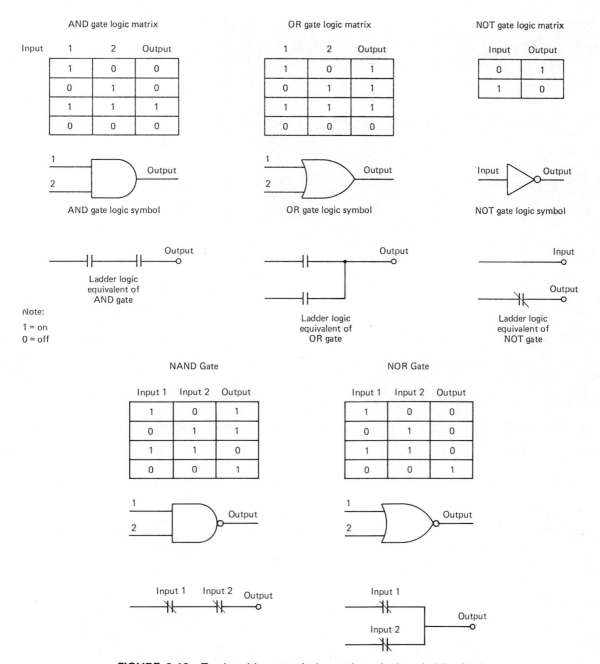

FIGURE 3-12 Truth tables, symbols, and equivalent ladder logic circuits for, AND, OR, NOT, NAND, and NOR logic functions.

is particularly useful for installation and troubleshooting.

Machine Diagrams and Plant Layout Diagrams

Some diagrams that you will use on the factory floor are designed to show you where components are located on the machine or where the machines are located in relation to the entire factory. Machine diagrams are quite useful when you have not worked on a machine before or if you are not completely familiar with the machine design. Figure 3-13a and 3-13b shows a picture of a plastics molding press and a diagram that would be used to indicate where parts are located on the machine. From this figure you can see that the diagram of the machine looks very similar to the picture of the machine. In fact, most machine diagrams are drawn to scale, so the representation should be nearly exact. The machine diagram will indicate the location of key components, such as hydraulic manifolds, transformers, electrical panels, and disconnects. These are all devices that will be shown in their relative location on the wiring diagram, but you must remember that the wiring diagram is a

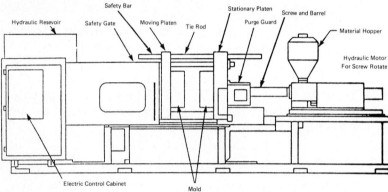

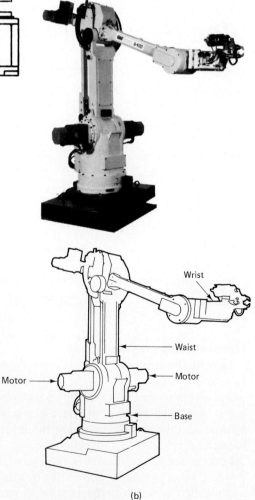

FIGURE 3-13 (a) Machine diagram of plastic molding machine with a picture of the machine to provide information about the location of components on the machine; (b) picture of a typical robot and a machine diagram used to show location of basic parts. [(a) Courtesy of NATCO, Inc; (b) courtesy of GMF Robotics.)

two-dimensional diagram that can show only directional relationships (right and left, and above and below). In the case of solenoids located on a hydraulic manifold, they will be shown to the far left side of the wiring diagram to indicate they are located to the left of the electrical panel, but the wiring diagram does not show how far to the left. When you see the machine diagram, you can see that the hydraulic manifold is located inside the last door on the left end of the machine. Many other details, such as height and depth, may also be shown with the machine diagram. In fact, some machine diagrams show the machine as a multiview diagram that shows top, side, and front views, with several auxiliary views that indicate specific sections of the machine.

The plant layout diagram is basically a top-view diagram that provides a floor plan of the entire factory. From this type of diagram you will be able to see that all the machines in the plant are numbered, which provides an easy way to locate them and a method of identification for record keeping. Each aisle in the factory is also identified, like streets in a city. This provides reference for all personnel in the plant when they must move about. The identification will also be useful for dispatching technicians when problems exist on automated systems, robots, or conveyors that may be installed across large portions of the plant. For example, a problem may occur in a large automation line, but you would need to know in what section of the plant the actual problem exists. This could be narrowed down to one or two grid locations on the layout diagram, which would allow all maintenance personnel to arrive at the correct area. The plant layout diagram will also identify locations of power distribution switch gear and transformers. Large sections of bus ducts will also be shown on this diagram, which will be useful for additional installation of new equipment or other times when plant expansion is taking place.

CONCLUSIONS

You have seen in this chapter that a variety of diagrams are provided and used on the factory floor for the installation, troubleshooting, and operation of equipment. Each diagram provides a different type of information presented in the manner that makes it most useful. For example, you learned that some diagrams, such as wiring diagram, plant layout diagrams, and machine diagrams, are presented to show the location of equipment and components. Some of these diagrams will also show wire routes and wire numbers so that specific conductors can be located.

Other diagrams, such as the ladder diagram and the ladder logic program in P/Cs, are provided to show the sequence of operation of equipment. The sequence of operation is provided by the machine designer so that a technician can determine the exact problem when a fault occurs. A diagram is also provided that lists all the components used in the electrical system. In this type of diagram you learned that part numbers, specification of voltage and current, and location of crib storage is specified so that replacement parts can be identified and quickly located.

You should also have learned that a large number of diagrams are needed to present all the information that is required to perform all work on the factory floor. If the diagram that you are using does not provide enough information, you should ask for associated diagrams and programs that may be helpful to you.

QUESTIONS

3-1. Explain how a JIC electrical symbol indicates what activates a switch.

3-2. Explain how the electrical symbol indicates an increase or decrease of air pressure, causing a switch to activate.

3-3. Explain the terms *normally open* and *normally closed*. (What is considered to be the normal condition?)

3-4. Explain why an abbreviation may be used with a symbol.

3-5. Explain why a number must be included with the abbreviation when identifying multiple limit switches in a diagram.

3-6. Explain the function of the wiring diagram in Figure 3-5.

3-7. Explain the function of the elementary (ladder) diagram in Figure 3-6.

3-8. Explain why an elementary diagram is sometimes called a ladder diagram.

3-9. Explain what a parts list is used for.

3-10. Explain why the programmable controller diagram must include documentation comments and an I/O cross-reference list.

3-11. Explain what a machine diagram and plant layout diagram are used for.

Power distribution   4

INTRODUCTION

A large amount of power is controlled and consumed in modern industry. This power must be produced at a utility power plant and distributed to the industrial site. It must also be distributed all around the factory at appropriate levels to power motors and control circuits. This causes the power distribution system in modern factories to be as large as some small towns or cities. In fact, in many of the processing factories where metal and plastic parts are produced or machined, the power consumption will be higher than that of most small towns or suburbs.

This causes the power distribution system, from the utility where the power is produced to the point on the factory floor where it is used in individual motors or heating elements, to be an important link in the motor control function. As a technician you will be responsible for installing new distribution, or altering existing distribution systems to accommodate new machines or processes. This involves selecting proper equipment to meet codes and standards, and determining proper sizes for this equipment. It will include the installation of switch gear and conductors, providing adequate safety devices (fuses and circuit breakers), and troubleshooting the system when it fails to operate correctly.

In this chapter we explain how electrical power is produced and where the production stations will be located. We also explain how this power is transmitted from the generating station to the substation at the factory. The distribution of the power after it has entered the building is also described, including the identification of vital pieces of equipment involved in the distribution. Methods for selecting proper sizes of factory distribution equipment and installation procedures are also covered. You will be presented with typical faults that occur and troubleshooting procedures that will help you locate the fault quickly and make repairs or replace equipment.

In this chapter we also provide information about power distribution so that you will understand safety considerations. You will learn the possible dangers that high voltage represents while it is destributed and procedures for working safely around it. Extensive information is provided about proper selection and sizing of safety devices so that the system, equipment, and personnel will be protected. You will also learn to measure the capacity of the power distribution system so that you can determine if it meets the current needs of the plant and if extra capacity is available for additional motors and controls. This will help you understand the capacity of the utility company's distribution system.

GENERATING ELECTRICITY WITH AN ALTERNATOR

All electrical energy that is used in industry today must be converted from some other energy source since it does not occur naturally. The most common form of conversion is the electrical generator. The electrical generator will produce a voltage when its shaft is rotated. It uses the principle of operation that a magnet passing a coil of wire will cause a current to flow in the coil of wire. The coil of wire in the alternator is called the *armature* and it is the stationary part of the machine. The magnetic field is produced by passing current through a small coil of wire called the *field*. Since the flux lines of the magnetic field must move pass the coils of wire in the armature to produce a current, the field is mounted on the rotating shaft of the alternator.

The amount of voltage that the alternator produces can be controlled in any one of several ways. One way to increase the amount of voltage at the armature is to spin the field faster, another way is to increase the number of turns of wire in the armature coil, and a third way is to increase the strength of the magnet that increases the flux lines in the field.

Since the speed of the rotating field will determine the frequency of the ac voltage, the main method of increasing the amount of voltage that the alternator produces is to increase the current flowing in the field, which controls the strength of the magnetic field and controls the number of flux lines.

Figure 4-1a shows the diagram of a simple alternator. From this diagram you can see that the armature is the stationary coil and the field is the rotating coil. As you know, the shaft is rotated to make the field's magnetic flux lines pass (cut through) the coils of wire in the armature. When this occurs, a strong current is induced (generated) into the armature coils.

Figure 4-1b shows the waveform of the induced current. From this diagram you can see that the waveform is sinusoidal. This occurs because the dc voltage causes a north and a south magnetic pole to be developed in the field coil. When the north magnetic pole passes a armature coil, the induced voltage will be positive. From the diagram you can see that when the magetic field begins to pass the armature coil, only a few lines of force are cut and the amount of voltage produced is small. This voltage will increase proportionately as the rotating field coil causes more of the field's magnetic flux lines to be cut, and the voltage will peak in the positive direction as the maximum number of flux lines cut across the armature coil. As the field continues to rotate, fewer flux lines are cut, until the positive magnetic field moves beyond the armature coil. On the diagram you can see that the voltage peaked when the maximum number coils were cut by the flux lines, and it decreases to zero as the magnetic field rotates away from the armature coil and cuts fewer lines. When the flux field is completely away from the coil and no lines are being cut, the voltage will return to zero. Notice that since the field magnet is rotated on a shaft, the flux lines cut through the armature coils in an arc, which causes the voltage to increase and decrease smoothly, which produces the sinusoidal waveform.

After the north (positive) magnetic field passes the armature coil, the south (negative) magnetic field begins to cut across the armature. When this

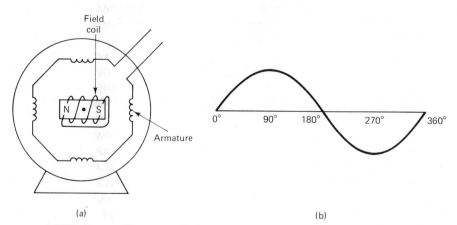

FIGURE 4-1 Diagram of an alternator and the ac sine wave that it produces.

occurs, a negative voltage begins to be induced into the armature. From the diagram you can see that the negative voltage will also increase, until it peaks in the negative direction, and again decreases to zero volts as the negative field moves past the armature.

This positive and negative oscillation occurs at the speed of the rotating field. This speed is regulated to 60 cycles per second (60 Hz) for all utility-generated voltages in the United States. The maximum amount of voltage that is generated is determined by the amount of dc current applied to the field. The actual amount of voltage generated will vary from alternator to alternator, and will be stepped up by set of transformers before it is sent out for distribution. In Chapter 6 we explain magnetic flux fields and ac and dc generating principles in greater detail. If you need more information about voltage generation, you should review that material.

THREE-PHASE VOLTAGE

The alternator is designed to produce as much power from the armature as possible. For this reason, three separate sets of armature coils are mounted in the generator, which will produce three separated voltages as the field rotates past them. These voltages are produced out of phase of each other by 120 electrical degrees. The three voltages are called *phases* and are identified as phase A, phase B, and phase C. They will carry phase shift all the way through the transmission system to the point in the factory where it is used.

To produce enough power for use in industrial and residential areas, the utility company must parallel several generators and generating stations together to increase the amount of current that is available. When these generators are connected together to make power available, it is called a *grid*. The grid allows extra power to be shifted from one area of the United States to another as the demand shifts. For instance, excess power can be generated with hydroelectric dams in areas where water is plentiful and the demand is not large. This power can be transmitted up to 1000 miles through the grid to areas of the country that could use the extra power. It is also possible to shift power across the grid to match load demands as they occur. For example, the peak demand period is early in the morning, when people are getting ready for work and commercial and industrial buildings are starting up for the day, and again in the evening when people come home for work and the sun is setting, which brings on a large number of appliances and lights. Since these peaks occur at a different times in each time zone, power can be shifted eastward for the morning peak and westward for the evening peak. This type of shifting is also possible when one area of the country has an extra-high demand due to extremely hot or cold weather, which causes air conditioners and heating equipment to use more than normal amounts of power. If another part of the country is not experiencing this weather, the extra power can be produced at other generating stations and put into the grid for use.

MEETING PEAK DEMANDS

When the peak demands for power are experienced on site at individual factories, it may be too expensive to draw the extra current from the power company. The power company bases its rate on the amount of power that is used and the peak demand. The peak demand is measured as the highest amount of power consumed over any 20-minute period. The peak demand is then converted into a penalty charge that is multiplied by the total usage for the entire monthly billing period. This means that if the factory uses 25 percent more power for 1 hour once during the entire month, a penalty rate is multiplied against the entire bill. Some utility companies use a 15-minute period, while others use a slightly longer time to base the peak measurement, but as you can see, the penalty is assessed against the total usage, so the penalty may result in an increase of 15 to 40 percent.

Factories use several methods to combat this penalty, which include controlling load shedding with a computer to turn off nonvital loads, such as extra lighting or ventilation, when people are not using the area. Other controls limit the number of motors that can be started at any one time, to keep the peak inrush to a minimum. Other companies use *peaking generators* that they have at their site. These generators use natural gas or fuel oil to produce the extra amount of power that is required for a short period. Since the factory owns and maintains the generating equipment and pays for its own fuel, the penalty rate for the extra demand can be avoided. The electric bills for many factories reach several hundred thousand dollars per month. As the high-demand penalty will increase this by thousands of dollars, peaking generators are easily seen to be cost-effective.

Utilities also use smaller generators at peak demand periods. These generators are also called peaking generators. Some hydroelectric generating stations use their excess power to pump extra water into peaking lakes to be released during peak demand periods. These lakes are rather small and do not provide the extra capacity naturally, so the extra water that is pumped into them allows the generating station to meet the peak demand without having to build more generating capacity.

DISTRIBUTING GENERATED POWER

To be useful, the power that is produced at a generating station must be distributed to all locations in a region. Some generating stations are located where water is plentiful because cooling water is essential in the process to return the steam that has passed the turbine wheel back to liquid form. The cooling water is used to condense the steam so that it can be pumped back to the boiler. Other power generating plants are located where large amounts of coal can be shipped into the site easily, or in the case of nuclear power plants, they may be located where populations are not so dense.

In each of these cases, the power is produced in large amounts where it is not being used, which means that it must be distributed. Figure 4-2 shows the path the power will take to reach a factory where it will be consumed. The first stage of the distribution system is the transformer substation at the generating site. This substation steps the voltage up to transmission levels. Some of the power from this substation (up to 35 percent) is returned to the generating plant for its operation. The rest of the power is transmitted out through wires connected to the large transmission towers that support the high-voltage wires. The power is usually transmitted at levels above 100,000 V. Some long-distance transmission lines use voltage levels up to 900,000 V.

A second substation is provided near the factory to step this large voltage down to levels around 2,000 to 4,000 V for transmission around the local commercial and residential areas. The voltage for commercial areas is set at a higher level than that for residential use.

A third substation may be provided at the factory site, where the voltage is reduced for transmission within the factory. If the factory does not use a large amount of power, one set of transformers are used instead of the substation. These transformers are part of the factory's power distribution and they provide the proper amount of voltage to the various areas of the factory. At various points the power can be distributed through service entrance equipment, lighting panels, motor control centers, and bus ducts.

The substation at the factory must provide switchgear, transformers, and switchboards to receive the utility high voltage safely and to step it down to the levels required for factory distribution. Figure 4-3 shows a power distribution diagram. As you can see from the diagram, there are several ways to receive and distribute the high voltage from the utility lines. As you know, this voltage may enter the factory substation at levels from 2.4 to 200 kV. The first set of devices inside the substation are the circuit breakers. They provide overcurrent protection, short-circuit protection, and a means of disconnecting the system when it must be worked on. Circuit breakers are available to protect circuits up to 3,000 A, and they will provide an interruption capacity up to 8,000 A rms (root mean square) symmetrical. The *interruption capacity* is the amount of current the circuit breaker can safely handle during a short circuit without damaging itself.

The next piece of equipment in the substation is the high-voltage switchgear. From the diagram you can see that once the main circuit is brought into the substation, it is divided into sections by the switchgear. The switchgear allows each circuit to be isolated and disconnected from all the other

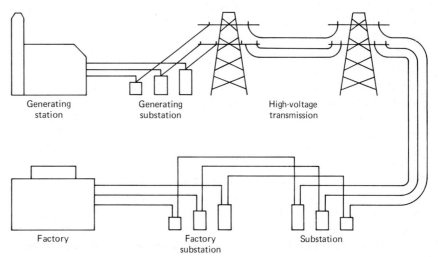

FIGURE 4-2 Diagram of a power distribution system.

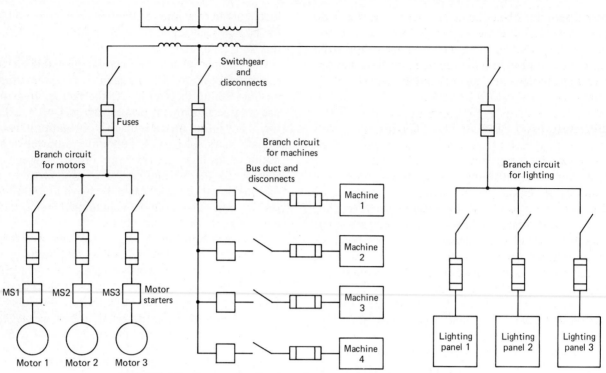

FIGURE 4-3 Power distribution diagram of a typical plant layout.

circuits for installation and maintenance work. This also allows for several different levels of voltage to be distributed throughout the plant. The high-voltage switchgear has voltage capacities up to 150 kV and 1200 A. In some systems it is preferable to use a step-down transformer prior to the circuit breaker and main switchgear so that the voltage rating of these devices can be reduced.

After the power comes through the high-voltage switchgear, it will be routed through step-down transformers, which will drop the voltage from the transmission level (up to 200 kV) down to in-plant transmission levels, which are usually less than 5 kV. The exact voltage level used for transmission in each factory will be different depending on the transmission distance and the amount of current that is needed. Higher current levels require larger wire sizes, which weigh more. This also causes the switchgear, circuit breakers, fuses, and terminal connectors to be larger than when higher voltages are used. Another consideration in the level of transmission voltage is the location of transformers. Some factories try to locate all their transformers in the substation area of the plant or in the powerhouse, so that they can be maintained more easily. This means that voltage is reduced at this point and the feeder circuits will be a little longer than if the transformers were located nearer the equipment where the power was being consumed.

After the voltage is stepped down by the transformers, it is routed through switchboards, load centers, motor control centers, and lighting panels. This equipment allows the power to be broken into feeder circuits and individual branch circuits. The feeder circuits group common types of loads with respect to their voltage specifications, current requirements, or applications. This strategy allows for selective protection with a combination of fuses and circuit breakers, and balancing the amperage load from phase to phase and circuit to circuit within the distribution system.

TRANSFORMERS

The transformer is the main device used in the power distribution system to step voltages up or down to levels that are desired. When the voltage comes from the armature of the ac synchronous generator, it may only be several thousand volts. This is not large enough for long-distance transmission, and the lower voltage level will cause the amount of current in the transmission to be very large. The transformer affects both the voltage and the current of the power that it handles.

Theory of Operation of Transformer

The transformer is made of two separate coils of wire that are located close to each other so that the flux lines generated by one coil will cut across the conductors of the second coil. The coil that receives the input voltage is called the *primary side* of the transformer, and the output coil is called the *secondary side*. The primary terminals of the transformer are labeled H1 and H2, and the secondary terminals are labeled X1 and X2.

When voltage is applied to the primary side of the transformer, a current will flow through that coil of wire. This current creates a strong magnetic field around each conductor in the coil. The magnetic field is made up of a high concentration of flux lines that move from the north pole to the south pole of the magnetic field. Since the voltage applied to the primary coil is ac voltage, it will oscillate in a sinusoidal waveform, which causes the voltage to peak positive, pass through zero volts, and then peak negative during each cycle. This causes the current to flow first in the positive direction through the coil and then in the reverse direction. This continual reversal of voltage will cause the magnetic field to build and collapse.

When the voltage peaks in the positive direction, the density of the flux lines in the magnetic field around each conductor will be at their maximum, and when the voltage waveform returns to zero volts prior to reversing, the strong magnetic field will quickly collapse. When the voltage reverses and peaks in the negative direction, the magnetic field will build again to its maximum and quickly collapse as the voltage returns to zero and begins to flow again in the positive direction. Each time the magnetic field builds in the primary coil and collapses, the flux lines will cut across the conductors of the secondary winding and cause an induced current to flow in it. When the magnetic field in the primary coil is positive, the flux lines will cause a negative voltage to be induced in the secondary when the field collapses, and when the field is negative the induced voltage will be positive. This means that the secondary voltage will be 180 degrees out of phase with the primary voltage.

Voltage Ratios

The amount of voltage that is induced in the secondary winding of the transformer is directly proportional to the ratio of the number of turns of wire in the primary and secondary coils. A formula is provided in Figure 4-4 that is used to calculate the KVA when the primary voltage and the amperage of the transformer are known. The amount of current available in the secondary can also be calculated, since it is indirectly proportional to the turns ratio.

The power available from the secondary of a transformer is measured in volt-amperes (VA). For

$$\text{Single-phase KVA} = \frac{\text{volts} \times \text{amperes}}{1000}$$

$$\text{Three-phase KVA} = \frac{\text{volts} \times \text{amperes} \times 1.73}{1000}$$

$$\text{Transformer efficiency} = \frac{VA_{out}}{VA_{in}} \times 100$$

FIGURE 4-4 Formulas used for calculating power ratings (kVA) and efficiency ratings for transformers.

single-phase circuits the voltage is multiplied by the amperage to determine the VA. If the value is over 1000, it is rated in kilovolt-amperes (kVA). If the transformer is connected to a three-phase circuit the power is determined from the calculation: volts × amperes × 1.73. The value 1.73 comes from the square root of 3. Again, if the value is greater than 1000, the answer can be divided by 1000 and the answer will be in kVA. These formulas are shown in Figure 4-4.

Step-Up and Step-Down Transformers

The number of turns of wire in the primary and secondary coils of wire will also determine if the transformer will be a step-up or step-down transformer. If the number of turns in the secondary winding is greater than in the primary side, the secondary voltage will be larger than the primary voltage and the transformer will be a step-up transformer. If the number of turns in the secondary winding is less than in the primary side, the secondary voltage will be less than the primary voltage and the transformer will be a *step-down transformer*. The voltage may be determined by the formula

$$\frac{V_1}{N_2} = \frac{V_2}{N_1}$$

Several other types of transformers are also used. They include an isolation transformer and an autotransformer.

The *isolation transformer* is designed so that the primary and secondary coils have the same number of turns, which means that the primary and secondary will have the same amount of voltage. This type of transformer is used to provide isolation between the primary circuit and the secondary circuit. It is often used where one of the circuits is grounded and the other is not.

Transformer Efficiencies

The efficiency of a transformer can be calculated and expressed as a percentage. VA_{in} is divided by VA_{out} and multiplied by 100. In all cases VA_{in} will be larger than VA_{out}, which means that the transformer has losses. These losses are called internal losses (W_i) and are caused by magnetic losses (W_m) and conductor losses (W_c). All of the losses add together to cause an increase in temperature in the core of the transformer.

The increase in temperature that is caused by the transformer losses is called *termperature rise*. The temperature rise in the transformer is listed as the increase in temperature above the ambient temperature in degrees Celsius. Normal temperature rise is usually listed as 40°C, and the transformer can be loaded to its maximum rating. When the temperature rise increases to 50°C, the transformer's load should be derated by 8 to 10 percent. If the temperature rise increases to 60°C, the transformer's load should be derated by 15 to 16 percent.

Typical Transformer Applications and Voltages

The high voltage that is used in transmission systems must be stepped down to use in all areas of a factory. The actual secondary voltage will depend on the rating of the equipment that will use the voltage. Several typical feeder circuits are shown in Figure 4-5, which also indicates the voltage for each circuit.

From Figure 4-5 you can see that the transformer is used to step down voltage for a three-phase branch circuit. You should remember that the voltage is stepped up when it is sent for transmission, and it must be stepped down to be usable for loads on the factory floor. Typical three-phase primary voltages found at the factory substation transformers are 13,800, 12,470, 12,000, 7620, 7200, 4800, 4160, and 2400 V. Typical voltages for motors and lighting are 600, 480, 277, 240, 208, and 120 V. The transformer that is selected will provide the step-down voltage in one or two steps. The transformer will also have taps that will account for minor increases or decrease in the voltage. The taps are designed for 2 and 2.5 percent above, and 2.5 and 4 percent below the rated voltage.

Since there are three transformer coils, they can be connected in either a wye or a delta configuration. In the first example voltage is received by the first transformer in the factory substation at 12,470 V. It is stepped down to 600 V to provide voltage for feeder circuits. The feeder circuit will provide voltage to a variety of transformers thoughout the plant, where it will be stepped down again for branch circuits. The next transformer will provide 240 V for branch circuits that will power three-phase motor loads. The primary and secondary side of this transformer is connected in delta configuration for power loads.

The second branch circuit transformer will also provide 240 V at its secondary, but this transformer is connected to provide only single-phase secondary with a neutral. This means that it will provide 240 V between L1 and L2, and 120 V between L1 and neutral and L2 and neutral. You can think of neutral as a return wire to the center tap of the transformer secondary. *It is not the ground wire*. The neutral is generally grounded in most single-phase systems, but it is not the ground wire. A separate ground wire is provided for this purpose. When you are working with the power distribution wires, the conductors that are used for ground wires will be colored green or remain bare. The neutral wire will be colored white, and the hot wires for L1, L2 or L3 will be other colors, such as black, red, and blue. Both the primary and secondary in this transformer will be connected in delta configuration for single-phase power loads such as motors and lighting loads. Several variations of this configuration are also shown beside this transformer which allow the transformer to respond to various load changes.

In some branch circuits a delta-connected secondary is required to provide 240 V for three-phase and single-phase loads, and a neutral is required to provide 120 V. The neutral is produced by a center tap in the second transformer winding, and it will provide 120 V between it and L1 or L3. If L2 is used with neutral, it will produce 208 V, which is unusable because it will cause the transformer load to become unbalanced and overheat. The 208 V, developed from 120 V of one phase of the transformer and half of the phase where the center-tapped neutral is connected, is called *wild leg voltage*. In this application the L2 conductors must be orange or marked with orange to indicate that the L2 and neutral connection will produce 208 V. Remember, wild leg voltage can be produced only when the secondary is connected in delta configuration with a center tap used to produce the neutral.

The third branch circuit transformer is connected as a delta primary and delta secondary and will provide 480-V three-phase voltage for motors and other loads, such as electrical heaters. It could also be connected as a delta primary and wye secondary and provide power for the same types of loads. The fourth branch circuit transformer is connected with a wye secondary to provide 277 V specifically for lighting. The secondary in this configuration provides a neutral that is connected to the wye point. The 277 V is provided between any line and neutral.

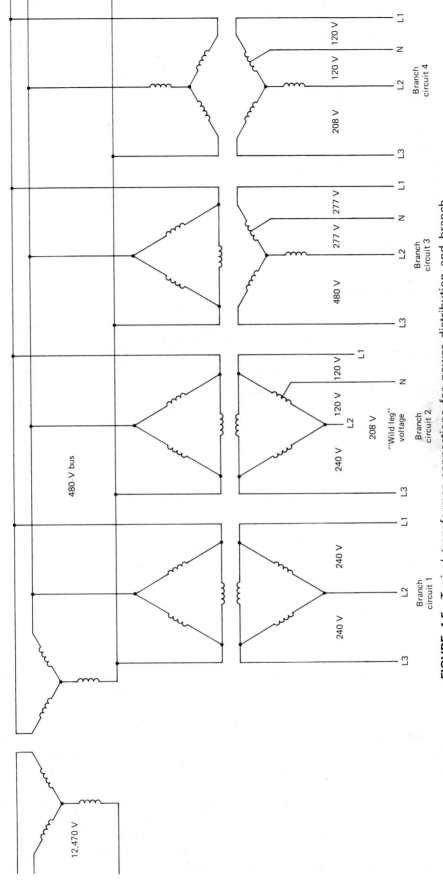

FIGURE 4-5 Typical transformer connections for power distribution and branch circuits. Additional transformer connections are provided. (Courtesy of Electrical Apparatus Service Association.)

TRANSFORMER CONNECTIONS

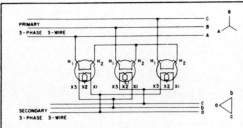

Y-DELTA FOR POWER

Often it is desirable to increase the voltage of a circuit from 2400 to 4160 volts to increase its potential capacity. This diagram shows such a system after it has been changed to 4160 volts. The previously delta connected distribution transformer primaries are now connected from line to neutral so that no major change in equipment is necessary. The primary neutral should not be grounded or tied into the system neutral since a single-phase ground fault may result in extensive blowing of fuses throughout the system.

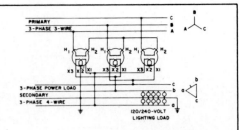

Y-DELTA FOR LIGHTING AND POWER

This diagram shows the connections for the Y-delta bank to supply both light and power. This connection is similar to the delta-delta bank with only the primary connections changed. The primary neutral should not be grounded or tied into the system neutral, since a single-phase ground fault may result in extensive blowing of fuses throughout the system. The single-phase load reduces the available three-phase capacity. This connection requires special watt-hour metering.

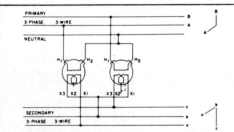

OPEN Y-DELTA

When operating Y-delta and one phase is disabled, service may be maintained at reduced load as shown. The neutral in this case must be connected to the neutral of the stepup bank through a copper conductor. The system is unbalanced, electro-statically and electro-magnetically, so that telephone interference may be expected if the neutral is connected to ground. The useful capacity of the open delta — open Y bank is 87 percent of the capacity of the installed transformers when the two units are identical.

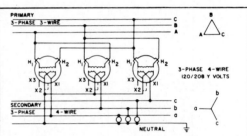

DELTA-Y FOR LIGHTING AND POWER

In the previous banks the single-phase lighting load is all on one phase resulting in unbalanced primary currents in any one bank. To eliminate this difficulty, the delta-Y system finds many uses. Here the neutral of the secondary three-phase system is grounded and the single-phase loads are connected between the different phase wires and the neutral while the three-phase loads are connected to the phase wires. Thus, the single-phase load can be balanced on three phases in each bank and banks may be paralleled if desired.

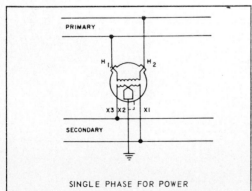

SINGLE PHASE FOR POWER

In this case the 120/240-volt winding is connected in series serving 240 volts on a two-wire system. This connection is used for small industrial applications.

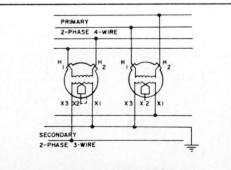

TWO-PHASE CONNECTIONS

This connection consists merely of two single-phase transformers operated 90° out of phase. For a three-wire secondary as shown, the common wire must carry $\sqrt{2}$ times the load current. In some cases, a four-wire or a five-wire secondary may be used.

FIGURE 4-5 Continued

TRANSFORMER CONNECTIONS

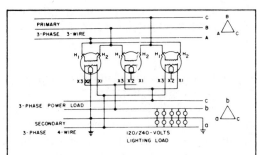

DELTA-DELTA FOR POWER AND LIGHTING

This connection is often used to supply a small single-phase lighting load and three-phase power load simultaneously. As shown in diagram, the mid-tap of the secondary of one transformer is grounded. Thus, the small lighting load is connected across the transformer with the mid-tap and the ground wire common to both 120 volt circuits. The single-phase lighting load reduces the available three-phase capacity. This connection requires special watt-hour metering.

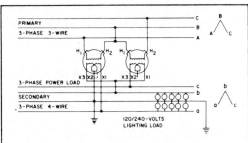

OPEN-DELTA FOR LIGHTING AND POWER

Where the secondary load is a combination of lighting and power the open-delta connected bank is frequently used. This connection is used when the single-phase lighting load is large as compared with the power load. Here two different size transformers may be used with the lighting load connected across the larger rated unit.

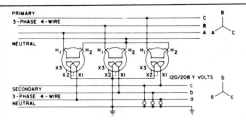

Y-Y FOR LIGHTING AND POWER

This diagram shows a system on which the primary voltage was increased from 2400 volts to 4160 volts to increase the potential capacity of the system. The previously delta connected distribution transformers are now connected from line to neutral. The secondaries are connected in Y. In this system the primary neutral is connected to the neutral of the supply voltage through a metallic conductor and carried with the phase conductor to minimize telephone interference. If the neutral of the transformer is isolated from the system neutral an unstable condition results at the transformer neutral caused primarily by third harmonic voltages. If the transformer neutral is connected to ground, the possibility of telephone interference is greatly enhanced and there is also a possibility of resonance between the line capacitance to ground and the magnetizing impedance of the transformer.

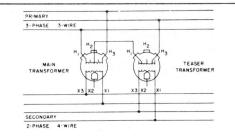

SCOTT CONNECTION – THREE PHASE TO TWO PHASE

In some localities, two-phase power is required from a three-phase system. The Scott connection is the most popular method of making this phase change. The secondary may be either three, four, or five wire. Special taps must be provided at 50 percent and 86.6 percent of normal primary voltage in order to make this connection.

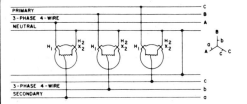

Y-Y AUTOTRANSFORMERS FOR SUPPLYING POWER FROM A THREE-PHASE FOUR-WIRE SYSTEM

When the ratio of transformation from the primary to secondary voltage is small, the most economical way of stepping down the voltage is by using autotransformers as shown. For this application, it is necessary that the neutral of the autotransformer bank be connected to the system neutral.

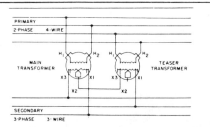

SCOTT CONNECTED TWO PHASE TO THREE PHASE

If it should be necessary to supply three-phase power from a two-phase system, the Scott connection may again be used In this case, of course, the special taps must be provided on the secondary side. In other respects, the connection is similar to the three-phase to two-phase transformation.

If it is desired to obtain the Scott transformation without a special 86.6 percent tapped transformer, it is possible to use one with a 10 percent or two 5 percent taps to approximate the desired value. It will introduce a small error of unbalance (overvoltage) which will require care in application.

FIGURE 4-5 Continued

The last two branch circuits provide 208 V at their secondary. Both of these systems use a wye-connected secondary, and the 208 V that is produced is usable because of the balanced load it will cause on the transformer. Do not confuse the 208 V produced by the wye-connected transformer with the unusable 208 V produced by a wild-leg delta connection. The first wye-connected transformer is used to provide three-phase for balanced three-phase loads such as motors and three-phase heating, while the other provides a neutral for 120-V loads. When the neutral is used, it is connected to the wye point. This means that three-phase loads can be connected to L1, L2, and L3, or single-phase loads can be connected to any two of the three lines. When low-voltage single-phase loads are used, they will be connected to any line and neutral. It is important to balance the current in each line of this type of transformer so that is does not overheat. This can be accomplished by moving single-phase loads from one line to another.

Each of these applications can also utilize transformers with multiple taps. The multiple taps allow a single transformer to be used for a wide variety of primary or secondary voltage requirements. Several other useful transformer connection diagrams are provided in Figure 4-5. These diagrams can be used for determining which transformers will be needed for a particular application.

Autotransformers

Other types of transformers are needed in motor control circuits to provide a small increase or decrease in secondary branch circuit voltage. For example, most factories try to standardize on one or two voltage levels throughout the plant for all motor and heating loads. This voltage is generally 480 and 208 V. At times a new piece of equipment such as a robot will be installed that requires a slightly higher voltage (such as 220 V). In this case, the robot may become damaged if it is operated at 208 V, so an autotransformer will be provided to produce the extra voltage. The autotransformer shown in Figure 4-6a may have one or two primary and one or two secondary windings. Each winding can produce 12, 16, or 24 V, which can also be combined to produce 32 and 48 V. Notice that the primary windings are marked H1, H2, H3, and H4, while the secondary windings are marked X1, X2, X3, and X4.

This means that the transformer can be connected as shown in Figure 4-6b and provide 220 or 244 V between terminal H1 and X2 or X4. As you can see, 208 of the volts are provided from the primary side, and the other 12, 24, or 36 is provided from the secondary. This also means that the load rating of an autotransformer is calculated only on the part of the voltage and current that is transformed through its secondary.

The autotransformer can also be connected to provide slightly lower voltage in applications where the supply voltage is 240 and the load needs 208 V. Figure 4-6c shows a diagram of this type of connection. As you can see, the autotransformer can be connected so that the secondary voltage is added to the primary, or so that the secondary voltage is subtracted from the primary. For this reason the transformer is also called a *buck–boost transformer*. The secondary of the autotransformer

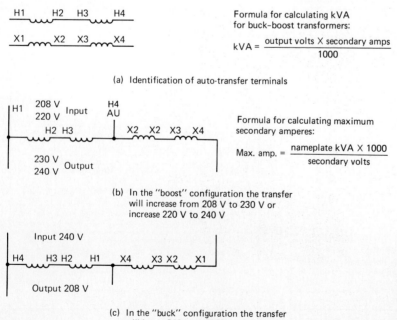

FIGURE 4-6 Autotransformer (buck–boost) connections that are used to slightly raise or lower voltages for specific loads.

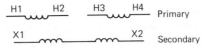

(a) Standard transformer terminal identification

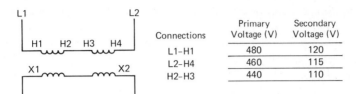

Connections	Primary Voltage (V)	Secondary Voltage (V)
L1–H1	480	120
L2–H4	460	115
H2–H3	440	110

(b) High-voltage connections for a control transformer

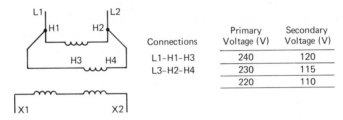

Connections	Primary Voltage (V)	Secondary Voltage (V)
L1–H1–H3	240	120
L3–H2–H4	230	115
	220	110

(c) Low-voltage connections for a control transformer

FIGURE 4-7 Control transformer terminal identification with wiring diagrams for connecting them to high and low voltages. (Photo courtesy of Eaton Corp./Cutler-Hammer Division.)

can also be used as a step-down transformer to provide 12, 16, 24, 32, or 48 V for control circuits.

Control Transformers

Control transformers are used to provide isolated low voltages that are used to energize solenoids, relay coils, and other small loads, such as indicator lamps. The control transformer is a step-down transformer that will accommodate the inrush current that is caused by electromagnetic loads being switched on and off. This type of transformer also provides some isolation from the spikes and transients that are found on normal power lines when large inductive loads such as motors and welding equipment are switched on and off. The transformer is also designed to provide a stable regulated output under a variety of fluctuations that are incurred on the primary side.

The control transformer is manufactured from coils of wire instead of using laminated sections that are found in larger transformers. A diagram of the control transformer is shown in Figure 4-7. From this diagram you can see that the secondary of the transformer has one coil with its terminals identified as X1 and X2. The primary winding is made of two separate windings that can be connected in parallel for 240, 230, or 220 V, or they can be connected in series for 480, 460, or 440 V. The secondary voltage will be 120, 115, or 110 V depending on which of the three primary volts are used. A table is provided in this figure to help you determine the connection needed for each application.

FUSES AND CIRCUIT BREAKERS

The motors, switchgear, transformers, and power distribution conductors must be protected against short-circuit current. The motors and other loads must also be protected against slow overcurrents or sustained overload that will allow them to overheat and become damaged. *Short-circuit current* is defined as any current that exceeds the normal full-load current of a circuit by 10 times. When short-circuit current occurs unprotected, any conductor or switchgear that is involved will be severely damaged by excessive magnetic forces as well as by extremely high heat levels that will melt most metal objects. Fuses, circuit breakers, and magnetic overloads can provide protection for motors and other loads against slow overcurrents, and fuses and circuit breakers can provide short-circuit protection.

Slow Overcurrents

The slow overcurrent develops from overloading devices or from malfunctions in motors and other loads. For example, if a motor-driven conveyor is overloaded, the motor will be required to draw extra current to try and move the load. If the motor is not protected, it will draw the extra current and begin to overheat. If the transformer that supplies voltage to the motor is also fully loaded, it will become overloaded when the motor draws the extra current. If the overload condition continues for 10 to 20 minutes, enough heat will be built up in the motor and transformer to cause the insulation in both of these devices to break down and deteriorate.

The same problem will occur if the bearings in a motor become dry and begin to wear. After the bearing has operated without any lubrication, it will begin to heat up and sieze on the motor shaft, which will in turn cause the motor to draw excessive current. This condition will cause the motor and transformer to overheat to a point where they are completely damaged because the overcurrent will continue as long as the motor is running.

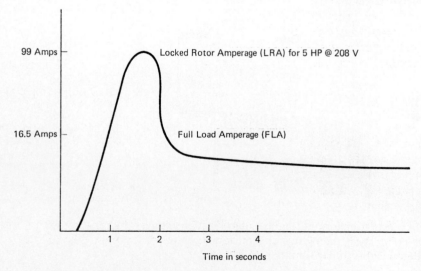

FIGURE 4-8 Graph showing inrush current caused by motors starting.

Protecting Against Slow Overcurrent

In all these examples the problem is caused by the increase in normal operating current to the point where damage can occur. Circuit breakers can sense either the heat or magnetic forces as they increase beyond the maximum safe level, and fuses and overloads can sense increased heat that the overload creates. Separation of these devices will cause a set of contacts or a conducting element to open any time the current increases above the safe level. This presents a problem with some loads, such as motors, that have a very large inrush current when they start. Figure 4-8 shows a graph of the inrush current caused by a motor starting. You can see that a 5-hp motor operating on 208 V will draw 16.5 A at full load. This motor will draw up to 99 A, which is six times the amount of full-load current, when it starts. This presents a problem in protecting against overcurrents, because a circuit breaker or fuse that is sized to protect the motor during full-load current (16.5 A) would trip when the motor is started, and if they are sized to allow the motor to start (99 A), they will not provide adequate protection when the motor is running at full-load current.

Several solutions to this probelm are available. One of them is a motor starter with heaters and overloads, and another is inverse-time circuit breakers. Each of these devices provides several minutes of time delay before they trip and take the motor off line. The operation of these mechanisms is explained in detail in Chapter 6. The theory of their operation involves allowing small overcurrents to exist for up to 4 or 5 minutes, and allowing larger overcurrents for less than 10 seconds. These times are based on the amount of time a specific overload can exist before a motor begins to sustain damage. The devices must be sized properly to provide adequate protection, and may be adjusted slightly once they are installed. The only problem that remains with the inverse-time circuit breaker and overloads is that they cannot sense a short circuit and open the circuit fast enough to provide interruption capacity.

Short-Circuit Currents

When a conductor from one potential comes in contact with the system ground or a conductor from another potential, a short circuit can occur. The short circuit provides little or no impedance to the power source, so that the current can rise to 100 times the full-load current levels in one or two cycles. This means that currents of 50,000 A or larger are possible if the current is allowed to build.

Figure 4-9a shows a graph of the current developed by a short circuit. Notice that the current continues to increase with each cycle of the ac voltage. As this current increases, it will build up powerful magnetic forces and tremendous amounts of heat energy that will cause the metal conductors and terminals to melt and explode.

A circuit breaker is an electromechanical device that requires approximately one-half an ac cycle to sense the short circuit and another half-cycle to trip its mechanical contacts. A graph of the short-circuit current that a circuit breaker will

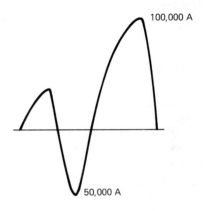

(a) Short-circuit current out of control

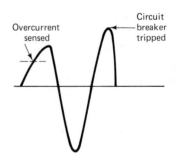

(b) Short-circuit current interrupted during 1 1/2 cycles by a properly sized circuit breaker

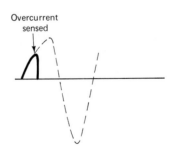

(c) Short-circuit current interrupted during first 1/2 cycle by a properly sized fuse

FIGURE 4-9 Graphs of short-circuit current protection provided by circuit breaker and fuses.

Adequate interrupting capacity and protection of electrical components are two essential aspects required by the 1984 National Electrical Code in Sections 110-9, 110-10, 230-98, and 240-1. The first step to assure that system protective devices have the proper interrupting rating and provide component protection is to determine the available short-circuit currents. The application of the point-to-point method permits the determination of available short-circuit currents with a reasonable degree of accuracy at various points for either 3φ or 1φ electrical distribution systems. This method assumes unlimited primary short-circuit current (infinite bus).

Basic Short-Circuit Calculation Procedure.

Procedure		Formulae	
Step 1	Determine transf. full-load amperes from either: a) Name plate b) Table 5-7-2 c) Formula	3φ transf. 1φ transf.	$I_{FLA} = \dfrac{KVA \times 1000}{E_{L-L} \times 1.73}$ $I_{FLA} = \dfrac{KVA \times 1000}{E_{L-L}}$
Step 2	Find transf. multiplier.	—	$\text{Multiplier} = \dfrac{100}{\text{Transf. \%Z}}$
Step 3	Determine transf. let-thru short-circuit current (Table 5-7-4 or formula).	—	†$I_{SCA} = \text{Transf.}_{FLA} \times \text{multiplier}$
Step 4	Calculate "f" factor.	3φ faults	$f = \dfrac{1.73 \times L \times I}{C \times E_{L-L}}$
		1φ line-to-line (L-L) faults on 1φ, center-tapped transformers	$f = \dfrac{2 \times L \times I}{C \times E_{L-L}}$
		1φ line-to-neutral (L-N) faults on 1φ, center-tapped transformers	$f = \dfrac{2 \times L \times I^*}{C \times E_{L-N}}$
		L = length (feet) of circuit to the fault. C = constant from Table 5-7-1. For parallel runs, multiply C values by the number of conductors per phase. I = available short-circuit current in amperes at beginning of circuit.	
Step 5	Calculate "M" (multiplier) or take from Table 5-7-3.		$M = \dfrac{1}{1 + f}$
Step 6	Compute the available short-circuit current (symmetrical) at the fault.		$I_{SCA \text{ at fault}} = I_{SCA \text{ at beginning of crk.}} \times M$

†**Note 1.** Motor short-circuit contribution, if significant, may be added to the transformer secondary short-circuit current value as determined in Step 3. Proceed with this adjusted figure through Steps 4, 5, and 6. A practical estimate of motor short-circuit contribution is to multiply the total load current in amperes by 4.

Example Of Short-Circuit Calculation.

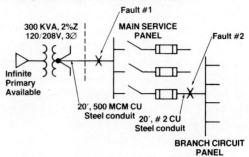

FAULT #1

Step 1 $I_{FLA} = \dfrac{KVA \times 1000}{E_{L-L} \times 1.73} = \dfrac{300 \times 1000}{208 \times 1.73} = 834A$

Step 2 $\text{Multiplier} = \dfrac{100}{\text{Trans. \%Z}} = \dfrac{100}{2} = 50$

Step 3 $I_{SCA} = 834 \times 50 = 41{,}700A$
At Transformer Secondary

Step 4 $f = \dfrac{1.73 \times L \times I}{C \times E_{L-L}} = \dfrac{1.73 \times 20 \times 41{,}700}{18{,}100 \times 208} = .383$

Step 5 $M = \dfrac{1}{1 + f} = \dfrac{1}{1 + 0.383} = .723$ (See Table 5-7-3)

Step 6 $I_{SCA} = 41{,}700 \times .723 = 30{,}150A$
Fault #1

FAULT #2

Step 4 Use I_{SCA} @ Fault #1 to calculate

$f = \dfrac{1.73 \times 20 \times 30{,}150}{4760 \times 208} = 1.05$

Step 5 $M = \dfrac{1}{1 + f} = \dfrac{1}{1 + 1.05} = 0.49$ (See Table 5-7-3)

Step 6 $I_{SCA} = 30{,}150 \times 0.49 = 14{,}770A$
Fault #2

Note: For simplicity, the motor contribution was not included.

*****Note 2.** The L-N fault current is higher than the L-L fault current at the secondary terminals of a single-phase center-tapped transformer. The short-circuit current available (I) for this case in Step 4 should be adjusted at the transformer terminals as follows:
At L-N center tapped transformer terminals
I = 1.5 × L-L Short-Circuit Amperes at Transformer Terminals
At some distance from the terminals, depending upon wire size, the L-N fault current is lower than the L-L fault current. The 1.5 multiplier is an approximation and will theoretically vary from 1.33 to 1.67. These figures are based on change in turns ratio between primary and secondary, infinite source available, zero feet from terminals of transformer, and 1.2 × %X and 1.5 × %R for L-N vs. L-L resistance and reactance values. Begin L-N calculations at transformer secondary terminals, then proceed point-to-point.

FIGURE 4-9 Continued

"see" before it opens a circuit is shown in Figure 4-9b. From this graph you can see that the short-circuit current may still reach 40,000 to 45,000 A before the circuit is fully opened. In some cases, the heat is so immense that it will actually weld the contacts of the circuit breaker together so that they cannot open even though the circuit breaker's trip mechanism has activated.

A fuse can sense the overcurrent as it begins to build and its link will melt before the current increases to a dangerous level. A graph of the protection that a fuse can provide during a short circuit is also shown in Figure 4-9c. From this graph you can see that the fuse element will open as soon as the current reaches the overcurrent level. This means that the short circuit fault will be sensed and opened in less than a half-cycle. The only problem with the single-element fuse is that it must be sized up to six times the full-load current level to allow the motor to start. This leaves two alternatives when providing short-circuit and overload current: fuses in combination with motor starters or circuit breakers, or the use of a dual-element fuse. Figure 4-9d shows example calculations for determining the amount of overcurrent.

Dual-Element Fuses

The dual-element fuse combines the best of overcurrent protection with short-circuit protection. Figure 4-10 shows cutaway diagrams of single-element and dual-element fuses. From this figure you can see that the single-element fuse is made of a single conducting element with several neck-down sections. The neck-down sections provide a point where the short-circuit current will be concentrated and cause the metal to melt. Since the element is made of thin metal, the larger the short-circuit current becomes, the quicker the element will melt.

Also shown is a cutaway diagram of a dual-element fuse. From this diagram you can see that the dual-element fuse provides a short-circuit element and an overload element. A short-circuit element is located on each end of the fuse, while the overcurrent element is located between them. The overcurrent element is made of a spring-loaded section with a heat absorber. The heat absorber has a small solder pot that anchors one end of the spring link. When the full-load current exceeds the specified limit, the solder will become increasingly warm. If the overload continues, the heat will increase sufficiently to allow solder to melt. When the solder melts, the trigger spring will fracture the calibrated fusing alloy and release the connector.

The overload element is sized to allow the fuse to sustain a 500 percent overload for approximately 10 seconds. If the overload is smaller, such as 300 percent, it will sustain it for approximately 60 seconds, and if it is larger, such as 800 percent, it will be cleared in approximately 1.5 seconds. This means that the small overloads will be allowed to continue for several minutes since the small heat

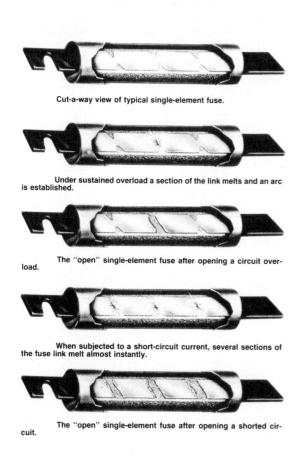

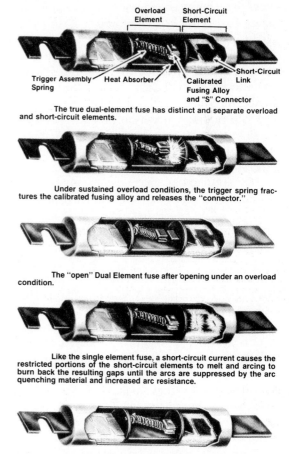

FIGURE 4-10 Cutaway diagrams of single-element and dual-element fuses. (Courtesy of Cooper Industries/Bussman Division.)

buildup will not damage the motor, and large overloads will be cleared quickly since they will cause heat buildup that will permanently damage the motor. This design allows the motor to draw high inrush current, up to 500 percent (5 times) of the full-load current for 10 seconds which is adequate time to allow the motor to start, and it will also allow the motor to develop extra horsepower to meet an increased load demand for several minutes, since the extra current that is drawn will not cause the fuse to open.

The dual-element fuse also provides another safety feature. This feature involves encasing the short-circuit element in silica sand. When the short-circuit current is applied to the short-circuit element, tremendous amounts of heat are built up while the element is melting. If this heat is allowed to build up, gases can be released when the metal is melted and cause the casing of the fuse to rupture. When silica is placed around the short-circuit element, it will absorb the extra heat and use it to melt the sand into a semiliquid state. Since the silica is forced to change state, it will absorb more heat than the reaction can produce, which results in the excessive energy being controlled without damaging the fuse or the hardware and enclosures that are used to mount the fuse.

The ability of the fuse to clear short-circuit currents safely is called its *interruption capacity*. The interruption capacity is listed as the maximum number of amperes that the fuse can safely clear. The interrupting capacity of modern current limiting fuses may be as large as 200,000 A. Fuses can also be used in the power distribution system for the expressed purpose of providing interruption capacity for protecting the system equipment and switchgear against large short-circuit currents.

Sizing Fuses According to the National Electrical Code®

When a power distribution system is designed, the fuses must be correctly sized and specified to provide protection for switchgear and conductors against short-circuit current and overload protection for motors. Figure 4-11 shows the number and types of fuses that are required in a power distribution system. This diagram shows fuses manufactured by Bussman Division of McGraw-Edison which will meet the specifications of the NEC. The electrical specification is the same regardless of brand name of fuse that is being used.

Fuses are selected on the basis of the application they are protecting. These applications include circuits with motors only, circuits with lighting loads only, circuits with a mixture of motors and lighting loads, and circuits with motors and power factor correction. Fuses are also specified for feeder and branch circuit applications. When fuses are used in conjunction with circuit breakers and motor starters they should be sized so that the fuse does not open before these resetting components.

From Figure 4-11 you can see that a wide variety of fuses must be available to meet the various loads that may be encountered. The fuse sizes have been standardized somewhat to ensure that loads of all sizes are adequately protected. Plug fuses are available in sizes that increase 1.15 percent starting at $\frac{3}{10}$ of an ampere. Cartridge fuses are available in sizes from $\frac{1}{10}$ to 600 A.

Types of Fuses

Several types of fuses are available for various applications in motor control systems. The two main types of fuses are cartridge and plug fuses. Figure 4-12 shows several cartridge and plug fuses in different sizes. Some of these fuses provide blades to make connections in the fuse holders, and other fuses provide connections at the neck of the fuse. The plug fuse comes in several sizes, based on the current rating of the fuse. A rejection feature is provided on cartridge, blade, and plug fuses. Examples of the rejection features are also shown in this figure. The rejection feature prevents a fuse with a larger current rating from being substituted for the fuse of a smaller size that was originally specified. This is accomplished by providing a matching rejection feature in the fuse holder with the feature that is provided for the fuse size. This means that each size of fuse has a different type of rejection feature, or it is located at a different point on the fuse body for each amperage rating.

Fuses for Solid-State Applications

Solid-state devices such as rectifier diodes, SCRs, inverters, and motor drives require current-limiting protection that can respond faster than traditional fuses. These devices are very sensitive to short-circuit current, so the fuse that is used to protect these devices should be able to sense the fault and clear it rapidly. Specialized fuses have been developed that have a high degree of current limitation for semiconductor applications.

Figure 4-13 shows several diagrams of the correct location for these fuses in both ac and dc circuits that have semiconductor components. From these diagrams you can see that it is important to locate fuse protection on both sides of solid-state devices since they can be damaged by short circuits that occur on the ac side of the circuit or the dc side of the circuit. The fuses used in these circuits must be sized according to the rms current rating of the circuit rather than the average current. This is required because the fuse must also protect against excessive heat buildup in the solid-state components caused by problems in the circuit. The nature of the solid-state circuit is to shift all current to the remaining components when one component fails. This means that if the circuit is overloaded and one component starts to fail before the others, it will ultimately shift its entire load to the remaining devices that are already overloaded. This tends to cause two or more solid-state components to fail when a fault occurs, if proper protection is not used. The solid-state fuses are designed to sense the overload and clear them before any components are affected. They will also afford proper protection to all remaining devices in a circuit if a solid-state device fails from internal breakdown. It is very important to size these fuses according to measured voltages and currents as well as specified values so that they can protect the circuit properly.

Cable-Limiting Fuses

Cable-limiting fuses are specialized fuses that are intended to protect cables from damage caused by short circuits. As you know, when a short circuit is allowed to occur, the current will cause both heat damage and magnetic distortion of all conductors that are in the faulty circuit. The cable limiter is specially designed to protect the cable against short-circuit current. The physical design of the fuse provides a termination point for the conductor and a hardware fitting so that it can easily by bolted to power distribution terminal strips. Figure 4-14 shows a picture of these specialized fuses and a diagram of where they should be located to protect the cables used in the power distribution system.

Troubleshooting and Testing Fuses

A fuse is the only component in a motor control system that is intended to be destroyed when it is doing its job properly. The main tendency for the troubleshooting technician is to test the fuses to find the one that has opened and then quickly replace it to put the equipment into production again. The fuse is in the circuit to protect the equipment from short circuits and overcurrents, but it is important to understand that the fuse is the only witness to the fault. This means that after a fuse is found to be faulty, you should check it carefully to determine whether it opened because of a short circuit or a slow overcurrent.

The first part of the troubleshooting sequence is determining which fuse is the cause of the open circuit. As you have seen in the power distribution circuit diagrams, fuses are located throughout the system to protect against problems. When you are called to test the circuit, you will be called because a machine or group of machines have stopped operating. This could be caused by several other problems that will be explained in other chapters, but in this case we will assume that one of the fuses has detected a fault and has opened to clear the fault.

You will need to take several voltage tests at various points in the circuit to determine which fuse has opened. The best way to test the fuses with voltage applied is to place the voltmeter probes on the top of fuses (where voltage comes in) from two different potential sources. The neutral or ground point in the system could also be used as a reference for one of the probes while the other probe is used to touch each fuse where power is incoming to determine if voltage is present. If voltage is present at the top of the fuse, it indicates that power is being supplied to the point. Next, the bottom of the fuse should be checked, and if voltage is not present at the bottom, the fuse is opened. If voltage is present at the bottom, one other test should be made to ensure that voltage is not backfeeding. This test requires that you place both terminals across each fuse in the box. If a fuse is open, it will read full voltage across it, and if the fuse is still in good condition, the voltage reading will be zero across it. This may confuse you at first, but Figure 4-15 shows how backfeeding can occur. If the circuit contains a motor starter or other switch that opens and isolates the load, the feedback problem will not occur.

From the diagram you can see that feedback can occur through the top set of fuses since the load that they are protecting is a transformer. In this case you should use both tests to find the open fuse. If two fuses are blown, you will find them with the test that checks for voltage at the top of the fuse in reference to ground or between phases. It should also be noted that if you test across each fuse when two or more fuses are blown, the test that uses ground or another phase as reference may confuse you because of the backfeeding problem. In this case the bad fuse will be found by reading voltage directly across each fuse.

As you can see, you must use a combination of each of these techniques to find all fuses that are blown. It is also best to check each fuse for continuity with an ohmmeter or test lamp after they have been removed from the circuit. *Do not try to test the fuse for continuity while power is applied, since the ohm-meter is a low-impedance meter and will cause*

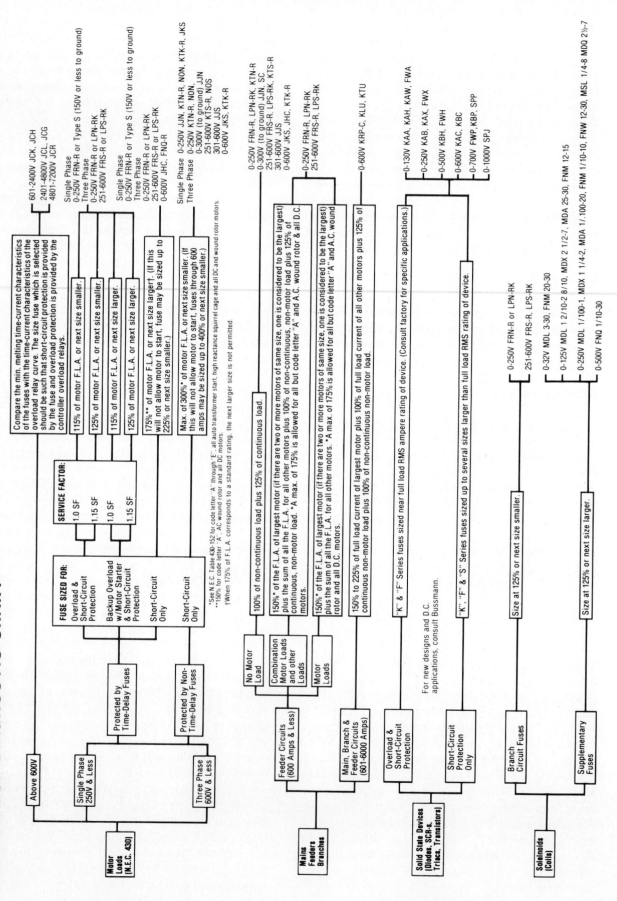

BUSS FUSE DIAGNOSTIC CHART

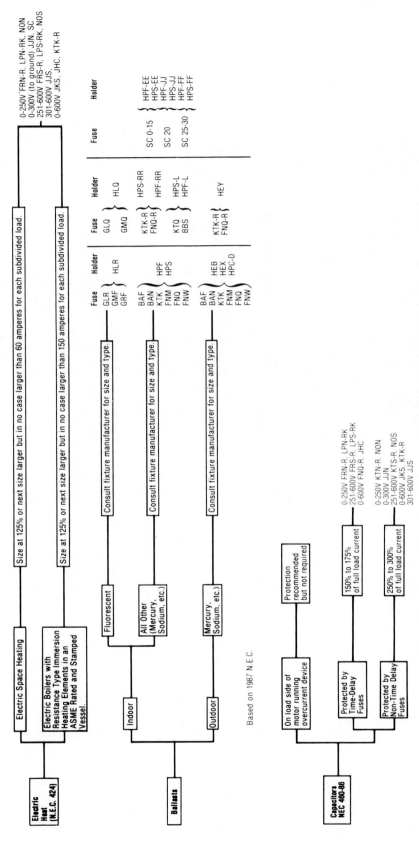

FIGURE 4-11 Selection chart for specifying fuse type and sizes. (Courtesy of Cooper Industries/Bussman Division.)

Buss Power Distribution Fuses

HI-CAP®
(Time-Delay)
KRP-C (600V)
601 to 6000A
200,000AIC
Current Limiting
UL Class L; CSA-HRC-L

The all-purpose silver linked fuse for both overload and short-circuit protection of high capacity systems (mains and large feeders). Time-delay (minimum of four seconds at five times amp rating for close sizing. Unlike fast acting fuses, pass harmless surge currents of motors, transformers, etc., without overfusing and any sacrifice of short-circuit current limitation (component protection). The combination use of 1/10 to 600 ampere LOW-PEAK dual-element time-delay fuses and 601 to 6000A KRP-C HI-CAP fuses is recommended as a total system specification. Easily selectively coordinated for blackout protection. Size of upstream fuse need only be twice that of downstream HI-CAP or LOW-PEAK fuses (2:1 ratio). HI-CAP fuses can reduce bus bracing; protect circuit breakers with low interrupting rating as well as provide excellent overall protection of circuits and loads.

LIMITRON®
(Fast-Acting)
KTU (600V)
601 to 6000A
Current Limiting
200,000AIC
UL Class L; CSA-HRC-L

Silver-linked fuse. Single-element units with no time-delay. Very fast-acting with a high degree of current limitation; provide excellent component protection. Particularly suited for protection of circuit breakers with lower interrupting ratings, and non-inductive loads such as lighting and heating circuits. Can be used for short-circuit protection only in circuits with the inrush currents. Must be oversized to prevent opening by the temporary harmless overloads with some sacrifice of current limitation. In motor circuits, must be sized at approximately 300% of motor full-load current and thus will not provide the overload protection of HI-CAP KRP-C fuses.

LIMITRON®
(Time-Delay)
KLU (600V)
601 to 4000A
Current Limiting
200,000AIC
UL Class L; CSA-HRC-L

10 seconds delay (minimum) at 500% of amp rating. Not as current limiting as KRP-C or KTU fuses.

LOW-PEAK®
(Dual-Element, Time-Delay)
LPS-RK (600V)
LPN-RK (250V)
1/10 to 600A
200,000AIC
Current Limiting
UL Class RK1; CAS HRC-I ("D")

High performance, all-purpose fuses. Provide the very high degree of short-circuit limitation of LIMITRON fuses plus the overload protection of FUSETRON fuses in all types of circuits and loads. Can be closely sized to full-load motor currents for reliable motor overload protection as well as backup protection. Close sizing permits the use of smaller and more economical switches (and fuses); better selective coordination against blackouts; and a greater degree of current limitation (component protection), LOW-PEAK fuses are rejection type but fit non-rejection type fuseholders. Thus, can be used to replace Class H, K1, K5, RK5 or other RK1 fuses.

FUSETRON®
(Dual-Element, Time-Delay)
FRS-R (600V)
FRN-R (250V)
1/10 to 600A
200,000AIC
Current Limiting
UL Class RK5; CSA-HRC-I ("D")

Time-delay affords the same excellent overload protection of LOW-PEAK fuses of motors and other type loads and circuits having temporary inrush currents such as caused by transformers and solenoids. (In such circuits, LIMITRON fuses can only provide short-circuit protection). FUSETRON fuses are not as fast-acting as LOW-PEAK fuses and therefore cannot give as high a degree of component short-circuit protection. Like LOW-PEAK fuses, FUSETRON fuses permit the use of smaller size and less costly switches. FUSETRON fuses fit rejection type fuseholders and can also be installed in holders for Class H fuses. They can physically and electrically replace Class H, K5, and other Class RK5 fuses.

LIMITRON®
(Fast-Acting)
KTS-R (600V)
KTN-R (250V)
1/10 to 600A
200,000AIC
Current Limiting
UL Class RK1; CSA HRC-I

Single-element, fast-acting fuses with no time-delay. The same basic performance of the 601-6000A KTU fast-acting LIMITRON fuses. Provides a high degree of short-circuit current limitation (component protection). Particularly suited for circuits and loads with no heavy surge currents of motors, transformers, solenoids, and welders. LIMITRON fuses are commonly used to protect circuit breakers with lower interrupting ratings. If used in circuits with surge currents (motors, etc.) must be oversized to prevent opening and thus only provide short-circuit protection. Incorporate Class R rejection feature. Can be inserted in non-rejection type fuseholders. Thus, can physically and electrically replace fast acting Class H, K1, K5, RK5, and other RK1 fuses.

ONE-
(General Purpose)
NOS (600V)
NON (250V)

1/10 to 600A
10,000AIC
Non Current Limiting
UL Class H
(K-5 in Sizes 1-60A)

With an inter 10,000 amp not considered ing, Class H are used in available short-Single-element fuses do not delay. The 1-a 50,000 AIC, Class K-5.

FIGURE 4-12 Examples of types of fuses and the rejection features that they provide. (Courtesy of Cooper Industries/Bussman Division.)

TIME

rupting rating of and generally current limit- ONE-TIME fuses circuits with low circuit currents. ONE-TIME incorporate time- 60A ratings have and are U.L.

SUPERLAG
(General Purpose)
RES (600V)
REN (250V)
1 to 600A
10,000AIC
Non-Current Limiting
UL Class H

Time-delay is excellent for a class H fuse; affords slower response to temporary overloads. After opening, SUPERLAG fuse links can be replaced and the fuse reused.

Plug Fuses
125V 10,000 AIC

FUSTAT Type S fuses have a size limiting feature which prevents "overfusing." Dual element construction provides the time delay necessary for motor running protection. Sizes from ¼ thru 30 amps.

FUSETRON Type T fuses are similar to Type S fuses except for the Edison (light bulb type) base.

Type W fuses are non-time delay, used with non-inductive loads.

HI-CAP®
(Time-Delay)
JHC (600V) UL Class J Dim.
CSA HRC-I

HI-CAP JHC fuses are similar to Class J LIMITRON fuses except they have the advantage of time-delay permitting them to pass temporary overloads without oversizing. Offer both backup motor overload and short circuit protection. JHC fuses are listed by CSA and therefore comply with NEC Section 90-6.

LIMITRON®
(Quick-Acting)
JKS (600V)
1 to 600A
200,000AIC
Current Limiting
UL Class J; CSA HRC-I

JKS LIMITRON fuses are basically the same as RK1 LIMITRON fuses except somewhat smaller in physical size (but considerably larger than Buss T-TRON fast-acting fuses). JKS fuses are single-element units with no time-delay and are thus best applied in circuits free of the temporary overloads of motor and transformer surges. The smaller dimensions of Class J fuses prevent their replacement with conventional fuses.

T-TRON®
(Fast-Acting)
JJS (600V) 1-800A
JJN (300V) 1-1200A
200,000AIC
Current Limiting
UL Class T; CSA HRC-I

The space-savers. Counterpart of the KTN-R/KTS-R LIMITRON fuses but only one-third the size; thus, particularly suited for critically restricted space. A single-element fuse; extremely fast-acting. Provide a high degree of current limitation on short-circuits for excellent component protection. Must be oversized in circuits with inrush currents common to motors, transformer, and other inductive components (will give only short-circuit protection). Commonly applied in electric heat circuits, load center, disconnect switches, meter, stacks, etc. The small size of T-TRON fuses permits them to be installed in panelboards and control centers for system upgrading when existing circuit breakers cannot safely interrupt larger available short-circuit currents.

Type SC®
300V
1-60A
100,000AIC
Current Limiting UL Class G
A high performance general-purpose branch circuit fuse for lighting appliance, and motor branch circuits of 300 volts (or less) to ground. Fuse diameter is ¹³⁄₃₂"; lengths vary with ampere rating from 1⁵⁄₁₆ to 2¼" (serves as rejection feature and thus prevents dangerous oversizing).

LIMITRON®
(Fast-Acting)
KTK-R (600V)
¹⁄₁₀ to 30A
200,000AIC
Current Limiting
UL Class CC; CSA HRC-I

U.L. listed for branch circuit protection. A very small, high performance, fast-acting, single-element fuse for protection of branch circuits, motor control circuits, lighting ballasts, control transformers, street lighting fixtures. . . . A diameter of only ¹³⁄₃₂" and a length of 1½" give cost and space savings. A grooved ferrule permits mounting in "rejection" type fuseholders as well as standard non-rejection type holders.

CC-TRON™
(Time-Delay)
FNQ-R (600V)
¼ to 7½ A
200,000 AIC
Current Limiting
UL Class CC

Ideal for control transformer protection. Meets requirements of NEC 430-72 (b) & (c) and UL 508. It's miniature design and branch circuit rating make it ideal for motor branch circuit and short circuit protection required by NEC 430-52.

Medium Voltage Fuses
R-Rated Fuses for Motor Circuits
  2400V
  4800V
  7200V
E-Rated Fuses for Potential Transformers
  2475V
  5500V
  8300V
  15,500V
E-Rated Fuses for Transformers & Feeder Protection
  2750V
  5500V
  8300V
  15,500V
Medium & High Voltage Links
  K, T, H, N Types

Buss Cable Limiters
UH Series
  (250V)   For Copper
  100 KAIC   or
K Series   Aluminum
  (600V)   Cable
  200 KAIC

Protect low voltage distribution networks and all types of service entrance cables. Totally self contained. UH Series is often used for residential applications.

K Series is used by utilities in downtown networks and for the protection of conductors between utility transformer & customer switchgear.

FIGURE 4-12 Continued

Fuses and Circuit Breakers   69

Typical Circuits

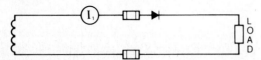

1. Single-Phase, Half-Wave

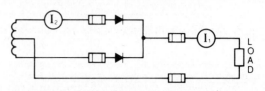

2. Single-Phase, Full-Wave, Center-Tap

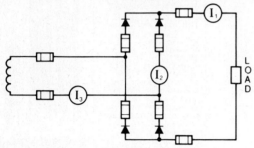

3. Single-Phase, Full-Wave, Bridge

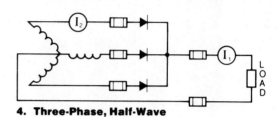

4. Three-Phase, Half-Wave

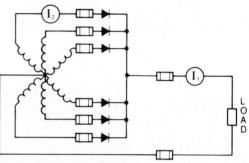

5. Three-Phase, Full-Wave

6. Six-Phase, Single-Wave

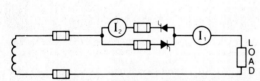

7. Single-Phase, Anti-Parallel, AC Control

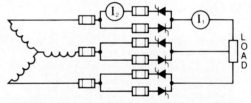

8. Three-Phase, Anti-Parallel, AC Control

Ratios Of Circuit Currents.

Crk. Diag.* No.	Relative Circuit Currents			
	$\dfrac{I_1 \text{rms}}{I_1 \text{average}}$	$\dfrac{I_2 \text{rms}}{I_1 \text{average}}$	$\dfrac{I_3 \text{rms}}{I_1 \text{average}}$	$\dfrac{I_2 \text{rms}}{I_1 \text{rms}}$
1	1.57	—	—	—
2	1.11	0.79	—	0.71
3	1.11	0.79	1.11	0.71
4	1.02	0.59	—	0.58
5	1.00	0.58	0.82	0.58
6	1.00	0.41	—	0.41
7	—	—	—	0.71
8	—	—	—	0.71

*For example, in Diagram No. 1, $\dfrac{I_1 \text{rms}}{I_1 \text{average}} = 1.57$

FIGURE 4-13 Diagrams of solid-state fuses are shown with the proper location and number of fuses to be used to provide protection for typical solid-state circuits. (Courtesy of Cooper Industries/Bussman Division.)

Application of Specialized Fuses and Limiters

Cable Limiters.
Cable limiters are distinguished from fuses by their intended purpose of providing only short-circuit response; they are not designed to provide overload protection. Typically, cable limiters are selected based on conductor size. They are available in a wide range of types to accommodate the many conductor sizes, copper or aluminum conductors, and a variety of termination methods. There are basically two broad categories of cable limiters:

1. 600 Volt or less rated—for large commercial, institutional and industrial applications.

Offset Bolt-to-Tube Terminals

Offset Bolt-to-Center Bolt Terminals

2. 250 Volt or less rated—for residential and light commercial applications.

Cable Limiters (250V).

In institutional and industrial systems, cable limiters are most often used at both ends of each cable on multiple cables per phase applications between the transformer and switchboard, as illustrated in the diagram and photographs below.

COMMERCIAL/INDUSTRIAL SERVICE ENTRANCE
(Multiple cables per phase)

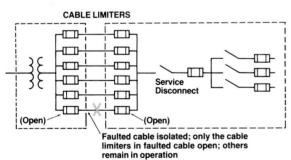

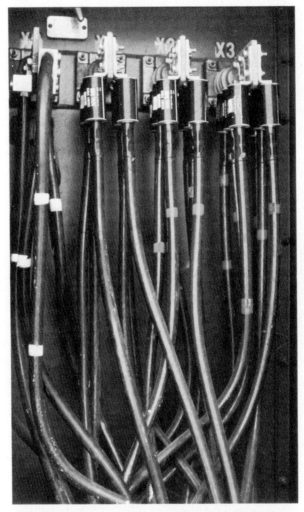
Buss Cable Limiters installed at secondary of 1500kva pad mounted transformer (six 500 MCM cables per phase).

Buss Cable Limiters installed at busbar of switchgear (six 500 MCM cables per phase from 1500kva pad mounted transformer).

FIGURE 4-14 Cable limiters fuses that are connected in line with cables to protect them from short-circuit damage. A diagram is also provided to show the proper location of these limiters. (Courtesy of Cooper Industries/Bussman Division.)

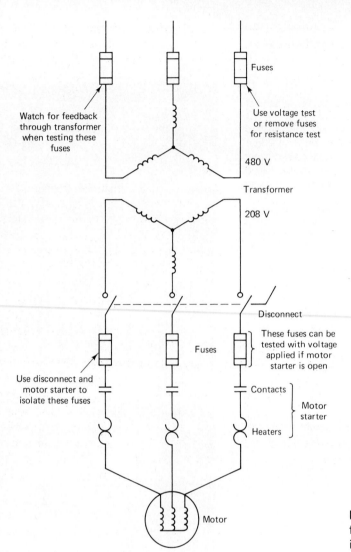

FIGURE 4-15 Methods of testing fuses to determine when a circuit is back feeding through a fuse.

a short circuit in any circuit that has power applied where you touch the probes.

Another problem that may confuse you with a blown fuse is caused when the fuse is subject to a slow overcurrent. In these cases, the link in the fuse is trying to pull away from the melting solder, and the solder in each fuse may not have melted enough to allow the spring to break away. If this occurs, the fuse will test good with voltage tests while it is in circuit and with the continuity test while it is out of circuit, yet the motors and loads will not be able to start when power is applied because sufficient current cannot pass through the partially melted solder to power the circuit. If you have tested the fuses and they seem to be good, yet the motor is not receiving sufficient power, you can recheck each suspected fuse. This time when you have the fuse out of its holder, tap one end of the fuse on the side of the machine or on the floor. This will cause the spring to break away if the solder is partially melted.

Some experienced technicians have learned to automatically change all the fuses in a circuit that has been involved in a slow overcurrent because the remaining fuses have been overheated and are probably stressed, which means that they may trip prematurely when the motor is started again.

It is always important to check each fuse that you find opened to see if it has blown because of a short circuit or a slow overcurrent. If the fuse is a plug type, its window will be black from the high current. If the window is clear but the fuse is open, shake the fuse and listen for solder to roll around, which indicates that the fuse has cleared a slow overcurrent.

If the fuse is a cartridge type, you may need to cut it in half with a bandsaw to determine if it opened from a short circuit or slow overcurrent. Once you have determined why the fuse has opened, you can look for the source of the problem. If the fuse has cleared a slow overcurrent, you should look for overload conditions on the motor.

You should also have an ammeter in place when the motor is started again. This will help you determine the extent of the overload and indicate when you have cleared the problem that is causing it.

If the fuse has opened because of a short circuit, you will have to begin disconnecting sections of the circuit until you test equipment indicates that the short circuit has been removed. At that point you will need to check more closely the circuit that has been disconnected for indications of a short circuit.

HARDWARE FOR POWER DISTRIBUTION

When power is brought into the substation and through the step-down transformers, it must be metered and controlled. The switchgear used for controlling the power as it is distributed around the plant must be housed in proper enclosures, and hardware that is used to connect these enclosures to the point of distribution must be matched to the application. This means that in some applications the conductors must be enclosed in metal conduit and enclosures that are raintight to keep moisture from causing damage. In other applications the power can be distributed through bus duct in the floor or overhead so that equipment can be connected with a minimum of work. You will need to understand the reason that each enclosure or other type of hardware is used in the distribution system, so that you can correctly install the proper equipment during initial installation or during replacement.

Metering Panels

When power is brought into the substation for distribution, it must be metered to determine the amount of voltage, current, wattage, power factor, and kVAR that are being used. Meters for sensing these variables can be mounted in a centrally located meter panel. Figure 4-16 shows a typical industrial meter panel. This panel provides a centralized location for installing and reading these meters. Each of the meter panels provide sockets that will accept a wide variety of meters.

Typical meters include voltage meters, current meters, wattmeters, and power factor meters. Since current must be measured in series, current transformers are used to provide the milliamperage equivalent of the actual current. These transformers look like doughnuts and are some times called *doughnut transformers*. The main conductor is inserted though the doughnut so that it can induce a current based on the actual current flowing through

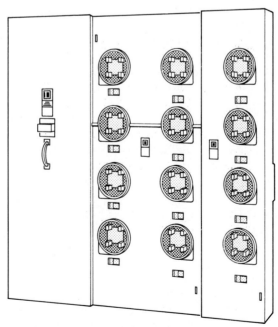

FIGURE 4-16 Metering panel used to mount a variety of voltage, current, wattage, kVAR, and power factor meters for utility companies and factory use. (Courtesy of Square D Company.)

Multi-socket,
1ϕ and/or 3ϕ indoor and rainproof
UL listed up to 100,000 rms symmetrical
short circuit current rating

the conductor. The voltmeter can be connected across each of the lines that are being measured.

The wattmeter takes the voltage from the voltmeter and current from the ammeter and indicates the amount of power that the system is using. This meter can also indicate the peak wattage that has been measured over any period of time. Since the utility company charges a penalty for excessive power factor, another meter will measure the apparent power on a kVAR (kilovolt-ampere reactance) meter. This value can be compared to the true power that is measured on the wattmeter and the power factor can be indicated.

Operators can use the power factor and demand meters to switch power factor correction equipment into the circuit or to drop off unnecessary loads to bring the peak power reading back below the maximum set point. These instruments may be installed by the utility company and sealed so that they cannot be tampered with. The utility company will read the meters once a month just like residential systems and assess the appropriate charges for the power that has been used. Other meters may be installed by the factory to determine energy costs on a batch or product basis. Since these costs change, they can be used constantly to determine the price of the product that is being manufactured.

Bus Duct Systems

Power can be distributed throughout the factory in a variety of ways, including routing cables inside conduits supported in cable trays that are mounted overhead, and in conduits buried in the floor. Other methods include using busbars mounted in bus ducts to distribute the power to locations where it will be used. The useful feature that bus duct provides is the ability to make multiple connections or taps to the busbars simply by attaching a disconnect that is protected with a fuse or circuit breaker.

The disconnect provides a switch that is used to disconnect power from that individual branch circuit without removing power from the entire distribution system. This allows the power to a single machine to be disconnected without bothering other circuits. The disconnect also provides a set of terminals where the conductors can be connected. The enclosure for the disconnect also provides a hardware connection for any conduits that are used between the disconnect and the machinery that will receive the power.

The bus duct and disconnects (taps) provide flexibility in the system, so that if machines are moved, the disconnect can be removed from the bus duct receptacle and moved to the new location. When a receptacle is not in use, it is covered to prevent material from entering the busbars and creating a short circuit. Figure 4-17 shows a typical busbar system enclosed in the bus duct. This figure also shows typical disconnects plugged into the bus duct. The busbar is a copper conductor that is large enough to carry several hundred amperes. The bus duct can have three or more busbars housed inside to provide proper voltage for three wire three-phase and four-wire three-phase voltage distribution systems. It is important to understand that once the bus duct is powered, the busbars inside the bus duct are extremely dangerous to work near since they are unprotected conductors that do not have any insulation covering them. If you or the tools that you are using come in contact with any of the bars, you will receive a severe or lethal shock. It is also important to understand that you can not allow any factory equipment, such as tow motors, forklifts, or

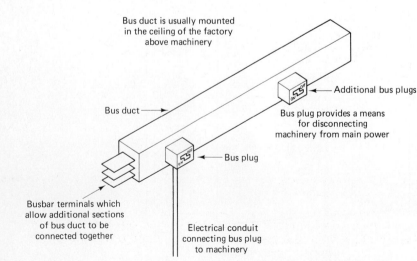

FIGURE 4-17 Bus ducts with busbars displayed so you can see their location. Notice the bus plugs that are used to tap into the power distribution system. (Courtesy of Square D Company.)

high-lift work platforms to come into contact with the bus duct because the sides may be dented to a point where the bus duct comes into contact with the busbar and creates a short circuit.

In some plants, exposed busbars are mounted high in the ceiling. Since these conductors are 40 or 50 ft in the air, they are assumed to be safe, but you you should not come into contact with them *at any time* because you can be killed by the extreme electrical shock they will cause. If you must work on the lighting system in the plant, be sure to work with several other technicians so they can be sure that you and your equipment do not come into contact with exposed busbars.

From Figure 4-17 you can see that the bus duct can be installed in the ceiling or under the floor and taps can be connected wherever power is needed for machine operation. Bus duct is typically used where many machines must be connected to power in a small area. Since it is rather expensive, it is not useful for long-distance cable runs that must cross the expanses of a large factory. In applications where longer runs are necessary, cable trays and wireways are preferred.

When conductors must be suspended overhead, a cable tray is generally used to support them. The cable tray allows easy access to all conductors and easy expansion when extra cables must be added to the system. Figure 4-18 shows a typical layout of cable tray with the fittings and accessories that are necessary for installation in all parts of the plant.

When conductors are supported in the cable tray, they are open to all types of environmental conditions. Sometimes the cable must be installed through parts of the factory where moisture and other liquids may penetrate the conductors and cause short circuits or deterioration of the shielding. In these applications wireways may be used to protect the conductors. Wireways are available in watertight, raintight, and oiltight versions. If the conductors must be subjected to weather conditions such as between the substation located outside the plant and the switchgear inside the plant, it must be enclosed in raintight or watertight wireways. The wireway is a square or rectangular duct that allows conductors to be installed inside. The top of the duct is removable so that conductors can be added or maintained. Gaskets are used to seal the wireway to maintain the watertight or raintight integrity. Tables are available that indicate the number of conductors of various sizes that can be carried in the duct without damage to other conductors because of heat buildup. When the wireway is filled to

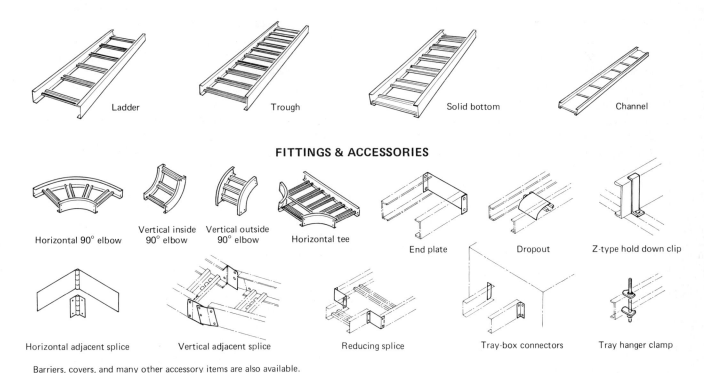

FIGURE 4-18 Cable tray and wireways that are used to support conductors that are used for distribution of power cables throughout the plant. (Courtesy of Square D Company.)

Hardware for Power Distribution 75

NEMA Type 1
Surface Mounting

Type 1 enclosures are intended for indoor use primarily to provide a degree of protection against contact with the enclosed equipment in locations where unusual service conditions do not exist. The enclosures are designed to meet the rod entry and rust-resistance design test. Enclosure is sheet steel, treated to resist corrosion.

NEMA Type 3R, 7 & 9
Unilock Enclosure
For Hazardous
Locations

This enclosure is cast from "copper-free" (less than 0.1%) aluminum and the entire enclosure (including interior and flange areas) is bronze chromated. The exterior surfaces are also primed with a special epoxy primer and finished with an aliphatic urethane paint for extra corrosion resistance. The V-Band permits easy removal of the cover for inspection and for making field modifications. This enclosure meets the same tests as separate NEMA Type 3R, and NEMA Type 7 and 9 enclosures. For NEMA Type 3R application, it is necessary that a drain be added.

NEMA Type 1
Flush Mounting

Flush mounted enclosures for installation in machine frames and plaster wall. These enclosures are for similar applications and are designed to meet the same tests as NEMA Type 1 surface mounting.

NEMA Type 3

Type 3 enclosures are intended for outdoor use primarily to provide a degree of protection against windblown dust, rain, sleet, and external ice formation. They are designed to meet rain **1**, external icing **2**, dust, and rust-resistance design tests. They are not intended to provide protection against conditions such as internal condensation or internal icing.

NEMA Type 4

Type 4 enclosures are intended for indoor or outdoor use primarily to provide a degree of protection against windblown dust and rain, splashing water, and hose-directed water. They are designed to meet hosedown, dust, external icing **2**, and rust-resistance design tests. They are not intended to provide protection against conditions such as internal condensation or internal icing. Enclosures are made of heavy gauge stainless steel, cast aluminum or heavy gauge sheet steel, depending on the type of unit and size. Cover has a synthetic rubber gasket.

NEMA Type 3R

Type 3R enclosures are intended for outdoor use primarily to provide a degree of protection against falling rain, sleet, and external ice formation. They are designed to meet rod entry, rain **3**, external icing **2**, and rust-resistance design tests. They are not intended to provide protection against conditions such as dust, internal condensation, or internal icing.

FIGURE 4-19 List of NEMA-type enclosures for indoor and outdoor industrial applications. (Courtesy of Allen-Bradley, Inc.)

capacity, another one must be used to carry the additional conductors.

Enclosures

Enclosures are required to house disconnects, motor starters, and other motor controls. The enclosures are available in a variety of designs that will protect the devices inside from all types of environmental conditions. Figure 4-19 shows a list of typical enclosure types as classified by the National Electrical Manufacturers Association (NEMA). Type 1 enclosures are designed for general-purpose application in indoor locations. This means that the enclosure should not be exposed to extreme conditions, such as excessive moisture. The type 2 enclosure is rated for drip-proof conditions that exist indoors. This means that some moisture may

NEMA Type 4X
Non-Metallic, Corrosion-Resistant Fiberglass Reinforced Polyester

Type 4X enclosures are intended for indoor or outdoor use primarily to provide a degree of protection against corrosion, windblown dust and rain, splashing water, and hose-directed water. They are designed to meet the hosedown, dust, external icing ☐, and corrosion-resistance design tests. They are not intended to provide protection against conditions such as internal condensation or internal icing. Enclosure is fiberglass reinforced polyester with a synthetic rubber gasket between cover and base. Ideal for such industries as chemical plants and paper mills.

NEMA Type 6P

Type 6P enclosures are intended for indoor or outdoor use primarily to provide a degree of protection against the entry of water during prolonged submersion at a limited depth. They are designed to meet air pressure, external icing ☐, and corrosion-resistance design tests. They are not intended to provide protection against conditions such as internal condensation or internal icing.

NEMA Type 7
For Hazardous Gas Locations
Bolted Enclosure

Type 7 enclosures are for indoor use in locations classified as Class I, Groups C or D, as defined in the National Electrical Code. Type 7 enclosures are designed to be capable of withstanding the pressures resulting from an internal explosion of specified gases, and contain such an explosion sufficiently that an explosive gas-air mixture existing in the atmosphere surrounding the enclosure will not be ignited. Enclosed heat generating devices are designed not to cause external surfaces to reach temperatures capable of igniting explosive gas-air mixtures in the surrounding atmosphere. Enclosures are designed to meet explosion, hydrostatic, and temperature design tests. Finish is a special corrosion-resistant, gray enamel.

NEMA Type 9
For Hazardous Dust Locations

Type 9 enclosures are intended for indoor use in locations classified as Class II, Groups E, F or G, as defined in the National Electrical Code. Type 9 enclosures are designed to be capable of preventing the entrance of dust. Enclosed heat generating devices are designed not to cause external surfaces to reach temperatures capable of igniting or discoloring dust on the enclosure or igniting dust-air mixtures in the surrounding atmosphere. Enclosures are designed to meet dust penetration and temperature design tests, and aging of gaskets. The outside finish is a special corrosion-resistant gray enamel.

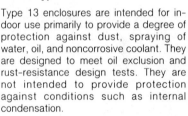

NEMA Type 12

Type 12 enclosures are intended for indoor use primarily to provide a degree of protection against dust, falling dirt, and dripping noncorrosive liquids. They are designed to meet drip ☐, dust, and rust-resistance tests. They are not intended to provide protection against conditions such as internal condensation.

NEMA Type 13

Type 13 enclosures are intended for indoor use primarily to provide a degree of protection against dust, spraying of water, oil, and noncorrosive coolant. They are designed to meet oil exclusion and rust-resistance design tests. They are not intended to provide protection against conditions such as internal condensation.

FIGURE 4-19 Continued

come in contact with the enclosure, but it is not approved where equipment must be washed down or steam cleaned daily.

Type 3 enclosures are designed for dust-tight, raintight, sleet-tight conditions that exist outdoors, and type 3R rainproof, sleet-resistant, and type 3S enclosures are designed for dust-tight, raintight, and sleet-proof conditions. These enclosures are intended for use where protection against falling rain, sleet, or dust. They are not intended to prevent condensation from forming on the inside of the enclosure or on internal components in the enclosure. If the application is located where windblown water or sleet may be encountered, a type 4 enclosure should be used.

The type 4 and 4X enclosure is designed to protect against windblown dust, rain, or water from direct-hose-down conditions. This enclosure is

Hardware for Power Distribution 77

specified by NEMA for watertight, dust-tight, sleet-resistant indoor and outdoor applications. The 4X enclosure also provides protection in environments where corrosion resistance is required. The covers of these enclosures have protective gaskets that provide a barrier against these conditions. All hardware for conduit must match the same requirements to maintain the integrity of the design throughout the installation.

NEMA type 5 enclosures are designed for dust-tight applications located indoors. This type of enclosure is designed to protect the switch and contacts from a buildup of dust that may prevent proper operation of the switches and starters that are enclosed.

Type 6 enclosures are designed to provide submersible, watertight, dust-tight, sleet-resistant indoor and outdoor applications. The submersible watertight specification allows the switch to be fully submersed in water to a limited depth and maintain watertight integrity. This type of enclosure also depends on gaskets to maintain the watertight condition. As with other enclosures, conduits and connectors must meet the same NEMA standard so that the entire installation maintains the stated protection.

Type 7 enclosures provide class I protection for group A, B, C, or D hazardous locations, with air-break protection for indoor locations. These enclosures are designed to prevent dangerous gases from penetrating the enclosure and coming into contact with the open parts of switches or other controls. This provides protection in explosive atmospheres. Class 1 locations are hazardous because flammable gases or vapors may be present in sufficient quantities to cause explosions.

Type 8 enclosures provide class I, group A, B, C, or D hazardous locations, oil immersed for indoor applications. This type of enclosure is very similar to the type 7 enclosure except that it will provide protection against oil immersion instead of air break. This allows the enclosure to be used where oil and machine coolants are used extensively.

Type 9 enclosures provide protection against class II, group E, F, or G hazardous locations and air-break for indoor applications. Class II locations are hazardous because of the presence of combustible dust in quantities sufficient enough to explode or ignite.

Type 10 enclosures are rated for all applications within mines. This type of enclosure must be able to protect switch gear against explosive conditions. Type 11 enclosures provide corrosion resistance and drip-proof, oil-immersion, protection for indoor applications. This type of enclosure is used where the vapors and fumes may be corrosive to switchgear or other motor controls that are mounted inside. The exterior of the enclosure is also resistant to corrosion from these fumes. This type of enclosure is also used for applications where machining operations are preformed.

Type 12 enclosures are intended for indoor applications to provide protection against dust, falling dirt, and dripping noncorrosive materials. They are not intended for use against direct spraying of these materials.

The last classification of enclosures is type 13, rated for oiltight and dust-tight applications. These enclosures provide protection against direct spraying of water, oil, and noncorrosive coolants.

Conduits and Hardware

Conduits must be used when conductors must be installed between enclosures and equipment operator panels. The conduits may be rigid or flexible to fit the application, and they will provide a degree of protection against physical damage or other environmental conditions. If the enclosure is rated for a specific classification, the conduit and fittings must have the same rating.

One type of conduit that is used extensively is called *rigid conduit*. This type of conduit requires preformed fittings to turn corners and match contours of machinery. These fittings include a variety of elbows and straight and 90-degree connectors for termination at enclosures and disconnects.

The National Electrical Code® (NEC) provides detailed specifications pertaining to the mounting and installation of rigid conduit and the number of conductors that can be carried by each size. Another type of conduit is *flexible conduit* or *electrical metallic tubing* (EMT). This type of conduit is capable of being bent and shaped in the field during installation. This eliminates the need for the majority of fittings, except for junction boxes, pull boxes, and pull elbows, which are used to pull conductors into the conduit after it has been mounted. Both rigid and EMT can be installed in concrete walls and floors, or they can be surface mounted on walls and machinery.

Another type of conduit is *armor cable*. It is very flexible and is sold under such trade names as Greenfield. This type of conduit is used where machinery may vibrate and destroy rigid and semirigid connections used by rigid conduit and EMT. This type of conduit is also available in a watertight jacket that allows it to be used in outdoor applications. It is extremely useful in conveyor applications and robot applications where connections between movable systems must be accomplished.

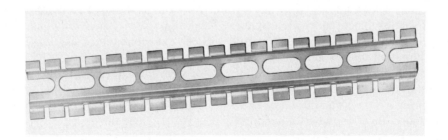

FIGURE 4-20 Example terminal blocks used in electrical panels. (Courtesy of Eaton Corp./Cutler-Hammer Division.)

Terminals and Wire Ties

When conductors are connected inside electrical panels, they are usually terminated at terminal blocks on switch gear or other motor controls. It is also common practice to group together all conductors that are used in motor control circuits and terminate them on a common terminal block. The conductors are usually separated by their application, such as ones connected to switches and ones connected to solenoids or relay coils. Figure 4-20 shows an example of typical channel-mounted terminal blocks. These blocks are mounted on a common channel that allows a wide variety of sizes to be mounted at the same location. The terminal blocks are matched to the size of the conductor that will be connected to it. The terminal blocks also provide an easy location for access to conductors during troubleshooting and testing procedures.

It is customary to identify each wire on the terminal block with a wire marker. These markers are available in the form of a tape that has a number, or letters and numbers, that can be added to the wire. An example of wire numbers is also shown in Figure 4-20. From this example you can see that these numbers can be applied by the technician at the time of installation, or they can be applied by a small imprinter that can burn the numbers into the coating of the conductor. It is important to maintain wire numbers since all conductors of similar voltages will be the same color.

After all conductors have been routed through the electrical panel, they should be secured. Several methods of securing the conductors in these panels are available. One method involves using a small plastic cable tray that has multiple slits that allow wires to be brought into it and removed from it at various points. When all connections have been made, a cover can be snapped into place over the tray to keep the conductors from falling out. Figure 4-21 shows an example of this type of cable tray. Another method of securing conductors in an electrical panel involves the use of cable ties. The ties are made of plastic and have a plastic strap that is inserted into a termination block and pulled tight. The block has a metal insert that allows the strap to move in only one direction, which allows it to be tightened. Once the strap is pulled tight, it must be cut loose to be removed. A example of these wire ties is also shown in Figure 4-21. The ties come in a variety of sizes and also have a number of support blocks with adhesive backs that allow them to be mounted on the surface of the panel. After the support block is mounted to the panel, the tie wrap can be routed through it and around the bundle of conductors to secure them into place.

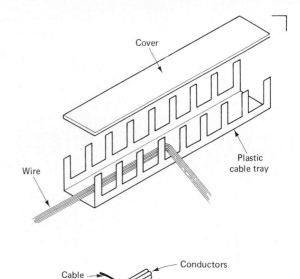

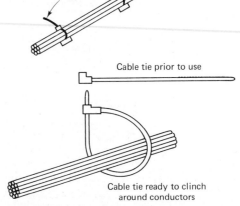

FIGURE 4-21 Cable tray and cable tie used to route and secure conductors in a panel.

Conductors

One part of the power distribution system that is involved at all points in the system is the conductor. Conductors can be made of aluminum or copper wire. Aluminum is generally used for long-distance distribution because of its lighter weight. Once the power is inside the plant, copper conductors are generally used for distribution. Copper conductors will be solid in busbar and busway applications, and they will be stranded conductors for all other applications.

The conductors are sized by the amount of amperage they can carry. A table of typical conductor sizes and their ampacities is provided in Figure 4-22. The standard used to determine the size of each conductor is called the *American wire gage* (AWG). These tables have been established by the National Electrical Code®, and they are used to select the proper size of conductor to carry the load.

Conductors are also classified by the type of insulation that is used as a cover. The cover also

Table 310-16. Ampacities of Not More than Three Single Insulated Conductors, Rated 0 through 2000 Volts, in Raceway in Free Air and Ampacities of Cable Types AC, NM, NMC and SE

Based on Ambient Air Temperature of 30°C (86°F).

Size AWG MCM	Temperature Rating of Conductor. See Table 310-13.								Size AWG MCM
	60°C (140°F)	75°C (167°F)	85°C (185°F)	90°C (194°F)	60°C (140°F)	75°C (167°F)	85°C (185°F)	90°C (194°F)	
	TYPES †TW, †UF	TYPES †FEPW, †RH, †RHW, †THW, †THWN, †XHHW, †USE, †ZW	TYPE V	TYPES TA, TBS, SA, AVB, SIS, †FEP, †FEPB, †RHH, †THHN, †XHHW*	TYPES †TW, †UF	TYPES †RH, †RHW, †THW, †THWN, †XHHW †USE	TYPE V	TYPES TA, TBS, SA, AVB, SIS, †RHH, †THHN, †XHHW*	
	COPPER				ALUMINUM OR COPPER-CLAD ALUMINUM				
18	....	....	....	14	....	....	....	....	
16	....	....	18	18	....	....	....	....	
14	20†	20†	25	25†	....	....	....	....	
12	25†	25†	30	30†	20†	20†	25	25†	12
10	30	35†	40	40†	25	30†	30	35†	10
8	40	50	55	55	30	40	40	45	8
6	55	65	70	75	40	50	55	60	6
4	70	85	95	95	55	65	75	75	4
3	85	100	110	110	65	75	85	85	3
2	95	115	125	130	75	90	100	100	2
1	110	130	145	150	85	100	110	115	1
1/0	125	150	165	170	100	120	130	135	1/0
2/0	145	175	190	195	115	135	145	150	2/0
3/0	165	200	215	225	130	155	170	175	3/0
4/0	195	230	250	260	150	180	195	205	4/0
250	215	255	275	290	170	205	220	230	250
300	240	285	310	320	190	230	250	255	300
350	260	310	340	350	210	250	270	280	350
400	280	335	365	380	225	270	295	305	400
500	320	380	415	430	260	310	335	350	500
600	355	420	460	475	285	340	370	385	600
700	385	460	500	520	310	375	405	420	700
750	400	475	515	535	320	385	420	435	750
800	410	490	535	555	330	395	430	450	800
900	435	520	565	585	355	425	465	480	900
1000	455	545	590	615	375	445	485	500	1000
1250	495	590	640	665	405	485	525	545	1250
1500	520	625	680	705	435	520	565	585	1500
1750	545	650	705	735	455	545	595	615	1750
2000	560	665	725	750	470	560	610	630	2000

	AMPACITY CORRECTION FACTORS								
Ambient Temp. °C	For ambient temperatures other than 30°C (86°F), multiply the ampacities shown above by the appropriate factor shown below.								Ambient Temp. °F
21-25	1.08	1.05	1.04	1.04	1.08	1.05	1.04	1.04	70-77
26-30	1.00	1.00	1.00	1.00	1.00	1.00	1.00	1.00	79-86
31-35	.91	.94	.95	.96	.91	.94	.95	.96	88-95
36-40	.82	.88	.90	.91	.82	.88	.90	.91	97-104
41-45	.71	.82	.85	.87	.71	.82	.85	.87	106-113
46-50	.58	.75	.80	.82	.58	.75	.80	.82	115-122
51-55	.41	.67	.74	.76	.41	.67	.74	.76	124-131
56-60	....	.58	.67	.71	....	.58	.67	.71	133-140
61-70	....	.33	.52	.58	....	.33	.52	.58	142-158
71-80	....	....	.30	.41	....	....	.30	.41	160-176

† Unless otherwise specifically permitted elsewhere in this Code, the overcurrent protection for conductor types marked with an obelisk (†) shall not exceed 15 amperes for 14 AWG, 20 amperes for 12 AWG, and 30 amperes for 10 AWG copper; or 15 amperes for 12 AWG and 25 amperes for 10 AWG aluminum and copper-clad aluminum after any correction factors for ambient temperature and number of conductors have been applied.

* For dry and damp locations only. See 75°C column for wet locations.

FIGURE 4-22 AWG wire sizes as determined by the National Electrical Code®; types of coverings that are used on conductors and their abbreviations. (Courtesy of National Electrical Code®.)

provides the voltage rating of the wire. Typical voltage ratings for conductors used in motor control applications include 300, 600, and 1000 V. The voltage and current ratings of the conductor should not be exceeded under any circumstances.

The type of covering is also listed in Figure 4-22. Abbreviations for each of these coverings are listed in the NEC tables and help determine the type of wire that should be selected for each application. The outer covering of the wire serves several purposes, including protecting the conductors from coming in contact with metal in the cabinet or other conductors. The cover also provides a location to stamp all specification data regarding the wire, including the voltage rating, AWG size, temperature specification, and type of covering.

Hardware for Power Distribution

QUESTIONS

4-1. Explain the term *power distribution*.

4-2. Why must you understand the power distribution system if you must install, troubleshoot, or repair industrial control systems?

4-3. Explain where the source for all electrical power is for a power distribution system.

4-4. Identify the location of the electrical generating plant in your area where the power for your building originates.

4-5. Name four sources of energy used to generate electricity.

4-6. State two ways the electrical utility company is able to meet peak demands on their generating system.

4-7. Explain how power is distributed from the generating facility to a factory where it is used.

4-8. List the main parts of a transformer and explain how each operates. Provide a sketch of the transformer windings and their identification.

4-9. Explain what is meant by the terms *step-up transformer* and *step-down transformer*.

4-10. Explain the term *transformer efficiency*.

4-11. Explain the terms *delta-wired transformer* and *wye-wired transformer*.

4-12. Explain the theory of operation of an autotransformer. Include a sketch of autotransformer windings and their terminal identification.

4-13. To what does the term *buck–boost transformer* refer?

4-14. What is a control transformer used for?

4-15. Explain the difference between short-circuit current and a slow overcurrent.

4-16. Explain the difference between a single-element fuse and a dual-element fuse.

4-17. Explain what a rejection feature is in regard to a fuse and the purpose it serves.

4-18. Explain the problem that may arise due to backfeeding when you must test for a blown fuse in a circuit. Explain how you would avoid the problem with feedback.

4-19. What is bus duct, and where would you find it?

4-20. Identify all the NEMA enclosure types and what they are used for.

4-21. Explain what a wire tie (cable tie) is used for.

Manual control devices 5

INTRODUCTION

Manual controllers are widely used in a large variety of industrial applications in the form of manual motor starters, drum switches, and disconnects. These applications include on/off controls with jog and reverse capabilities for pumps, compressors, fans, conveyors, grinders, drill presses, mixers, choppers, meat cutters, textile looms, and woodworking and metalworking machines. Manual controls are also useful for noninductive loads such as resistive heat and lighting. Sometimes disconnects are used for both on/off control and protection of motors. The disconnect switch was introduced in Chapter 4, where it was used primarily for power distribution applications. Some additional information about the disconnect being used as a manual control is provided in this chapter.

Earliest manual controls were simple knife switches. The knife switch had a movable part and a stationary part. Figure 5-1 shows a typical knife switch used for single-phase motor control. The stationary part of the switch provides two isolated terminals where the field wires are connected. Line 1 is connected to one of the terminals and the motor is connected to the terminal. In addition to providing a location to mount the field wires, the terminal also provides a V slot where the movable part of the switch will seat when it is closed. The movable part of the switch is called the knife blade and it is hinged at one end. From the diagram you can see that the knife blade will seat tightly into the V slot at each terminal. When the knife blade seats into the V slot, it will make electrical contact with each terminal and provide a low-resistance path for electrical current. The knife blade is made from copper and has a nonconductive handle applied so that the switch can be safely operated manually. Early knife switches were made durable enough to carry starting current safely for small motors and other small loads. If the motor was switched on and off frequently, the copper part of the knife blade would heat up and begin to warp. Constant use of the switch may also cause arcing at the point where the knife blade makes contact with the V slot. The arcing would eventually deteriorate the copper and cause resistance to increase, which would cause additional heat on the V slot and the switch would warp and become faulty. To combat this wear, the point where the switch makes contact can be plated with a special alloy so that it will not pit when arcing occurs.

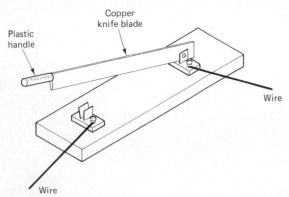

FIGURE 5-1 Diagram of knife switch used for simple manual motor control.

The knife switch has evolved from the simple switch that was used for earliest motor control to the point where it is used in modern disconnect switches. The knowledge gained from problems with early switches has been used to design contacts that are used in modern motor starters. This chapter will help you understand the operation of various manual controls as well as applications for which they are well suited. Wiring diagrams are provided to show distinct differences between various starters, and as an aid during field wiring and troubleshooting. Procedures for selecting the proper type of control and enclosure are also detailed. This includes methods of sizing the controls to provide adequate protection for the motor loads and noninductive loads being controlled. Installation of manual controls is also detailed. These procedures will be useful during initial installation or when the control is replaced. The final part of this chapter includes troubleshooting and diagnostic procedures for these controls. It also shows these controls wired directly to motors for speed control and reversing assumes that you understand principles of dc and ac motors. If you need to review information pertaining to methods of reversing motors and changing motor speeds, see Chapters 10 and 11.

DISCONNECT SWITCHES

The simple type of knife switch has evolved into the modern disconnect for motor control circuits. The disconnect switch was introduced in Chapter 4, where it was used as a disconnect for power distribution systems. The information in this chapter shows applications where the disconnect can be used as a motor control to switch equipment where the motor may be operated for long periods without being turned off. A ventilation fan is an example of this type of application. The disconnect switch will meet most electric code requirements since it provides a means to disconnect power from the motor, and it can also be fused to provide motor overload protection. Figure 5-2 shows a picture of a two-pole disconnect switch. The electrical symbol for the disconnect is also shown. The disconnect is available as a single-pole, two-pole, or three-pole switch. The single-pole type is used for single-phase 110-V circuits, the two-pole type is used for single-phase 208/220-V circuits and dc circuits, and the three-pole switch is used for three-phase ac circuits.

The disconnect is rated for the maximum voltage and maximum current to which the contacts will be subjected. If the disconnect is a fusible type, the fuses and the disconnect must be sized to match the load. Since the disconnect will be used to disconnect and protect a motor without any other controls, the fuses must be a dual-element type and capable of allowing the motor to start while it pulls locked-rotor amperage while protecting it at its full-load amperage rating. Another useful feature that the disconnect provides is lockout protection. The disconnect can be locked in the off position with a padlock for safety or security reasons.

In most applications the disconnect will be used primarily to satisfy the National Electrical Code® (NEC) article pertaining to motor protection. This article requires that for safety, all motor-driven equipment have a disconnecting means within easy access, which provides the operator with a way to safely disconnect power to motorized equipment. The disconnect also provides a method to lock out the switch when it is turned off. This enables personnel to work on the machinery safely.

MANUAL MOTOR STARTERS

Several types of manual motor starters are used in a variety of applications for starting, jogging, reversing, and stopping motors. The starters can be used for control of single-phase and three-phase ac and dc motors. Motor starters also provide the motor protection against overcurrent, which also provides a means of safely cycling a motor on and off numerous times per hour without overheating or damaging the switching mechanism. The manual motor starter can also provide a feature that allows the motor to restart automatically after a loss of power. A picture and electrical diagram for a typical manual motor starter are shown in Figure 5-3. The motor starter is usually installed in an enclosure where its pushbutton or toggle switch is accessible to an operator.

Manual motor starters are generally used for controlling and protecting smaller motor loads or noninductive loads. Like the disconnect switch, the

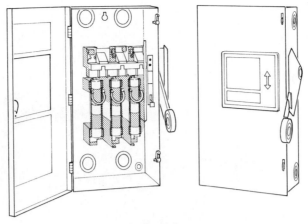

Fused disconnect

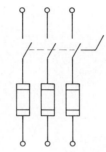

Diagram of fused disconnect

FIGURE 5-2 Disconnect switch with electrical diagram. (Photo courtesy of Square D Company.)

starter is available with a single-pole switch for single-phase 110-V loads and a two-pole switch for 230-V loads up to 3 hp. Three-pole starters are available for loads up to 10 hp. Other types of manual starters are available for applications that require motors to be reversed and for two-speed motors. An autotransformer type is also available for reduced-voltage starting applications.

Figure 5-4 shows a diagram of a typical manual motor starter. The starter has three major parts: the base assembly, the movable contact carrier, and the operator assembly. The base assembly provides the mounting holes and plastic assembly to mount the stationary part of the contact assembly. The stationary part of the contact assembly provides a terminal point, which is a means of securing the field wiring for the line and load wires. The movable contact carrier contains the movable contacts and spring assembly as well as any undervoltage protection, and the operator assembly provides the means to activate the manual starter. The operator assembly can use pushbutton switches, selector switches, or toggle switches to activate the switch. The function of each of these parts and their operation are explained so that you will be able to install, troubleshoot, and replace major parts of the starter, or the complete device.

Operator Assembly, Base Assembly, and Contact Carrier

The base assembly for the manual motor starter is the largest part of the starter. It is made of a durable nonconductive plastic material that will house the stationary part of the contact assembly for the starter. It also provides a base on which to mount the operator assembly. The operator assembly is the part of the manual starter that activates the contact assembly. The operator assembly is available with a variety of switches. A pushbutton switch is available with the buttons flush mounted or recessed. Recessed pushbuttons are for applications where the buttons may be activated accidentally. Flush-mounted pushbuttons are available for applications where the switches must easily be activated by operators who must wear gloves or have other problems depressing the switches. Both the pushbuttons and contacts in the starter are of the maintained type, since the starter is activated manually. This means that when the start pushbutton is depressed, the starter will remain in the energized mode until the stop pushbutton is depressed. When the motor is deenergized by the stop button, the starter will remain in that mode until the start button is depressed again.

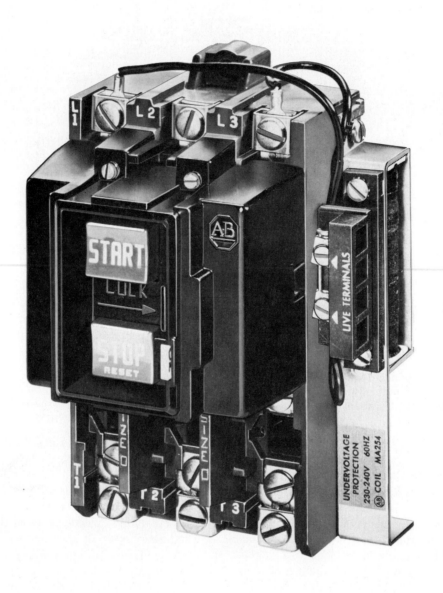

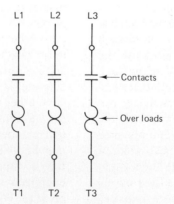

FIGURE 5-3 Manual starter with electrical diagram. (Photo courtesy of Allen-Bradley, Inc.)

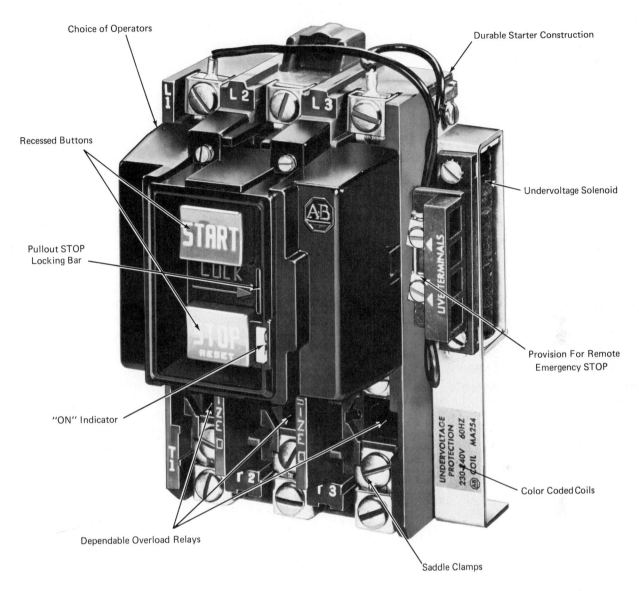

FIGURE 5-4 Exploded view diagram of a typical manual motor starter. (Courtesy of Allen-Bradley, Inc.)

The toggle switch operates similarly to a single-pole switch used to control residential lighting. This type of switch is usually used for smaller starters. When the toggle switch is switched to the on position, the starter contacts will close and energize the motor. The toggle switch is also a maintained type of switch and will remain in the position to which it has been switched.

The selector switch operator is generally used for applications where the control has two or more positions. This type of switch is useful for applications where three functions are needed, such as hand, auto, and off; or forward, reverse, and off. Figure 5-5 shows examples of each of these types of operators. Some manufacturers may use different names for these operators, but the function is the same.

The base assembly also provides a plastic barrier between each set of contacts. This barrier prevents arcing between sets of contacts that are of different potential, such as between L1 and L2. A cover called an *arc hood* is placed above the contacts to prevent arcing to anything outside the starter. Sometimes the potential at a set of contacts may become very large if a motor does not start or if a short circuit occurs. In these cases, a large arc may be drawn to any metal, such as an enclosure, and the arc hood prevents the arc from damaging any part of the starter assembly.

The contacts in the starter can be designed to

Manual Motor Starters

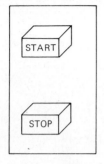

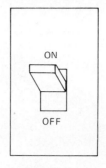

(a) Pushbutton switch (b) Selector switch (c) Toggle switch

FIGURE 5-5 Types of switch operators for manual motor starters.

operate in one of several ways. Figure 5-6 shows several of these designs. One type of design uses contacts that are flat (Figure 5-6a). Notice that both the stationary and movable parts of the contacts are shown in this diagram. The stationary part of the contacts consists of two isolated contacts that have wiring terminals provided to connect field wiring to their terminals. The movable contacts are mounted so that they will align perfectly with stationery contacts when the movable contact carrier is depressed. The movable contact carrier is designed to prevent the contacts from twisting as they seat with the stationary contact. The contacts in the movable carrier actually float in the carrier to allow exact alignment when they seat with the stationary contact. A spring is used to provide pressure on the contacts as they close so that they will not bounce, rebound, or vibrate.

Notice that two contact points are used for each set of contacts. This arrangement is called *double-break contacts*. Double-break contacts allow larger currents and voltages to be switched on and off without causing severe arcing or overheating. The movable contact assembly also provides a snap action mechanism that causes the contacts to make (close) and break (open) quickly, which limits arcing. The contacts are manufactured from a silver alloy. Cadmium and cadmium oxide are mixed with the silver to make the best material for conduction and to prevent contact pitting. The part of the contacts that come together to provide a circuit for current flow are made to withstand arcing from voltage and current surges that occur when the contacts are closing or opening. The metal on the tip of the contacts tends to melt and transfer from one of the contacts to the other during arcing. This reaction, very similar to arc welding, will leave a pit in one of the contacts and a deposit in the shape of a hill on the other contact. On earlier contacts sets, the pitting was so common that procedures for dressing or cleaning the contacts were provided. Modern motor starters use new metal alloys that resist pitting, and the small amount of arcing that takes place will actually help clean the contact surface. The newer contacts are made from silver alloys and do not require cleaning or dressing. In fact, any time the contacts are filed or dressed with sandpaper, they tend to lose more of their plating and begin deteriorating faster. The coatings on these modern contacts resist oxidation and can withstand continual operation. If the contacts must be cleaned, use a burnishing tool.

Another contact design used in motor starter contacts allows contacts to be mounted on a slight angle. This type of contact arrangement is shown in Figure 5-6b. The contacts are mounted at an angle to provide tension on them when they are closed. This arrangement allows the contacts to mate more

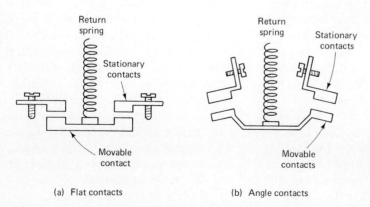

(a) Flat contacts (b) Angle contacts

FIGURE 5-6 Typical contact arrangements used in manual motor starters.

88   Chap. 5 / Manual Control Devices

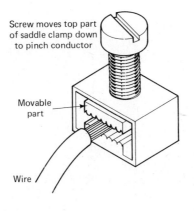
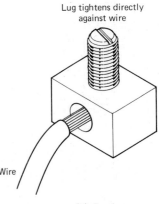

FIGURE 5-7 Saddle clamp and lug terminal connectors.

(a) Saddle clamp   (b) Box lug

easily during the travel as they come together. The contacts will operate very similarly to the flat contacts.

The contacts are rated according to the amount of current that will pass through them safely. The maximum amount of voltage that can be applied to the contacts is determined by the contact rating as well as the rating of the base assembly. Typical voltage ratings of manual starters are 115, 200 to 230, and 460 to 570 V ac and 125 and 250 V dc. The manual starter's current rating may be identified in amperes, wattage, or horsepower. Typical current ratings are usually shown as a horsepower rating. Figure 5-8 shows typical ac and dc horsepower ratings for manual starters. You may need to review the horsepower-to-amperage conversion formula to determine exact current ratings.

Make sure that both the voltage and current rating for the starter exceed the actual voltage and current that is controlled to the load. The manual motor starters are also identified by size. The sizes are shown in Figure 5-8. Notice that size 00, called "double ought," is the smallest. The sizes for manual starters include 00, 0, and 1. The amount of current that each size of starter can control is dependent on the load voltage.

For example, you can see from Figure 5-8 that a size 1 starter can safely control a 3-hp motor at 115 V, 7.5 hp at 200 to 230 V, and 10 hp at 460 to 575 V. The reason for this difference is that motors draw less amperage at higher voltages and the contacts can withstand a specific maximum amount of current safely as long as the voltage rating is not exceeded. For example, a 3-hp motor at 115 V will draw 34 A, a 7.5-hp motor at 230 V will draw 40 A, and a 10-hp motor at 460 V will draw 14 A. Each of these loads can be controlled safely by a size 1 starter.

The contacts in manual motor starters are "normally open" contacts. The number of sets of contacts is be determined by the application for the switch. Figure 5-9 shows the diagram for several types of manual motor starters. The motor starter in Figure 5-9a with a single set of normally open contacts is used for single-phase 110-V circuits. The motor starter is required only to break (open and close) the line side of the circuit. The neutral side of

Single-phase motors

NEMA Size	Full-Voltage Starting	
	115 V	230 V
00	1/3	1
0	1	2
1	1	3
1½	3	5
2	—	7½
3	—	15

Polyphase motors

NEMA Size	Full-Voltage Starting		
	200 V	230 V	460 V, 575 V
00	1½	1½	2
0	3	3	5
1	7½	7½	10
2	10	15	25
3	25	30	50
4	40	50	100
5	75	100	200
6	150	200	400
7	—	300	600
8	—	450	900
9	—	800	1600

FIGURE 5-8 Maximum horsepower ratings for NEMA-size starters. (Courtesy of Electrical Apparatus Service Association and Allen-Bradley, Inc.)

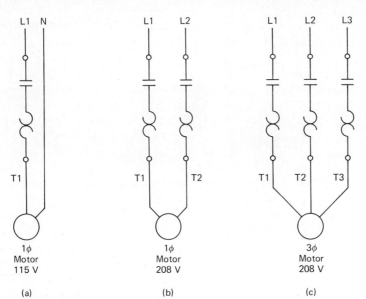

FIGURE 5-9 Diagram of one-, two-, and three-pole motor starters.

the motor circuit should not be broken. The motor starter in Figure 5-9b that has two sets of normally open contacts is used for 208/220 or 480-V ac single-phase applications or dc voltage applications. In single-phase ac voltages, both of the hot lines should be broken by the contacts. The motor starter in Figure 5-9c uses three contacts to break each of the three-phase lines.

Overload Device

The manual motor starter provides a device to detect and prevent motor overloads. Any motor is capable of drawing current in excess of its rated load. This condition is called an *overload*. It can occur because of bad bearings, or from trying to turn a load that has become too large, such as a conveyor or a mixer being overloaded. When the motor draws too much current, its insulation will become overheated and begins to break down and deteriorate to the point of damaging the motor. Another damaging problem may occur even though the motor current is controlled within tolerance when a motor begins to overheat from starting and stopping too frequently.

The overload device in the manual motor starter is designed to check for too much current and open the motor starter circuit before the motor windings become damaged. Several types of overload devices are used in modern motor starters. Figure 5-10a shows pictures and diagrams of single-pole and three-pole overload devices.

From Figure 5-10 you can see that there are two basic types of overloads. The type shown in Figure 5-10a uses a solder pot to sense the heat that is built up from excessive motor current. This type of control is called *solder pot overload* or *melting alloy thermal overload,* and it has several main parts: the solder pot, the heater (a wire that produces heat when current flows through it), the ratchet shaft with rewind spring, the overload contacts with return spring, and the reset button. From the diagram you can see that the ratchet shaft is mounted in the solder pot. The ratchet shaft has spring tension applied to it that tries to make it rotate. Since the shaft is secured in solder and the solder is solidified, it cannot turn. The ratchet has its teeth beveled so that they catch the end of the contact actuator lever. The rewind spring on the ratchet shaft keeps pressure on the contact actuator lever so that it cannot move. When the contact actuator lever is secured in place by the ratchet, it keeps pressure on the contacts, which keeps them closed. When the contacts are closed, they compress a very strong spring. This spring is constantly exerting pressure to try and open the contacts. Any time the solder becomes soft, the ratchet shaft will become free from the solder and the rewind spring will cause the ratchet to spin. When the ratchet spins, the contact actuator lever is released and the contact return spring forces the contacts open.

The overload contacts are connected in series with the motor. The heater is also connected in this series circuit. This means that any current that flows to the motor must also pass through the heater. The heater is physically mounted so the heat that it generates is conducted directly to the solder pot. When the motor draws normal current the heater does not generate enough heat to melt the solder in the solder pot. But when the motor draws excessive current, the heater generates enough heat to cause the solder to begin to melt and release the ratchet shaft. When the ratchet shaft spins the contact lever is released and the contact return spring forces the overload contacts open. Since the motor is connected in series with the overload contacts, the motor is deenergized.

The motor starter overload can be reset by pressing on the reset button. The reset button rotates the ratchet shaft so that it rewinds the shaft spring. The contact actuator lever is allowed to return to its reset condition, where it is again resting against the ratchet teeth. If the reset button is pressed too soon, the solder will not be solidified enough to hold the ratchet shaft, and the shaft spring will again spin the ratchet and allow the contacts to open. Sometimes the overload must be allowed to cool down for several minutes before it can be reset.

Since the overload contacts should open only when the motor is actually drawing too much current, the overload heater must be sized to the exact amount of current the motor is drawing during full-load conditions. The heater and overload unit are manufactured as a single unit so that they are easy to size and install. This also makes the overload tamper-proof.

The overload and heater are widely used for motor protection because they can be reset without changing any parts. This allows the electrician to determine the cause of the overload and reset the motor starter after the condition has been cleared. Sometimes the condition that is causing the motor to overheat is not readily apparent, so the electrician will reset the motor starter and observe the motor when it is energized again. This may include taking a full-load motor amperage reading with an ammeter. If the motor continues to trip the overload device, the electrician must determine if the motor is faulty or if the load has changed in some way. This test procedure requires the load to be disconnected from the motor to see if it is still drawing overload current. If the overload disappears, the load is at fault. If the overload continues, the motor is at fault.

The procedure for correctly sizing motor starter heaters to specific motor and load applications is very important. The heater cannot be expected to protect the motor if it is not sized properly. The sizing procedure involves determining the motor horsepower, the full-load motor amperage, and the temperature conditions around the motor and motor starter. This method of sizing the heaters is included in the installation procedures provided later in this chapter.

The second type of overload device is called a *bimetallic thermal overload relay* (Figure 5-10b). This type of overload relay is not used as often as the solder pot type of overload, but you should understand its operation. The bimetallic thermal overload uses a U-shaped bimetal strip as the heater for the thermal overload. The bimetal strip has an additional variable-resistance element that will allow the heater to be precisely adjusted to the full-load motor current. The bimetal heater is wired in series with the motor. When the motor is drawing normal current the heater will become slightly warm, but the heat is not enough to cause the U-shaped element to change position. If the motor begins to draw excessive heat, the U-shaped element begins to move and the contacts begin to open. When the element has moved enough, the contacts will snap open crisply and deenergize the motor. The overload is designed to snap the contacts open so that they do not arc excessively and wear out prematurely.

The bimetal type of overload can be converted in the field for manual reset or automatic reset. If the overload is converted for automatic reset, the contacts of the U-shaped element are allowed to return to the closed state when the element cools. Automatic reset allows motors and controls that are mounted in a remote or inaccessible location to be restarted without being serviced. A set of alarm contacts are provided to energize an indicator lamp to alert maintenance personnel that an overload condition is occurring. The alarm contacts may also be used where continuous processes are controlled. In an application such as continuous pouring of glass, the process should not be interrupted. In this application, if an overload is sensed, the circuit is wired so that the motor contacts cannot deenergize the motors. Instead, the alarm contacts will energize an indicator lamp which will warn the operator that an overload exists. If the overload continues for several minutes, maintenance personnel are called. The motors are deenergized at the end of the pouring cycle and checked for faults. In some applications the process is so expensive that some damage to the motor may be acceptable so that the pouring process can be completed.

Some applications using manual starters can also use dual-element fuses to protect the motor against short-circuit current and slow overcurrents. The inverse time characteristics of the dual-element fuse were explained in Chapter 4. You may need to review this information to better understand the methods of motor protection and control that are available.

Undervoltage Protection

Motor starters also are available that can provide undervoltage protection. This type of control actually provides loss of voltage protection as well as detecting small undervoltages. This type of

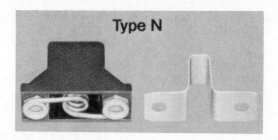

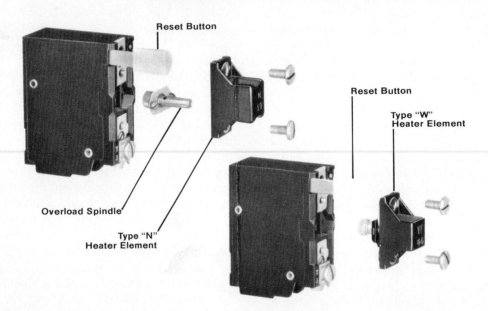

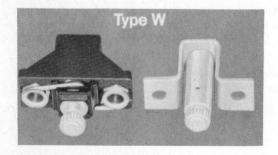

INSTRUCTIONS FOR SELECTION AND INSTALLATION

Conversion From Type N To Type W Eutectic Alloy Overload Relay Heater Elements

1. Remove each existing Type N heater element and each overload spindle (2 screws).
2. Consult index for selection of heater and refer to table covering the Bulletin number of the controller. Select the corresponding Type W heater element for the Type N being replaced. **Note:** For new installations or when motor full load current is known, refer to "Heater Element Selection Tables" in the catalog.
3. Install the Type W heater element(s) selected and secure with the two mounting screws.
4. Reset relay by depressing reset button.

Note: Single pole overload relay is shown. Install a Type W heater element in each active pole of a 2 or 3 pole block style overload relay in the same manner.

Bulletin Number Index	Table Color Code
705 with Block Style Overload Relays	Table 3
705 with Side Mounted Overload Relays	Table 1
706	Table 4
707	Table 4
709	Table 1
712	Table 2
713	Table 1
715 with Block Style Overload Relays	Table 3
715 with Side Mounted Overload Relays	Table 1

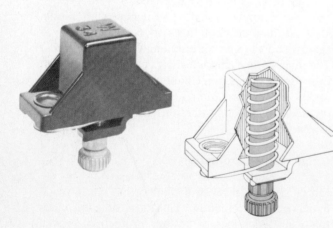

FIGURE 5-10 Old- and new-style overload devices with instruction and tables for selection or conversion. (Courtesy of Allen-Bradley, Inc.)

1. Bulletin 705, 709, 713 and 715 With Side Mounted Overload Relays

N Element Number	W Element Number					
	Size 00	Size 0	Size 1	Size 2	Size 3	Size 4	Size 5
NN4	W10	W10	W10				
NN3X	W11	W11	W11				
NN3	W12	W12	W12				
NN2X	W13	W13	W13				
NN2	W14	W14	W14				
NN1X	W15	W15	W15				
NN1	W16	W16	W16				
N1X	W17	W17	W17				
N1	W18	W18	W18				
N2X	W19	W19	W19				
N2	W20	W20	W20				
N3X	W21	W21	W21				
N3	W22	W22	W22				
N4X	W23	W23	W23				
N4	W24	W24	W24				
N5	W25	W25	W25				
N6	W26	W26	W26				
N7	W27	W27	W27				
N8	W28	W28	W28				
N9	W29	W29	W29			W29	
N10	W30	W30	W30			W30	
N11	W31	W31	W31			W31	
N12	W32	W32	W32			W32	
N13	W33	W33	W33			W33	
N14	W34	W34	W34			W34	
N15	W35	W35	W35			W35	
N16	W36	W36	W36			W36	
N17	W37	W37	W37			W37	
N18	W38	W38	W38			W38	
N19	W39	W39	W39			W39	
N20	W40	W40	W40			W40	
N21	W41	W41	W41			W41	
N22	W42	W42	W42			W42	
N23	W43	W43	W43			W43	
N24	W44	W44	W44	W44			
N25	W45	W45	W45	W45			
N26	W46	W46	W46	W46			
N27	W47	W47	W47	W47			
N28	W48	W48	W48	W48			
N29	W49	W49	W49	W49			
N30	W50	W50	W50	W50			
N31	W51	W51	W51	W51			
N32		W52	W52	W52			
N33		W53	W53	W54	W53		
N34		W54	W54	W55	W55		
N35		W55	W55	W56	W56		
N36		W56	W56	W57	W58		
N37		W57	W57	W58	W59	W60	
N38		W58	W59	W60	W61		
N39			W60	W61	W61	W62	
N40			W61	W62	W63		
N41			W62	W63	W64		
N42			W63	W64	W64	W65	
N43				W65	W65		
N44				W66	W66	W67	
N45				W67	W67		
N46				W68	W68	W69	
N70				W69	W69	W70	
N71					W70	W71	
N72					W71	W72	
N73					W72	W73	
N74					W74	W74	
N75					W75	W75	
N76					W76	W76	
N77						W78	
N78						W79	
N79						W81	

2. Bulletin 712

N Element Number	W Element Number					
	Size 0	Size 1	Size 2	Size 3	Size 4	Size 5
NN4	W10	W10				
NN3X	W11	W11				
NN3	W12	W12				
NN2X	W13	W13				
NN2	W14	W14				
NN1X	W15	W15				
NN1	W16	W16				
N1X	W17	W17				
N1	W18	W18				
N2X	W19	W19				
N2	W20	W20				
N3X	W21	W21				
N3	W22	W22				
N4X	W23	W23				
N4	W24	W24				
N5	W25	W25				
N6	W26	W26				
N7	W27	W27				
N8	W28	W28				
N9	W29	W29			W29	
N10	W30	W30			W30	
N11	W31	W31			W31	
N12	W32	W32			W32	
N13	W33	W33			W33	
N14	W34	W34			W34	
N15	W35	W35			W35	
N16	W36	W36			W36	
N17	W37	W37			W37	
N18	W38	W38			W38	
N19	W39	W39			W39	
N20	W40	W40			W40	
N21	W41	W41			W41	
N22	W42	W42			W42	
N23	W43	W43			W43	
N24	W44	W44	W44			
N25	W45	W45	W45			
N26	W46	W46	W46			
N27	W47	W47	W47			
N28	W48	W48	W48			
N29	W49	W49	W49			
N30	W50	W50	W50			
N31	W51	W51	W51			
N32	W52	W52	W52	W53		
N33	W53	W53	W54	W54		
N34	W54	W54	W55	W55		
N35	W55	W55	W56	W56		
N36	W56	W56	W57			
N37	W57	W57	W58	W59		
N38	W58	W59	W60			
N39		W60	W61	W61		
N40		W61	W62	W62		
N41		W62	W63	W63		
N42		W63	W64	W64		
N43			W65	W65		
N44			W66	W66		
N45			W67	W67		
N46			W68	W68		
N70			W69	W69		
N71				W70		
N72				W71		
N73				W72		
N74				W73	W74	
N75					W75	
N76					W76	
N77					W78	
N78					W79	
N79					W80	

3. Bulletin 705 and 715 With Block Style Overload Relays

N Element Number	W Element Number					
	Size 00	Size 0	Size 1	Size 2	Size 3	Size 4	Size 5
NN4	W10	W10	W10				
NN3X	W11	W11	W11				
NN3	W12	W12	W12				
NN2X	W13	W13	W13				
NN2	W14	W14	W14				
NN1X	W15	W15	W15				
NN1	W16	W16	W16				
N1X	W17	W17	W17				
N1	W18	W18	W18				
N2X	W19	W19	W19				
N2	W20	W20	W20				
N3X	W21	W21	W21				
N3	W22	W22	W22				
W4X	W23	W23	W23				
N4	W24	W24	W24				
N5	W25	W25	W25				
N6	W26	W26	W26				
N7	W27	W27	W27				
N8	W28	W28	W28				
N9	W29	W29	W29				W29
N10	W30	W30	W30				W30
N11	W31	W31	W31				W31
N12	W32	W32	W32				W32
N13	W33	W33	W33				W34
N14	W34	W34	W34				W35
N15	W35	W35	W35				W35
N16	W36	W36	W36				W36
N17	W37	W37	W37				W37
N18	W38	W38	W38				W38
N19	W39	W39	W39				W39
N20	W40	W40	W40				W40
N21	W41	W41	W41				W41
N22	W42	W42	W42				W42
N23	W43	W44	W44				W43
N24	W44	W45	W45	W45			
N25	W45	W46	W46	W46			
N26	W46	W47	W47	W47			
N27	W47	W48	W48	W48			
N28	W48	W49	W49	W49			
N29	W49	W50	W50	W50			
N30		W51	W51	W51			
N31	W51	W52	W52	W52			
N32		W53	W53	W53	W54		
N33		W54	W54	W54	W55		
N34		W55	W55	W55	W56		
N35		W56	W56	W56	W59		
N36		W57	W57	W58	W60		
N37		W58	W58	W59	W61		
N38		W59	W59	W60	W62		
N39			W60	W61	W62		
N40			W61	W62	W63	W63	
N41			W62	W63	W64		
N42			W63	W64	W65		
N43				W65	W66	W66	
N44				W66	W67	W67	
N45				W67	W68	W68	
N46				W68	W69	W69	
N70					W70	W71	
N71					W71	W72	
N72					W73	W73	
N73					W74	W74	
N74					W75	W75	
N75					W76	W76	
N76					W78	W78	
N77						W79	
N78						W81	
N79						W82	

4. Bulletin 706 and 707

N Element Number	W Element Number				
	Size 0	Size 1	Size 2	Size 3	Size 4	Size 5
NN4	W10	W10				
NN3X	W11	W11				
NN3	W12	W12				
NN2X	W13	W13				
NN2	W14	W14				
NN1X	W15	W15				
NN1	W16	W16				
N1X	W17	W17				
N1	W18	W18				
N2X	W19	W19				
N2	W20	W20				
N3X	W21	W21				
N3	W22	W22				
N4X	W23	W23				
N4	W24	W24				
N5	W25	W25				
N6	W27	W27				
N7	W28	W28				
N8	W29	W29				
N9	W30	W30				W29
N10	W31	W31				W30
N11	W32	W32				W31
N12	W33	W33				W32
N13	W34	W34				W34
N14	W35	W35				W35
N15	W36	W36				W35
N16	W37	W37				W36
N17	W38	W38				W37
N18	W39	W39				W38
N19	W40	W40				W39
N20	W41	W41				W40
N21	W42	W42				W41
N22	W43	W43				W42
N23	W44	W44				W43
N24	W45	W45	W45			
N25	W46	W46	W46			
N26	W47	W47	W47			
N27	W48	W48	W48			
N28	W49	W49	W49			
N29	W50	W50	W50			
N30	W51	W51	W52			
N31	W52	W52	W53	W54		
N32	W53	W54	W55			
N33	W54	W55	W56			
N34	W55	W56	W57			
N35	W56	W57	W58			
N36	W57	W58	W59			
N37	W58	W59	W60			
N38	W59	W60	W61			
N39	W60	W61	W62	W62		
N40		W62	W63	W63		
N41		W62	W63	W64		
N42		W63	W64	W65		
N43			W65	W66		
N44			W66	W67		
N45			W67	W68		
N46			W68	W69		
N70				W70		
N71				W71	W71	
N72				W72	W72	
N73				W73	W74	
N74				W74	W75	
N75				W76	W77	
N76					W79	
N77					W80	
N78					W81	

Manual Motor Starters

protection is required by OSHA and ANSI for many applications where a motor must be reset manually if the motor stops for any reason. You should recall that since the motor starter uses a manual switch operator such as a pushbutton or toggle switch, the contacts are maintained and they will stay in the condition (open or closed) where they were last actuated. For example, in drill-press or wood-cutting applications, severe damage could occur if the motor stops for any reason and is suddenly restarted. This could occur if power was lost for a few seconds during a power outage or lightning storm. The loss-of-voltage device will trip the motor starter contacts to their open position even though the start pushbutton is depressed or if the toggle switch is in the on position. This trip action requires someone to come to the motor starter and manually reset the start button to allow the motor to restart. When the operator comes to restart the motor, a check can be made of the condition of the work in the cutter blade so that the work can be repositioned prior to restarting the motor, to prevent damage to the machine. If the machine were allowed to restart on its own, the material in the cutter blade might be out of position and become damaged, or the cutter head might become jammed and damage the machine or malfunction and cause an accident.

The undervoltage sensor can also detect substantial drops in voltage caused by large motors and spot-welding loads cycling on and off. The lowering of voltage may also occur during a brownout condition. This may occur during peak usage periods in large industrial and metropolitan areas such as during lengthy warm periods in the summer. During this time, residential usage may cause utility supplies to drop 10 to 15 percent. This lowering of voltage will cause motors to overheat. If the motor is susceptible to overheating, the motor should be protected by an undervoltage monitor. In these cases the motor starter would be deenergized by the undervoltage sensor even though power was not lost completely. The motor starter would remain deenergized until manually reset and the undervoltage condition cleared.

Figure 5-11 shows a diagram that indicates the location and operation of the undervoltage device. You can see that a continuous-duty solenoid has been added to the manual motor starter. The solenoid is wired directly across the line-side terminals, such as L1 and L2. This allows the solenoid to be energized any time that power is applied to the motor starter. The first diagram shows the motor starter with line voltage present and the motor starter switched off. In this condition the motor starter's stop button, which is a maintained type of switch, has been depressed. Since power is applied to the line side of the motor starter, the undervoltage solenoid plunger is in the up position.

You must understand that the start and stop bottons on the manual starter are both maintained switches. This means that whichever switch is depressed, it will stay in that position until the other switch is depressed. The two switches are mounted on a toggle actuator that is similar to a children's "seesaw." When the start button is depressed, the mechanism pushes the stop button to the off position. When the stop button is depressed, it will automatically cause the start button to move to the off position. This action maintains the switch that is depressed until the other switch is activated.

The second diagram shows power applied and the start button depressed. In this case the "on" indicator on the front of the motor starter is showing, which indicates that the contacts are closed. and the motor would be energized and running. Notice that the undervoltage solenoid's plunger is still in the up position because power is still applied to the line side of the motor starter.

The third diagram shows the undervoltage device when it has detected a loss of voltage. Notice that the start button is still depressed because the motor starter was energized when the loss of voltage occurred. The undervoltage solenoid's plunger is now in the down position because it has detected a loss of voltage. The plunger will remain in the down position even if supply voltage returns, until the starter is manually reset. You would be able to determine that the undervoltage device has tripped because the starter is deenergized and the motor is no longer running even though the start button remains depressed. The "on" indicator would also be retracted so that it was not visible.

The fourth diagram shows the motor starter after it has been reset. To reset the starter after the undervoltage device has detected the loss of voltage, the stop button must be firmly depressed. The motor can then be restarted at any time by again depressing the start button. Remember that the undervoltage solenoid will not be returned to the up position if power has not returned to the line side of the motor starter, and the starter cannot be reset.

The electrical diagram in this figure shows an emergency stop switch added to the undervoltage monitor circuit. Since the undervoltage solenoid is activated electrically, a remote stop switch with normally closed contacts can be connected in series with the undervoltage solenoid. Any time the stop switch is activated, the contacts will open and deenergize the undervoltage solenoid. Once the solenoid is deenergized, the motor starter must be manually reset, just as if a voltage loss has been

Manual Starting Switches With Undervoltage Protection

OPERATION OF UNDERVOLTAGE SOLENOID

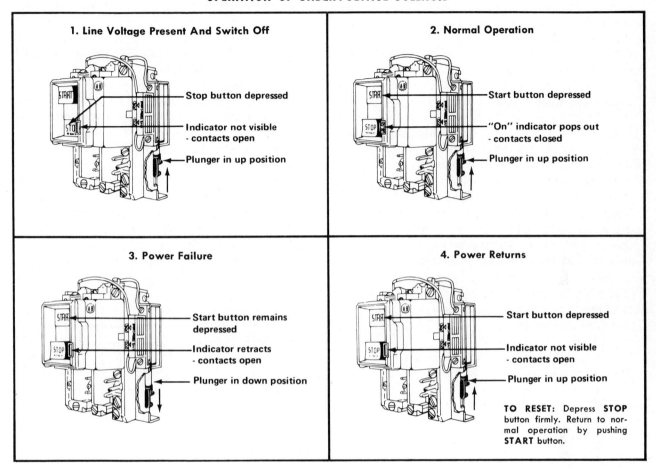

TYPICAL WIRING DIAGRAM

(See Applicable Codes and Laws)

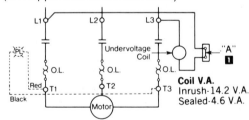

1 Remove Jumper "A" To Connect Remote Stop Operator Wires To Vacated Terminals.
NOTE: Jumper Not Available on devices in NEMA Type 7 & 9 Enclosure.

FIGURE 5-11 Mechanical diagrams of a manual motor starter with undervoltage protection. (Courtesy of Allen-Bradley, Inc.)

detected. You should remember that additional stop switches can be added to this series circuit as needed. If any of the switches is depressed, current to the solenoid will be interrupted and the motor starter would need to be reset manually.

Types of Manual Starters

Several types of manual starters are available for a variety of applications. These starters provide protection for single-phase and three-phase ac motors and dc motors. Figure 5-12 shows an electrical diagram for each of these types of starters. In Figure 5-12a a single-phase, single-pole starter is shown. This starter is used to control fractional-horsepower motors that operate on 115 V. Notice that only L1 has a set of contacts and an overload device. In accordance with the National Electrical Code® the neutral line is not allowed to have contacts or be interrupted.

The starter shown in Figure 5-12b is a two-pole starter used to control single-phase motors that operate on 208 to 220 V. Notice that both lines L1 and L2 have a set of contacts to disconnect power from the motor, but only one overload is required. The reason only one overload device is required is that the motor current that flows to the motor through L1 must return through L2. Since this creates a series circuit, the single overload can sense any increased current and protect the motor. Figure 5-12c shows a two-pole starter with a pilot light connected across the load-side terminals. In this configuration, the pilot lamp will be illuminated any time the motor is energized. The pilot lamp is mounted on the cover of the motor starter enclosure so that it is visible at a distance. Another variation of the two-pole motor starter connects a selector switch and pilot device in series with L1. When the selector switch is in the hand position, the motor is connected directly to L1 and will run any time the starter is energized. When the selector switch is moved to the auto position, the pilot device is connected in series with L1. This means that the motor will run only when the starter is energized and the operating condition on the pilot device has be reached. In the case of the temperature switch shown in the diagram, the motor will run any time the temperature exceeds the temperature set point on the switch. This control is wired to control a cooling fan which will be energized any time the temperature exceeds the set point. This type of application could utilize any type of pilot device as long as its contacts matched the current and voltage

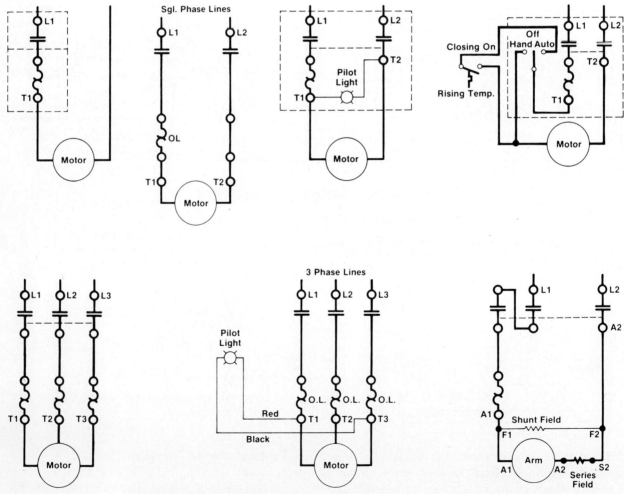

FIGURE 5-12 Diagrams for single-phase, three-phase, and dc motor starters. (Courtesy of Allen-Bradley, Inc.)

96   Chap. 5 / Manual Control Devices

requirements of the motor. In another typical application for this type of control circuit a level switch is used to control a sump pump. Any time the level in the sump reaches the set point, the pilot switch closes and energizes the pump motor. The motor starter provides overload protection for the pump motor in this application. The diagram in Figure 5-12d shows a single pole switch utilized where only one line of the source voltage is broken.

The motor starter in Figure 5-12e shows a three-pole switch. This type of starter is used to control three-phase motors. Notice that an overload is provided for each of the three lines. The dashed lines below the three sets of contacts indicate that the three contacts operate as a set. This means that each of the three contacts will always be in the same condition, either open or closed. Figure 5-12f shows a three-pole starter with a pilot lamp connected across the load-side terminals T1 and T3. This lamp will be illuminated any time the motor is energized.

Figure 5-12g shows a dc motor starter. Notice that only one overload is required since the two motor lines make a series circuit. You should also notice that the shunt field is wired directly across the two motor starter terminals marked A1 and A2. This configuration places the shunt field directly across the applied voltage so that the motor will see maximum starting torque.

Any starter may have a pilot light assembly added after it has been installed. The pilot light should always be wired to the load-side terminals to indicate the status of the motor starter. Any time the motor starter's contacts are energized, the pilot light will be illuminated, indicating that the switch is operating correctly and that the motor is energized. If the motor starter is deenergized for any reason, the indicator lamp will also be deenergized. This allows an operator to determine the status of a machine from some distance away. This is useful in a noisy environment when an operator is responsible for keeping several machines operating at the same time. The lamps are available with red or green lenses to indicate that the motor is energized or deenergized.

Some applications require that the motor starter be capable of being locked in the off position. Several styles of manual motor starters are available with a locking guard that allows the handle to be locked in the off position. This provides security when the system must be deenergized for critical operations or maintenance. This also ensures that only qualified personnel may operate the system.

Two-Speed and Reversing Manual Motor Starters. Manual starters are also available for two-speed motor applications. Figure 5-13 shows the diagram for a two-speed motor controlled by a manual motor starter. Notice that the two-speed motor starter has two separate sets of contacts. One of these sets is for the low-speed winding and the other set is for the high-speed winding. These contacts are controlled by a toggle or selector switch which has three positions: high speed, low speed, and off. A mechanical interlock is used to prevent both high-speed and low-speed contacts from being closed or energized at the same time. This interlock automatically toggles the two sets of

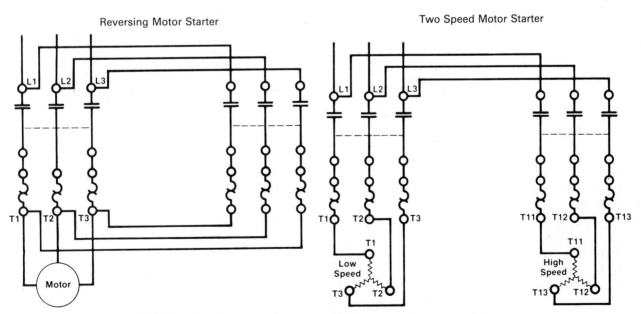

FIGURE 5-13 Diagrams for two-speed and reversing manual starters. (Courtesy of Allen-Bradley, Inc.)

contacts to opposite conditions. This type of starter is useful for applications where two-speed fans are used.

Another specialized type of manual motor starter is a reversing starter. A diagram of an application using a reversing starter is also shown in Figure 5-13. This type of starter also utilizes two separate sets of contacts. One set of contacts allows the motor windings to receive voltage directly from the line-side terminals, which will cause the motor to run in the forward direction. The other set of contacts will reverse two phases of the three-phase voltage line voltage for a three-phase motor, or reverse the start winding for single-phase motors. If the motor starter is for reversing dc motors, the second set of contacts will reverse the polarity of the line-side voltage for dc series type and reverse the field windings for shunt and compound dc motors. In some dc motor applications resistors are added to protect the motor when it is reversed. An example diagram for each starter and motor type is shown in Figure 5-13. If you need more review of reversing ac and dc motors, refer to Chapters 10 and 11.

Manual Reduced-Voltage Starters. Another specialty type of manual starter allows motors to be started by reducing voltage to decrease the amount of line current that is drawn when a motor is started. This type of control uses an autotransformer to allow larger current to flow in the secondary winding of the autotransformer and in the motor winding while current is reduced in the supply line. The autotransformer provides the function of a step-down transformer. By decreasing the secondary voltage to the motor, it will draw less current in the primary side of the circuit. The smaller primary current will keep the overall current demand lower so that the switch gear and other control components will last longer. It will also provide a savings in the electric bill by keeping the demand factor lower.

The autotransformer type of control provides better torque during motor starting than a resistive type of reduced voltage starter because resistors decrease the starting current as well as the starting voltage. You should remember that when a motor loses current, it will also lose torque.

To understand fully the principals of motor starting and the need for reduced-voltage starting that is used in this type of control, you may need to review motor principles in Chapter 11. The autotransformer reduced-voltage starter uses a three-position switch for normal operation. The electrical diagram in Figure 5-14 shows the location of this switch. When the operator is ready to start the motor, the switch is moved to the start position. In this position the switch connects the three-phase supply voltage lines to terminals L1, L2, and L3. Terminals L1 and L3 are each connected to their own transformer at the terminal marked 100. L2 is connected to the transformer terminal marked 0. Notice that the 0 terminal on the two transformer windings are jumpered and connected to terminal T2. T2 is connected directly to the motor without using any contacts. Terminals T1 and T3 are connected to the 65 percent tap on their respective winding. When the switch is in the start position, the contacts connect T1 and T3 to the motor windings.

In the starting configuration the motor receives reduced voltage. Since this voltage is reduced through a transformer, the current through the transformer is increased. The net effect is that the motor receives reduced voltage with increased current, whereas the three-phase supply actually uses less current. This allows the motor to be started without causing the demand meter to register a large current demand.

After the motor begins to accelerate, the operator can turn the switch to the run position. This usually occurs when the motor reaches near full speed. When the switch is changed to the run position, the contacts disconnect the motor from the transformer and reconnect it directly across the line voltage. Notice that the overloads are in this circuit and will be used only when the starter is in the run position. This configuration keeps the overloads out of the starting circuit, where current will be large and tend to overheat them.

Any time the operator wants to stop the motor, the motor starter switch is moved to the stop position. This action disconnects the motor windings from voltage provided by either running or starting circuitry. An undervoltage solenoid is used to protect the motor from trying to restart if a power failure occurs or if the motor is deenergized at any time. One or more remote stop switches can be connected in series with this solenoid to provide safety for personnel who work near the motor.

An ammeter may also be installed on the motor starter to provide the operator with an indication of when the motor has reached sufficient speed to change the switch from start to run. A time-delay relay is also available to prevent the undervoltage solenoid from tripping if voltage dips for a few seconds while other motors are starting. If the brownout or low-voltage condition exists longer than the setting on the time-delay relay, the undervoltage relay will trip and protect the motor.

This type of control uses air-break contacts or oil-immersed contacts. These contacts are rated for

*Size A, NEMA Type 1
General Purpose Enclosure
Air break construction
Catalog Number 646-LAB71
(Door Removed)*

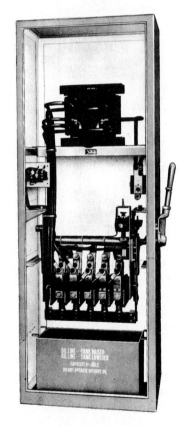

*Size D, NEMA Type 1
General Purpose Enclosure
Oil-immersed construction*

(Door Removed)

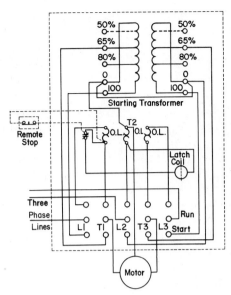

FIGURE 5-14 Diagrams of autotransformer reduced voltage starters. (Courtesy of Allen-Bradley, Inc.)

the high interruption capacity required for starting larger motors. You should note that oil-immersed contacts are rated only for NEMA type 1 general-purpose application and are not usable where explosive conditions exist. The oil-immersed contacts utilize heavy copper contacts that work well in oil. The oil helps to keep the contacts cool over long periods of operation where motors are cycled frequently. The duty cycle for this type of control is very critical. A duty cycle based on setting the autotransformer taps at 65 percent with an inductive load where the current is three times the full-load current would allow one start every 4 minutes, up to a total of four starts. Then the motor must be rested for 2 hours. The time required for the motor to start and reach running speed should not exceed 15 seconds at any time. Since these are maximum ratings, the motor can easily be started two or three times over a longer period without damage, or started several times and then allowed to remain running for a while. Remember, it is when the motor is being started that the large inrush current creates enormous amounts of heat in the motor and starter mechanism. In industrial applications, it is important to identify the number of starts that a motor is expected to have during normal use; the motor-starting equipment can be sized accordingly.

Installing the Manual Motor Starter

One of the steps in the process of installing a manual motor starter is sizing the starter and overload to match the motor load. Figure 5-15 shows

Manual Motor Starters **99**

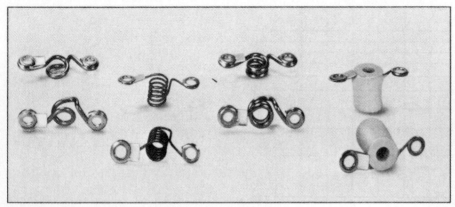

FIGURE 5-15 Types of heaters and a specification table used to size heaters for motor protection. (Courtesy of Allen-Bradley, Inc.)

TABLE 5	
Heater Type No.	Full Load Amps.
P1	0.17
P2	0.21
P3	0.25
P4	0.32
P5	0.39
P6	0.46
P7	0.57
P8	0.71
P9	0.79
P10	0.87
P11	0.98
P12	1.08
P13	1.19
P14	1.30
P15	1.43
P16	1.58
P17	1.75
P18	1.88
P19	2.13
P20	2.40
P21	2.58
P22	2.92
P23	3.09
P24	3.32
P25	3.77
P26	4.16
P27	4.51
P28	4.93
P29	5.43
P30	6.03
P31	6.83
P32	7.72
P33	8.24
P34	8.9
P35	9.6
P36	10.8
P37	12.0
P38	13.5
P39	15.2

pictures of heaters and a table that is used to size heaters to motors. The selection process takes several conditions into consideration in choosing a proper heater for each application. One of these conditions is the ambient temperature at the location at which the motor and the starter will be located. Another condition is the service factor of the motor being controlled. The final condition is the motor's rated full-load amperage.

The table in Figure 5-15 shows heaters by part number and the amount of full-load motor current that each will protect. If the motor has a service factor of 1.15, select the heater size nearest the full-load current rating listed on the motor name plate. This table assumes that the motor temperature rise will not be over 40°C (104°F). Sometimes the motor's full-load amperage will not be listed on the heater table, or special conditions will be encountered that will require you to adjust slightly the selection from the list. If the motor's full-load amperage does not exactly match one of the heaters on the list, select the next-higher-rated heater. If the temperature will be higher at the motor starter than where the motor is mounted, you should also select the next-higher-rated heater. If the temperature will be lower at the motor starter than where the motor is mounted, select the next-lower-rated heater. If the motor has a service factor of 1.0 or if it is rated for continuous duty, select a heater that is one rating smaller than is listed on the chart. After the motor is in operation, be sure to check the actual amperage that the motor is pulling to determine that the motor is not overloaded and that it is pulling slightly less than its rated nameplate value.

Another step in the installation procedure involves selecting a proper enclosure for the motor starter. Enclosures are rated according to the environment from which they are protecting the controller. Sometimes a consideration is to keep out elements such as acid fumes or moisture that may damage the control. Other times a consideration is to keep explosive fumes away from the contacts of the control where sparks or arcing occur. The National Electrical Manufacturers Association (NEMA) has classified each of these conditions and written specifications that rate the enclosures for all electrical equipment. Figure 5-16 shows the classification type for motor starter enclosures. This table applies to enclosures for both manual motor starters and full-voltage starters. Notice that some of the enclosure types are specified for outdoor use, while other are rated for indoor use. You must also understand that the enclosure will not meet the intended specifications if the proper type of hardware is not used. This means that a type 4X enclosure will not be watertight if regular thin-wall conduit is used instead of a watertight conduit. Be sure that the location where the starter will be mounted is specified on the installation blueprint so that the proper enclosure and hardware can be selected.

Most enclosures are available in both flush-mounted and surface-mounted enclosures. Figure

NEMA enclosures for starters

TYPE	ENCLOSURE
1	General purpose—indoor
2	Dripproff—indoor
3	Dusttight, raintight, sleet tight—outdoor
3R	Rainproof, sleet resistant—outdoor
3S	Dusttight, raintight, sleetproof—outdoor
4	Watertight, dusttight, sleet resistant—indoor & outdoor
4X	Watertight, dusttight, corrosion resistant—indoor & outdoor
5	Dusttight—indoor
6	Submersible, watertight, dusttight, sleet resistant—indoor & outdoor
7	Class I, group A, B, C or D hazardous locations, air-break—indoor
8	Class I, group A, B, C or D hazardous locations, oil-immersed—indoor
9	Class II, group E, F or G hazardous locations, air-break—indoor
10	Bureau of mines
11	Corrosion-resistant and dripproof, oil-immersed—indoor
12	Industrial use, dusttight and driptight—indoor
13	Oiltight and dusttight—indoor

FIGURE 5-16 NEMA classification for motor starter enclosures. (Courtesy of National Electrical Manufacturers Association and Electrical Apparatus Service Association.)

NEMA Type 1
Flush Mounting

NEMA Type 1
Surface Mounting

FIGURE 5-17 Flush-mounted and surface-mounted motor starter enclosures. (Courtesy of Allen-Bradley, Inc.)

5-17 shows an example of each type. Each enclosure is available with pushbutton or toggle operators mounted on the cover. This provides the widest possible choice of equipment, to meet the requirements of any application.

After the proper heater and enclosure have been selected, the motor starter is ready to be installed. This procedure should be accomplished with the aid of a field wiring diagram. You should also use a blueprint that shows the exact location to mount the starter and enclosure. The enclosure should be mounted and all interconnecting conduits should be installed before you begin to wire the control. In some cases, the starter will need to be removed from the enclosure so that it can be mounted properly. After the enclosure is mounted correctly, the starter can be wired in accordance with the field wiring diagram. If the control is a single-pole single-phase starter, it will be easier to connect the field wires if the starter is removed from its enclosure. After the field wires have been secured properly, the starter can be remounted into the enclosure and the cover can be replaced.

If the control is a larger starter, it will be easier to connect the field wiring with the starter mounted in the enclosure. This procedure will also allow the wires to be properly routed in the motor starter enclosure so that they will not rub or chafe against any metal parts or inhibit the motor starter parts from operating correctly. Be sure to apply field

Manual Motor Starters 101

wiring markers to any wires that are connected to the motor starter. These markers should remain on the wires permanently for any diagnostics or troubleshooting procedures that may be required if the motor starter is suspected of being faulty at a later date.

After you have completed the installation procedure and have all the field wires connected, the supply power should be applied and the motor should be tested. If the motor starter has an undervoltage solenoid, be sure to test it by denenergizing power at the nearest disconnect. The motor starter should be in the on position and the motor should be running for this test. The pilot-lamp indicator can also be tested during this procedure if it is being used. It is important to remember that the quality of the installation will determine how well the motor starter will operate over its operational life.

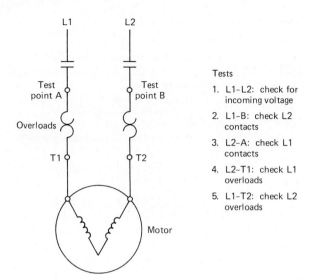

FIGURE 5-18 Troubleshooting test points for a manual motor starter.

Troubleshooting the Manual Motor Starter

At some time you will be required to troubleshoot a manual motor starter that is suspected of being faulty. It is important to remember what basic parts are used in the motor starter and to be able to identify them from the field wiring diagram. In most cases you will be called to troubleshoot a motor that has malfunctioned, and you will find that the motor is controlled by a manual motor starter. This means that you must troubleshoot the complete motor circuit and not pick on parts of the system to test randomly. Some electricians have found by experience that the motor starter overload will be the most likely problem in this circuit. Other typical problems will be the loss of supply voltage or a faulty motor. This means that these components can be tested first and the majority of the time the problem can be quickly found and repairs can be made.

At other times it will be necessary to make the diagnostic tests in an orderly fashion to isolate the problem. The test points are shown in Figure 5-18. This test procedure should be used if the circuit fault is not readily apparent or if the motor does not restart when the motor starter is reset. The first test in this procedure should be for supply voltage across the terminals marked L1 and L2. If the starter is for a three-phase application, the test should be across all three of the supply voltage lines, two at a time. If the supply voltage is dc, be sure to set the meter for dc voltage. If voltage is not present at these terminals, it means that the supply voltage has been lost and you should check the disconnects and fusing in the motor starter circuit.

If voltage is present across these terminals, proceed to the next test point. One of the voltmeter probes should be left on the L2 terminal during the next test. This will provide a point of reference for the test. The other probe should be moved to test point A (between the contacts and the overload). If voltage is not present at this point, it means that the motor starter contacts for line 1 are not passing power. Check the contacts and replace them if necessary. If voltage is present at this terminal, it means that the L1 contacts are passing voltage correctly. Continue the test by reversing the voltmeter probes. Leave one of the probes on terminal L1 and place the other probe at test point C. If voltage is not present, it indicates that the L2 contacts are faulty. Check these contacts and replace them if necessary. If voltage is present at this test point, it indicates that the L2 contacts are operating correctly. If the starter has three sets of contacts, leave one probe on the L1 terminal and duplicate this test at the point between L3 contacts and its overload. If the starter has an undervoltage or loss of voltage solenoid, be sure that it is energized. If the undervoltage solenoid is not energized, you will need to check for a remote stop button that may have been activated or that may be malfunctioning. Remember, the undervoltage solenoid may be tripped because of a previous power failure or low-voltage condition that has since cleared. Try to reset the motor starter by first pressing the off switch and then pressing the on switch. If voltage is present at test points A and B, continue the procedure. Leave one of the meter probes on terminal L2 and move the other probe to terminal T1. If voltage is not present at this point, it indicates that line 1 heater is open. Allow the motor starter to cool down for several minutes and try to reset the overload mechanism. If voltage is present at the T1 terminal, reverse the voltmeter probes and leave one of the probes at terminal L1. Touch the other voltmeter probe to terminal T2 and check for voltage at this point. If voltage is not present at this point, the L2 overload is open. If voltage is present at this point, you can assume that the starter is operating correctly and providing voltage to the motor. Additional tests should be made on the motor to determine the exact fault. You may need to repair or replace the motor to make the circuit operational again. You may need to review the information on troubleshooting motors that is listed in Chapters 10 and 11 to locate the fault that exists in the motor.

It should also be noted that some technicians prefer to start this test procedure at the motor terminals and work their way back to the motor starter and disconnect. Each of these methods of testing the circuit is acceptable. Another point that you should remember is that the contacts in most starters can be replaced rather than replacing the complete starter. You will need to check the parts catalog for your starter to see if replacement contacts are available. If they are available, they should be kept in stock so that they can be changed if your troubleshooting procedure proves they are faulty. Replacing faulty contacts is faster and less expensive than changing the complete starter. Instructions will be provided with the contacts to help you remove the faulty one and install the new ones properly.

MANUAL DRUM SWITCHES

Manual drum switches are used for reversing single-phase and three-phase ac squirrel-cage motors and dc shunt and compound motors. The drum switch is also available for controlling multispeed motors with two to four speeds. This type of switch is capable of controlling 5-hp ac motors at 200 to 300 V, 7.5-hp ac motors at 460 to 575 V, and 2-hp dc motors at 115 to 230 V. If larger motors are used, a reversing magnetic starters should be used in place of the drum switch.

One application of the drum switch is for reversing motors such as reversing overhead hoists that are used to move heavy objects such as press dies and molds. The hoists must have the ability to reverse a motor quickly to handle these loads safely. Other uses for the drum switch include controlling multispeed motors in applications such as blowers, mixers, compressors, pumps, tumblers, and special machine tools.

The drum switch can be used to control these types of motors directly, but you must understand that it does not include any type of motor overcurrent protection or undervoltage protection. This means that the motor must be protected by a fused disconnect or magnetic starter. Several options are also available with most drum switches to provide interlocks with the magnetic starter to protect the motor against these perils.

Figure 5-19 shows a photograph and switching diagrams of a reversing drum switch. Notice that the switch has three sets of contacts with six separate terminals. The contacts are numbered 1

FIGURE 5-19 Picture and diagrams of reversing drum switch. (Courtesy of Eaton Corp./Cutler-Hammer Division.)

through 6 and are switched with a three-position selector. These positions include forward, reverse, and off. In the reverse position, each of the terminals on the left side makes contact with the terminal directly opposite to provide three separate circuits. This connects 1 to 2, 3 to 4, and 5 to 6. When the switch is in the forward position, contacts 1 and 2 are connected to the terminal directly below it in the diagram. This connects 1 to 3 and 2 to 4. Terminal 5 remains connected to terminal 6 just as when the switch was in the reverse direction. In the off position, each of the six terminals is isolated from each other.

Figure 5-20 shows several diagrams of a single-phase ac motor, a three-phase ac motor, and dc shunt and compound motors being reversed by a drum switch. These diagrams show the switch in its most commonly used applications. Notice in the diagram for a single-phase ac motor (Figure 5-20a) that the start winding must be reversed for the motor's rotation to be reversed. The start winding is connected to terminals 2 and 5, and the run winding is connected to terminals 1 and 4. Single-phase voltage L1 and L2 are connected to terminals 6 and 2. Terminals 3 and 5 are jumpered. When the drum control is switched to the forward position, terminal 1 is connected to terminal 3, terminal 2 is connected to terminal 4, and terminal 5 is connected to terminal 6. In this configuration L1 voltage is applied to terminal 5 of the start winding and L2 voltage is applied to terminal 8 of the start winding. This allows the motor to run in the forward direction. When the switch is reversed, the drum switch contacts change so that terminal 1 is connected to terminal 2 and terminal 3 is connected to terminal 4. This action applies L1 voltage to terminal 8 of the start winding, and L2 is connected to terminal 5 of the start winding. By reversing the applied voltage to the start winding, the motor will now run in the reverse direction. When the switch is changed to the off position, L1 voltage is disconnected from the motor, which would deenergize it. If the motor were connected to 110 V, the neutral line would be connected where L2 is. This would allow the drum switch to reverse the motor, and when the switch is in the off position only the L1 voltage is disconnected. The neutral line remains connected to the motor at all times so that the switch will meet the

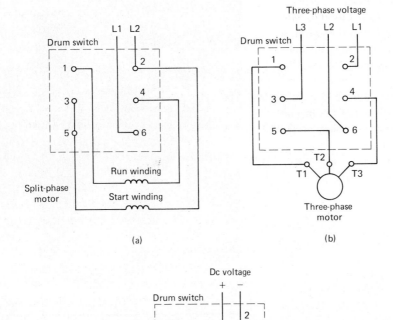

FIGURE 5-20 Diagrams of (a) split-phase, (b) three-phase ac motors, and (c) dc motors connected to drum controllers.

National Electrical Code®. Remember: the NEC states that the neutral line should never be opened by a switch or disconnect.

In Figure 5-20b the drum switch is connected to reverse rotation for a three-phase motor. Since the shaft rotation of a three-phase motor may be reversed by reversing the applied voltage to any two of the three motor terminals, the drum switch can accomplish this easily. When the switch is moved to the reverse position, L1 is connected to T1, L2 is connected to T2, and L3 is connected to T3. When the drum control is switched to the forward connection, L1 is connected to T3, L3 is connected to T1, and L2 remains connected to T2. By reversing the T1 and T3 connects, the motor will rotate in the opposite direction. When the control is switched to the off position, L1 and L2 are disconnected from the motor terminals, which will deenergize the motor. You should remember that two of the three supply voltage lines must be disconnected to deenergize a three-phase motor. Wherever possible, all three of the lines should be disconnected.

In Figure 5-20c the drum control is used to reverse the rotation of a dc shunt and a dc compound motor. The principle used to reverse the direction of rotation of a dc motor is to reverse the voltage polarity to the field winding. The field winding is wired to the control like the start winding of a split-phase ac motor. This means that when the switch is changed to the reverse position, the field winding is reversed to the power supply. In a compound motor, both the series field and the shunt field are reversed.

Changing Motor Speeds with a Drum Switch

The drum switch can also be used to change motor speeds. This function is useful where a motor may require several speeds, such as a multispeed grinding or other machine tool applications. Figure 5-21 shows the diagram of this type of switch. The multispeed drum switch is available with three, five, six, or nine sets of contacts. The speed of a three-phase motor may be varied with the torque remaining constant or with the torque allowed to vary.

Installing and Troubleshooting the Drum Switch

Drum switches can be installed in surface-mounted enclosures or cavity enclosures. The cavity enclosure allows the switch to be flush mounted on an operator's panel. The drum switch also provides a

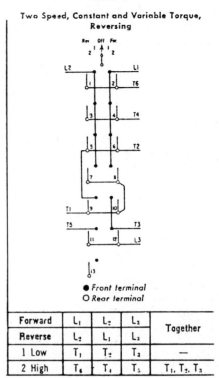

FIGURE 5-21 Drum switch used to control two-speed and four-speed motors. (Courtesy of Allen-Bradley, Inc.)

choice of handle types. These include the pistol grip, the tear drop, and the knob and pointer.

The first part of the installation procedure involves the selection of the type of drum switch, based on the motor control application and the type of motor. These conditions will determine the number of sets of contacts the switch will require. The second part of the procedure is to connect the motor and line voltage wires to the drum switch according to the wiring diagram. Be sure that power to the line wires has been deenergized at the disconnect. This will allow the wires to be handled without any danger of electrical shock.

After the field wires are connected, the disconnect can be closed to energize the supply voltage lines. The switch should be operated through all the selections to ensure that the motor is operating correctly. If any of the switch modes do not operate correctly, the troubleshooting procedures should be used.

The drum switch should be troubleshooted in the same way as other switches or pilot devices. This means that since the switch opens or closes one or more sets of contacts, each of the sets can be tested. The reversing drum switch has three sets of contacts that are connected in two different configurations when the switch is operated. A switch diagram must be used to determine the terminals that are connected in each mode. A matrix provided at the bottom of this type of diagram indicates which contacts are closed when the handle is in the forward, reverse, or off position. An ohmmeter is used to test these contacts if the switch is removed from the circuit. Remember that the ohmmeter will show continuity through motor windings if the motors remain connected to the switch. These readings are confusing, because they can tend to show that a set of contacts are closed even though they are open. This occurs because the motor winding provides a path for the ohmmeter current to backfeed instead of flowing through the contacts. This causes the technician to assume that the contacts are operating correctly and a fault in the switch contacts may be overlooked. In general, this means that a continuity test will be useful only if the switch has been disconnected from both the motor wires and the supply voltage wires.

Since you must disconnect the wires from the switch to make an ohmmeter test, it will be faster to make a voltage test with power applied to the circuit and switch. A voltmeter should be used to make these tests. In this case, one of the meter probes should be attached to one of the supply voltage lines while the other probe is moved from terminal to terminal to test for an open set of contacts. This test will show exactly where voltage is lost in the drum switch circuit. As with other electrical circuits, the motor or supply voltage may be at fault rather than the drum switch. Therefore, the complete circuit should be tested rather than selecting individual components and testing then at random until the problem is found. After the fault is found, the bad components should be replaced and the switch should be tested after the circuit is put back into commission.

QUESTIONS

5-1. Explain one problem that occurs when a knife switch is used to switch large current loads.

5-2. Explain where you will see the knife switch in manual controls used in industry today.

5-3. Draw the symbol for a disconnect and a fused disconnect. Explain the operation of the disconnect switch.

5-4. Give two purposes for which a disconnect switch is used in industrial equipment.

5-5. Explain why equipment that uses electrical power must be locked out at the disconnect when repair work is in progress. (Remember: it is acceptable to allow power to be applied to control circuits while you are making specific voltage tests.)

5-6. Draw a diagram that shows the electrical and mechanical operation of a manual motor starter. Be sure to identify all parts on the starter.

5-7. State two uses for manual motor starters.

5-8. Explain how manual motor starters are identified by size. List the sizes of starters that are available.

5-9. Explain the function of the arc hood on a manual starter.

5-10. Explain why double-break contacts are used on some starters.

5-11. Explain what you should do if you find exceedingly dirty contacts in a starter.

5-12. Explain why contacts should never be filed.

5-13. If you are using a size 2 starter to control a machine at 208 V. What would the allowable amperage be? If this starter is used on a machine connected to 480 V, what would the allowable amperage be?

5-14. Explain when you would use a single-pole, a two-pole, and a three-pole motor starter. Draw an electrical diagram of a motor connected to each of these starters.

5-15. Draw the electrical symbol of the overload used on a motor starter.

5-16. Explain the operation of the overload and heater on a manual starter.

5-17. Explain how you would reset a tripped overload on a motor starter.

5-18. Explain why an overload may not be able to be reset immediately after it has tripped.

5-19. Explain the operation of an automatically resetting overload.

5-20. Provide examples of the use of an automatically resetting overload and of a manual resetting overload.

5-21. Explain how undervoltage protection operates on a manual motor starter.

5-22. Draw an electrical diagram of a manual motor starter using undervoltage protection.

5-23. Draw an electrical diagram of a manual motor starter used to reverse the direction of rotation of a motor.

5-24. Draw an electrical diagram of a manual motor starter used for a two-speed motor.

5-25. Explain the operation of reduced-voltage manual motor starter. Use an electrical diagram to assist your explanation.

5-26. Explain why the duty cycle of an autotransformer reduced-voltage starter is critical.

5-27. List the types of enclosures provided by manufacturers that adhere to NEMA standards.

5-28. Explain electrical tests you would use to test a drum switch.

5-29. Draw an electrical diagram of a drum switch.

5-30. Explain which contacts are connected when the drum switch is in each of its three positions.

5-31. Draw a diagram of a dc motor, an ac single-phase motor, and an ac three-phase motor connected to drum switches.

5-32. List two applications for which you would use a drum switch.

Magnets, solenoids, and relays 6

INTRODUCTION

In this chapter we explain the theory of magnetic properties. We also discuss the way basic magnetic principles have been incorporated into modern motor control devices such as relays and contactors. It is very important to understand these principles because they are also used in many other applications, including ac and dc electric motors and magnetic motor starters. In this chapter we follow electromagnetic principles as they are applied to solenoids, relays, and contactors. Theory-of-operation and electrical diagrams for these devices are also explained. Typical applications, with installation and troubleshooting procedures, are included. In the final part of this chapter we discuss methods of interfacing solenoids, relays, and contacts to programmable controllers and robots. These applications show relays and contactors used as both input and output devices to the programmable controller and robot. When you have completed this chapter, you will be able to utilize and repair these devices when you encounter them in industrial motor control systems.

MAGNETS

Magnet is the name given to material that has an attraction to iron or steel. This material was first found naturally about 4000 years ago as a rock in a city called Magnesia. The rock was called *magnetite* and was not really usable at the time it was discovered. Later it was found that pieces of this material could be suspended from a wire and would always orient themselves such that the same ends always pointed toward the earth's poles. Scientists soon learned from this phenomenon that the earth itself is magnetic. At first they used this material in compasses. It was many years later that the forces caused by two magnets attracting or repelling could be utilized as part of a control device.

As scientists gained more knowledge and equipment became available to study magnets more closely, a set of principles and laws evolved. The first of these discoveries showed that a magnet has two poles. These poles, which are determined by the electronic makeup of the magnet, are called the north and south poles. When two magnets are placed end to end so that similar poles are near each

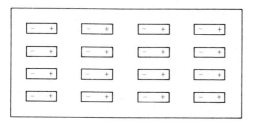

Strong magnet; magnetic dipoles all in alignment

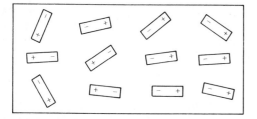

Weak magnet; magnetic dipoles misaligned

FIGURE 6-1 Diagram of alignment of magnetic domains for a strong and weak magnet and nonmagnetic material.

other, the magnets repel each other. It does not matter if the poles are both north or both south, the result is the same. When the poles are opposite, the two magnets attract each other. These concepts are called the *first and second laws of magnets*.

When sophisticated laboratory equipment became available it was found that this phenomenon is due to the basic atomic structure and electron alignments. When the atomic structure was studied, it was found that the atoms were grouped in regions called *domains*. In material that was not magnetic or magnetized, the alignment of the electrons in the domains was random and usually followed the crystalline structure of the material. In material that was magnetic, the alignments in each domain were along the lines of the magnetic field. Since each domain was aligned exactly like the ones next to it, the magnetic forces were additive and became much stronger. In material where the magnetic forces were weak, it was found that the alignment of the domains was not exactly along the magnetic field lines. The more closely this alignment was to the magnetic field lines, the stronger the magnet would become. Figure 6-1 shows diagrams of the domains of material for a strong magnet and a weak magnet. Notice that the alignment of the magnetic domains is related directly to the strength of the magnet.

You may be more familiar with a typical bar magnet. This type of magnet, made of soft iron that has been magnetized, is shown in Figure 6-2. From this figure you can see that the magnet is in the shape of a bar, and it has its north and south poles identified. The magnet produces a strong magnetic field from the flux lines that move from the north pole to the south pole. These flux lines are invisible, but they are still very powerful. A simple experiment will allow you to see the configuration of the invisible flux lines as they encompass the bar magnet. A piece of clear plastic film such as the plastic sheets used for overhead transparencies should be placed over a bar magnet. Make the plastic as flat as possible and sprinkle iron filings over it. The filings will be attracted by the flux lines as they extend from the north pole to the south pole. Since the flux lines begin at one pole and stretch to the other, the highest concentration of flux lines will be near the poles. The iron filings will also concentrate around

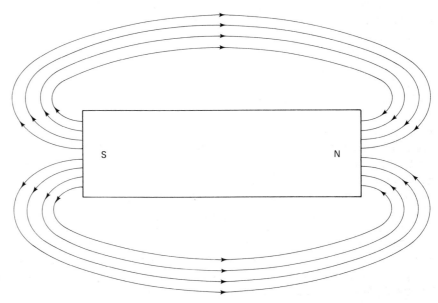

FIGURE 6-2 Bar magnet showing location of flux lines.

Magnets

the poles, but a definite pattern of flux lines can be seen. If an overhead projector is available, the image of the flux lines can be projected onto a projector screen or blackboard so that they can be seen more easily. The number of flux lines around a magnet is related directly to the strength of the magnet. A stronger magnet will have more flux lines than a weaker magnet. The strength of a magnet's field can be measured by the number of flux lines per area. Since the strength of a magnet's field is based on the alignment of the magnetic domains, the number of flux lines will increase as the alignment of the magnetic domains increases.

Some material, such as alnico and Permalloy, make better magnets than iron, since the alignment of their magnetic domains remains consistent even after repeated use. These permanent magnets are useful in many types of controls, especially in motors. Permanent magnets have several drawbacks for use in some applications. One of these problems is that the magnetic force of a permanent magnet is constant. This means that if something is attracted to a magnet, it will remain so until it is physically removed from the force of the flux lines. Another problem with a permanent magnet's flux field being constant is that it cannot easily be made stronger or weaker if circumstances so require.

ELECTROMAGNETS

In the early nineteenth century, scientists began working with simple electric cells (batteries). When a wire conductor was connected to the terminals of the battery, current began to flow. If the wire was placed near a pile of iron filings while current was flowing through it, the filings were attracted to the wire. This occurrence caused scientists to investigate the magnetic properties of a current-carrying conductor. Figure 6-3 shows several diagrams indicating the location of magnetic flux lines around conductors. Again, several simple experiments may be set up to show these principles. Insert a current-carrying conductor through a piece of cardboard. Place iron filings around the conductor on the cardboard. The filings will settle around the conductor in concentric circles showing where the flux lines are located. As the amount of current is increased, the number of flux lines will also increase. The flux lines will also concentrate closer and closer to the wire until the current reaches saturation. When the flux lines reach the saturation point, any additional increase of current in the wire will not produce any more flux lines. Figure 6-3a shows the location of flux line around conductors.

Figure 6-3b shows the wire connected to a dry

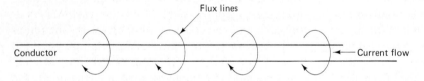

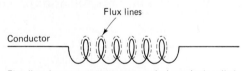

(a)

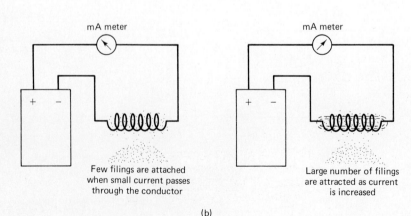

(b)

FIGURE 6-3 Diagram showing the magnetic principles of current-carrying conductors.

cell battery and placed near a pile of iron filings. A variable resistor can be added to the circuit to increase or decrease the current. When the current is set to a minimum, the wire will attract only a few filings. As the current is increased, the number of filings the wire will attract also increases until the current causes the wire to reach saturation. When the saturation point is reached, any additional current flowing in the wire will not produce additional flux lines.

In a later experiment, scientists found that by wrapping the wire into a helix coil, the strength of the flux lines was compounded. As the number of coils was increased, the strength of the magnetic field increased. It was also found that when the current in the wire was interrupted and forced to zero, the strength of the magnetic field was minimal. In this way the magnetic field could be switched on and off by switching the current in the conductor on and off.

In another experiment it was found that the strength of a magnetic field is increased by placing a soft-iron core into the helix coil. The farther the core is inserted into the helix, the stronger the flux field becomes. When the core is removed completely from the coil, it is considered to be an air coil and the magnetic field is at its weakest point. When different materials are used as the core, the strength of the field can also be varied somewhat. In some cases, when soft iron is used as the core, it becomes permanently magnetized, and when current is switched off in the wire, the magnetic flux lines remain. When powder iron is used for the core material, the flux lines are not as strong, but no flux lines remain when current is switched off.

Since the magnetic flux lines in a coil of wire are caused by electric current flow, this type of magnet is called an *electromagnet*. As the electromagnet was studied further, the movement of the flux lines could be determined. Figure 6-4 shows several diagrams that locate the flux lines. Figure 6-4a shows the direction the flux lines travel around a straight current-carrying conductor. The direction of the flux lines is dependent on the direction of the current flow. For this reason a method of marking the conductor to indicate the direction of current flow has been standardized. A dot or a cross (X) is used to mark the conductor. In Figure 6-4a you should notice that a dot is used to indicate that current is flowing toward the observer. The flux lines in this diagram shows their flow is in a clockwise motion. In Figure 6-4b, a cross is used to indicate that current is flowing away from the observer and the flux lines are shown moving in the counterclockwise direction.

Figure 6-4c shows another way to show the directions of the flux lines, the direction of current flow, and which pole is the north pole. This method is called the *left-hand rule*. From this diagram you can see that you need to know either the direction of current flow or the direction of the flux lines. Normally, the direction of current flow is easy to determine with a voltmeter. Place your left hand on the wire with your thumb pointing in the direction of the current flow. After this is accomplished, you will notice that your fingers will be wrapped around the conductor in the direction of the flow of the flux lines. When the straight conductor is looped in a coil or helix, the same rules can be applied. In simple magnetic theory for solenoids and relays, the

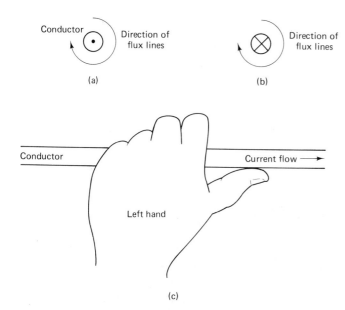

FIGURE 6-4 Diagrams used to show the directions of flux lines around straight and coiled conductors; example of left-handed rule.

direction of the flux lines and current flow is not really important. When magnetic theory is used to explain the operation of transformers, motors, and generators, the direction of flux lines and the field strength are vitally important.

In addition to determining the flow of the flux lines in different types of electromagnets, it is important to understand the methods of controlling its strength. The strength of an electromagnet can be increased by:

1. Increasing current in the conductor
2. Increasing the number of coils of wire
3. Changing the type of core

Once you understand these basic principles, the operation of solenoids, relays, and contactors can easily be explained. This information will allow you to troubleshoot and repair these devices with a minimum of effort.

SOLENOIDS

One of the simplest applications of an electromagnet is as a solenoid. In this application the coil of wire is mounted in a fixed position and its core is allowed to move freely. Figure 6-5 shows the locations of components you can use to demonstrate solenoid action. When no current is applied to the coil of wire, gravity allows the core (bolt) to move to the bottom of the coil. As current is applied a magnetic field is quickly built in the coil. This field is strong enough to pull the core up from the table toward the middle of the coil of wire. The magnetic field is concentrated in the middle of this coil, and this causes the core to center itself there. When current is removed from the coil of wire, gravity will again cause the core to drop out of the coil. Another occurrence in this experiment is a change in the amount of current flow. The coil requires a maximum amount of current to pull the movable core to the center of the coil. This current is called *pull-in* current. When the core centers itself in the coil, the amount of *current* drops off dramatically. This current is called *holding current* or *sustaining current*. Generally, the pull-in current is 5 to 10 times larger than the holding current. This action is called *solenoid action*. The holding current is smaller in that the amount of energy needed to make the core move is minimal because it is centered and no further movement is required. As the core moves into the coil, the electrical impedance of the coil increases to a point where current flow is minimal. The impedance of the coil is inversely related to the portion of the core that is inside the coil. When the maximum amount of the core is inside the coil, the impedance is at its minimum, and when the minimum amount of the core is in the coil, the impedance is at its maximum.

You can easily duplicate this experiment to see the basic principles. You will need to locate a solenoid coil that has fairly heavy wire. (*Never apply power to the solenoid coil when the metal*

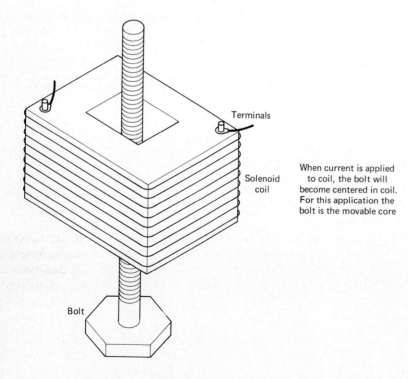

FIGURE 6-5 Diagram showing basic solenoid action.

112   *Chap. 6 / Magnets, Solenoids, and Relays*

core has been removed, since it will draw exceedingly large currents and burn up.) Next you will need to locate a bolt that is ½ in. in diameter and approximately 2 to 3 in. long. Be sure that the bolt is made of iron or steel.

Place the bolt on a table so that it is standing up. Hold the coil so that the tip of the bolt is inserted into it, and begin to apply dc power. Increase the power until the magnetic field pulls the bolt up and centers it in the coil. When you have the voltage set to a value that has the bolt centered in the coil, leave the adjustment alone and turn the power supply's power switch off. When the power is turned off, the bolt will drop out of the coil to the table again. Be sure to set the bolt on end again so that its tip is just inside the coil.

Now you can simulate the action a solenoid produces by turning the power supply on and off several times. Every time the solenoid is energized, the bolt will be pulled directly into the center of the solenoid, and every time power is removed, the bolt will drop out of the coil and fall back to the table. While you are energizing and deenergizing the coil, you should notice the amount of force the bolt has when it centers into the coil. If the voltage is increased slightly or the number of turns in the coil is increased, the bolt will have more power when it centers itself. You will also notice that after repeated operation, the coil will begin to heat up since the bolt (core) is not sized correctly for the coil, and the coil is not sized for the power supply.

Adding a Plunger to the Solenoid to Make a Valve

Figure 6-6 shows an example of the electomagnetic coil used to move the plunger in a solenoid valve. In this application the movable core is connected to a plunger in a solenoid valve which controls the flow of air. When no current is flowing in the coil, there is no magnetic field and a small spring keeps the movable core (plunger) in its closed position. In this position the plunger closes off the air valve. When current is applied to the coil, a strong magnetic field is produced which will attract the movable core to the center of the coil. This field is strong enough to pull the movable core to the center of the coil. Since the plunger for the solenoid valve is attached to the end of the movable core, the air valve is opened. Any time that current is deenergized from the solenoid coil, the magnetic field is reduced and the small spring will cause the movable core to return the plunger to the closed position. Whenever the valve is required to be open, current is applied to the coil and the magnetic field pulls the core with its plunger to the open position. These valves can be used to control the flow of air or fluids such as water or hydraulic fluid. A one-way solenoid is usually used to control on/off flow of water or air when an exhaust port is not needed. The valve body is usually unidirectional (one way). This means that the inlet and the outlet sides are marked so that the fluid flow will not prevent the solenoid plunger from operating correctly. If the fluid flow is in the wrong direction through the valve, it may not allow the valve to open or close correctly. In some cases the valve will operate, but the increased force required to open the valve will cause the coil to overheat.

Solenoid valves are classified by the number of inlet and outlet ports that are controlled. The valve shown in Figure 6-6 is a simple on/off solenoid. Other types of valves are more complex, such as the two-way valve. This valve is so named because it has one inlet and one outlet port. The valve can connect the inlet to the outlet port, or the inlet port can be blocked. When this type of valve is used to control a pneumatic cylinder the valve has one inlet port, one exhaust port, and two outlet ports. In one position the inlet is connected to the outlet marked A, and outlet B is connected to the exhaust port. When the valve is switched to the

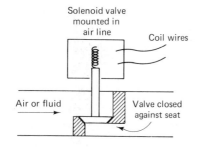

(a) Solenoid valve with no current applied; noitce that spring keeps valve closed

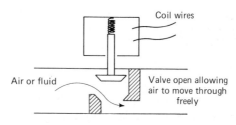

(b) Solenoid valve in the energized position; magnetic field pulls valve to the open position

FIGURE 6-6 Diagram of an inline solenoid valve.

4-Way, 5-Port, 2-Position Single & Double Solenoid Valve 1/4 & 3/8 Inch Ports

Application
These valves are used to operate double acting cylinders. They may also be used as 2-way or 3-way valves by plugging appropriate ports. Valves are actuated by a maintained electrical signal.

Mounting
These valves are designed for inline or stacking mounting. Side by side mounting may also be used, with each valve individually piped.

Operation
Single Pressure At Inlet Port 1:
De-energized position — Solenoid operator #14 de-energized. Pressure at inlet port 1 connected to outlet port 2. Outlet port 4 connected to exhaust port 5.
Energized position — Solenoid operator #14 energized. Pressure at inlet port 1 connected to outlet port 4. Outlet port 2 connected to exhaust port 3.

Dual Pressure:
May be used for dual pressure service with pressure at ports 3 and 5 (higher pressure at 5) — Exhaust at port 1. (Must be ordered as dual pressure.)

Overrides:
Non-locking manual overrides are standard.

Application
These valves are used to operate double acting cylinders. They may also be used as 2-way or 3-way valves by plugging ports. Valves are actuated by a "momentary" electrical signal alternately on each of the two solenoids.

Mounting
These valves are designed for inline or stacking mounting. Side by side mounting may also be used, with each valve individually piped.

Operation
Single Pressure At Inlet Port 1:
Solenoid operator #14 energized last. Pressure at inlet port 1 connected to outlet port 4. Outlet port 2 connected to exhaust port 3. *Solenoid operator #12 energized last.* Pressure at inlet port 1 connected to outlet port 2. Outlet port 4 connected to exhaust port 5.

Dual Pressure:
May be used for dual pressure service with pressure at ports 3 and 5 (higher pressure at 5) — Exhaust at port 1. (Must be ordered as dual pressure.)

Overrides:
Non-locking manual overrides are standard.

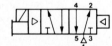 #14 Operator End / #12 Operator End

 #14 Operator End / #12 Operator End

FIGURE 6-7(a-b) Examples of four-way single solenoid and double solenoid valves. (Courtesy of Parker Hannifin Corporation.)

other position, the inlet is connected to port B and port A is connected to the exhaust port. This type of valve is energized with the solenoid coil in one direction and returns with spring pressure to the other position.

Three-Way Solenoid Valves

The valve shown in Figure 6-7a is a four-way, two-position valve. It is important to understand all the things that the pneumatic diagram is indicating.

Solenoids Manifolded Together

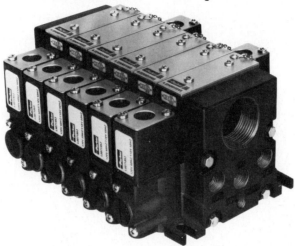

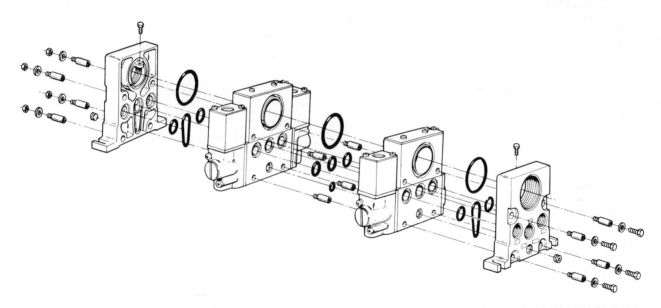

FIGURE 6-7c Pictures of a four-way air solenoid valves and a diagram of air valves connected in a manifold. (Courtesy of Parker Hannifin Corporation.)

You should notice that this valve has five ports on the right end, identified as 1, 2, 3, 4, and 5. The diagram also shows a similar set of ports on the left end of the valve that are not identified. Since this valve has only five ports, the connections shown on the right side of the valve indicate the internal connections of the valve when it is deenergized and the return spring has shifted the plunger. The diagram on the left side of the valve shows the internal connections of the valve when the solenoid is energized. *Remember that every diagram will show the connections for the valve when it is energized and when it is deenergized. This will cause the valve to appear to have twice the number of ports that it actually has.*

When this valve is operated as a two-way valve, air is supplied to port 1, port 2 is connected to the device that is to receive the air, and ports 3, 4, and 5 are plugged. This application is quite useful as a simple on/off valve to control air supply to a system. It is also useful in simple applications such as suction-cup grippers in robots where air is applied to a venturi manifold. When air is allowed to flow through the venturi, suction will be produced at the cup. The amount of load that can be lifted by the suction cup will be determined by the amount of air pressure and the size of the cup. When airflow to the venturi is shut off, suction will be lost and the cup will drop the part. This allows activation of the suction cup to be controlled by the solenoid valve.

A three-way valve is required when a device such as a spring-return air cylinder requires an

Solenoids 115

HAND OR FOOT ACTUATED AIR VALVES

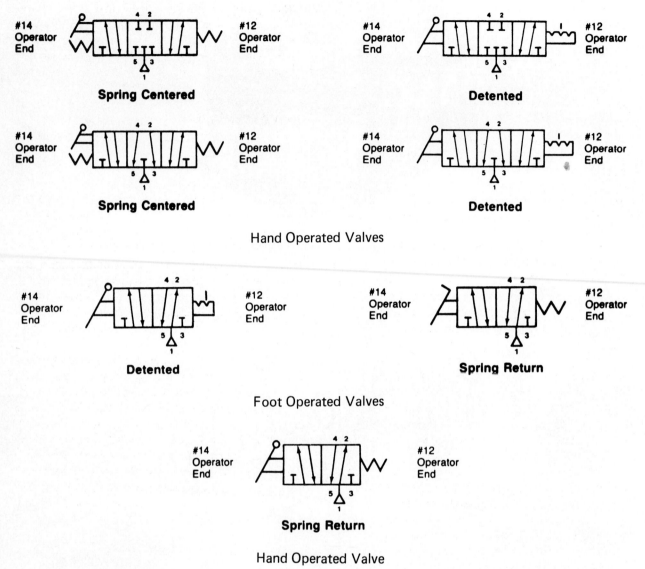

FIGURE 6-7d Examples of hand and foot actuated air valves. (Courtesy of Parker Hannifin Corporation.)

exhaust port to allow the spring return to operate correctly. For this explanation, port 1 will be used as the air inlet, port 2 will be used as the outlet, and port 3 will be used as an exhaust. Notice that the solenoid (coil) is shown on the left end of the valve as a box with a diagonal line through it. When the solenoid coil is deenergized, this diagram shows the return spring pushing the plunger to the left, and when the coil is energized, the magnetic field will shift the plunger to the right.

In the deenergized state, port 1 will be connected to port 2. This means that the device connected to port 2 will receive air when the coil is not powered. When the valve is powered, the device connected to port 2 will be connected to port 3, which is allowed to bleed to atmosphere as an exhaust port. As indicated before, this type of valve could be used to control a spring-returned air cylinder (piston) that pushes parts off of a conveyor line when it is activated (extended). When the air supply is removed from the cylinder, the return spring will cause the cylinder to retract. It is important to note that the cylinder must have a way to exhaust the air from the cylinder when the spring is used to retract the cylinder. This valve is called a three-way valve since it has an inlet, an outlet, and an exhaust port. The valve can connect the inlet to the outlet port, or the inlet port can be blocked and the outlet can be connected to an exhaust port. If a double-acting cylinder is used instead of a spring-return cylinder,

116   Chap. 6 / *Magnets, Solenoids, and Relays*

a four-way valve is required to provide proper operation.

Four-Way Solenoid Valves

An example of the four-way double solenoid valve is shown in Figure 6-7b. From the diagram you can see from the slash mark across the box at each end that this valve has two solenoid coils. The ports on this valve are identified as 1, 2, 3, 4, and 5. Again you should notice that two diagrams are shown inside the valve, which represent the connection of the ports when the valve is energized and deenergized. The diagram on the left side of the valve indicates the connections for the valve when coil 14 is energized. As in previous diagrams, the ports are identified in only one side of the valve diagram.

It will be easier to understand the operation of this type of valve if you look at the diagrams of the valve connected to a double-acting air cylinder. In this condition, the return coil 12 moves the plunger so that port 1 is connected to port 2, and port 4 is connected to port 5. Since port 1 has supply air connected to it, supply air will be sent to port 2 and port 4 will allow the front side of the cylinder to bleed to exhaust through port 5 of the valve which will cause the cylinder to extend its rod. It is important to remember that the cylinder requires a method of exhausting air from the opposite side of the cylinder that is pressurized, or the cylinder rod will not be able to move.

When the solenoid 14 is energized, the plunger will shift to the left; port 1 will be connected to port 4, while port 2 will be connected to port 3. In this position, air pressure will be applied to the front side of the cylinder by the connection of ports 1 to 4, and the rear port of the cylinder will be exhausted by the connection of port 2 to port 3. This configuration will cause the cylinder rod to retract. You will also see the location of O-rings used as seals for the plunger.

Multiple Air Solenoid Valves Mounted on a Manifold

Another way you will find the various types of valves is when they are combined with similar valves and mounted on a manifold. Figure 6-7c shows a photograph of the previous types of solenoid valves mounted on a manifold. An exploded-view diagram is also provided that shows how these valves are mounted to share a common air supply. The manifold also reduces the installation cost of mounting multiple valves. This type of arrangement provides a more compact arrangement of the valves, which saves space. The manifold can be assembled to mount any number of single-solenoid and double-solenoid two-position or three-position valves that have similar valve body styles. When you install, troubleshoot, or repair the valves on a manifold, they should be treated as individual valves. This means that the only part of the system that they have in common is the supply air and the exhaust. All other aspects of the valve operation must be viewed as actions of an individual valve. If one of the valves is found to be defective, that valve may be replaced without damage to the rest of the valves on the manifold.

Double Solenoid and Manually Operated Valves

Another version of the four-way valve uses a second solenoid coil to cause the plunger to return, instead of a spring. The internal connections and operation of the valve are identical to those of the spring-return valve except that the return is caused by a second solenoid. In this diagram you will also notice that the spring symbol has been replaced by a second solenoid symbol.

This type of valve can also be operated manually with a foot pedal or a hand lever. Figure 6-7d shows example diagrams of these types of valves. You should notice that the valve body is the same as that of the electric-operated valve. The only difference in the valve and its diagram is that the valve is manually operated in one direction and a return spring is used to cause the valve to operate in the other direction. These valves can be integrated with electrically controlled valves throughout any system.

Applications for Solenoid Valves

Solenoid coils can be powered by either ac or dc voltage. Typical voltages include 24, 120, and 220 V ac and 24, 120, and 220 V dc. These voltages allow the solenoid to match any voltage that is presently being used in the control system or that is being used for other machine controls. It also allows the solenoid to be controlled by any available pilot device. Figure 6-8a shows a sketch and electrical diagram of a float switch that controls the level of water in a tank. The water flow is controlled by an ac solenoid wired in series with the float switch. As water is used and the level in the tank drops below the low-level set point, the float switch closes and energizes the solenoid and allows water to flow into the tank. When the water level in the tank raises above the high level set point, the float switch is opened and the water solenoid is deenergized. This cycle continues to keep the water level automati-

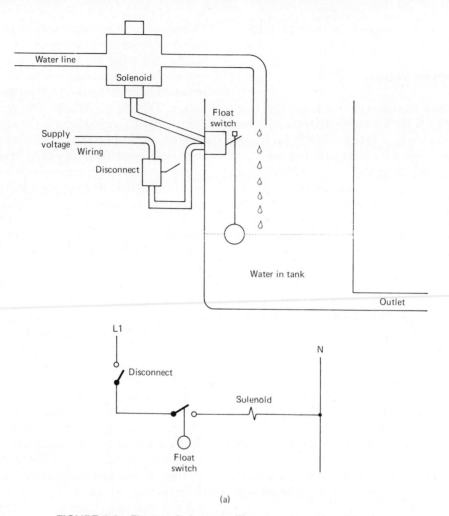

FIGURE 6-8 Float switch controlling a water solenoid valve.

cally between the high and low set points as water is used from the tank. This system uses water from a city water source so that pressure is maintained without operating a water pump. This allows the level to be controlled with only a solenoid valve.

Another application uses multiple solenoid valves to control the distribution of granular plastic material for plastic presses. Figure 6-8b and 6-8c shows a sketch and electrical diagram of this application. This system uses several silos for storing the plastic raw material in granular form. The silos can be filled from train cars that are pulled alongside them outside a factory. Material travels from the silos to each plastic press through an air conveyor. The air conveyor consists of a system of 3-in. tubing that feeds raw plastic material to each of the plastic presses on the factory floor. This allows raw material to reach each press automatically from overhead instead of supplying each machine individually from cardboard containers. Each plastic press has a small hopper mounted on it that will store enough raw material to last 20 to 30 minutes. The air conveyor system monitors the level of each hopper and adds material whenever it is needed. Solenoid valves are mounted in each tube where it feeds into each machine's hopper. A master timer controls the pattern of filling the hoppers in sequence. This timer allows each hopper 3 minutes for adding material. Each hopper has a pilot switch which is used to control the level of the material. This pilot switch is connected in series with the contacts of the master timer. If the hopper is low on material, the low-level switch is closed. When the master timer energizes this circuit for its allotted 3 minutes, the solenoid is energized and material is allowed to flow into the hopper. When the material reaches the high-level limit the level switch opens the circuit and deenergizes the fill solenoid. If the 3-minute cycle runs out prior to the hopper becoming full, the solenoid will also be deenergized and the hopper will have to wait 15 minutes for another cycle time. If any of the presses is not operating or using material at a fast rate, the level switch may prevent the solenoid from energizing during each timing cycle.

The sequential part of the system can be enhanced or controlled by a sequencer switch,

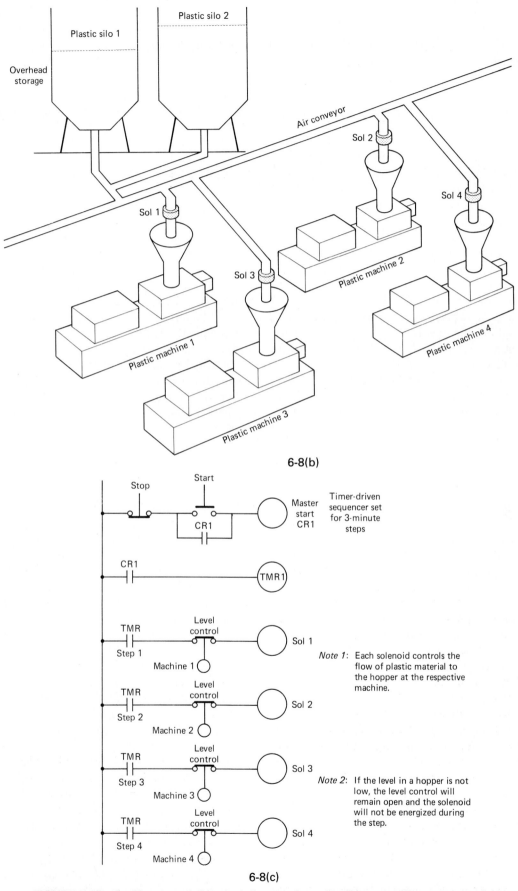

FIGURE 6-8(b-c) Sketch and electrical diagram for solenoids controlling raw material supply to plastic injection molding machines.

control relays, or with a programmable controller. A timer is used at this point in the book since other devices have not yet been explained. After you have learned more about these control devices, you may wish to review this application.

Installing and Troubleshooting the Solenoid

Solenoids are easy to install electrically. The simplest solenoids have only two wires, which can be connected without regard to polarity. The most important part of the installation procedure is to be sure and match the voltage rating for the solenoid to the amount of voltage available in the supply circuit. This means that if the supply circuit is 110 V ac, the solenoid should also be rated for 110 V ac. If the solenoid is being specified for a new circuit, the voltage for this circuit should be specified to match the voltage of other machines in the area. This practice allows the fewest possible number of parts to be kept in stock instead of stocking solenoids for every voltage.

The first step in the installation procedure is to ensure that the solenoid valve body is installed properly. This may involve personnel from other trades. In some shops you will be responsible for installing the electrical and plumbing parts of the valve. It is important to remember that some solenoids are location sensitive. This means that some solenoids will be marked with an arrow indicating the direction in which the valve should be mounted. These valves are sensitive to location because they rely on gravity to aid in opening or closing the valve. If the valve is mounted in the opposite vertical position or in an incorrect horizontal position, the solenoid will not operate correctly or may even become damaged from excessive heat. After the plumbing part of the valve is installed, the solenoid coil can be placed over the plunger and the field wires can be connected. Use approved methods for making these connections. In some applications a set of electrical terminals will be used. In other applications wire nuts may be used to make the connection. Be sure to check any applicable codes or standards for making these connections. All electrical connections should be returned under the cover of the solenoid connection box. *Do not apply voltage to the solenoid coil unless it is mounted on the valve plunger.*

After the solenoid coil is mounted on the valve plunger and the electrical connection have been made, power may be applied to the circuit for test purposes. Remember that the solenoid valve plunger provides high impedance to the coil circuit to keep current to an acceptable level. If power is applied to the coil when it is not on the plunger, the amount of impedance is not sufficient to limit current through the coil and the excessive current will burn the coil open. Remember that the coil for the solenoid is wound with very small wire gage to provide enough turns for the coil to have the proper amount of magnetism. Since the wire gage is so small, any excessive current will cause it to burn open.

Troubleshooting the Solenoid Valve

When you are testing the solenoid for proper operation, it is possible to hear the solenoid's plunger move when the valve is energized. In some cases you can also feel the plunger move or "click" as the valve is energized. If the valve is controlling the flow of a fluid, it is also possible to check visibly for the presence of this flow.

If you energize a solenoid valve and it does not activate its plunger, the valve must be tested as two separate parts, the mechanical part and the electrical part. Remember that it is possible for the coil to receive the proper amount of electrical current to magnetize it, but the plunger may be malfunctioning mechanically. This may happen if the valve is corroded or if fluid pressure is too great for the valve.

If the valve is controlled by an electrical control circuit, the complete circuit must be tested. Do not check the solenoid randomly because you assume that it is malfunctioning. If you were to test the solenoids shown in Figure 6-7a-b, you could use the voltage drop test. This test requires power to be applied to the circuit and all control devices to be energized. A voltmeter should be used to make these tests, and Figure 6-9 indicates the test points with number sets. One of the voltmeter probes is identified with the letter A and other probe is

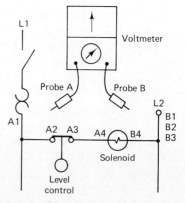

FIGURE 6-9 Sketch and electrical diagram showing positions for voltmeter probes to test for a faulty solenoid valve.

identified with the letter B. The locations for the probes at the first test points are marked A1 and B1. The points for the second test are marked A2 and B2. Notice that since the test point for probe B is the same for the second and third tests, B3 identifies the same point as B2.

The first test point is indicated on the diagram to check for voltage at the supply lines (L1 and L2). If voltage is not present at these terminals, you have found the problem. Continue looking for problems, such as a blown fuse or an open disconnect.

If voltage is present at these terminals, continue testing for voltage directly across the two solenoid terminals. If voltage is present at the supply terminals but is not present at the solenoid terminals, an open has developed in the switch or wiring. Continue testing for the open circuit by leaving one of the voltmeter probes on L2 and move the other probe to the input side of the float switch. If voltage is missing at this point, the wire between the supply terminal and the switch is open. Replace this wire and retest the circuit. If voltage is present at this terminal, move the voltmeter probe to the output terminal of the switch. If voltage is missing at the output terminal of the switch, it indicates that the contacts in the switch are open or broken. Make repairs as required and try to operate the circuit again. Remember, the switch may be open because the water level may be too high. In this case the switch is operating correctly and you will have to lower the water level for the solenoid to be energized.

If voltage is present at the output side of the switch, it indicates that the switch is operating correctly and you should move the meter probe to the terminal in the diagram that is on the left side of the solenoid. Leave this meter probe in place and move the right-side terminal to the L2 supply terminal. If voltage is not present when the meter probes are placed directly across the solenoid terminals, but it is when the probes are placed on the left side of the solenoid and L2, it indicates that the wire connecting the solenoid to L2 is open. Replace this wire and again try to operate the solenoid.

In some cases it is possible to test the solenoid without using a voltmeter. This can be accomplished using one of several methods. One way to determine if a solenoid is magnetized is to place a metal screwdriver tip near the solenoid core. If the solenoid coil is magnetized, it will also attract the screwdriver. If the coil is not magnetized, the screwdriver will not be attracted. Another method of determining if the solenoid is magnetized is to loosen the locking nut on the coil and try to pull the coil off the plunger. If the coil is magnetized, the magnetic field should exert a pulling force on the coil. *This method will damage the coil if you pull too hard and the coil comes off the plunger.* For this reason you must be very careful while you are pulling on the coil so that it does not come completely off the plunger. The preferred method of testing a solenoid is to use a voltmeter. This method can be used to test for solenoids that are powered by either ac or dc voltage.

Another way to determine if the coil is producing a magnetic field is to see if it is drawing any current. If the coil is powered by ac current, you can use a clamp-on ammeter. Since the coil will draw such a small amount of current, you may need to wrap the wire that supplies voltage to the coil around the ammeter claw several times to multiply the amount of current. If the coil is drawing current, you can assume that it is operating correctly and that the problem is in the air part of the valve.

Another quick way to test the solenoid is to use a screwdriver tip to press against the actuator on the end of the valve to cause the plunger to shift. If the valve can be shifted manually with the screwdriver, causing the system to operate correctly, you can assume that you have proper airflow and that the problem is in the electrical part of the valve.

If you shift the valve and the system does not respond, you can assume that the problem is in the air system or that the load is not functioning properly. Remember, the valve only routes the air to the proper port. If the valve is controlling a cylinder to move a load, the load must be free to move so that the cyclinder can be activated. If the load is jammed or the air pressure is too low, the valve will actuate and route air to the proper port, but the load will not move. If the load does not move when you activate the valve manually, the problem may be caused by too low air pressure or there may be a problem with the load. After you test the air pressure and are sure that it is sufficient, the load should be inspected to ensure that the cylinder can move it.

RELAYS

Another application of solenoid action involves a relay. The relay uses solenoid action to move a set of electrical contracts from the open to the closed position, or from the closed to the open position. The basic principles involved in the operation of a relay have been included in the design of contactors and motor starters. There are several variations of the solenoid principal used to operate relays. The simplest of these actions is shown in Figure 6-10a. From this diagram you can see that the relay has

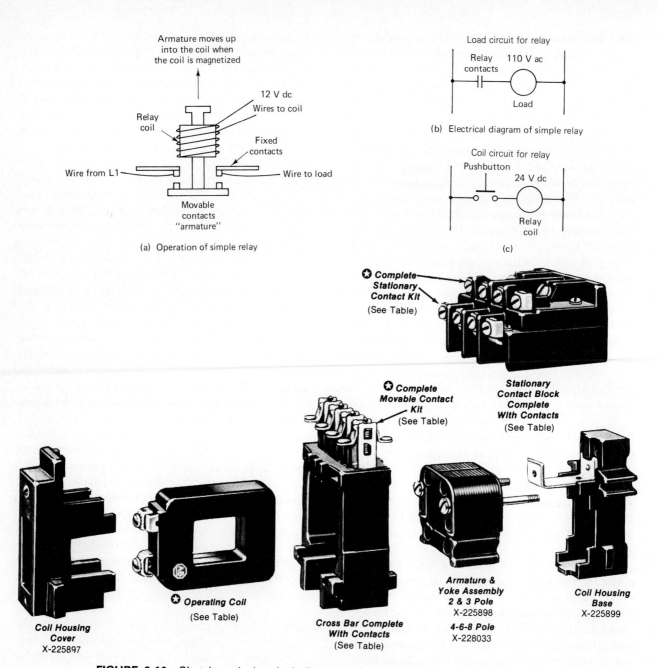

FIGURE 6-10 Sketch and electrical diagram of a simple relay; exploded view of single relay. (Photo courtesy of Allen-Bradley, Inc.)

several more parts than those in the simple solenoid valve. The relay is basically a set of contacts that are moved in one direction by the solenoid coil. The solenoid coil is generally referred to as the "coil" when it is used in a relay. This helps prevent confusion between solenoids and relays, as solenoids usually refer to an electrically operated valve. The relay coil uses the same magnetic principles as those used by the the solenoid. When it is energized, the magnetic field is strong enough to cause the movable core to be pulled into the center of the coil. The movable part of the relay is called the *armature*. In this diagram you can see that the armature is the part of the relay that makes the contacts move from the open position to the closed position.

When the coil is deenergized, the magnetic field is lost and gravity pulls the armature down out of the center of the coil. When the armature returns to its normal condition, the contacts are allowed to return to their open position. The contacts in the relay can pass current to control an electrical load just like the contacts of any other switch. *The most important point to understand about the relay is that the electrical circuit that energizes the coil is not connected electrically with the contacts. The*

122   Chap. 6 / Magnets, Solenoids, and Relays

contacts are moved by a magnetic force, so they are electrically isolated from the coil voltage.

The electrical diagram for this relay is shown in Figure 6-10b. Notice that the coil is energized by 12 V dc and the load is energized by 110 V ac. The two power supplies are not connected in any way. The coil simply acts as the operator to pull the contacts closed. This may seem like unnecessary additional equipment since a single-pole pushbutton switch could perform the same function as the relay in controlling current to the lamp. But you must remember that relays, contactors, and motor starters have many other complex functions besides controlling current to a lamp.

The important aspect of a relay that must be kept in mind is that the coil requires relatively small amounts of current to energize the magnet and pull the contacts closed. This means that small pilot switches can be used in the coil circuit to control much larger currents in the contact's circuit. This also allows the pilot switches to be located a long distance from the relay since their current is rather small and voltage drops are not a problem. This allows relays to be placed near the loads they are controlling, and the pilot switches used for control can be mounted together on an operator's panel. This was called remote control in early control circuits.

Another feature of being able to control the relay with a small amount of coil current is that this current can be controlled by microprocessors. This is one of the main advantages of a programmable controller and robot. The microprocessor in these controllers can control enough current to energize the coils of relays, contactors, and motor starters. The small amount of current required to energize the coil also allows the contacts of one relay to control current to the coil of another relay. In this type of arrangement the relay is called a *control relay*. In fact, one relay can be the permissive for power to all the other relays in the control circuit. Since additional pilot switches can be added to each relay, the control circuit can become quite complex.

Applying AC Voltage to a Relay Coil

The solenoids and relay coils discussed so far have all been energized with dc voltage. Since the current in a dc current flows in only one direction, the magnetic field becomes quite strong as soon as current begins to flow through the coil. When current is deenergized, the magnetic field that was developed in the coil collapses rapidly.

When ac current is used to energize a coil, the magnetic field will follow the sine wave of the applied voltage. When the ac sine wave is at the 0-degree mark, that ac current will be at 0. As the sine wave moves to the 90-degree mark, the current will increase from its minimum amount to its maximum amount. When the sine wave moves to the 180-degree mark, the current diminishes from the maximum amount back to zero. At this time the magnetic field builds to its maximum strength and then collapses.

The ac current continues to the 270-degree mark, where it builds from zero to maximum in the negative direction. The magnetic field also follows this pattern as it starts to build up in the opposite polarity. Since the polarity of the current has been reversed, the magnetic field around the coil also reverses. Even though the magnetic field is reversed, its net affect is that it will be strong enough to keep the armature in the center of the coil. As the current moves from the 270-degree mark to the 360-degree mark it returns from its maximum negative amount to zero and the magnetic field diminishes and collapses. This cycle continues as the ac current follows the sine-wave pattern. You should remember that the sine wave is produced as the ac current is generated.

From this diagram you can see that since the ac current is a zero part of the time it builds to a peak, the strength of the magnetic field will be an average based on the frequency of the ac current. Another problem that occurs when ac current is used to energize the coil is that the armature (movable core) receives induced current. The induced current causes eddy current, which causes the armature to heat up. These problems have caused the design of the ac relay to be modified slightly.

Using a Laminate Steel Core

The first modification that is used to make the relay coil suitable for ac circuits is to construct the core from laminated sheets of metal instead of from a solid piece of iron. Figure 6-11 shows a sketch of a laminated armature assembly. Notice that this diagram shows the thickness of a a single piece of the laminated material. These single pieces are stacked on each other and riveted to form a component that is approximately the same size as a solid core. The laminations are isolated from each other with a thin film of insulation which stops eddy current when induction occurs.

In another modification the armature assembly is designed so that the laminated sections have the coil permanently mounted to it. The laminated sections are shaped like the letter E, and the coil of wire is placed over the center section. This is the

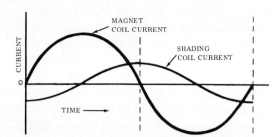

FIGURE 6-11a Diagram that shows the effect of the shading coil on coil current.

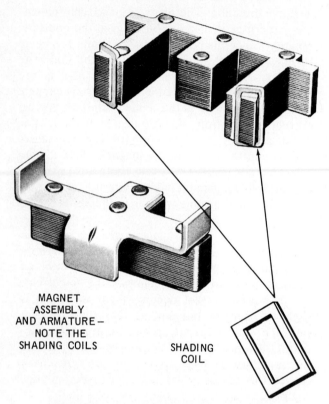

FIGURE 6-11b Location of the shading coil. (Courtesy of Square D Company.)

stationary part of the component, called the *magnet assembly*. The movable part of this assembly is a flat section of laminated material called the *armature*. When the coil is energized, the magnet assembly builds a strong magnetic field that draws the armature tight to the laminated legs. The middle leg of the E section is slightly shorter than the two outside legs. This provides a small air gap between the middle leg and the armature.

The air gap serves a special purpose. When the magnetic field is deenergized a small amount of residual magnetism always remains. This residual magnetism is strong enough to keep the armature attracted to the magnet assembly. The air gap in the middle leg of the assembly breaks up any residual magnetism and allows the armature to fall free any time the magnetic field is deenergized.

Another change to the coil assembly was required because the rise and fall of the magnetic field caused the armature to chatter or vibrate. The armature would start to drop away from the magnet assembly during every cycle when the ac current would return to zero at the 180-degree mark. During this time, the magnetic field would also decrease to zero and the magnetic pull would not be strong enough to hold the armature. When the current began to increase again, the magnetic field would become strong enough to pull the armature tight again. The action of the armature falling away from the magnet assembly and then being pulled tight again caused the relay to vibrate.

One way to cure the vibration was to insert a shading coil around the two outside legs of the laminated section. Figure 6-11 shows the exact placement of this coil. The shading coil is a single turn of conducting material such as copper or aluminum that is placed directly on the end of the magnet assembly. The shading coil is magnetized by induced current from the laminated section. Since the magnetism is induced it will be 90 degrees out of phase with the flux field in the laminated sections. The net effect of this arrangement is that the current in the shading coil is out of phase with the current in the magnet's coil. This creates a magnetic pull that is out of phase with the coil's magnetic pull. This means that as the coil's magnetic pull diminishes to zero, the magnetic pull created by shading coil is reaching its peak. The small amount of magnetic pull created by the shading coil is strong enough to keep the armature secured to the laminated sections and prevents the relay from vibrating. The shading coil also prevents the coil from becoming overheated from the vibrations.

Types of Magnet Frame and Armature Assemblies

Several types of magnet frame and armature assemblies are used to operate relay contacts. Examples of each of these are shown in Figure 6-12. The first type of assembly is called a *clapper*. This type of assembly has the coil mounted on a stationary laminated magnet assembly. The armature is hinged at one end. A small spring is attached to the tail of the armature to keep the contacts in the open position. When the coil is energized, the laminated sections become magnetized and pull the armature tight to it. The motion of the armature causes the contacts to be actuated to the closed position.

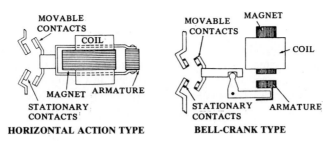

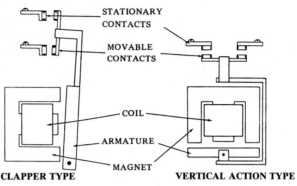

FIGURE 6-12 Types of armature assemblies. (Courtesy of Square D Company.)

When the coil is energized, its magnetic pull is strong enough to overcome the spring tension. But when the coil is deenergized, the return spring pulls the armature away from the magnet assembly. This type of assembly can also be mounted so that gravity will cause the contacts to return the contacts to the open position.

The second diagram shows a *vertical action assembly*. This type of actuator keeps the magnet assembly stationary and the armature is allowed to move in an upward motion when the magnetic field is formed. The armature has one half of the contact assembly mounted directly to it so that when it is moved by the pull of the magnetic field, the contacts are forced closed. When the coil is deenergized, gravity will allow the armature to drop and open the contacts. Single and double sets of contacts can be actuated by this type of assembly.

The third type of assembly is called a *horizontal action assembly*. This assembly has the laminations and coil mounted permanently. The armature is mounted in through the coil in the horizontal plane and the contacts are mounted at one end. The other end of the armature is shaped like the letter T. A return spring is located in the coil where the armature will compress it when the coil is magnetized. When the coil is energized, the T end of the armature is attracted to the laminated coil. This motion causes the armature to move in the horizontal direction and push the movable contacts close against the stationary mounted contacts. When armature is moved to the point where it causes the contacts to close, it also compresses the return spring. When the coil is deenergized, the magnetic field collapses and the return spring forces the armature back to its original position, where it is slightly out of the coil. This action causes the contacts to open. This type of relay is position sensitive and should be mounted so that the armature will operate in the horizontal direction.

The last type of armature assembly shown in Figure 6-12 is the *bell-crank assembly*. The coil assembly is mounted so that it is stationary. One half of the contact set is also mounted so that it will not move. The armature assembly is mounted directly below the coil assembly so that it will be pulled up against it when the coil is energized. The bell-crank assembly consists of a ball and socket. The ball is on the end of an L-shaped lever, and the socket is part of the contact carrier. The L-shaped lever is mounted on a pin so that when the armature is pulled against the coil assembly in the vertical direction, the ball moves the contact carrier in the horizontal direction. This action causes the movable contacts to be pressed against the stationary contacts to complete the electrical circuit. When the coil is deenergized, the weight of the armature assembly causes it to fall away from the coil. When this occurs, the lever pulls the movable contacts away from the stationary contacts and opens the

Relays 125

circuit. This type of assembly is also used where horizontal movement is required, such as the starter motor used in your car. When the starter is energized, the bell-crank mechanism engages the starter motor drive gear against the gear mounted on the flywheel. When the starter motor begins to turn, the flywheel is also turned. When the starter motor is deenergized, the bell-crank pulls the starter motor gear away from the flywheel. This allows the gears to be engaged when the starter motor is turning and to be disengaged when the engine is running.

This type of assembly is used in relays to help eliminate contact bounce when the contacts are closed. When the contacts are mounted directly to the armature, they are subject to bouncing when the coil is first energized and the armature is pulled into place. When the contacts are mounted on the end of the bell-crank assembly, the contacts are isolated enough to prevent bounce.

Relay Coil Specifications

The coils used on solenoids, relays, and contactors have specific electrical requirements that must be met for proper performance. These specifications are generally listed in manufacturers' literature so that you can select the correct coil for your application (Figure 6-13). It is important to remember that relay coils are available to be energized by either ac or dc current. A variety of coils are also available to match existing control circuit voltages. Typical voltages include 6, 12, 24, 32, 48, 110–115, 120, 208, 220, 240, 277, 380, 440, 480, 550, and 600 V ac, and 6, 10, 12, 24, 32, 48, 120, and 240 V dc.

Other specifications are used to classify coils. These include the *inrush current* and *sealed current ratings*. *Inrush current* is the amount of current the coil will draw when it is first energized and *sealed current* is the amount of current the coil will require to maintain the magnetic field. This current is rated in volt-amperes (VA). Inrush current tends to be 5 to 10 times the amount of current needed to maintain the magnetic field. Inrush current is larger than sealed current for several reasons. First, the armature is not inside the coil initially, so the coil impedance is minimum, which causes current to be maximum. Another reason that inrush current is large is because the armature is at rest when current is first applied to the coil. The coil requires a maximum amount of current to begin to get the armature to move. Once the armature begins to move, the amount of energy required to keep it moving decreases. This occurs as the armature moves into the coil and changes the coil's impedance.

The combination of these two factors allows inrush current to subside to the minimum level required to maintain the magnetic field. A typical coil may be rated for 800 VA at inrush and 80 VA when it is sealed. The VA may be developed by a 120-V coil drawing 6.66 A at inrush and 0.66 A when sealed. If the coil voltage were 480 V, the inrush current would be 1.66 A and the seal current would be 0.166 A. Each of these coils would develop the same amount of magnetic force required to move the armature.

Seal-in voltage is also specified in some manufacturers' literature. This specification is important for armature assemblies that use gravity to return the armature to its deenergized position. You must remember that gravity is also pulling on the armature even when the coil is energized. This means that a minimum amount of voltage is required to keep the armature tightly seated against the pole pieces. If the seal-in voltage drops too low, the armature is allowed to drop away from the pole faces of the magnet assembly. When this occurs the magnetic field is weakened and the contacts are allowed to draw back slightly from their closed position, where they will begin to arc and pit. In these application where the armature is moving in the vertical direction, the seal-in voltage must be greater than the pickup voltage. Usually, this is accomplished by sizing the coil correctly for the circuit voltage. In applications where the armature moves in the horizontal direction, the seal-in voltage is not a real concern.

Another specification that is usually listed for coils is the pickup voltage. *Pickup* voltage is the minimum amount of voltage required to pull the armature from its deenergized position. This value is important when the voltage on the factory floor drops significantly as motors are started. It is also important when the power distribution system on the factory floor is at its maximum or when it is slightly overloaded. These conditions cause the supply voltage to be lower than it normally would be. In some areas the voltage can also be lowered during a condition known as a brownout. When the voltage drops significantly, it may not be strong enough to provide the amount of VA required to pull the armature. If this is a problem in your applications, it is recommended that the next smaller size of coil be used. This means that a 208-V coil could be used on a 220-V system, or a 110-V coil could be used on a 125-V system. Substitutions should not be considered where the next smaller coil would cause more than a 20 percent overvoltage.

Another specification that is important to magnetic coils is the dropout voltage. The *dropout*

REPLACEMENT AC MAGNET COILS

Equipment To Be Serviced					Coil Prefix or Class and Type	Hz	24 Volts	110-115 Volts	120 Volts	208 Volts	220 Volts	240 Volts	277 Volts	380 Volts	440 Volts	480 Volts	550 Volts	600 Volts	Coil Volt-Amperes In-rush	Sealed	Price
Device	Size	Type	Poles																		
	20 Amp.	L	2-6		31071-400	60/50	23/24	⋯/44	44/45	50/52	△/53	53/54	55/56	⋯/60	⋯/62	62/63	⋯/65	65/66	150/140	30/30	
		L	8-12		31071-408	60/50	23/24	⋯/44	44/45	50/52	△/53	53/54	55/56	⋯	⋯/62	62/63	⋯/65	65/66	180/170	35/35	
		LL	All		31071-411	60/50	23/24	⋯/44	44/45	50/51	△/53	53/54	55/56	⋯	⋯	⋯	⋯	⋯	150/140		
		LX	2-4		31071-400	60/50	23/24	⋯/44	44/45	50/52	⋯/53	53/54	55/56	⋯/60	⋯/62	62/63	⋯/65	65/66	150/140		
		LX	6-12		31071-408	60/50	23/24	⋯/44	44/45	50/52	⋯/53	53/54	55/56	⋯	⋯/62	62/63	⋯/65	65/66	180/170		
Coils for Present Design Magnetic Contactors and Starters		PC 3	All		9998PC3C	60/50	24C/24B	110A	110A	208C	220A	220A	277C	380B	440A	440A	⋯	600C	23/23	4.9/4.9	
		PD, PE (Series B)	All		9998PD2C	60/50	24C/24B	110A	110A	208C	220A	220A	277C	380B	440A	440A	⋯/65	600C	50/50	7.3/7.3	
		PF	All		9998PF1C	60/50	24C/24B	110A	110A	208C	220A	220A	277C	380B	440A	440A	⋯	600C	224/216	16.6/18.9	
		PG	All		9998PG1C	60/50	24C	110A	110A	208C	220A	220A	277C	380B	440A	440A	⋯	600C	394/381	27.2/30.5	
		PJ	All		9998PJ1C	60/50	⋯	110A	110A	208C	220A	220A	277C	380B	440A	440A	⋯	600C	700/635	45/48	
		PK	All		9998PK1C	60/50	⋯	110A	110A	208C	220A	220A	277C	380B	440A	440A	⋯	600C	1262/1233	54.7/58.1	
	00	SA▲ (Series B)	All		9998SAC-	60/50	23	△/45	45	52	△/54	54	55	59	⋯/62	62	△/65	65	165	33	
					9998SAD-	60	⋯	Dual Voltage Coils: 115/230V—01						120/240V—02		240/480V—04			165	33	
	0, 1, 1P & 30 Amp.	SB, SC & SM	All		31041-400	60/50	20/22	△/42	42/43	48	△/51	51/53	52	56/57	△/60	60/61	△/62	62/64	245/232	27/26	
					31041-402	60/50		Dual Voltage Coils: 115/230 50 Hz.						120/240—02 60 Hz.		240/480—04 60 Hz			245/232	27/26	
	2 & 60 Amp.	SD & SP	2 & 3		31063-409	60/50	16/17	△/38	38/39	44	△/47	47/48	49	53/54	△/57	57/58	△/60	60/61	311/296	37/36	
			2 & 3		31063-411	60	16/17	Dual Voltage Coils: 120/240—02						240/480V—04					311	37	
			4 & 5		31063-400	60/50	16/17	△/38	38/39	44	△/47	47/48	49	53/54	△/57	57/58	△/60	60/61	438/429	38/37	
	3 & 100 Amp.	SE & SQ	2 & 3		31074-400	60/50	16/17	△/38	38/39	44	△/47	47/48	49	53/54	△/57	57/58	△/60	60/61	700/678	46/47	
			4 & 5		31091-400	60/50	16/17	△/38	38/39	44	△/47	47/48	49	53/54	△/57	57/58	△/60	60/61	1185/1260	85/89	
	4 & 200 Amp.	SF & SV	All		31091-400	60/50	16/17	△/38	38/39	44	△/47	47/48	49	53/54	△/57	57/58	△/60	60/61	1185/1260	85/89	
	5 & 300 Amp.	SG & SX	All		31096-400	60/50	⋯	△/09	09/10	15	△/18	18	19	21/22	△/24	24	△/29	29/30	2970/2970	212/250	
	6 & 7	SH & SJ																			
	400, 600 & 800 Amp.	SY, SZ, SJ (Elect. Held)	2-3			60/50		△/09	09	15	△/18	18	19		△/24	24	△/29	29	1530/1250		
		SY, SZ, SJ (Mech. Held)	2-3		31104-418				Coil Part Number 31104-400-50 (All System Voltages)												

△ Use next higher voltage, 60 hertz coil.
▲ Use on Type S Series B devices only.
NOTE: Coil Selection Table for Obsolete Design Magnetic Contactors and Starters is listed on Page 719.

FIGURE 6-13 Coil specifications. (Courtesy of Square D Company.)

voltage is the level that applied voltage must be reduced to before the armature is released from the coil's magnetic pull. This level is typically 5 to 10 percent lower than seal-in voltage. This value is very important in areas that are subject to voltage fluxuations or where low-voltage conditions may occur. If the voltage drops below this level, even for one or two cycles, the armature will begin to drop away from the coil and the contacts will begin to open. If this occurs frequently, the relay will chatter or vibrate and the contacts will wear out prematurely. If this is a problem with a relay installation, it can usually be remedied by selecting a coil that has a slightly lower voltage rating for pull-in and dropout. If the problem is critical in an application where multiple relays are mounted in several panels, an alternative would be to use a multitap transformer or autotransformer to raise the control voltage slightly.

Magnetic Coil Variations

Since coils for solenoids, relays, contactors, and motor starters are manufactured in a large variety of voltages, styles, and shapes, a set of rigid standards has been written by the National Electrical Manufacturers Association (NEMA). These standards deal with the specifications discussed previously and the amount of variance in voltage that is permissible. These standards require that a coil be capable of operating safely in environments where the voltage is up to 110 percent of the rated voltage, or as low as 85 percent of the rated voltage. These standards ensure that coils will withstand temperature rises that occur when the voltage is 10 percent too high and that the armature will be picked up and sealed in when voltages are 15 percent lower than the rating.

If the control circuit voltage is too high, the coil will overheat and cause the coil's insulation to break down. This will eventually lead to a short circuit in the coil, which will destroy it. The heat buildup is a direct effect of increasing VA. The increase in VA can be calculated by multiplying the applied voltage by the impedance of the coil. Since the coil's impedance is rather constant when the armature is pulled inside the coil, any increase in voltage will cause a corresponding increase in the coil's temperature. It is important to remember that the temperature begins to build from the inside of the coil to its outside. This makes it difficult for the coil to radiate excessive heat and also means that the coil has been overheated for some time by the time you notice the temperature rise by feeling the outside of the coil.

Another problem regarding the coil receiving excessive voltage is that the armature will be pull into the coil with a much greater force. This causes the movable contacts to be slammed against the stationary contacts with too much force. When this occurs repeatedly, the contact assembly will become warped and damaged and the contacts will no longer seal correctly. This action also causes the contacts to wear prematurely.

Relay Contacts

Contacts are the part of the relay that act like a switch. The coil causes the contacts to move from the open to the closed position, or from the closed to the open position. When the contacts close, the electrical circuit is completed and the load is energized. Since the coil circuit determines when the contacts are energized, it is called the *control circuit*. The contacts complete the circuit to the load, so this circuit is called the *load circuit*.

The contacts for relays and contactors can be normally open or normally closed. The word "normally" in this case indicates the position of the contacts when no power is applied to the coil. The contact's position when the coil is deenergized may be determined by either gravity or a return spring. Figure 6-14 shows the electrical symbol for both normally open and normally closed contacts. An operational sketch is also shown to help you see the difference between the normally open and normally closed contacts. The normally open contacts are held open by a return spring. When the coil is

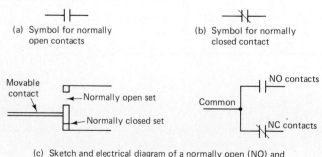

FIGURE 6-14 Electrical symbol and operational sketch for normally open and normally closed relay contacts.

energized, the magnetic field pulls the relay armature and causes the contacts to close.

In the relay with normally closed contacts, the energized coil pulls the armature and opens the set of contacts. When the coil is deenergized, the return spring pulls the contacts to their closed position. Since the contacts return to their normal position when power is removed from the coil, it is also referred to as the *fail-safe position*. This means that the contacts will return to the normal position any time power is lost in the control circuit. This is important in some circuits where the control circuit is powered from a different source from the load circuit. Generally, normally open or normally closed contacts are selected to make the circuit logic function properly.

Contacts are identified by the number of poles and the number of throws. The number of contact poles refer to the number of sets of contacts that are controlled by one operator. The operator in this sense refers to the armature assembly that moves and causes the contacts to be activated. The term *throw* indicates the number of circuits that each set of contacts can control. Figure 6-15 shows an example of a single-pole, single-throw contact and a single-pole, double-throw contact. Notice that the single-throw contact has only one circuit. The contact can either energize or deenergize the motor in that circuit. The second diagram in this figure shows an example of a double-pole contact. In this arrangement the movable contact is normally closed with the top permanent contact and is normally open with the bottom permanent contact. When the coil is deenergized, the motor in the normally closed circuit will be energized and running. The motor connected to the normally open set of contacts is deenergized at this time.

When the coil is energized by the selector switch, the movable contact changes position and closes the lower circuit and opens the upper circuit. This causes the conveyor's A motor to deenergize and the conveyor's B motor to energize. In this type of relay, one common contact is used for both the normally open and the normally closed circuits. The diagram for this type of contact is also shown in the figure. The contacts are identified as NC for normally closed, NO for normally open, and C for common. Even though these are single-throw, double-throw contacts, they are generally known as NC, NO, and common. Notice that the contacts are also identified by abbreviations. These abbreviations, commonly used to identify relay contacts, are shown with the diagrams.

Another set of terms is used to discribe contact types. The terms *single break* and *double break* refer to the type of contact set used. The single-break contact has only one movable contact point. When the contact opens or closes, only one point is involved. This means that current is interrupted at only one point in the circuit. The double-break set of contacts uses contact points to complete the circuit. When the contacts are opened, the circuit is broken at two points, which spreads the arc caused by the contacts opening across two contacts rather than one. The contacts for the double-break set form a bridge, which also provides the contacts with more surface area, which tends to keep them cooler during repeated operation.

You should remember from earlier switch contacts that the term *break* refers to the time when the contacts are opening. Some manufacturers refer to the time when contacts are closing rather than when they are opening. In this instance the contacts are called *single make* and *double make*.

Diagrams for Relay Contacts. Relay contacts can be shown in several different ways. They can be shown as a wiring diagram, showing the basic parts in relation to where they are located on the component, or as an elementary diagram, which

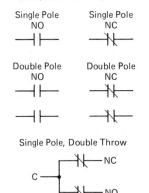

SUPPLEMENTARY CONTACT SYMBOLS

	SPST, N.O.		SPST N.C.		SPDT		TERMS
	SINGLE BREAK	DOUBLE BREAK	SINGLE BREAK	DOUBLE BREAK	SINGLE BREAK	DOUBLE BREAK	SPST - SINGLE POLE SINGLE THROW
							SPDT - SINGLE POLE DOUBLE THROW
	DPST, 2 N.O.		DPST, 2 N.C.		DPDT		DPST - DOUBLE POLE SINGLE THROW
	SINGLE BREAK	DOUBLE BREAK	SINGLE BREAK	DOUBLE BREAK	SINGLE BREAK	DOUBLE BREAK	DPDT - DOUBLE POLE DOUBLE THROW
							N.O. - NORMALLY OPEN
							N.C. - NORMALLY CLOSED

FIGURE 6-15 Identification of sets of contacts with common abbreviation. (Courtesy of Square D Company.)

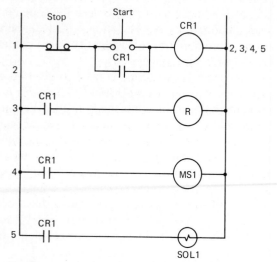

FIGURE 6-16 Wiring diagram and elementary (ladder) diagram of a four-pole relay.

is the way they would show up on a ladder diagram or logic diagram. When components are shown in a ladder diagram, the contacts and coil will be shown in respect to the circuit in which they are used. For example, the coil will generally be shown in the control part of the circuit and the contacts are usually shown in the load diagram. Other diagrams may be shown to explain specific operations of the relay, or to explain product differences. These diagrams are usually developed by the manufacturer and may not conform to any standard.

Figure 6-16a shows an example of a four-pole relay shown as a wiring diagram. This diagram shows the coil and contacts in the exact location where they would be found on the relay. Notice that the two contacts are shown on the right side of the coil and two contacts are shown on the left side of the coil. This diagram also shows that each set of contacts has its terminal identified as line side at the top and load side at the bottom. The line side is identified by the letter T for "terminal" and the load side is identified by the letter L for "load." The first set of contacts is marked as L1 and T1 and the second set is marked L2 and T2. Each of the other two sets are identified in a similar fashion to indicate they are set 3 and 4. This diagram also shows the coil terminals, but does not differentiate between the two terminals. This diagram also shows the location of these terminals at the top and bottom of the relay. When you look at the component and need to locate these terminals, they will be located exactly where they are shown in this diagram. In other words, this diagram is similar to a picture of the relay.

Figure 6-16b shows this relay in an elementary diagram. This diagram is also called a ladder diagram since each of the circuits in this diagram is connected to the two outside rails, which gives the diagram the overall appearance of a wooden ladder. The two outside rails represent the two lines of supply voltage. They are identified as L1 and N for line 1 and neutral. The amount of voltage is also marked to indicate that the power source for this circuit is 120 V ac. If the supply voltage for this circuit were dc or a different amount of ac voltage, it would be indicated at the top of the diagram. Each line on the diagram is also called a *rung* since it gives the appearance of the rungs of a ladder.

This type of diagram sometimes confuses students who are trying to read diagrams for the first time. In the ladder diagram all the components are located in the diagram in respect to the total operation of the circuit. This means that the diagram is laid out from left to right and top to bottom to show the sequence of operation.

This diagram is used for troubleshooting since it indicates the sequence of operation. This means that the coil may be on one rung of the diagram and the contacts they control located on several other rungs. In this diagram you must remember that the contacts are shown on the rung of the diagram where they occur in the machine's operation. If a set of contacts is near the top of a diagram, they probably occur early in the sequence of operation. If the contact set is near the bottom of the diagram, they probably occur later in the sequence of operation. This is sometimes confusing since a set of contacts may be shown several rungs away from the coil, and the contacts are always shown close to the coil in a wiring diagram. This creates the illusion that the contacts are too far from the coil to be operated by the magnetic field. *Remember: the ladder diagram is drawn to indicate sequence of operation. When the coil is energized, all its contacts are operated, regardless of which rung they are shown in.*

Since a diagram may contain more than one relay, it is important to identify each set of contacts with respect to which coil operates it. Notice that this control relay is identified as CR1 and each of its set of contacts is also identified as CR1, which indicates that coil CR1 controls the operation of these contacts. If additional relays were shown in the diagram, they would be identified as CR2, CR3, and so on. The contacts that are controlled by each of these relays would also carry the same identification. This means that when you see a set of contacts in a ladder diagram, you must check its identification to determine which coil controls it. In larger diagrams, the number of rungs may require several pages or blueprint sheets to show the complete diagram. In these diagrams, each rung of the diagram is numbered. When a coil is shown in the diagram, each line where its contacts are used is indicated as a number beside the coil. This allows the troubleshooting technician to determine the number of contacts the coil controls. This also shows other parts of the machine sequence that is controlled by this relay coil.

The contacts may also be identified with a rung number that is marked just below the contact symbol. The number indicates the rung where this contact's coil is located in the diagram. The contact is also identified with the relay identifier such as CR1 at the top of the contact symbol. This provides the troubleshooter with two ways of determining which coil operates this set of contacts and on which line of the diagram the coil can be located. Figure 6-17 shows several commonly used relays with typical combinations of normally open and normally closed sets of contacts.

The ladder diagram is also the design used for programmable controllers. The display on the programmable controller terminal will show relay contacts similar to the ladder diagram shown in this figure. In Chapter 16 we explain the similarities between hard-wired relays and contacts and coils in the programmable controller. Some manufacturers call the contacts and coils programmed in their programmable controller relay, *ladder logic*.

Convertible Contacts

Figure 6-18 shows contacts that can be converted in the field for use as normally open or normally closed contacts. This feature allows one type of relay to be stocked as replacement parts on the factory floor. If the relay that requires replacement has several sets of normally open contacts and several sets of normally closed contacts, the repair technician can adjust the contacts at the time of installation. From the diagram you can see that the return spring can be under the contacts to make the set operate as a normally open set, or the contacts can be turned over and the return spring can be placed on top of them so that they will operate as a normally closed set.

Contact Ratings

Contacts have ratings that can be used so that they can be matched for specific applications. Each relay or contactor can operate any type of contact. This means that the same relay type may operate contacts with a different current rating. These ratings are also useful in selecting a relay for a specific application, or when a relay requires replacement.

Figure 6-19a shows a typical specification sheet offered by manufacturers for selecting relays. Figure 6-19(b) (labelled A-G) shows examples of these relays. The main ratings for contacts involve the size of the electrical load the contacts can safely control. Since the load can be rated in amperes or horsepower, both ratings are usually provided. The

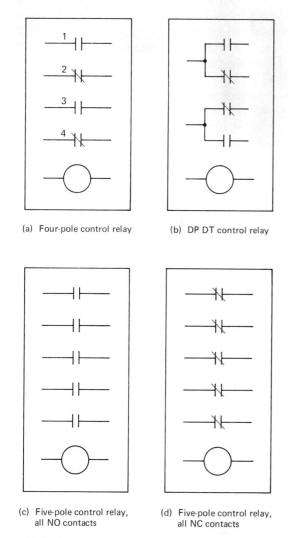

(a) Four-pole control relay
(b) DP DT control relay
(c) Five-pole control relay, all NO contacts
(d) Five-pole control relay, all NC contacts

FIGURE 6-17 Diagram of relay contacts commonly used for motor control circuits.

Relays 131

ADDING or CONVERTING CONTACT CARTRIDGES HAVING "SWINGAROUND" TERMINALS

General Instructions (Specific cases below.)

1.1 Adding a contact cartridge:

As received, accessory cartridges are in the normally open mode with terminal screws adjacent to N.O. symbols. If normally closed mode is desired, convert contact as indicated in Step 1.2 below. When cartridges are inserted, the terminal screws must face the front. The clear cover may face either side. **Do not install more than 8 N.C. contacts per relay.** When installing one cartridge, locate it at an inner pole position. When installing 2 cartridges, locate both in inner or outer (balanced) positions.

1.2 Converting a contact to its alternate mode (N.O. ⇌ N.C.):

Withdraw an assembled cartridge for replacement or conversion by inserting the blade of a suitably-sized screwdriver under a terminal screw pressure plate. Slide cartridge out. See Figure 2. Back the terminal screws out of the cylindrical nuts a sufficient amount (approximately 2 turns for a fully-tightened screw) to permit rotation of each screw and nut assembly to its alternate position. See Figure 3.

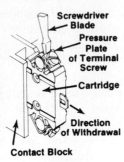

FIGURE 2

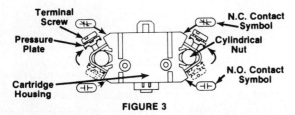

FIGURE 3

FIGURE 6-18 Reversible contacts. (Courtesy of Allen-Bradley, Inc.)

other important rating provided for the contact is the maximum load voltage.

The maximum voltage rating should not be exceeded. This rating is provided for both ac and dc application. The maximum voltage rating is determined by the design of the relay or contactor. When multiple contacts are used for three-phase applications, you should remember that each of the three lines may be the return for the others. This means that there is always a potential difference between any two lines of the three-phase that is equal to twice their individual voltage. This large potential difference between the phases tends to draw arcs between the contacts. The relay and contactors are designed to prevent this arcing between the contacts. If the maximum allowable voltage is exceeded, arcs may be drawn from one set of contacts to another when the contacts are opening and closing.

Some manufacturers list the voltage rating in volts per pole. This means that the contacts are rated by the amount of voltage each pole will control. If the contacts are used in a single-phase 115- or 230-V system, each pole will control 115 V with respect to neutral. When the contacts are used in a three-phase 230-V system, each contact will control 230 V. Since the potential between contacts on the 115-V system is smaller, these contacts can be used to control a larger load than when the load operates on a higher voltage. Remember that some motors may be capable of operating on either one of two voltages. These motors are known as dual-voltage motors. When a 1-hp motor is operated on 115 V, it will draw 7.2 A. When the same motor is operated on 230 V, the motor will draw 3.6 A. The difference in the amount of current draw will allow the same relay or contactor to be used in both applications.

Another problem that occurs when the maximum voltage rating is exceeded is premature contact wear. The excessive voltage causes the contacts to pit more easily and wear the special coating off the contacts. Notice that the load for the contacts must be derated if the relay or contactor is used to jog or plug a motor. These conditions require the motor to be turned on and off frequently or require reverse-polarity voltage to be applied directly to the motor to stop it. During these applications the contacts tend to overheat.

Another rating listed in this specification sheet is the maximum current rating or horsepower rating for the contact. The current ratings of the contacts actually determine their size. The larger the current rating, the larger the contacts must be. The current rating is also listed for inductive and noninductive loads. You should notice that the same-size contacts can be used for a larger noninductive load than when they are used to control inductive loads. This accounts for the large inrush current that inductive loads must handle for a few cycles when a motor is started. It also provides extra capability for the inductive spike that occurs when inductive loads are deenergized. Sometimes the contact's current rating is listed as full-load amperes (FLA) or locked rotor amperes (LRA). This indicates that the contacts are rated for an inductive load.

Noninductive loads are primarily resistive in nature. These types of loads usually consist of heating elements and lighting fixtures. When these loads are energized and deenergized, the current remains rather constant. This allows the contact's noninductive load rating to be slightly larger than its inductive load rating.

Other ratings that are listed on some specification sheets are the make and break currents. When the contacts close (make) the inrush current may be 5 to 10 times the normal operating current. When the contacts open (break) the inductive spike may cause current to rise to two to three times the full-load current for several cycles. The make and

Contact Ratings

Maximum AC Contact Rating Per Pole 50 or 60 Hertz					
NEMA Rating Designation	Maximum Voltage	Amperes		Continuous Carrying Current (Amperes)	Voltamperes	
		Make	Break		Make	Break
A600 A300	120	60	6	10	7200	720
A600 A300	240	30	3	10	7200	720
A600 —	480	15	1.5	10	7200	720
A600 —	600	12	1.2	10	7200	720
— B300	120	30	3	5	3600	360
— B300	240	15	1.5	5	3600	360

Maximum DC Contact Rating Per Pole					
NEMA Rating Designation	Maximum Voltage	Amperes		Continuous Carrying Current (Amperes)	Voltamperes	
		Make	Break		Make	Break
P300	125	1.1		5	138	
	250	0.55		5	138	

Coil Current Data

Type of Relay	60 Hertz		50 Hertz		25 Hertz	
	Sealed VA	Inrush VA	Sealed VA	Inrush VA	Sealed VA	Inrush VA
BR	32	135	34	130	20	55
BRM	0	90 (Latch) 25 (Unlatch)	0	81 (Latch) 30 (Unlatch)	—	—
BX	20	109	20	96	12	39
BXL	20	109	20	96	12	39
C — 2 Pole	12	35	12	33	7	12
C — 4 Pole	15	60	14	58	12	29
CL — 2 Pole	6	17	6	16	4	8
CL — 4 Pole	10	31	10	31	9	20
D	25	102	25	94	14	33

Type of Relay	60 Hertz		50 Hertz		25 Hertz	
	Sealed VA	Inrush VA	Sealed VA	Inrush VA	Sealed VA	Inrush VA
N — Bulletin 700	23	134	24	120	—	—
NM — Bulletin 700	0	148 (Latch) 19 (Unlatch)	0	157 (Latch) 18 (Unlatch)	—	—
R RM — Bulletin 700	6.5	—	6.5	—		

Type of Relay	Sealed VA	Inrush VA	Latch VA	Unlatch VA
N — Bulletin 700DC	9.75	145	—	—
NM — Bulletin 700DC	—	—	146	28
R RM — Bulletin 700DC	6.5	—	6.5	—

Standard Coil Voltages and Frequencies

Type of Relay	Standard Frequencies Hertz	Standard Voltages												
C	60	6	12	24	32	48	64	—	120	208	—	240	—	480	—	600
	50	6	12	24	32	48	64	110	—	208	220	—	440	—	550	—
BR BX BXL	60	6	12	24	32	48	64	—	120	208	—	240	—	480	—	600
	50	6	12	24	32	48	64	110	—	208	220	—	440	—	550	—
BRM	60	—	—	—	—	—	—	—	120	208	—	240	—	480	—	600
	50	—	—	—	—	—	—	—	—	208	220	—	440	—	550	—
CL D	60	6	12	24	32	48	64	—	120	208	—	240				
	50	6	12	24	32	48	64	110	—	208	220					
N NM ■ N with Pneumatic Timer N with Solid State Timer R RM	60	—	—	24	32	48	64	—	120	208	—	240				
	50	—	—	24	32	48	64	110	—	208	220	—				
R with Solid State Timer	DC	—	—	24	32	48	64	—	115-125	—	—	230-250	—	—	—	—

FIGURE 6-19 Specification sheet for selecting and sizing relays.

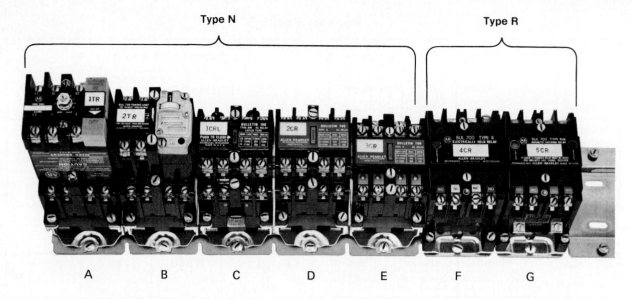

FIGURE 6-19b Various types of relays available for industrial applications. A, Add-on type solid state time delay relay; B, Add-on type pneumatic time delay relay; C, AC latching relay; D, AC standard relay; E, AC standard relay; F, Electrically held relay with sealed contacts; G, Magnetic latching relay with sealed contacts. These relays are mounted directly on a mounting strip. (Courtesy of Allen-Bradley.)

break ratings of contacts are not as critical for noninductive loads since they do not cause current surges.

Other ratings include the operating time for the contacts to make or break. Typical make times range from 4 to 20 milliseconds (ms), while time to drop out may take 6 to 15 ms. These times will be vitally important in high-speed switching applications, where the relay must open and close its contacts at high frequencies. These applications include interfaces with programmable controllers, robots, and other solid-state devices. Some solid-state motor controls require relays to interface them to other motor controls. These controls will require current ratings in the milliamperage range. The voltage requirement for these small current applications will be 5 V dc for TTL circuits and up to 12 V dc for CMOS devices.

Types of Relays

Relays are available with a variety of types of contacts for many types of applications. These types of contacts include fast-acting contacts for logic circuits and interface to solid-state controls or programmable controllers, power contacts for switching currents up to 30 A, sealed contacts such as reed contacts for power and logic applications, latched and unlatched contacts for logic applications and control circuits, and auxiliary contacts that can be added to a relay after initial installation.

A complete family of small relays is available for *fast-acting* applications. Figure 6-20 shows pictures and specifications for these types of specialty relays. Since they must operate at very high speeds, they generally have very small contacts with minimal current-carrying ability. The speed of operation for these types of relays is around 5 to 10 ms. From this figure you can see that this type of relay is usually used to interface standard motor controls to microprocessor-driven motor controls. It is also useful for interfacing controls from robots with controls of other systems, such as programmable controllers. This type of interface is generally required when the input and output signals of the robot switch (control) the negative voltage of the control device instead of the positive voltage. Since these relays are usually used in conjunction with other printed-circuit-board components, they also provide a means of being mounted into a printed circuit board, or on a socket that has been mounted on the circuit board.

From the specification sheet in this figure you can see that the relay is available in a large variety of coil voltages. These include dc from 3 to 48 V. These relays also provide a large variety of contact ratings, up to 1 A at 24 V dc and up to 0.3 A at 110 V ac for noninductive loads. Notice that other contact and coil specifications are also listed to provide you with all pertinent data for these relays. This information is necessary for selecting the proper relay for your application for initial installations and as replacement parts.

Another special type of relay is made for *high-current* applications. These relays can be sealed or open relays with a variety of contact

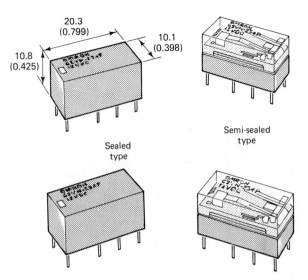

CONTACT RATINGS

Item	Type	G2VN-237P, G2VN-237PH, G2VN-234P, G2VN-234PH		G2VN-287P, G2VN-287PH, G2VN-284P, G2VN-284PH	
	Load	Resistive load (p.f. = 1)	Inductive load (p.f. = 0.4; L/R = 7 ms)	Resistive load (p.f. = 1)	Inductive load (p.f. = 0.4; L/R = 7 ms)
Rated load		110 VAC, 0.3 A 24 VDC, 1 A	110 VAC, 0.2 A 24 VDC, 0.3 A	110 VAC, 0.3 A 24 VDC, 1 A	110 VAC, 0.2 A 24 VDC, 0.3 A
Carry current		2 A (1 A)			
Max. operating voltage		125 VAC, 125 VDC			
Max. operating current		2 A (1 A)			
Max. switching capacity		60 VA, 30 W	20 VA, 10 W	60 VA, 30 W	20 VA, 10 W
Minimum permissible load (reference value)		100 mVDC 10 μA (10 mVDC 10 μA)		1 VDC 1 mA	

NOTE: The values in parentheses are of the sealed type.

COIL RATINGS

Classification	Item / Rated voltage	Rated current (mA)	Coil resistance (Ω)	Coil inductance (ref. value) (H)		Must operate voltage	Must dropout voltage	Maximum voltage	Power consumption (mW)
				Armature OFF	Armature ON	% of rated voltage			
Standard type	3 VDC	120	25	0.013	0.013	75 max.	10 min.	Approx. 130	Approx. 360
	5 VDC	72	69	0.054	0.054				
	6 VDC	60	100	0.09	0.09				
	9 VDC	40	225	0.27	0.28				
	12 VDC	30	400	0.57	0.6				
	24 VDC	15	1,600	3.7	4.3				
	48 VDC	7.5	6,400	17	19				
High-sensitivity type	3 VDC	50	60	0.096	0.083	75 max.	10 min.	Approx. 200	Approx. 150
	5 VDC	30	167	0.3	0.27				
	6 VDC	25	240	0.45	0.41				
	9 VDC	16.7	540	1.1	1.1				
	12 VDC	12.5	960	2.1	2.0				
	24 VDC	6.3	3,840	9.7	9.7				
	48 VDC	3.3	14,400	52	48				

NOTE: 1. The coil resistance is measured at a coil temperature of 23°C (73°F) with tolerances of ±10°.
2. The performance characteristic data are measured at a coil temperature of 23°C (73°F).

FIGURE 6-20 Fast-acting relays for printed-circuit-board installation. (Courtesy of Omron Electronic, Inc.)

arrangements. From Figure 6-21 you can see that the contacts are available to carry loads up to 20 A. Other models of this type of relay are capable of controlling up to 30 A. Notice that the voltage rating is listed as 250 V ac and 125 V dc maximum. When the higher voltages are used, the amount of current that can safely be controlled must be derated to 10 A ac or 7 A dc for inductive loads and 15 A ac or dc for resistive loads. This rating allows the relay to be used in a wide variety of applications that involve controlling larger currents such as fractional-horsepower motors and small electric heating

CONTACT RATINGS

Item \ Load	Resistive load (p.f.=1)	Inductive load (p.f.=0.4, L/R=7msec)
Rated Load	220 VAC 15A 24 VDC 15A	220 VAC 10A 24 VDC 7A
Carry current	20A	
Max. operating voltage	250 VAC, 125 VDC	—
Max. operating current	20A	—
Max. switching capacity	3,300VA 360W	2,200VA 170W
Minimum permissable load (reference value)	5 VDC 100mA	

COIL RATINGS

	Rated voltage (V)	Rated current (mA) 50Hz	Rated current (mA) 60Hz	Coil resistance (Ω)	Coil inductance (ref. value) (H) Armature OFF	Coil inductance (ref. value) (H) Armature ON	Must operate voltage	Must dropout voltage	Maximum voltage	Power consumption (VA, W)
AC	6 12 24 50 100 120	254 126.5 63 30.5 15 12.6	217 108 54 26 13 10.8				80 max.	30 min.	110	Approx. 1.3VA
DC	6 12 24 48 100 110	200 89 50 25 12 9		30 135 480 1,920 8,300 12,300				10 min.		Approx. 1.2W

NOTE: The rated current and coil resistance are measured at a coil temperature of 68°F with tolerances of ±15% for DC rated current and +15%, −20% for AC rated current, and ±15% for rated coil resistance.

CHARACTERISTICS

Contact resistance	30mΩ max.
Operate time	AC: 20msec max., DC: 20msec max.
Operating frequency	Mechanically: 1,800 operations/hour Under rated load: 1,800 operations/hour
Insulation resistance	100MΩ min. (at 500 VDC)
Dielectric strength	2,000 VAC, 50/60Hz for 1 minute (1,000 VAC between non-continuous contacts)
Vibration	Mechanical durability: 10 to 55Hz; .06" double amplitude Malfunction durability: 10 to 55Hz; .06" double amplitude
Shock	Mechanical durability: approx. 100G's Malfunction durability: approx. 20G's
Ambient temperature	Operating: +14 to +131°F
Humidity	45 to 85% RH
Service life	Mechanically: 10,000,000 operations min. (at operating frequency of 18,000 operations/hour) Electrically: See "CHARACTERISTIC DATA."
Weight	Approx. 1.55oz.

NOTE: The data shown above are of initial value.

FIGURE 6-21 Power relays. (Courtesy of Omron Electronic, Inc.)

elements. Sometimes these relays are used to energize large coils for hydraulic valves or large motor starters. Notice that the specification sheet also provides the approximate number of operations that this relay can perform in its life. This specification is important for determining the useful life of the relay. Other useful information is provided in the specification sheet, such as ambient temperature rating, humidity rating, weight, and size. This information is necessary when you must select the proper relay for the environment of the application. Size and weight information is useful where multiple relays must be housed in one electrical cabinet. This provides information required to select the cabinet size. Vibration and shock specifications are also useful in selecting the proper relay in an environment where excessive machine vibration is present. In some machines, such as stamping presses and plastic injection molding machines, the repetition of hydraulic cylinders pressing mold or die cavities together causes severe shocks and vibrations that may cause relay contacts to vibrate or malfunction.

Another type of relay is a *general-purpose relay*. Figure 6-22 shows a picture and specifications for the general-purpose relay. This type of relay can be used as a control relay or to control power to fractional-horsepower motors. This type of relay can control currents up to 10 A for ac and dc resistive loads and up to 6 A ac and 7.5 A dc for inductive loads. Notice that the contacts for these relays are also rated in watts or volt-amperes (VA). The VA rating is higher because it takes into consideration apparent power of inductive loads.

FIGURE 6-22 General-purpose relays. (Courtesy of Omron Electronic, Inc.)

The coil for this type of relay is also rated for control circuits powered by 6 to 240 V ac or 6 to 110 V dc. Minimum operating voltage, minimum voltage for drop out, and maximum applied voltage are listed to determine pull-in and seal-in voltages.

Some special-purpose applications require the coil to be latched once it has been energized. This type of application allows the coil to be energized with a pulse of current. When the coil energizes and moves the contacts to the energized position, a special mechanical latch holds the relay in the energized state even when coil voltage is interrupted. This relay has a second coil to deenergize the contacts. The coil used to deenergize the contacts is called the *reset coil* and the coil used to latch the relay is called the *set coil*. The reset coil unlatches the mechanical latch and allows the contacts to return to their normal position. The reset coil is not rated for continuous duty, so the signal to energize it should be of short duration.

This type of coil is said to have memory. This means that the pulse that is received by the latching coil energizes the contacts and and the latching mechanism. The coils stays in this position even though the coil pulse is removed.

Figure 6-23 shows a picture, diagrams, and specifications for this relay. It is very useful in interfacing circuits with solid-state components or in motion control circuits where the signal may exist for only a few seconds. Other applications include sequential events in machine control where a relay must be latched while the machine is in a particular step and then must be unlatched when the machine moves to the next step.

From the specifications shown in this figure you can see that the coil can be powered by either ac or dc current. You should also notice that the reset coil requires much less power to operate since it only needs to deenergize the latching mechanism. Once the latching mechanism is released, the return spring will pull the contacts back to their normal position.

Other specialized relays include relays specially designed as master control relays. This type of relay meets the National Electrical Code® (NEC) requirement for all control circuits that utilize programmable controllers or other microprocessor controllers. This type of relay is used to interrupt all power to the programmable controller inputs and outputs in the case of microprocessor failure. It also has relay contacts that are capable of controlling all of the current used in the input and output components in hard-wired circuits.

In either of these applications, the master control relay contacts are placed in series with both lines of the supply voltage. This allows one signal to

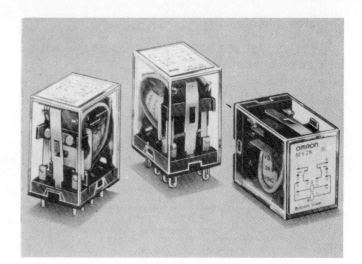

CONNECTING SOCKETS
Same as the Standard Type.

Terminal arrangement/Internal connections (bottom view)

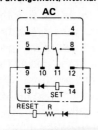
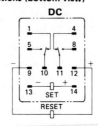

NOTES:
1. R is a resistor for ampere-turn compensation, and is incorporated in the relays rated at 50 VAC or above.
2. Pay attention to the polarity of the set and reset coils, as incorrect connection of positive and negative terminals will result in malfunctioning of the relay.

Contact Data

Type	MK1P-UA (SPDT) MK2P-UA (DPDT)		MK3P-UA (3PDT)		MK1EP-UA MK2EP-UA MK3EP-UA	
Load	Resistive load (p.f. =1)	Inductive load (p.f. =0.4: L/R=7msec)	Resistive load (p.f. =1)	Inductive load (p.f. =0.4: L/R=7msec)	Resistive load (p.f. =1)	Inductive load (p.f. =0.4: L/R=7msec)
Rated load	220 VAC 5A 24 VDC 3A	220 VAC 2A 24 VDC 2.5A	220 VAC 3A 24 VDC 2A	220 VAC 1.2A 24 VDC 1.5A	230 VAC 10A 28 VDC 10A	230 VAC 6A 28 VDC 7.5A
Contact material	Ag	Ag	Ag	Ag	AgCdO	AgCdO
Carry current	5A	5A	3A	3A	10A	10A
Max. operating voltage	500 VAC 250 VDC	500 VAC 250 VDC	500 VAC 250 VDC	500 VAC 250 VDC	500 VAC 250 VDC	500 VAC 250 VDC
Max. operating current	5A	5A	3A	3A	10A	10A
Max. switching capacity	1100VA 72W	440VA 60W	660VA 48W	260VA 35W	2300VA 280W	1400VA 210W
Minimum permissible load	1 VDC 10mA	1 VDC 10mA	1 VDC 10mA	1 VDC 10mA	5 VDC 100mA	5 VDC 100mA

Coil Data

	Rated voltage	Rated current (mA)		Coil resistance (Ω)	Coil inductance (ref. value) (H)		Pickup voltage	Dropout voltage	Maximum voltage	Power consumption (VA,W)
		50Hz	60Hz		Armature OFF	Armature ON		% of rated voltage		
AC	6	404	360	5.3	0.028	0.041	30% min.	80% max.	110% max.	Approx. 2.3 VA
	12	202	180	21.5	0.115	0.165				
	24	98	88	88	0.422	0.678				
	50	43.6	39	390	1.95	3.2				
	120	20.2	18	2,300	10.5	16.4				
	240	10.3	9.2	9,300	33.0	63.9				
DC	6	255		23.5	0.14	0.23	10% min.			Approx. 1.5 W
	12	126		95	0.56	0.87				
	24	56		430	2.82	4.46				
	48	29.5		1,630	10.99	16.52				
	110	11.5		6,800	36.2	54.4				

NOTE: The rated current, coil resistance and inductance are measured at a coil temperature of 23°C with tolerances of ±20%.

FIGURE 6-23 Picture and diagram of latching relays. (Courtesy of Omron Electronic, Inc.)

relay coil to deenergize all rungs of the logic circuit. This type of relay is also useful when designing a control system with zones or for automated systems that are so large that parts of the system need to be deenergized for maintenance while other sections must remain operational.

Another specialized relay is controlled by pneumatic or solid-state time-delay elements. The time-delay elements can be added to the relay by being mounted on the front of the contact operator. A typical pneumatic time-delay element is shown in Figure 6-24. From this figure you can see that the amount of time delay can be adjusted by setting the wheel on the front of the time-delay element. Adjusting the wheel changes the rate that air is allowed to escape from the bellows in the control. The relay's contact operator is pressing on the bellows, keeping the contacts from reaching full travel until all the air is removed. This type of time-delay element can be mounted on the relay to make it operate in the delay on or the delay off mode. In the delay on mode, the relay coil is energized and the contacts begin forcing air out of the bellows. Since the removal of all air takes time, the closure of the relay contacts can be delayed. When the delay module is mounted on the relay in the off delay mode, the contacts are closed immediately when the coil is energized. When the coil is deenergized, the return spring tries to open the contacts, but the time-delay element is full of air and slow bleeds down. While the bellows is loosing its air it prevents the contacts from opening, which creates the time-delay off.

The time-delay element can be mounted on these special relays, making them time-delay relays. This allows a relatively inexpensive relay to be converted to a time-delay device. It also allows the time-delay element to be mounted after the relay has been operating. Sometimes after a system has been installed and is being tested, it becomes apparent that a small time delay is required for the system to operate more efficiently or safely, so a time-delay element can be added.

The electrical symbol for the time-delay contacts is also shown in this figure. Notice that the delay contacts are shown with the time-delay symbol, which is an arrowhead pointing down for time delay off or the tail of an arrow pointing up, indicating time delay on. Since the time-delay module can be added to a relay with open or closed contacts, the symbols for on-delay-timed open, and off-delay-timed open are also shown. Timing diagrams and additional time-delay applications are provided in Chapter 9, where in-depth information about the various types of timers and their theory of operation is presented.

Installing the Relay

The installation of a relay involves two basic operations. The first operation involves mounting the relay to the panel board or location where it will operate. The second step involves using a field wiring diagram to connect the field wiring.

Relays may be mounted in electrical control cabinets and as stand-alone devices. The most used method of mounting is the flange mount. Figure 6-25 shows two basic types of sockets. This type of mounting utilizes one or two U-shaped slots on either side of the relay base or at the top and bottom of the relay base. Since a slot is used, one of the mounting screws can be installed prior to mounting the relay. After the relay has be mounted so that the slot slides under one of the mounting screws, the other screw can be installed. After both screws have been installed, they can be tightened to mount the relay securely. All field wiring connections are made directly to the terminals on the relay.

When multiple relays must be mounted into an electrical panel, a mounting rail can be used. The rail is mounted securely into the back of the electrical control panel. After the rail is mounted, each relay can be inserted onto the rail and secured in position with a mounting screw. The rail allows multiple relays to be mounted easily in the electrical panel so that the relay terminals are easily accessible during troubleshooting activities.

Other types of mounting involve sockets for the relay to be mounted into. Figure 6-25 shows two basic types of sockets that are usually used. The round base shown is also called the *tube base socket*. The "Tube base" refers to the octal base socket used widely for Vacuum tubes in early radios and televisions. The other type of socket is the *square base socket*. An example of square base socket is also shown in Figure 6-25. You can see from the diagram that the socket terminals are laid out in the shape of a square, which gives this socket type its name.

When a socket type of relay is mounted in a panel, a hold-down clip must be used to prevent the relay from coming loose. Several examples of clips are shown in Figure 6-26. The clips are secured at the base of the relay and snap over the top of the relay to hold it in place. To remove the clip, a screwdriver is inserted between the clip and the side of the relay and turned slightly.

The socket relay is very useful in high-production or automation applications where downtime must be limited. In these applications, faulty relays must be removed and replaced as quickly as possible. The socket relay can be pulled from its socket and replaced with a minimum of effort.

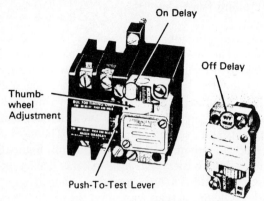

Pneumatic timer is supplied with two mounting screws for attaching timer to the Type N or Type NM relays.

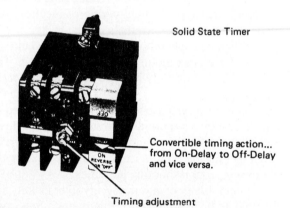

CONVERTING PNEUMATIC TIMING OPERATION — To
convert timing unit from "On-Delay" operation to "Off-Delay" operation or vice versa loosen two captive screws (A) on the die cast plate. Remove the die cast plate from the relay and rotate 180°. Replace the die cast plate in rotated position and tighten screws. Type of operation is indicated at the top of the die cast plate.

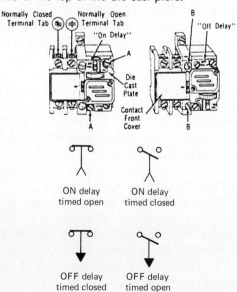

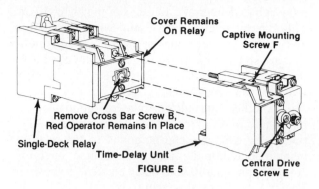

TIME DELAY UNIT

Note: Reset time — 75 milliseconds minimum.
A time delay unit may be added to a single deck relay only.

To Add a Time-Delay Unit to Single-Deck Relay:

1. On the relay, remove and discard screw B, Figure 5, from the center of the red operator. **Do not remove red operator or cover.**
2. Orient time-delay unit as in Figure 5.
3. Engage central drive screw E in hole of red operator, and thread this screw carefully through the red operator and into drive yoke within the relay while guiding the time-delay unit into keyed position. Tighten screw E until firmly seated (14-20 in.-lbs. torque).
4. Tighten the two mounting screws F securely (8-12 in.-lbs. torque).
5. Establish desired delay mode. Convert mode, if needed, as described in following section. Mode is shown at lower midpoint of timing assembly, Figure 6.

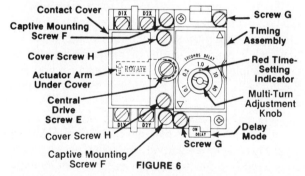

To Convert Delay Mode (On-Delay ⇌ Off-Delay):

1. Loosen the two captive slotted head screws G, Figure 6, identified by the symbol △, which secure the timing assembly.
2. Remove and invert timing assembly.
3. Reposition timing assembly and tighten screws G securely (8-12 in.-lbs. torque).
4. Verify delay mode marking.

To Convert Time-Delay Contacts (N.O. ⇌ N.C.):

1. Remove the contact cover by loosening the two captive screws H, Figure 6.
2. Rotate actuator arm 90° clockwise.
3. Convert contact as in Step 1.2.
4. Return actuator arm to original position.
5. Reassemble contact cover and tighten screws H securely (8-12 in.-lbs. torque).

To Set Time Delay (Timer must be mounted on relay):

1. Adjust to time setting desired using calibration of approximate scale. Full traverse of the red time-setting indicator, Figure 6, requires about 6 revolutions of the knob. Do not turn knob to move red indicator beyond end limits of scale.
2. Verify actual timing.
   On-Delay: Push in and hold central drive screw E.
   Off-Delay: Push in and release central drive screw E.
   Adjust if necessary.

FIGURE 6-24 Pneumatic time-delay elements added to a relay. (Courtesy of Allen-Bradley, Inc.)

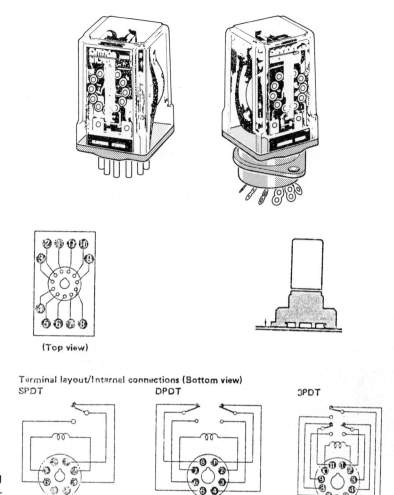

FIGURE 6-25 Socket mounting bases. (Courtesy of Omron Electronic, Inc.)

Another feature that allows the socket relay to be changed quickly is that all field wiring connections are made to the terminals on the socket base. This means that no field wiring connections need to be made when a socket relay is replaced. This feature also means that any field wiring connections only need to be made at the time of installation. The field wiring connections at the socket can be made at screw terminals or permanently soldered. The soldered connections require more time for installation, but provide added security in that they will not come loose as a screw might.

Several other types of terminal connections are also available for making field wiring connections easy when a relay must be replaced. A spade connector may be attached to the end of the field wire, and it can be pushed on the spade terminal on the relay. When the relay must be replaced, the spade terminals can quickly be removed from the relay terminals and returned to the new relay. This allows the relay to be exchanged rapidly and easily. Another type of terminal that is commonly used on a relay is a screw terminal. Some of these terminals allow a ring terminal or eye terminal to be placed under the screw and secured. The other type of screw terminal uses a saddle clamp where the wire is placed under the clamp and the screw is tightened until the clamp secures the wire tightly. After all field wiring connections have been made, the relay circuit should be tested for proper operation. If any of the circuits do not operate correctly after the initial installation, the troubleshooting procedures listed in this chapter should be used.

Interfacing Relays to Programmable Controllers and Robots

The relay can be used as an output driver or as an input signal for programmable controllers and robots. The signal that is sent and received by these controllers is usually less than 1 A and rather small dc voltages, in the range 5 to 24 V. Some robots also use the positive voltage line as the common and put all switching devices such as contacts and pilot

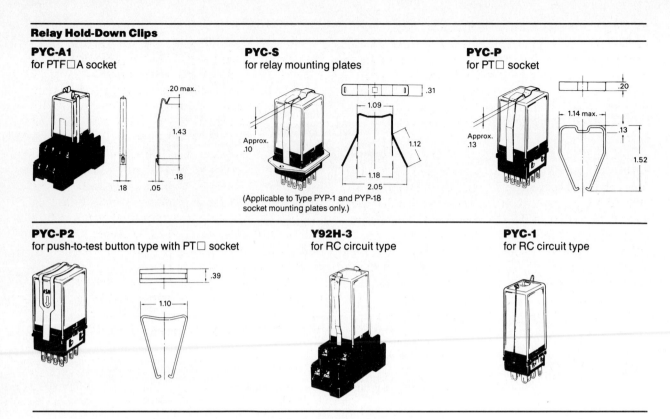

FIGURE 6-26 Types of hold-down clips for socket-mounted relays. (Courtesy of Omron Electronic, Inc.)

devices in the negative line. If other devices that must be interfaced to these robots use the negative line as the common, a relay must be used to isolate these two systems when they are interfaced. The small output signal is also too small to power motors or heating load directly, so a relay is very useful to interface programmable controller and robot outputs to larger loads.

Figure 6-27 shows an electrical diagram of several relays connected to the output bus of a robot and to ac and dc output modules from a programmable controller. In the diagram showing the relay coils connected to the programmable controller output modules, notice that only the positive line of the supply voltage is connected to the dc module and only L1 of the ac supply voltage is connected to the ac output module. The negative side of the dc and the neutral side of the ac is used as common for all the relay coils.

One side of the relay is connected to the output terminal screw and the other side is connected to common. When the programmable controller's logic energizes address 1, a small signal is sent to the output module and that address is energized. When the module is energized, the circuit at address 1 switches on and dc positive voltage is sent to that relay coil. The light-emitting diode (LED) behind the status indicator for address 1 is also energized, so that it glows. This allows the troubleshooting technician to determine from a distance if the output is energized.

If address 10 is energized, the programmable controller sends a small signal to address 10 on the ac output module and energizes it. When address 10 becomes energized, L1 voltage is sent to that relay coil and the status indicator is energized to show that output 10 has been powered.

If a status indicator is not glowing, that address is not energized. This is very useful in troubleshooting the system quickly. The electronic circuitry inside the output modules is explained in detail in Chapter 16. At this point the internal operation of the module is not important. Instead, you can assume that if the status indicator is energized, the relay coil should be receiving the proper amount of voltage. It is possible for the status indicator to be faulty and not be glowing even though power is being sent to the coil. This does not occur often and methods to troubleshoot this condition are presented in Chapter 16.

It is important to note that the relay's contacts are not shown in this diagram. The contacts are shown in the ladder diagram, where they are controlling the system's operation. The contacts will be identified with the same number that is listed on the coil that controls them. On some diagrams numbers will be listed to the right of each coil. These numbers indicate the ladder diagram rung numbers where the relay's contacts can be found. Some diagrams also use a "/" with the rung number to indicate that the contacts found on that rung will be normally closed.

The other diagram in this figure shows four relay coils connected to the output bus of a robot. The robot's output bus has room for 16 outputs, but only four are shown in this diagram. Notice that the output bus is powered by 24 V dc. In this system the negative side of the voltage is being switched on and off while the positive side of the power supply is used as the common. Each relay coil is connected to a specific address on the robot I/O bus (input/output bus). The other side of each relay is connected to the common side of the power supply. Since the common for this circuit is the positive side of the power supply, these wires may be connected to the common terminals on the robot bus or connected together and one wire jumpered to the positive terminal of the power supply.

An individual relay coil is energized when the robot microprocessor sends a signal to energize its address. When the I/O bus receives the signal, the negative side of the power supply is switched on that address and the relay coil becomes energized. Some robots provide small status indicators for each address, while other robots require the status to be displayed on an address screen as a 1 to indicate that the output is energized and as a 0 to indicate that the output address is deenergized. The output can also be tested with a voltmeter to determine if it is energized.

Since the robot I/O bus uses solid-state parts to switch the voltage, it is very sensitive to voltage surges or inductive spikes. It is important to remember that a relay coil emits an inductive spike of

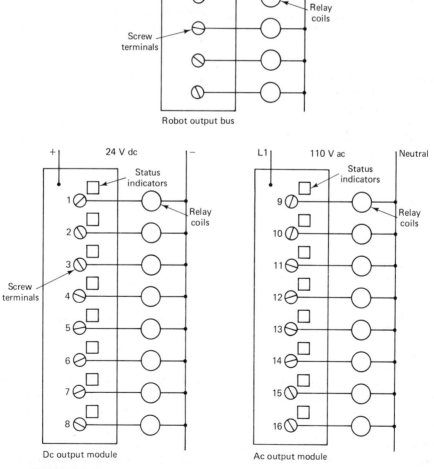

FIGURE 6-27 Diagram of relay coils connected to programmable controllers and robots.

voltage when it is deenergized. This spike is out of phase with the applied voltage so it can damage the dc solid-state component with the reverse-polarity voltage. To prevent this damage, a diode is placed across the coil in the reverse-bias direction. When the reverse-polarity voltage from the spike is encountered, it will be routed back into the coil, where it will be absorbed. A resistor can be added with the diode if the surge is exceptionally strong. This action prevents the surge voltage from reaching any of the solid-state components in the robot. If this is a problem with the programmable controller output module, the same components can be added to it. An example of this surge protection circuit is shown connected to the relay coil at the first robot address.

Notice that the relay's contacts are not shown in this diagram either. Like the diagram for the programmable controller, the contact is shown in the ladder diagram for the system. In some cases, a solenoid valve or other output load will be connected to the robot output bus instead of a relay coil. In this case the output is the entire component and it will not have any other parts, such as contacts.

Relay Contacts as Inputs

In some motor control circuits, the contacts from a control relay or general-purpose relay are used to control an input signal to a programmable controller or to a robot. This type of interface is not used as often as the relay coil is used as an output. The relay contacts can be used to interface two incompatible signals. This may be required when you have a robot that has a 24-V dc I/O bus and several of the input signals are powered from a 110-V ac source. In this application a relay coil could be connected to the 110-V circuit and the contacts could be wired in series with the 24-V input bus. This type of application would generally arise when a robot is installed into an existing motor control system where all the controls are rated for 110-V applications.

Another condition that requires relay contacts to interface two incompatible signals occurs when the output of a robot must be sent to a programmable controller as an input signal. If the output signal is sent from the robot, the robot's power supply is used. The input module of the programmable controller may need to have its input signal coming from its own isolated power supply. In this application, the robot output is sent to the coil of a relay, and the contacts of that relay are wired in series with the input module and its isolated power supply. This allows the robot power supply to be completely isolated from the programmable controller's power supply. A solid-state relay or optocoupler could also be used to isolate these signals, but the general-purpose relay is a reliable choice.

Figure 6-28 shows an electrical diagram of the interface between the robot and the programmable controller. The robot is using the dc positive terminal as the common of its circuit and the programmable controller is using the dc negative as its common. This also presents a problem in trying to connect these two signals to each other directly.

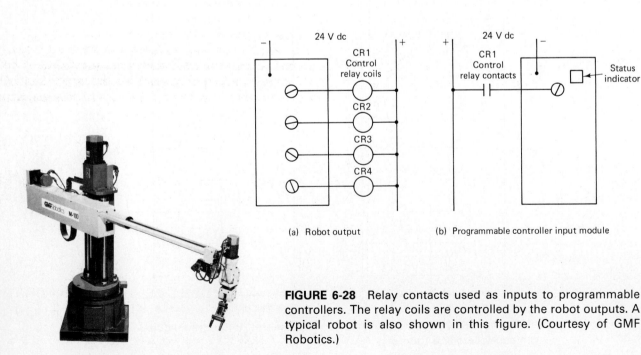

FIGURE 6-28 Relay contacts used as inputs to programmable controllers. The relay coils are controlled by the robot outputs. A typical robot is also shown in this figure. (Courtesy of GMF Robotics.)

Another application where a relay could be used is to interface two signals when one of the signals is coming from a microprocessor-controlled device and needs to be sent as an input to another microprocessor-controlled device, such as a robot. Industrial robots generally have all their inputs rated at the same voltage; 24 V dc is a typical voltage for a robot I/O bus. If a TTL or CMOS signal is being sent from another microprocessor-controlled system to the robot I/O bus that is expecting a 24-V signal, a small relay will be required to complete the interface. The coil of the relay would be rated for the TTL or the CMOS voltage and the contacts would be wired in series to a robot input address. When the small signal is energized, the coil would close the contacts and the 24-V signal would be sent to the robot input address.

Another time a relay should be used to interface two incompatible signals is when a small dedicated type of programmable controller is used. It is important to remember that some small programmable controllers do not provide a choice of input modules. They come standard with all inputs either ac or dc voltage. If you have a small programmable controller with dc inputs, you will need to build interface circuits for any ac voltage signals. The same would be true if your programmable controller had ac inputs and you needed to interface a dc input signal.

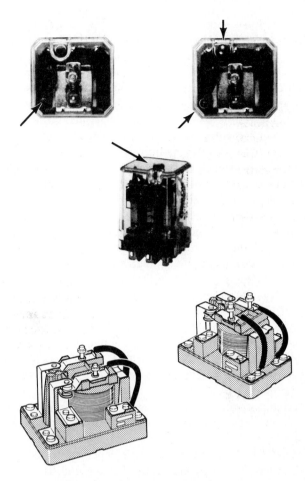

FIGURE 6-29 Methods of checking a relay to determine if its contacts are closed.

Troubleshooting the Relay

When a circuit containing a relay malfunctions, the relay will need to be troubleshooted in two sections to determine where the fault exists. The first test to make when troubleshooting a relay is to determine if the coil has the contacts energized. Some relays provide an operational indicator. An example of this indicator is shown in Figure 6-29. The indicator consists of a small button that protrudes through the top of the relay. This button is popped up when the relay is energized and is more visible than when the relay coil is deenergized. This allows the troubleshooting technician to determine the condition of the coil without using a test meter.

Another way to determine if the relay coil has energized its contacts is to observe the movement of the contacts. This is possible when the relay is an open type such as the one shown in Figure 6-29, or when the relay is enclosed in a clear plastic cover. The relay contacts are visible in these types of relays and the technician can observe their operation and determine the state of the coil. Some sealed relays provide a indicator LED or lamp that will glow to show that the relay coil has been energized.

The position of the relays contacts can also be determined by operating them manually. Figure 6-29 also shows a manual actuator button that allows the relay contacts to be operated manually. When the button is depressed, it simulates the action of the coil and causes all normally open contacts to close and all normally closed contacts to be opened.

If the circuits that the relay is controlling become energized when the button is pressed, you can assume that the coil is not energized. If nothing changes in the relay circuits when the button is pressed, you should assume that the relay coil already had the contacts energized. To confirm your suspicions, you should use a voltmeter to make additional tests.

The first test that should be made with the voltmeter is a check for the proper amount of voltage across the coil terminals. Remember that if the voltage is not the proper amount, the relay coil cannot pull the contacts to their energized position. If no voltage is present, all the components in the coil circuit should be tested to locate the open circuit. Use the test used in Chapter 5 to locate an open pilot device or open wire in a control circuit. This test consists of checking for applied voltage at the supply voltage terminals, then leaving one meter probe on the common terminal and testing the terminals of each component in the order in which they are connected in the circuit. Any point in the circuit where voltage is lost indicates the open circuit. If this occurs at one of the pilot devices, be sure to check and see if the device is operating correctly and protecting the circuit against a condition it is supposed to protect it against.

When you are sure that the proper amount of voltage is applied directly across the two coil terminals, you can assume that the coil should be energized. If the contacts do not move, you should suspect that the relay coil is defective, which means that it is opened. The fastest way to test the coil for an open is to disconnect power and remove one of the wires to the coil and test for continuity. The continuity test should be made with an ohmmeter that has several scales. Do not be misled by looking for a specific amount of resistance, since the amount of coil resistance will depend on the size and length of wire that is used in the coil. Instead, you are looking for an infinite reading on the ohmmeter when the highest range is used. A test lamp could be used, but an ohmmeter will differentiate between a high-resistance coil and an open.

If this test indicates that the coil is opened, the complete relay or only its coil should be replaced. If the relay coil is replaceable, follow the directions provided with the replacement part and complete the operation as quickly as possible. If the relay is the sealed type, the entire component must be replaced as a unit. It might be faster to replace the entire relay even if the coil can be replaced, to get the equipment back into commission as soon as possible. The type of application and the cost of downtime will help you decide which method will be best for you.

If this coil or others near it in a multiple-relay installation have burned out often, you should suspect an overvoltage condition or excessive surge voltages. If the source of the surge voltage or overvoltage can be determined, it should be corrected. If the source cannot be determined, surge suppression devices or snubbers should be installed across the supply voltage lines for the coil voltage, or across each individual relay coil.

If the relay coil is operating correctly, each of the contact circuits should be tested individually. Remember: it is possible for the relay coil to energize each set of contacts, and individual contacts can be faulty. This may occur because one of the sets of contacts is not seating properly or the contact point may be pitted from excessive arcing. It is also possible to lose supply voltage to one set of contacts and not another. This occurs frequently when the contact sets receive voltage from separate sources. Remember: it is highly possible for one fuse or one circuit breaker of a set to open. This may also occur when one set of contacts is powered with ac voltage and another set is powered with dc voltage.

If you suspect that one or more of the sets of contacts are faulty, use the voltage loss test to locate the open point in the contact circuit. If the result of this test indicates that voltage is not passing through the relay contacts, you should inspect the contacts physically. It is also possible for the contacts to become twisted in the contact carrier and not seat properly.

If a set of contacts is found to be faulty, change the contacts without changing the entire relay. This is possible if a contact maintenance or replacement kit is available. The contact maintenance kit supplies new contacts and springs to repair the relay contacts. If a contact kit is not available, or if the relay is a sealed type, the entire relay must be replaced as a unit. Again it must be noted that the entire relay might be replaced as a unit to save downtime and the faulty relay can be repaired later in the shop and returned to stock.

It is also important to be aware of several other conditions that may make you assume that a set of contacts is defective. If the relay uses a plug-in socket for mounting, it is possible that another troubleshooting technician has tested the circuit before you have arrived and exchanged the suspect relay. Sometimes when socket-mounted relays are suspected, technicians will exchange the relay with one that is close by in the electrical panel. Since the relay looks the same physically, it will be swapped with the suspect one. Sometimes the relays do not contain the same normally open and normally closed contact arrangement and a new problem has be entered into the system. If you suspect that relays have been swapped, the remedy is to locate the field wiring diagrams and check out all the terminals and field wiring connections.

After any relays have be replaced, the entire circuit should be tested to be sure that the system is operating correctly. It is possible that more than one component became faulty or out of adjustment after the system was tested.

QUESTIONS

6-1. Explain how the alignment of the dipoles in a magnetic material affects the strength of a magnet.

6-2. Explain three ways in which the strength of an electromagnet can be increased.

6-3. Explain which of the three ways that you mentioned in Question 6-2 is most often used to vary the strength of electromagnets in motors and relay coils. Why?

6-4. Draw a diagram of a solenoid and explain how it works.

6-5. Explain why a iron core is used inside a solenoid coil.

6-6. Draw a diagram of an in-line solenoid valve and explain its operation.

6-7. Draw a diagram of a two-way and a three-way solenoid valve and explain their operation.

6-8. List three applications each for one-, two-, and three-way solenoid valves.

6-9. Explain how you would troubleshoot a faulty solenoid valve.

6-10. Explain the term *plugging* a solenoid valve.

6-11. Draw a diagram of a simple relay and explain its operation.

6-12. Explain the function of a shading coil. Where would you expect to find a shading coil?

6-13. State the three types of aramature assemblies used for relays.

6-14. List typical voltages used to power relay coils.

6-15. Explain the terms *inrush current* and *sealed current* with regard to relay coils.

6-16. Explain why a latched relay that is powered with low voltage would keep its contacts closed if they were closed manually with a screwdriver.

6-17. Explain the function of the coil in a relay.

6-18. Explain the function of the contacts in a relay.

6-19. Draw the electrical symbol for a relay with one set of normally open contacts and one set of normally closed contacts.

6-20. Explain the terms *poles* and *throws*.

Contactors and motor starters 7

INTRODUCTION

The contactor is similar to the relay in that it has a coil and contacts, but it is more closely related to the magnetic motor starter because of its applications and size. In fact, some models of magnetic motor starters are made by adding a set of overloads to a contactor. At one time the distinction between relays and contactors was based strictly on size. The *relay* is classified as a set of contacts rated less than 20 A that are controlled by a coil. A *contactor* was classified as a set of contacts rated more than 20 A controlled by a coil. Today, this classification becomes less clear because power relays are available with contacts rated up to 30 A. Another reason this classification is no longer valid is that a relay and a contactor can utilize solid-state components instead of using a coil. This caused the new classification to include any device that could energize and deenergize an electrical circuit repeatedly. The contactor and relay are now distinguished more by their applications rather than only their size.

The relay is generally used for general-purpose switching applications and light-duty inductive and resistive loads. It is also used for control or logic purposes in electrical circuits. The contactor is used for motor starting applications and to control lighting, heating, transformer, and capacitive loads.

The operation of the contactor is explained in this chapter. The basic design is discussed and the function of all major parts is provided. Additional information is provided regarding NEMA sizes, types of auxiliary contacts, and typical applications and circuits. Installation, preventive maintenance, and troubleshooting procedures and detailed.

The last part of this chapter continues this discussion of the difference between a contactor and a magnetic motor starter. The motor starter has provisions for possible motor overload protection. Since the motor starter is the most widely use motor control, a detailed explanation of its basic parts and operation is provided. Additional information is given regarding typical applications for the basic motor starter, including wiring diagrams. A detailed comparison of traditional motor starter specifications and European motor starter standards is also provided. Since you have not yet studied solid-state controls, a detailed comparison of a magnetic motor starter and a solid-state motor starter is postponed to Chapters 12 and 14.

CONTACTORS

Contactors are being used to control motors and other types of noninductive loads, such as industrial heating and lighting loads. The contactor is designed similar to the relay in that it has a magnetic coil that controls the operation of one or more sets of contacts. The contactor can also have any number of auxiliary contacts rated slightly less than the main line contacts added to the main contactor frame. The contactor is available with mechanical or electrical holding and latching mechanisms.

Basic Parts of the Contactor

Figure 7-1 shows a picture and exploded-view diagram of a typical size 1 contactor. From the diagram you can see that it has six basic parts. The largest part of the contactor is the base. It has several other pieces molded or fastened to it permanently. These parts include the mounting plate, arc hood, and stationary contacts with field wiring terminals.

The mounting plate provides the method of mounting the contactor to an electrical panel. It also provides the mounting platform for the contactor base and keeps all of the moving and stationary parts properly aligned. The base is generally made of high-quality die-cast metal so that it can easily absorb and dissipate heat from the coil. The base also provides tapped threaded pads for all the additional parts of the contactor to be mounted to securely. Since the additional parts, such as contact carrier and coil, are mounted to the base with screws, they can easily be removed and replaced when they become faulty.

The stationary contacts and field wiring terminals are mounted with screws to the top of the base. Since these contacts are also mounted with screws, they can be changed as part of the contact assembly in the event that they wear prematurely.

The arc quencher is mounted on the stationary contacts, where they can absorb any arc drawn between the movable and stationary contacts as the contactor opens and closes them. This shield also prevents dangerous arcs from coming out of the contacts to other metal parts, such as the electrical cabinet.

The contact carrier serves several vital purposes. The first purpose is to provide a platform for the movable contacts. It also provides a housing for the coil and magnet assembly. When the coil is energized the magnetic field pulls the contact carrier upward until the movable contacts seat against the stationary contacts. This assembly allows the contactor to have only one movable part when it is fully assembled.

The contact carrier also provides a place to mount the armature and retainer spring. The molded coil is held in place in the middle of contact carrier by the magnet yoke and clips. The coil cover is then mounted to the base with screws that prevent the coil from falling out of the contact carrier. The coil cover provides enough room for the contact carrier to be pulled up when the coil is energized. Since the contact carrier is the only moving part for the contactor, the operation is rather simple and reliable. The magnet is made with a permanent air gap that allows the contact carrier to drop out freely and prevent residual magnetism.

The contacts are the double-break style and are made of solid cadmium oxide silver, which provides high conductivity and resistance to welding and pitting. The field wiring terminals provide a means of connecting field wiring to the two sides of the stationary contacts. When the double-break movable contact segment is seated to the stationary contacts, a conductive bridge is made across them, allowing current to travel from the line terminal to the load terminal.

Types of Contacts and Protection Features

The contactor is available with a variety of contacts for different applications. The contactor is also available with any number of poles up to five. When the contactor is purchased, the number of main contacts must be specified. If the contactor is used for dc circuits, a single-pole normally open contactor is available. Up to five normally open main contacts are available for ac loads.

The contacts are made from cadmium oxide and silver to resist corrosion and wear due to arcing. This material also helps ac arcs extinguish themselves. Several types of protection are built into the contactor to prevent damage due to arcing. These features include arc shutes, arc traps, and blowout coils. The arc shutes and arc traps are provided on dc contactors to route the arc that is produced when the contacts are opened or closed, away from damaging the contact surface. You should remember that dc loads produce larger arcs when contacts are opened or closed because dc current flows in one direction. Ac arcs are not as excessive since the ac current is sinusoidal. The ac current changes direction during every cycle and returns to zero at the 180-degree point in the cycle. This action tends to keep to a minimum the arc caused by contacts opening and closing.

Since the dc current flows in only one direction, the magnetic field can become excessive and large arcs are caused when the contacts are opened

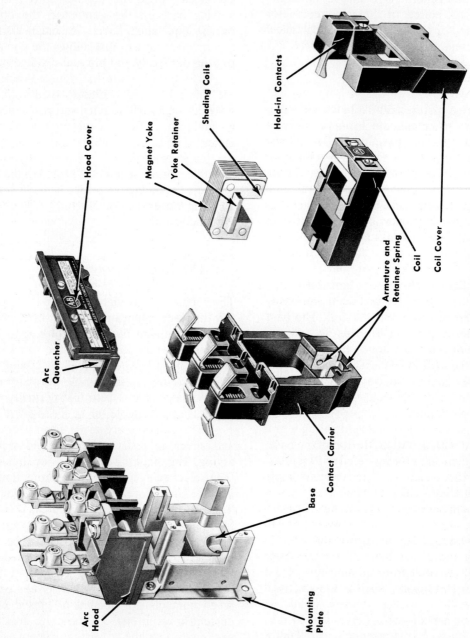

FIGURE 7-1 Exploded-view diagram of typical contactor. (Courtesy of Allen-Bradley, Inc.)

or closed. The arc will begin when the contacts start to open and continue to grow as the distance between the contacts continues to grow. The arc will be at its largest when the contacts are fully opened. The arc that is developed when the contacts are opened tends to be much larger than the arc that is produced while the contacts are closing.

The arc produced when the contacts are closing is caused by inductive loads pulling locked-rotor amperage (LRA). The locked-rotor amperage, also called *inrush current*, tends to be very large while the motor is coming up to running speed. This current will be at its largest when the contacts are initially closed.

The contactor provides arc shutes and arc traps to keep these arcs in check. The idea of these features is to limit the arc as much as possible and to shunt any residual arc to a position in the contactor away from the contact surface. The contactor also provides protective covers over the contacts to prevent the arc from coming out from the contacts and hurting someone. The arc shutes and arc traps provide a location in the contactor where the arc can be extinguished without damaging the contactor. These features are called *arc supression* and are provided in one means or another in all contactors. It is important to remember that you should not operate a contactor under load with any of these protective devices removed or damage may occur. When the contactor is inspected during preventive maintenance, it is important to see that the arc supression features of the contactor are operational and undamaged.

Contact Sizes

Contactors are rated by NEMA according to the size of load that they can handle safely. The load rating for a contactor is based on the size of the contacts. Figure 7-2 shows a NEMA rating for contactors. The smallest contactor is a size 00 and the largest is a size 9. This table shows the load rating as a maximum continuous current or as a maximum horsepower. Notice that these ratings include the operating voltage, and indicate whether the supply voltage is single phase or three phase.

Figure 7-3 shows some of these contactor sizes. These photographs provide an idea of the relative size of these contactors. The size 00 is about the size of a control relay and can carry a maximum of 9 A in a continuous load. The size 0 is also relatively small in that it can carry a maximum load of 18 A. You should remember that these amperage capabilities are also within the range of a power relay. This is the point where some confu-

NEMA Size	Continuous Amp. Rating	600 VOLTS MAXIMUM			
		Volts	Maximum Horsepower Rating [2] Full load current must not exceed the "Continuous Ampere Rating"		Maximum Horsepower Rating For Plugging Service [1]	
			Single Phase	3 or 2 Phase	Single Phase	Three Phase
00	9	120	1/3	3/4	—	—
		208	—	1½	—	—
		240	1	1½	—	—
		480	—	2	—	—
		600	—	2	—	—
0	18	120	1	2	½	1
		208	—	3	—	1½
		240	2	3	1	1½
		480	—	5	—	2
		600	—	5	—	2
1	27	120	2	3	1	2
		208	—	7½	—	3
		240	3	7½	2	3
		480	—	10	—	5
		600	—	10	—	5
2	45	120	3	—	2	—
		208	—	15	—	10
		240	7½	15	5	10
		480	—	25	—	15
		600	—	25	—	15
3	90	120	—	—	—	—
		208	—	30	—	20
		240	—	30	—	20
		480	—	50	—	30
		600	—	50	—	30
4	135	120	—	—	—	—
		208	—	50	—	30
		240	—	50	—	30
		480	—	100	—	60
		600	—	100	—	60
5	270	120	—	—	—	—
		208	—	100	—	75
		240	—	100	—	75
		480	—	200	—	150
		600	—	200	—	150
6	540	208	—	200	—	150
		240	—	200	—	150
		480	—	400	—	300
		600	—	400	—	300
7	810	208	—	300	—	—
		240	—	300	—	—
		480	—	600	—	—
		600	—	600	—	—
8	1215	208	—	450	—	—
		240	—	450	—	—
		480	—	900	—	—
		600	—	900	—	—
9	2250	208	—	800	—	—
		240	—	800	—	—
		480	—	1600	—	—
		600	—	1600	—	—

[1] An example is plug-stop or jogging (inching duty) which requires continuous operation with more than five openings per minute.

[2] **Non-Motor Loads** — When contactors are required to switch non-motor loads such as lighting circuits, ovens, transformer primaries, etc., use the Bulletin 702L contactor.

Interrupting Capacity—The double-break contacts will carry rated loads continuously, are designed for full voltage starting and are capable of interrupting six times the current corresponding to the horsepower rating.

Coil Performance — Continuous operation at 110% of the rated voltage; and pick-up at 85% of rated voltage.

FIGURE 7-2 NEMA rating for contactors. (Courtesy of Allen-Bradley, Inc.)

SIZE 1
7 1/2 HP 240 V
10 HP 600 V

SIZE 3
30 HP 240 V
50 HP 600 V

SIZE 4
50 HP 240 V
100 HP 600 V

SIZE 5
100 HP 240 V
200 HP 600 V

FIGURE 7-3 Pictures of contactors to show relative sizes. (Courtesy of Allen-Bradley, Inc.)

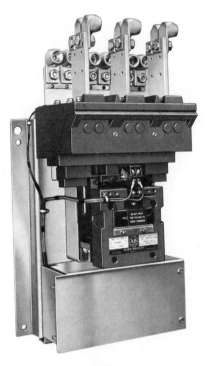

SIZE 6
200 HP 240 V
400 HP 600 V

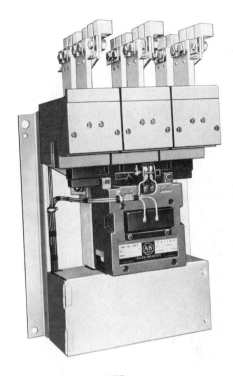

SIZE 7
300 HP 240 V
600 HP 600 V

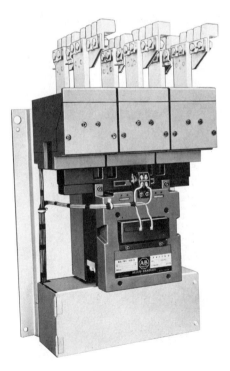

SIZE 8
450 HP 240 V
900 HP 600 V

SIZE 9
800 HP 240 V
1600 HP 600 V

FIGURE 7-3 Pictures of contactors (*Continued*)

sion comes from when relays and contactors are classified strictly by size.

The ratings for each of the contactors include the largest horsepower load that can be controlled by plugging. The smallest contactors are not rated for plugging and should be limited to continuous operation.

Auxiliary and Special-Purpose Contacts

The main contactor can have a variety of contacts added to it for special purposes. Auxiliary contacts can be added to the frame of the main contactor so that they can be operated by the contactor's coil. When the coil is energized, the contact carrier will also activate the auxiliary contacts. Figure 7-4 shows auxiliary contacts added to a contactor. A diagram is also provided to show the location of these contacts and how they are operated. An electrical diagram is also shown in this figure. This diagram will be used in all wiring diagrams and ladder diagrams to indicate circuits that utilize auxiliary contacts.

Auxiliary contacts can be normally open or normally closed. Some types are field convertible so that they can be converted from normally open to normally closed, or vice versa. In the diagram and on the contactor the types of contacts that are used will be identified clearly so that you will be able to troubleshoot the contactor.

The auxiliary contacts are also rated by NEMA. The auxiliary contacts for a size 0 contactor are rated for 18 A. The auxiliary contacts for a size 1 is 27 A and the size 2 is 45 A. These contacts are heavy enough to control fairly large loads. Other types of auxiliary contacts are rated for pilot duty only. These types of contacts are usually used as sealing or maintaining contacts. They may also be used as control contacts to energize other lines of the circuit.

Other contacts are available to be used as alarm contacts. These contacts are usually wired to an indicator lamp. When the contactor is energized, the indicator lamp will be energized through the alarm contacts. When the contactor is deenergized the alarm contacts will deenergize the indicator lamp. The alarm contacts can be wired as a set of normally closed contacts. In this type of application, the indicator lamp would become energized when the contactor is deenergized.

Mechanical holding or interlock contacts are available for use with most contactors. The mechanical interlock operates like a latch. Figure 7-5 shows an electrical diagram of a contactor with a mechanical interlock. When the coil is energized, the contact carrier is pulled up so that movable contacts seat to the stationary contacts. When the contact carrier is pulled up as far as its stroke will allow, a mechanical interlock is tripped and keeps the contact carrier in this position even when current is interrupted to the coil. The reset or unlatch coil must be energized for the interlock to be released.

This figure also shows NEMA ratings for the

Auxiliary Contacts
Bulletin 500 Line contactors and starters can accommodate a total of eight auxiliary contacts. Four of these contacts can be used without increasing panel space. Each contact easily snaps into place and is held firmly to guard against dislodging by vibration. Contacts are identified with a N.O. or N.C. symbol which is clearly visible from the front of the starter. The contacts are bifurcated and positioned in a vertical plane for reliability.

Electrical diagram of auxiliary contacts

FIGURE 7-4 Picture of auxiliary contacts on a contactor and diagram showing their operation. An electrical diagram is also provided to show the auxiliary contacts with the contactor. (Courtesy of Allen-Bradley, Inc.)

154   Chap. 7 / Contactors and Motor Starters

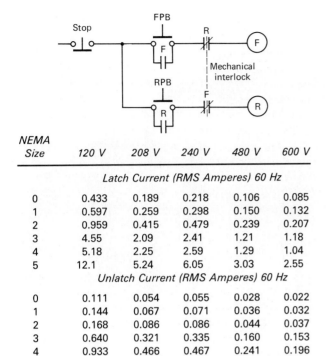

NEMA Size	120 V	208 V	240 V	480 V	600 V
	Latch Current (RMS Amperes) 60 Hz				
0	0.433	0.189	0.218	0.106	0.085
1	0.597	0.259	0.298	0.150	0.132
2	0.959	0.415	0.479	0.239	0.207
3	4.55	2.09	2.41	1.21	1.18
4	5.18	2.25	2.59	1.29	1.04
5	12.1	5.24	6.05	3.03	2.55
	Unlatch Current (RMS Amperes) 60 Hz				
0	0.111	0.054	0.055	0.028	0.022
1	0.144	0.067	0.071	0.036	0.032
2	0.168	0.086	0.086	0.044	0.037
3	0.640	0.321	0.335	0.160	0.153
4	0.933	0.466	0.467	0.241	0.196
5	1.22	0.625	0.628	0.317	0.258

FIGURE 7-5 Electrical diagram of interlock used with a contactor. NEMA specifications are also provided for the latch and unlatch coils shipped. (Courtesy of Allen-Bradley, Inc.)

should also notice that the coil current will be smaller as the voltage is increased.

Coils for Contactors

The coils for all the NEMA starters are available in a variety of voltages. This allows the contactor to be used in most any ac or dc control circuit. Figure 7-6 provides a table that shows the ac voltage on which these coils will operate. It also lists the pull-in and seal-in currents for each of these coils. Notice that ac coils are available for 120-, 208-, 240-, 480-, and 600-V circuits. Another table in this figure shows the typical operating time it takes for the coils to operate each of the NEMA contactors. These data are useful in selecting the proper contactor and coil for original installation or as a replacement part.

Figure 7-7 shows a picture of a typical encapsulated coil. These coils are made by tightly winding small wire on a core. The coil is then dipped in thermoset epoxy and terminals are provided for field wiring connections. Most manufacturers use color codes to identify the voltage rating for each coil. Some manufacturers use a different color thermoset epoxy for each voltage, while others use a color patch on the side of the coil to make the identification. All coils will be clearly marked with the voltage rating and the operating frequency.

The design of a dc coil assembly in slightly different from that of an ac coil assembly. The dc coil is built around a solid steel core which can build

latch coil and unlatch coil. From this table you can see that the reset coil is smaller than the main coil since it only needs to pull the interlock mechanism and allow the contact carrier to drop out. You

COIL CURRENTS

NEMA Size	No. of Poles	Inrush Current (Amps.) 60 Cycles					Sealed Current (Amps.) 60 Cycles				
		120V	208V	240V	480V	600V	120V	208V	240V	480V	600V
00	1-2-3	0.50	0.29	0.25	0.12	0.07	0.12	0.07	0.06	0.03	0.02
0	1-2-3-4	0.88	0.50	0.44	0.22	0.17	0.14	0.08	0.07	0.04	0.03
1	1-2-3-4	1.54	0.89	0.77	0.39	0.31	0.18	0.10	0.09	0.04	0.04
2	2-3-4	1.80	1.04	0.90	0.45	0.36	0.25	0.14	0.13	0.06	0.05
3	2-3	4.82	2.78	2.41	1.21	0.97	0.36	0.21	0.18	0.09	0.07
3	4	5.34	3.08	2.67	1.33	1.07	0.39	0.23	0.20	0.10	0.08
4	2-3	8.30	4.80	4.15	2.08	1.66	0.54	0.31	0.27	0.14	0.11
4	4	9.90	5.71	4.95	2.47	1.98	0.61	0.35	0.31	0.15	0.12
5	2-3	16.23	9.36	8.11	4.06	3.25	0.81	0.47	0.41	0.20	0.16
6	2-3	0.62	Current Shown is The Very Small AC Current Passing Through Coil of Control Relay. Any Standard Duty 120V Control Station May Be Used In This Circuit.				0.082	Current Shown is The Very Small AC Current Passing Through Coil of Control Relay. Any Standard Duty 120V Control Station May Be Used In This Circuit.			
7	2-3	0.62					0.082				
8	2-3	0.62					0.082				
9	2-3	1.2					0.16				

OPERATING TIME

Size	Approximate Operating Time in Milliseconds 3 Pole Contactors	
	Pick-up	Drop-out
00	28	13
0	29	14
1	26	17
2	32	14
3	35	18
4	41	18
5	43	18
6	88	40
7	88	45
8	118	94
9	118	84

FIGURE 7-6 Tables showing coils with available voltages for contactors and typical operating times for coils to energize contactors. (Courtesy of Allen-Bradley, Inc.)

Coils

Thermoset Epoxy—Bulletin 500 Line coils are hot pressure molded in thermoset epoxy to protect against mechanical damage and harmful environments.

Identification—Coils are ink stamped with the coil number, voltage and frequency which are clearly visible from the front of the starter. The ink stamp is color coded for easy identification. The left coil terminal is also stamped with the last three digits of the coil number which can be used to determine voltage and frequency.

Coil Shunt Plate—The shunt plate is designed to retard the magnetic flux until the voltage applied reaches the "pick-up" voltage. The flux generated is then sufficient to overcome the shunt plate and close the contacts with snap action. This guards against the contacts partially closing which could result in insufficient contact pressure or possibly welded contacts.

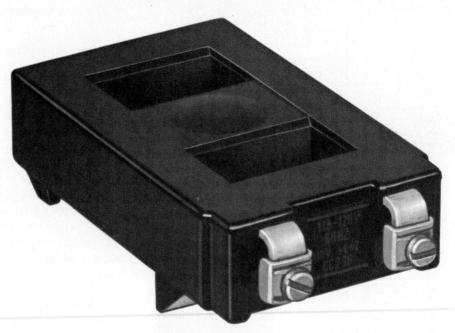

Thermal Cutout—Each coil is provided as standard with an exclusive thermal cutout, designed to open on excessive currents or misapplied voltage. This temperature sensing device protects against the coil insulating material melting and causing irreparable starter damage.

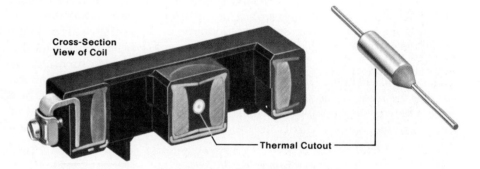

Magnet

The high efficiency magnet has a permanent air gap. Pole face wear cannot affect the air gap and cause magnetic sticking due to residual magnetism. Short stroke and cushion mounting reduce impact and increase life. Magnets are corrosion resistant; each lamination is phosphate coated, the magnet is assembled and epoxy impregnated, and the pole faces are phosphatized after grinding.

FIGURE 7-7 Picture of a typical encapsulated coil and diagram showing built-in thermal cutout and energized and deenergized inductors. (Courtesy of Allen-Bradley, Inc.)

extremely strong magnetic fields. The dc coil also has built-in shunt plates that retard the magnetic flux that builds in the coil until the applied voltage increases to the minimum coil pickup value. When the applied voltage reaches the coil pickup value, the flux will quickly build and the magnetic field will snap the contact carrier into place. This action causes the contacts to come together in a snap action, which limits the arc across the contacts. It also prevents the contacts from pitting and wearing when the contacts are energized.

The coils that are used in ac circuits are made from simple encapsulated coils of wire. The major difference between the ac and dc assembly is in the design of the laminated sections used in the ac coil assembly. The ac coil utilizes laminated sections instead of the solid steel coil used in dc coils. The laminated sections use a design that is similar to the type used in ac relay coils. The contactor coil assembly requires an air gap to allow the contactor to break loose cleanly when the coil is deenergized. The laminated sections provide protection against eddy current buildup and residual magnetism, just as in a relay coil.

Some coils provide an internal thermal cutout that is built into the coil when it is manufactured. The thermal cutout is located in the middle of the coil where heat initially builds up when the coil begins to overheat. This cutout will open the coil circuit and deenergize the coil, which protects it from overheating. When the coil cools down the cutout will reset and allow current to flow again. Since the cutout resets itself automatically, this type of coil should be used only with contactors that are wired for manual reset. This means that the contactor is controlled by a momentary-type start button. The start button is sealed with a set of auxiliary contacts, and when the thermal cutout deenergizes the coil circuit, the auxiliary contacts drop the circuit out until the pushbutton is pushed manually. The coil must be cool enough for the cutout to reenergize or the start button will not energize the coil again.

This type of protection will prevent the coil from burning out due to overheating. This could be caused by the contact carrier jamming and not allowing the armature to center itself in the coil. When this occurs, the coil's impedance remains low and the coil will draw excessively large currents and cause the coil to overheat. The coil could also overheat if the applied voltage became too large. This could occur if the utility voltage increased due to minimal loading on its generator. Coil voltage could also become too large if the wrong-size coil was used. This may occur when a coil or complete contactor is replaced by one of incorrect size.

Contactor coils can also be protected from excessive voltage transients with a surge suppression device. These devices are manufactured small enough to be mounted directly across the coil terminals. Like the coil, the surge suppression device is encapsulated in epoxy plastic to withstand rugged factory floor environment. These devices must be matched with the coil voltage and may not be available for the larger coil voltages from all manufacturers.

Applications for Contactors

Contactors can be used for a variety of motor starting applications. These applications include starting and jogging single-phase and three-phase ac motors and dc motors of all sizes. Since the contactor is available in sizes up to a NEMA size 9, it can be used to start motors as large as 1600 hp.

Figure 7-8 shows the contactor used in two-wire and three-wire motor starting circuits. The motor in each of these diagrams could be a three-phase motor, or a single-phase motor. The two-wire circuit is so named because only two wires are connected to the coil (Figure 7-8a). This means that the coil circuit is one large series circuit. Any pilot devices are located in this series circuit. The circuit in this diagram shows a float switch in series with the coil. The float switch will close its contacts any time the level of water in the tank in which it is mounted drops below the predetermined set point. When these contacts close, current is passed to the contactor's coil. When the contactor is energized, the contactors contacts are energized and the motor will begin to run. Since the motor is used in a pumping application, it will be driving a water pump from its shaft. When the water level in the tank reaches the predetermined high set point, the float

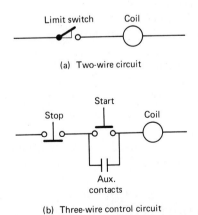

FIGURE 7-8 Examples of two-wire and three-wire control systems.

switch contacts will be opened and the coil will be deenergized and allow the contacts to drop out and stop the pump.

Notice that the contactor does not provide overcurrent protection for the motor. This means that overcurrent protection must be provided by another means. In a discussion later in this section we explain the method of adding overloads to the contactor. You will notice that the motor in this application is protected with a fused disconnect. The fuses in the disconnect are the dual-element type, which allows the motor to start with large inrush current and yet protect the motor at the full-load current level under normal operating conditions. The fuses will blow and protect the motor any time the full-load current exceeds a safe limit. The fuses also provide excellent short-circuit protection for this application.

The diagram in Figure 7-8b use a three-wire control circuit to control a contactor. The contactor in this application is controlling a motor that is driving a conveyor. Notice that a set of auxiliary contacts are used in conjunction with a momentary-type start/stop buttons. When the start button is pressed, its contacts close as long as the button is depressed. Current is passed through the start button to the stop button. As long as no one is pressing on the stop button, its contacts will remain closed and pass current to the contactor's coil. The contactor's coil will energize and pull its contacts closed. The normally open auxiliary contacts will also close when the coil is energized. These contacts are in parallel with the start button contacts and will seal them in when they are close. The auxiliary contacts provide an alternative route for current to pass the normally open start button contacts. This means that current will flow through the auxiliary contacts around the start button after it has been released and has returned its contacts to the normally open position. This allow adequate current to reach the coil through the auxiliary contacts and keep it energized.

When the stopped button is depressed, all current to the contactor's coil will be interrupted and the coil will deenergize and allow all the contacts to drop out. When the auxiliary contacts drop open, they will keep the circuit deenergized until the start button is depressed manually.

A variety of safety devices can be added in series with the stop button contacts. If any of these safety devices sense an unsafe condition, their contacts will open and the circuit will be deenergized just as if the stop button had been depressed. Since the auxiliary contacts will drop out the seal-in circuit, the circuit will have to be restarted manually by pressing the start button.

Jogging a Motor with a Contactor. The contactor can also be used to control motors for jogging and plugging applications. The jogging circuit is shown in Figure 7-9. From this circuit you can see that the major change is in the control circuit. The jog button is part of the start/stop station. In this application the jog button is a "roto-push" selector pushbutton and a normally closed stop pushbutton. When the stop/start pushbuttons are used for normal starting and stopping the start button utilizes the normally open momentary contacts. When the start button is depressed, the relay coil is energized and auxiliary contacts at terminals M2 and M3 in parallel with the start pushbutton seal it in. This operation is exactly like the three-wire control circuit discussed previously.

When the circuit is used to jog the motor, you want the motor to operate only when the jog button is depressed. When you take your finger off the jog button, the motor should turn off. The jog button is used for single button control. This allows the operator to move the motor shaft slightly to cause a small amount of machine motion. If a normal start/stop station were used, the operator would have to press the start button to energize the motor and then quickly press the stop button to deenergize the motor when it had moved enough.

The basic principle of the jog circuit allows single-button control by preventing the auxiliary contacts from sealing in the circuit. The jog/start button has two sets of starting contacts, marked A and B. When the button is in the normal starting position, the A set of contacts are closed when the start button is depressed, and the B set of contacts, which are in series with the auxiliary contacts, are closed at all times.

When the roto selector is switched to the jog position, the B set of contacts are forced open at all times, which disables the seal-in circuit. The A set of contacts operate as a normal momentary set of contacts. When the start pushbutton is depressed, the coil is energized, and the auxiliary contacts close, but the current used to seal the circuit is interrupted by the normally open set of B contacts. This means that the coil will remain energized as long as the start pushbutton is depressed. Whenever the push button is released, current to the coil is deenergized and the coil drops the main load contacts out, which stops the motor.

The contactor must be rated for jogging to be used in this type of application. The reason for the special rating is that the motor will draw locked rotor amperage (LRA) every time the motor is started. Since this current is 5 to 10 times the normal running current, the contactor will tend to

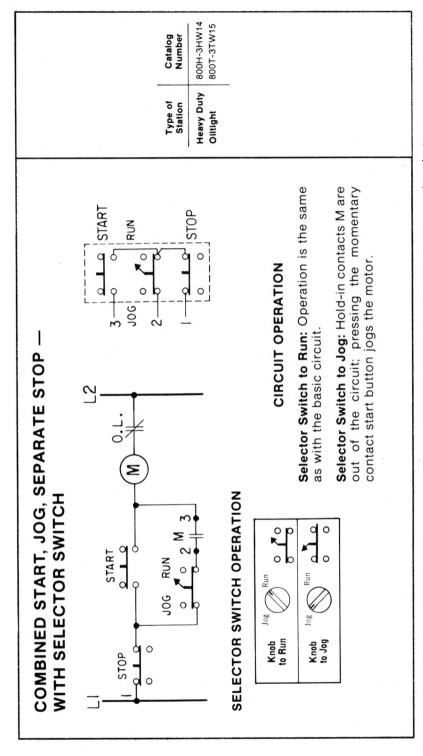

FIGURE 7-9 Wiring diagram and ladder diagram showing a roto push selector pushbutton to control a jogging application. (Courtesy of Allen-Bradley, Inc.)

overheat from starting too often. The contactors that are rated for jogging are manufactured to handle the extra heat. Some manufacturers derate the load capacity of their normal contactor when it is used for jogging.

Reversing a Motor with a Set of Contactors. A motor can be reversed using two contactors. The contactors can be interlocked electrically or mechanically. One of the contactors is used to control the motor in the forward direction and the second contactor is used to control the motor in the reverse direction. Figure 7-10 shows an example of this type of control circuit. When the forward pushbutton is depressed, the forward contactor coil is energized, which pulls in the forward contacts and applies power to the motor in the forward direction. The forward contactor has one set of normally open auxiliary contacts connected in parallel with the forward pushbutton to seal it in and a set of normally closed auxiliary contacts wired in series with the reverse contactor's coil. Each of these sets of contacts is identified with the letter F to indicate that they will be energized by the forward contactor coil.

The second contactor is used to apply reverse voltage to the motor to cause it to operate in the reverse direction. When the reverse pushbutton is depressed, the reverse motor starter will not become energized because the normally closed set of contacts controlled by the forward contacts will be energized to the open position, which disables the reverse coil circuit.

This means that the stop pushbutton must be pressed first to deenergize the forward contactor. When the forward contactor is deenergized, the normally closed contacts controlled by the forward contactor will return to the closed position and allow the reverse contactor circuit to become operational if the reverse pushbutton is depressed. When the reverse pushbutton is depressed, the reverse contactor coil will become energized and the motor will begin to operate in the reverse direction. The normally closed set of auxiliary contacts that are controlled by the reverse contactor will disable the forward contactor's coil. This prevents the forward contactor and reverse contactors from coming on the same time, which would cause a direct short circuit across the two contactors.

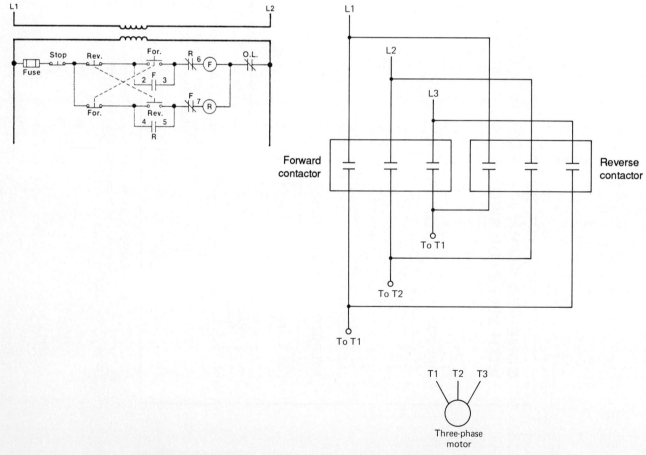

FIGURE 7-10 Wiring diagram showing two contactors used for reversing a motor.

160 Chap. 7 / *Contactors and Motor Starters*

The reversing contactor can be used for applications where the motor must be capable of operating in the forward and reverse directions, such as elevators and lifts. This circuit could also be used for reversing conveyors or machine tool applications.

Using a Contactor for Plugging a Motor. Another special motor control application is called *plugging*. This involves applying full reverse voltage to a motor while it is running in the forward direction. The reverse voltage causes a reverse torque to be applied to the motor armature and causes the shaft to stop rotating in the forward direction and begin to turn in the reverse direction. A switch is used to deenergize the motor before it actually begins to run in the reverse direction. In this way, plugging action will actually work as a method to brake or stop a motor quickly. The reversing contactors shown in Figure 7-10 can be used for plugging if the contactors are rated for this type of service. The plugging switch is actually a rpm switch that determines when the rpm is slowing down to zero. A timer can also be used to estimate the time it takes the motor to decelerate the motor to zero rpm. At that point the plugging switch deenergizes the circuit completely, which stops the motor. If the switch does not deenergize the motor at the correct time, the motor will actually begin to run in the reverse direction rather than stop.

In Chapter 14 we discuss plugging circuits for dc and ac motors in much more detail. We also explain the operation of other specialized motor control circuits, such as jogging and inching with regard to the changes that must occur inside the motor windings.

Since the contactors do not provide any overload protection, fuses must be used to protect the motor. In the second part of this chapter we explain the addition of magnetic and thermal overloads added to the contactor. When the overload is an integral part of the contactor, it is called a motor starter. When the overload is added to the contactor, it is still considered a contactor. This allows the user the flexibility to protect the motor while providing complete control.

Using a Contactor to Control Lighting and Heating Loads. The contactor is well suited for controlling lighting loads and other noninductive loads. Lighting loads such as incandescent filaments used in tungsten lights can be controlled with contactors. The contactor allows the lights to be controlled from a remote location. It also allows multiple switches for controlling the lighting system, which is very useful in large industrial areas.

The contactors can also be used for ballast lighting used in mercury arc lighting. These types of lighting systems present an inductive load during starting which may cause some arcing when the contactor is energized. Contactors for these application provide three, four, or five poles for specialized lighting circuits. The contactor can control individual lights or complete circuits. Figure 7-11 shows a circuit with the contactor being used to control an industrial lighting system. The contactor's coil is also available for 277-V applications. This allows the coil to use the same voltage that is used for the main power. Any type of pilot device can be used to activate the control circuit. This includes pushbutton or selector switches. In some applications such as commercial lighting in stores or public buildings, key lock switches are used to energize the contactor. This prevents the lights from being turned off or on accidentally.

Heating Applications and Switching Capacitive Loads. The contactor is very well suited for controlling electric heating loads. Since unlike motors, resistive heaters do not need overload protection, they can be controlled by a contactor and protected by a fused disconnect. The single-

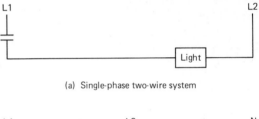

(a) Single-phase two-wire system

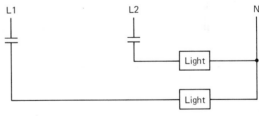

(b) Single-phase three-wire system

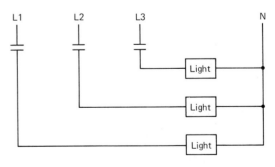

(c) Three-phase four-wire system

FIGURE 7-11 Wiring diagram showing a contactor used to control industrial lighting circuits.

element fuse in a disconnect is sufficient to meet all NEC code requirements. A circuit breaker could also be used as branch circuit protection and provide protection for the heating elements.

Since the contactor does not require overload protection, it can use a two-wire or three-wire control circuit with thermostatic pilot switches to switch the coil off and on. The contactor is useful for controlling single-phase or three-phase heating systems since they are available with a variety of poles.

The contactor is also rated for inductive, resistive, and radiant types of heating systems. Figure 7-12 shows a wiring diagram of a single-point controller. The controller uses a thermocouple to provide accurate temperature control in an industrial furnace. The controller allows the user to input a set point and an amount of high and low differential that it will use as operating parameters. The controller will switch off the contactor any time the furnace temperature rises above the set point. It will switch the contactor back on when the temperature falls below the set point. In this application the set point is 600°F. The high differential is set at 10°F and the low differential is set at 5°F. This causes the controller to energize the contactor's coil any time the temperature is below 595°F and to deenergize it any time the temperature rises above 610°F. The 15°F between 595°F and 610°F becomes a dead band for the controller. Any time the temperature has reached 610°F, the contactor is deenergized until the temperature falls below 595°F. Once the temperature falls below 595°F the contactor is energized again and will remain energized until the temperature reaches 610°F. The controller selects the amount of time of a 10-second timing cycle that the contactor coil will remain energized. This function allows the contactor to be energized a percentage of the 10-second timing cycle that is proportional to the amount of error the temperature is from the set point. This means that if the temperature is 5°F or more below the set point, the controller will cause the contactor to be energized for the entire 10-second cycle time. In this case the contactor will remain energized continually. When the temperature gets closer to the set point, the controller energizes the coil for less time out of the 10-second cycle. For example, if the temperature is 597°F, the controller would see that it is 40 percent below the set point (using the 5°F low differential as a reference). The controller would respond by allowing the contactor to remain energized 4 seconds out of every 10-second timing cycle.

Since the contactor would be energized 4 seconds of every 10 seconds, the contactor would switch on and off 6 times a minute. This rate would cause the contactor to switch on and off 360 times an hour. At this rate the contactor would be switching off and on 8640 times a day and over 3 million times each year. You can see from this application that the contactor would have to be rated for the extensive amount of switching. Since the heating load does not generate a large amount of inrush current, the contacts tend to wear rather well.

The contactor is especially well suited for switching large industrial capacitive circuits use for power factor correction. Since the contactor's coil can be energized by rather small voltages and currents, it can easily be interfaced with a microprocessor power factor control. The amount of leading or lagging phase shift can be programmed into the controller and when the power factor exceeds the set point, large banks of capacitors can be switched into the circuit by the contactors to correct it. This type of application only requires that sets of contacts be switched open or closed at the appropriate times as directed by the controller unit. If the power factor can be controlled within the limits specified, the utility bill could be reduced by thousands of dollars a month in some cases.

NEMA and IEC Standards for Contactors

There are over 30 manufacturers of contactors from North America and around the world. Several associations and commissions have agreed upon sets of tests to provide some standardization of electrical contactors. The major association for North America is the National Electrical Manufacturers Association (NEMA), and the commission for the rest of the world is International Electro-technical Commission (IEC). Some manufacturers refer to the IEC as the European standards. The IEC makes recommendations for manufacturers to test products and publish technical data for product comparisons. NEMA has established standards for electrical products and test specifications for equipment. Many of these specifications and standards have been adopted by other agencies, such as the Underwriters' Laboratory (UL).

The IEC and NEMA do not perform testing like UL does, but they do help establish test parameters. The net effect of these organizations is that they help engineers and technicians compare and select the correct-size product for the application. The IEC was established in England in 1906 and has been making recommendations through 42 member nations since that time. NEMA was established in 1926 and presently have 550 members, representing manufacturers of all types of electrical products.

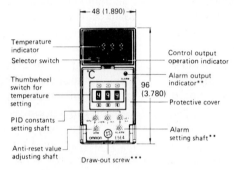

NOTE:
* The types without PID action are not provided with these constant setting shafts.
** The types with SHIFT set function are provided with the SHIFT output operation indicator and SHIFT setting shaft in lieu of the ALARM output operation indicator and ALARM setting shaft, respectively. The types without alarm or shift set function are provided with neither the indicator nor the setting shaft.
*** The front panel and internal mechanism can be drawn out by turning this screw counter-clockwise.

■ CONNECTIONS

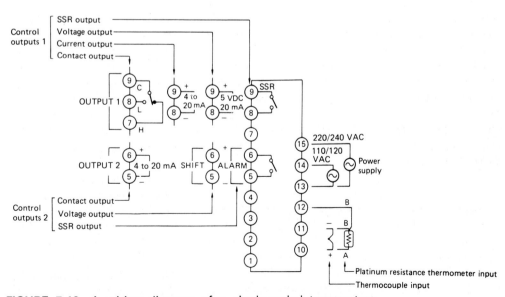

FIGURE 7-12 A wiring diagram of a single-ended temperature control used to switch a heating contactor on and off to provide temperature control to a industrial furnace. (Courtesy of Omron Electronic, Inc.)

Contactors are rated by both of these groups in different ways. The IEC does not specify sizes for individual contactors, but they do provide some recommendations of test results for life expectancy of the electrical contacts and mechanical mechanism. When comparing NEMA-rated contactors and IEC contactors you will find that the IEC type is smaller than a NEMA contactor with the same load rating. This size difference came from conditions in the two regions of the world when the components were originally manufactured. When the contactors were originally designed between the time of WWI and WWII, manufacturing material was in short supply and very expensive in Europe. For this reason, IEC contactors tend to be much smaller and more compact. By comparison, raw material was not a problem in North America, so their contactors were made for durability and life expectancy. These factors tended to make the contactor larger than was necessary for minimum operational standards.

Since original IEC contactors were smaller and more compact, they have maintained these basic sizes for current models. When North American manufacturers designed contactors to be used in European cabinets, they found that the larger NEMA-style contactors were not accepted since they caused cabinet sizes to become too large. On the other hand, many engineers were afraid that the smaller IEC-style device would not stand up to the operational standards that they were used to.

One other major difference between IEC contactors and NEMA contactors relates to the way they interact with European and North American motors. North American motors are designed to withstand heat buildup caused by overloading and continual starting, stopping, or jogging. The European-style motors tend to overheat more rapidly, which causes excessive wear on the contacts. This means that overload components must be sized to protect the motors from rapid heat buildup.

Figure 7-13 shows the EIC utilization categories. From this table you can see that the contactors are rated according to the type of load application, such as AC-1 noninductive or slightly inductive loads, or AC-3 for squirrel-cage motors for normal running operations. Contactors designed for dc applications are also identified on this chart. These devices are identified as DC-1 through DC-5.

Figure 7-14a shows a table with the IEC test load specifications. From this table you can see that each of the ac and dc categories of devices is shown. The notes at the bottom of this table help explain the ratings shown in the table. The following abbreviations are used in the table; I_e is rated operational current, U_e is rated operational voltage, I is for current when the contacts are closing (make), U is for voltage before make, U_r is for recovery voltage, and I_c is for current when the contacts open (break).

The important specification from this table is the amount of current the contactor can safely handle during testing. This value is listed in column I/I_e. *Note that these ratings are for testing purposes only; they are not intended as continuous load conditions.* These ratings are used as guidelines for manufacturers to design contactors and for personnel who must select contactors for original equipment (OEM) or as replacement parts. For example, the AC-1 contactor can be expected to handle safely 100 percent of its rated current load. The AC-2 contactor can handle 2.5 times or 250 percent of its rated load, and the AC-3 and AC-4 can handle up to six times or 600 percent of its rated loads during the test cycle. The AC-3 and AC-4 must safely handle larger currents than their ratings for short periods, such as when motors are started, jogged, or reversed. Since the AC-1 contactor is not used for inductive loads, it should never be subjected to inrush currents since it is not rated for that type of application.

Notice that the operational voltage and the recovery voltage U_r/U_e is derated 17 percent for the AC-3 contactor. The reason for this is that when a motor is operating at full load, a back EMF (counter

Utilization Categories

IEC recommendations state contactor usage in utilization categories as shown in the following table.

Category		Typical Applications
ac	AC-1	Non-inductive or slightly inductive loads, resistance furnaces.
	AC-2	Slip-ring motors: Starting, plugging ❶.
	AC-3	Squirrel-cage motors: Starting, switching off motors during running. Starting, switching off the motor only after it has come to its full speed.
	AC-4	Squirrel-cage motors: Starting, plugging ❶, inching ❷. Plugging or inching duty which is making and breaking the motor load at locked rotor current levels.
dc	DC-1	Non-inductive or slightly inductive loads, resistance furnaces.
	DC-2	Shunt-motors: Starting, switching off motors during running.
	DC-3	Shunt-motors: Starting, plugging ❶, inching ❷.
	DC-4	Series-motors: Starting, switching off motors during running.
	DC-5	Series-motors: Starting, plugging ❶, inching ❷.

❶ By plugging, is understood stopping or reversing the motor rapidly by reversing motor primary connections while the motor is running.
❷ By inching (jogging), is understood energizing a motor once or repeatedly for short periods to obtain small movements of the driven mechanism.
Note — The application of contactors to the switching of rotor circuits, capacitors or tungsten filament lamps shall be subject to special agreement between manufacturer and user.

AC-3 Starting, switching off the motor only after it has come to its full speed.
AC-4 Relates to plugging or inching duty which is making and breaking the motor load at locked rotor current levels.

FIGURE 7-13 Utilization categories for IEC contactors. Comparison of IEC and JIC components. (Courtesy of Eaton Corp./Cutler-Hammer Division.)

TABLE VI. PRODUCT COMPARISON IEC TYPE AND NEMA TYPE

	ISSUE	IEC TYPE	NEMA TYPE
1	Starter Size	Smaller / horsepower (rating)	Larger / horsepower (rating)
2	Starter Price	Lower price / horsepower (rating)	Higher price / horsepower (rating)
3	Contactor Performance	Electrical life — 1 Million AC-3 Operations 30,000 AC-4 Operations typical when tested per IEC 158-1	Electrical life typically 2.5 to 4 times higher than equivalently rated IEC device on the same test *
4	Contactor Application	Application Sensitive — greater knowledge and care necessary	Application easier — fewer parameters to consider
5	Overload Relay Trip Characteristics	Class 10 typical — designed for use with motors per IEC recommendations calibrated for 1.0 service factor motors	Class 20 typical — designed for use with motors per NEMA standards calibrated for 1.15 service factor motors
6	Overload Relay Adjustability	Fixed heaters. Adjustable to suit different motors at the same horsepower (heaters not field changeable)	Field changeable heaters allow adjustment to motors of different horsepowers
7	Overload Relay Reset Mechanism Characteristics	RESET/STOP dual function operating mechanism typical	RESET ONLY mechanism typical
		Hand/Auto Reset Typical	Hand Reset Only
8	Fault Withstandability	Typically designed for use with fast acting, current limiting European fuses	Designed for use with domestic time delay fuses and circuit breakers

* Statement valid only for the devices at the lower horsepower ratings. IEC type and NEMA type devices become more performance equivalent as device horsepower ratings increase.

FIGURE 7-13 Utilization categories for IEC contactors (*Continued*)

EMF) is generated by the motor windings and causes the net voltage to be approximately 17 percent less. This allows the contacts to carry the load when contacts are opening without design changes.

Life and Load Curves for Contactors

Another important specification for contactors is the life expectancy of the electrical contacts and the mechanical assembly. One way of rating these conditions is by the number of expected cycles the component is expected to complete successfully during its lifetime. Figure 7-14b shows a graph that indicates the expected life of contactors. From this table you can see that the data listed on the bottom are in reference to an AC-3 contactor. Notice that the AC-3 contactor's contacts are expected to have 1.15 million cycles and 10 million mechanical cycles during its life. Remember that an AC-3 contactor is rated for general-purpose motor applications.

Contactors may be rated for severe applications such as plugging, jogging, inching, and other applications that cause the motor to see inrush or LRA current repeatedly. When the contactor is used in this type of application, its life expectancy is reduced to 30,000 cycles.

Additional information on IEC recommendations for overload devices and motor starters are provided at the end of this chapter. You can also obtain specific information and product literature about a particular product from your electrical component vendor. Most equipment representatives will provide you with assistance in making equipment comparisons for selecting the proper components for your applications.

Contactors **165**

Verification of the number of on-load operating cycles
Conditions for making and breaking corresponding to the
several utilization categories ❶ ❹

Category		Value of the rated operational current	Make			Break		
			I/I_e	U/U_e	$\cos \varnothing$ ❷	I_c/I_e	U_r/U_e	$\cos \varnothing$ ❷
ac	AC-1	(All values)	1	1	0.95	1	1	0.95
	AC-2	(All values)	2.5	1	0.65	2.5	1	0.65
	AC-3	$I_e \leq 17A$	6	1	0.65	1	0.17	0.65
		$I_e > 17A$	6	1	0.35	1	0.17	0.35
	AC-4	$I_e \leq 17A$	6	1	0.65	6	1	0.65
		$I_e > 17A$	6	1	0.35	6	1	0.35
			I/I_e	U/U_e	L/R ❸ (ms)	I_c/I_e	U_r/U_e	L/R ❸ (ms)
dc	DC-1	(All values)	1	1	1	1	1	1
	DC-2	(All values)	2.5	1	2	1	0.10	7.5
	DC-3	(All values)	2.5	1	2	2.5	1	2
	DC-4	(All values)	2.5	1	7.5	1	0.30	10
	DC-5	(All values)	2.5	1	7.5	2.5	1	7.5

I_e Rated operational current
U_e Rated operational voltage
I Current made
U Voltage before make
U_r Recovery voltage
I_c Current broken

❶ In ac the conditions for making are expressed in r.m.s. values, but it is understood that the peak value of asymmetrical current, corresponding to the power-factor of the circuit, may assume a higher value
❷ Tolerance for cos ∅: ±0.05
❸ Tolerance for L/R: ±15%
❹ Information extracted from IEC Publication 158

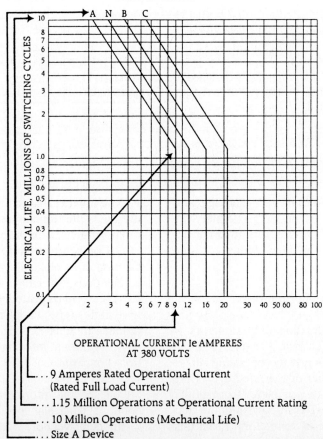

FIGURE 7-14 Test load specifications and life expectancies for IEC contactors. (Courtesy of Eaton Corp./Cutler-Hammer Division.)

Installing, Interfacing, and Troubleshooting Contactors

The contactor must be installed in some type of NEMA enclosure or electrical panel. NEMA enclosures are available in a wide variety of types for each of the different types of environments. The most common types of these enclosures include waterproof, explosion-proof, and general-purpose types. A full list of enclosure types is provided in Chapters 2 and 5. The start/stop buttons are generally mounted inside the enclosure, with the buttons protruding through the cover for easy access.

Smaller contactors can be installed and mounted in much the same manner as relays. Refer to Chapter 6 if you have questions regarding installation of these types of devices. The larger contactors, such as sizes 7 to 9, should be treated similarly to the large switchgear devices discussed in Chapter 3.

The smaller contactors can be interfaced to robots and programmable controllers for motor starting and heating loads in a manner similar to that for relays. The programmable controller allows a wide variety of control strategies to be used, such as process control algorithms for heating circuits and power factor control for switching capacitors.

Contactors can also be tested and troubleshooted using the procedures listed for relays in Chapter 6. Since contactors utilize contacts and a coil, their operation is very similar to that of relays. The major differences between contactors and relays relate to size and applications. Since you only need to determine if the coil is energized and the contacts or operating correctly, the test used for troubleshooting relays can be used to troubleshoot contactors. Additional information regarding troubleshooting contactors is provided at the end of this chapter, with information about motor starters.

MAGNETIC MOTOR STARTERS

The magnetic motor starter is one of the most widely used motor controls. The starter provides operational contacts such as the contactor and additional built-in overcurrent protection. The overcurrent protection is in the form of thermal or magnetic overloads. An overload can be added to a contactor for motor protection, or it can be considered an integral part of the motor starter. In reality, the same overload that is built into the motor starter is also available for use on a manual motor starter and can be added to a contactor to provide motor protection.

In this section of the chapter we explain the operation of electrical and mechanical components of a magnetic motor starter. The operation of the overloads will also be explained. Typical applications and circuits will be presented as well as methods of selecting the proper size of motor starter and overloads. In the last part of this section we explain installation and troubleshooting procedures that you will be able to utilize in the classroom and later on the factory floor while you are on the job.

Figure 7-15 shows a photograph of a typical magnetic motor starter, also called a magnetic starter. A magnetic starter looks very similar to a contactor; in fact, it has many of the same components: contact assembly, contacts, coils, and auxiliary contacts. The contact assembly operates the same as the assembly in the contactor. When the coil is energized the contact assembly is pulled up so that the armature is centered in the coil. This upward action causes the movable contacts to press closed to the stationary contacts. When the armature assembly is pulled up into the coil, the pole faces on the armature are pulled tight to the upper pole face. A small air gap remains between the pole faces so that the magnetic field will be broken when the coil is deenergized.

Since the armature assembly with the contact carrier is the only movable part of the magnetic starter, it can be pulled straight up and has a very short vertical stroke. This means that the coil can snap the contacts closed, which causes a very small amount of contact wear.

Parts of the Magnetic Motor Starter

It is important that you be able to recognize the basic parts of a magnetic motor starter. This is required since you will have to install, troubleshoot, and repair the starter. For example, if you need to test the coil and later change it, you will need to know where the coil is and how the contactor comes apart. This is also true if you need to check the contacts and later change them.

Figure 7-16 shows a cutaway view of a motor starter, showing all the parts of a magnetic starter as they come together and operate as a starter. From this diagram you can see that the base is attached to the mounting plate. The mounting plate is used to mount the starter to the electrical panel or enclosure. The stationary contacts are mounted in the top of base. They have the field wiring terminal points for both the line-side terminals and the load-side terminals.

The overloads are also built into the base. This is the part of a magnetic starter that makes it different from a contactor. The overloads can be

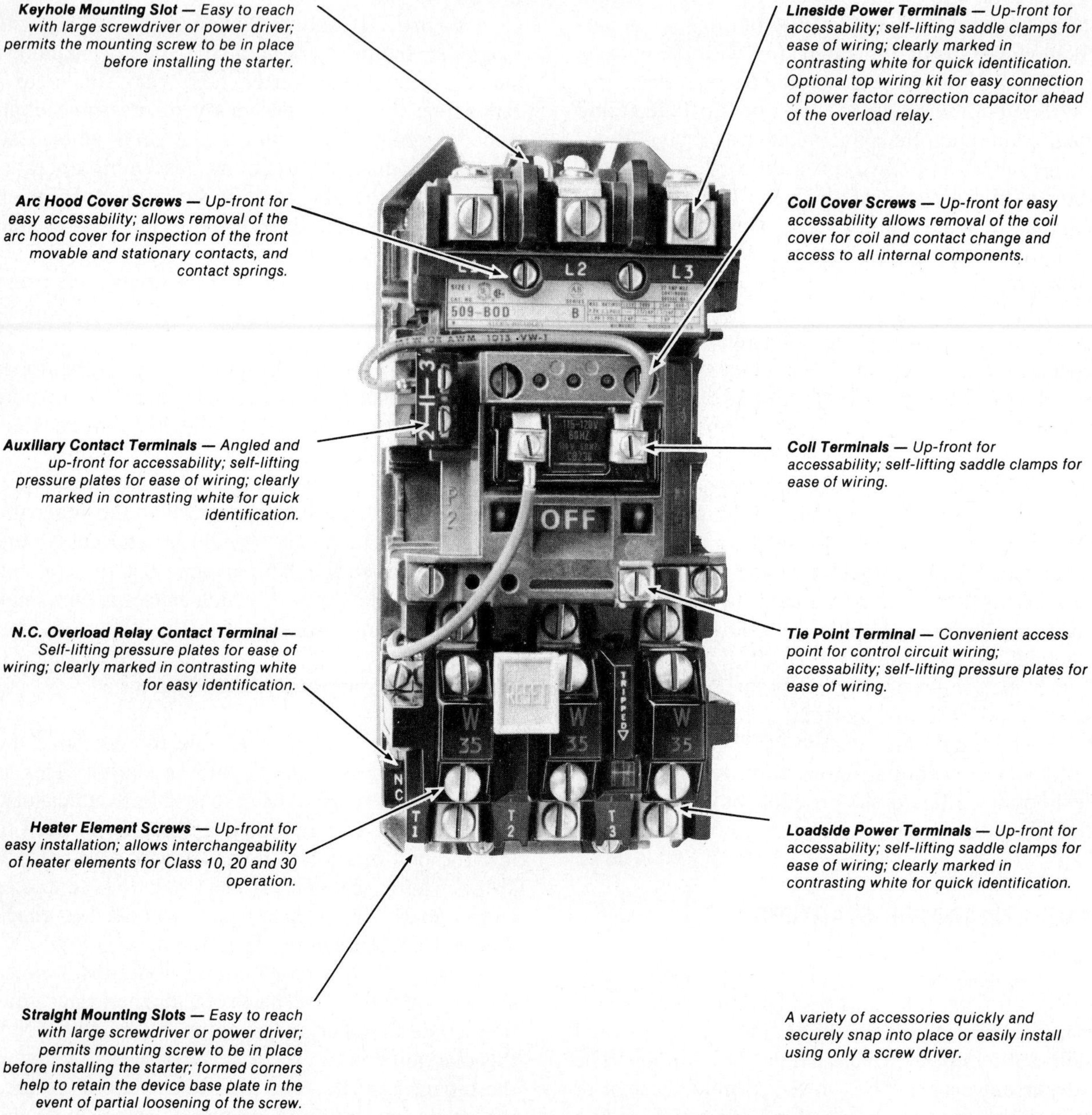

Keyhole Mounting Slot — Easy to reach with large screwdriver or power driver; permits the mounting screw to be in place before installing the starter.

Arc Hood Cover Screws — Up-front for easy accessability; allows removal of the arc hood cover for inspection of the front movable and stationary contacts, and contact springs.

Auxiliary Contact Terminals — Angled and up-front for accessability; self-lifting pressure plates for ease of wiring; clearly marked in contrasting white for quick identification.

N.C. Overload Relay Contact Terminal — Self-lifting pressure plates for ease of wiring; clearly marked in contrasting white for easy identification.

Heater Element Screws — Up-front for easy installation; allows interchangeability of heater elements for Class 10, 20 and 30 operation.

Straight Mounting Slots — Easy to reach with large screwdriver or power driver; permits mounting screw to be in place before installing the starter; formed corners help to retain the device base plate in the event of partial loosening of the screw.

Lineside Power Terminals — Up-front for accessability; self-lifting saddle clamps for ease of wiring; clearly marked in contrasting white for quick identification. Optional top wiring kit for easy connection of power factor correction capacitor ahead of the overload relay.

Coil Cover Screws — Up-front for easy accessability allows removal of the coil cover for coil and contact change and access to all internal components.

Coil Terminals — Up-front for accessability; self-lifting saddle clamps for ease of wiring.

Tie Point Terminal — Convenient access point for control circuit wiring; accessability; self-lifting pressure plates for ease of wiring.

Loadside Power Terminals — Up-front for accessability; self-lifting saddle clamps for ease of wiring; clearly marked in contrasting white for quick identification.

A variety of accessories quickly and securely snap into place or easily install using only a screw driver.

FIGURE 7-15 Identification of main parts of magnetic motor starter. (Courtesy of Allen-Bradley, Inc.)

CUT-AWAY STARTER SAMPLE

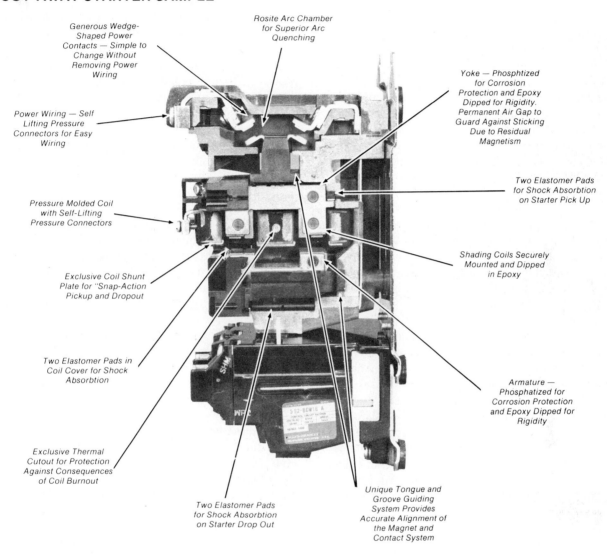

- Vertical lift gravity drop out design.
- Armature, contact carrier and contacts lift as a "single-piece" in a short vertical stroke.
- No pivot points, drop out springs or linkages.
- Elastomer pads reduce shock and contact bounce to extend mechanical and electrical life.
- Coil and yoke are keyed for proper reassembly.
- Domed surface on top of coil permits self alignment of the pole faces.

FIGURE 7-16 Cutaway diagram of a magnetic motor starter. (Courtesy of Allen-Bradley, Inc.)

attached externally to the contactors discussed earlier in this chapter, but they are built into the molded plastic base assembly. You should notice that the location of the overloads on the front of the starter allows them to be inspected and changed easily. This is necessary since you must size the heaters on the overload to the running current of the motor and sometimes they must be exchanged for the next larger or smaller size while the starter is installed in its enclosure or panel.

The other parts of the starter that are part of the base assembly are the built-in arc hoods specially molded into the plastic to prevent the arc from one set of contacts from getting near any other set of contacts or other part of the starter. You can also see that the contacts are covered by a hood cover mounted on the front of the contacts with mounting screws. The hood cover contains the arc quencher, which helps extinguish an arc as soon as it begins to develop.

A set of mounting terminals are provided on the mounting plate for the auxiliary contacts to be attached. This location is immediately to the left of the base. It allows the auxiliary contacts to be mounted in close proximity to other component parts of the starter so that interconnecting wires can be kept as short as possible.

Another component shown in this diagram is the contact carrier, which holds the movable set of contacts. Notice that each of the contacts have a spring mounted under it to help it remain tightly seated when it is closed against the stationary set of contacts. The contact carrier also houses the magnet assembly. The magnet assembly consists of the coil, which is molded in epoxy, and the laminated magnetic yoke, which has a shading coil attached. The yoke is placed through the coil and the coil is then snapped into the contact carrier. After all the parts are assembled into the contact carrier, the carrier is mounted in the base section of the starter, where it can move in the vertical direction to cause the contacts to close and open when the coil is energized. It pulls the magnet yoke upward, where it presses against the movable contacts and causes them to seat against the stationary contacts.

It is also very important that you be able to recognize the parts of the starter with the symbols used to represent them in the various types of diagrams. This is a very important concept to understand since you will be using the wiring diagram and electrical ladder diagram to install and troubleshoot the magnetic starter. This means that your mind's eye, the picture in your mind, must show the physical component when you see the symbol for it on an electrical diagram. You must also visualize the operation of these parts when you are trying to troubleshoot various components, such as the coil or contacts.

Wiring Diagram

Figure 7-17 shows a wiring diagram used to represent the magnetic motor starter. It is important that you learn to recognize this group of symbols as a magnetic starter when they appear on an electrical diagram. The best technicians and troubleshooters are quickly able to recognize the symbols on a diagram as a operating component and its individual parts, and then transfer this concept to the physical part when they must locate them for making electrical tests or change parts.

Each of the component parts of a starter is identified in the diagram. It is important to remember that the symbols in the diagram are placed to show the actual location where you would find them if you were looking at the front of the diagram. This diagram is presented as a map of where you would expect to find the parts on the starter. Notice that each electrical connection (terminal) is represented by a small circle. The three terminals at the top of the diagram are marked L1, L2, and L3. Since they are shown at the top of the diagram, that is where you would find them when you are looking at the starter.

The next symbol, which you will find directly under the line terminals, represents the load contacts. There are three sets of normally open load contacts and one set each of normally open and normally closed auxiliary contacts.

Notice that the load contacts are drawn with a heavy dark line to represent that they carry the large motor or load current and the auxiliary contacts are drawn with a thiner line to shown that they carry control current. In some diagrams the auxiliary contacts are drawn with a different color, but they will always be shown with thinner lines. This is another example of trying to show the physical relationship of components in a wiring diagram. Since the load contacts are physically much larger than the auxiliary contacts, the load contacts are always represented on the wiring diagram as being larger.

Since the load contacts are shown directly under the terminals marked L1, L2, and L3, this is exactly where you would expect to find them physically on the starter. The auxiliary contacts are shown on the left side of the starter, and the normally open pair are designated as being mounted on the rear pair. In Figure 7-17 you will notice that the auxiliary contacts are located on the left side of the starter. There are four terminals on the auxiliary contacts; two are mounted on the top of the contact

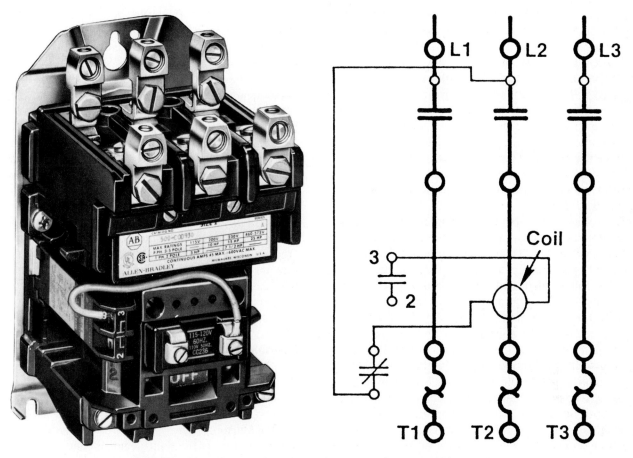

FIGURE 7-17 Electrical wiring diagrams of a magnetic starter. (Courtesy of Allen-Bradley, Inc.)

assembly and two are located at the bottom of the assembly. The front terminal on the top and bottom make up the normally closed set and the rear terminals on the top and bottom make up the normally open set.

Three terminals are identified directly below the load contacts. The small circle on the diagram indicates that there is a terminal screw located at this point on the starter. The heaters for each overload are shown directly below these terminals. The heaters are part of the overload assembly, so they are identified by the letters OL.

In the diagram three terminals marked T1, T2, and T3 are shown connected to the overload heaters. These terminals are the point where the wires from the load (motor) are connected. *Remember: the terminals marked L1, L2, and L3 identify the terminals where three-phase line voltage is connected and the terminals marked T1, T2, and T3 are where the load is connected.* This convention has been adopted for use on all North American motor starters.

The normally closed overload contacts are shown between the heater in line 1 and the heater in line 2. This diagram shows one set of overload contacts. Other starters will use one set of OL contacts is provided for each OL assembly.

Since the coil is physically located in the middle of the starter, in this diagram it is shown in the middle of the starter as a circle with two terminals. Note also that there are three small wires shown in this diagram. Two of the wires are connected to the coil. One connects the coil to the overload contacts and the other is connected to the auxiliary contacts. The third small wire connects the other side of the normally closed overload contacts to the terminal point at L2. These wires are connected to the starter when it is manufactured. This allows for simple field wiring since a pilot device such as a start/stop station can be connected between L1 and the top terminal on the normally open auxiliary contacts to complete the field wiring during installation.

Ladder Diagram

The magnetic starter can also be shown in an elementary (ladder) diagram. You must remember that the ladder diagram shows the sequence of operation for the entire circuit rather than show

Magnetic Motor Starters **171**

location of components such as the wiring diagram. The most important point to remember is that the ladder diagram is generally shown as two distinct sections. The top of the diagram usually contains the control circuit and the bottom of the diagram usually shows the load or motor circuit. The control circuit is generally shown in the top of the diagram because it must be energized first in the sequence of operation before the load contacts close and energize the motor. The load contacts are part of the load circuit with the motor, and the coil will be shown in the control circuit with the control devices.

Figure 7-18 shows two diagrams of a magnetic motor starter connected to a three-phase motor as a ladder diagram. The control circuit shows a start/stop button connected in series with the starter coil. The coil is identified by the letter M or MS for motor starter. If more than one motor starter is shown in the diagram, the coils will be identified as MS1, MS2, MS3, and so on. It is also important to remember that the load contacts and auxiliary contacts will be identified by the same letters used to identify the coil. The same mark is used to indicate which contacts the coil will activate when it is energized. This is extremely important if additional contacts are used in the circuit, because sometimes these contacts will be shown on a line of the diagram that is some distance away from the coil that activates them. In fact, sometimes the contacts will be shown closer to another coil in the circuit, which may cause some confusion if the contacts and coils are not clearly identified.

The ladder diagram should be read as a sequence of operation. This means that the start button must manually be pressed closed momentarily to allow control circuit voltage to flow to the starter coil and energize it. When the coil is energized, its magnetic field closes the three sets of load contacts, which allows three-phase voltage to pass to the motor. As you know, this voltage also flows through the overload in each line, so that the motor can be protected against overcurrents.

When the armature pulls up to close the line contacts, it also closes the auxiliary contacts. Since these contacts are wired in parallel around the start button, they will provide an alternative path for current when they are closed. This action seals the circuit in, so that the coil remains energized even when the start button is allowed to return to its normally open position. When the stop button is depressed, current to the coil is interrupted and the coil becomes deenergized. This allows the contacts to drop out and deenergize the motor. The auxiliary contacts that are used to seal in the coil circuit are also allowed to open, which breaks the seal circuit.

This requires that the start button be depressed again to reenergize the circuit.

Another way the circuit can become deenergized is if one of the overloads senses an overcurrent condition. When this occurs, the overload contacts are opened and will remain open until they are manually reset. This causes the coil current to be deenergized and the line contacts will drop open to stop the motor. If the heater element in the overload cools down, the coil will not become energized again until the overload is reset and the start button is depressed again.

If additional pilot devices or control relays are used to design circuit logic, they will be placed in the control circuit. Examples of these logic circuits are shown throughout this book. You should always remember when you are looking at a diagram that wiring diagrams are trying to show location of components and ladder diagrams are trying to show sequence of operation.

Overload Assembly

The overload assembly is considered to be an integral part of the magnetic motor starter. The overload is very similar to the overloads used on manual starters. Some manufacturers use the same overload for both their manual and magnetic starters. In this section we review the operation of the overload assembly in terms of its use in magnetic starters and the interconnections between the overload contacts and the magnetic starter's coil.

Figure 7-19 shows a photograph and diagram of a typical overload assembly. You can also refer to Figure 7-15 and see the location of the overload. Figure 7-19 shows a closer view of the overload and a diagram of the heater assembly. From this figure you can see that the overload assemblies are connected side by side so that they will fit under the line contacts in the starter. You should also notice that each overload has two terminals. The top terminal is where the overload is connected to the terminal for the line contacts and the bottom terminal is used to connect the field wires for the motor. From the picture of the overload assembly you can also see the location of each heater. A close-up picture and diagram of a heater is also provided, in which you can see the ratchet gear at the end of the heater that is inserted into the heater assembly.

Figure 7-20 shows several diagrams that animate the action of the overload assembly as it heats up from an overcurrent. Figure 7-20a shows the operation of a eutectic alloy overload. Remember from previous discussions of overloads that eutectic means low-melting-temperature solder. From this diagram you can see that the overload has

3-Phase Starters

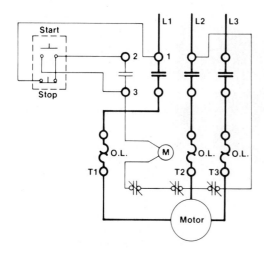

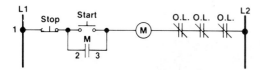

**Bulletin 509
Size 00**

Standard wiring with START-STOP push button station

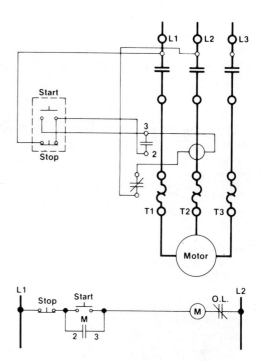

**Bulletin 509
Sizes 0, 1, 2, 3 & 4**

Standard wiring with START-STOP push button station

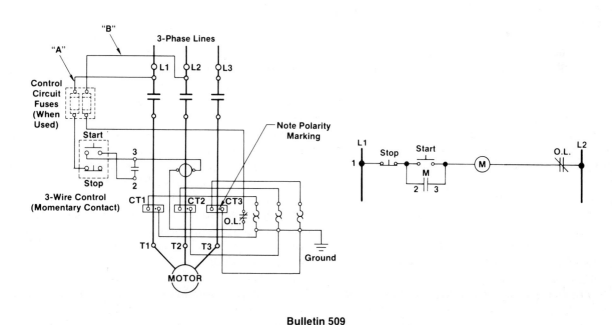

**Bulletin 509
Size 5**

Standard wiring with START-STOP push button station. Current transformers are utilized thereby allowing the use of Size 1 overload relays.

FIGURE 7-18 Ladder diagrams of a magnetic motor starter. (Courtesy of Allen-Bradley, Inc.)

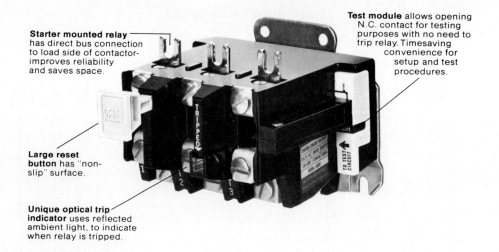

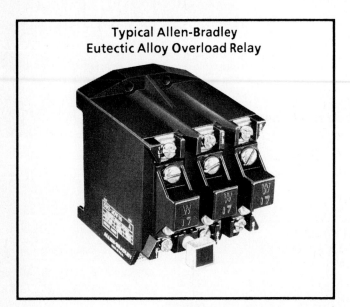
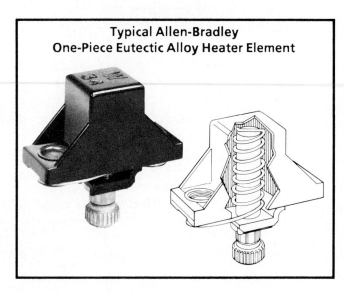

FIGURE 7-19 Picture and diagram showing the basic parts of an overload assembly. (Courtesy of Allen-Bradley, Inc.)

four parts: the ratchet and shaft, the pawl that is connected to the end of the pivot, the overload contacts and actuator, and the heating element.

Design of the Ratchet, Pawl, and Pivot

The operation of the overload depends on the design of the ratchet and pawl and how they actuate the pivot. An exploded-view diagram of these components is shown in Figure 7-20b. The ratchet is built into the end of the overload shaft and has a series of teeth that are similar to the teeth in a common gear. These teeth are set on the shaft so that they provide a tight pocket for the pawl to rest in between any two consecutive teeth.

The other end of the shaft is surrounded by the eutectic alloy (solder) where it is prevented from turning as long as the solder is solidified. A spring is connected to the pivot to try and turn it in a counterclockwise direction. The pawl is located on the end of the pivot in such a way that it will make contact with the teeth of the ratchet. When the overload is in the reset condition, the spring on the pivot is applying pressure against it so that the pawl is trying to spin or turn the shaft in the clockwise direction. Since the solder is solidified, it will prevent the shaft from turning.

When the overload's heater element senses an excess of current, extra heat is generated from the element to the eutectic alloy (solder) and begins to melt it. When the solder becomes soft enough, the spring pressure on the pivot is strong enough for the pawl to turn the ratchet and shaft. When the ratchet spins about one-fourth of a turn, it moves far

174 *Chap. 7 / Contactors and Motor Starters*

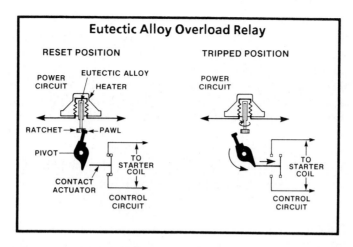

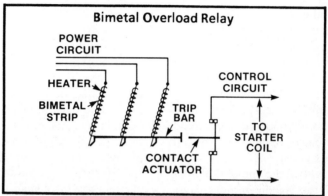

PHASE LOSS SENSITIVITY

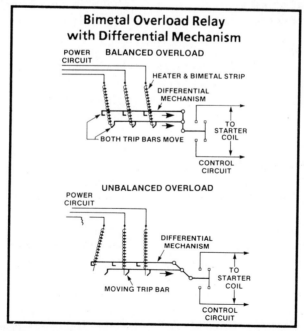

FIGURE 7-20 Diagrams showing the operation of the overload assembly. (Courtesy of Allen-Bradley, Inc.)

Magnetic Motor Starters **175**

enough to release the pivot and pawl. When the pivot and pawl move this slight distance, they will activate the overload contacts and open them.

The reset button on the overload assembly presses the pivot and pawl back into their original position. If the solder has solidified around the shaft again, the overload mechanism can be reset. If you try to reset the overload too quickly after it has tripped, the solder will still be soft and the spring pressure on the pivot will cause the trip action again and keep the overload contacts open.

Protecting against Loss of One Phase

The three-phase voltage supplied to a motor through a motor starter may lose one phase. This condition will cause the motor to overload the other two phases. This means that two of the three overload assemblies will build up heat rapidly and cause the overload contacts to trip. Figure 7-20c shows the location of a drive bar that is used to trip all the sets of overload contacts when any one of the overload heaters senses an overcurrent. This could also occur if the motor became faulty and developed an open in one of its coils.

The drive bar is located at the end of the pivot for each overload. If any one of the heaters melts the solder around its shaft, it will activate its pivot and pawl assembly. When the pivot moves, it will also move the drive bar, which will cause the other two pivots to break loose from their ratchet teeth. This arrangement ensures that if any of the overloads senses excessive current and is activated, the drive bar will also be moved and cause the other two overloads to trip.

Sizing the Heater Element for Motor Protection

The heater is a coil of precisely measured resistive wire. The amount of resistance in the wire is calculated so that the element will give off an exact amount of heat for a given current. This ensures that the heater element will heat the eutectic alloy in the overload assembly to an exact temperature to melt its solder and release the shaft and pawl. Since each motor size will draw a different amount of current, a variety of different-sized heater elements must be available to provide the proper amount of heat to melt the eutectic alloy in the assembly when the motor draws too much current. If the heater is too small for the application, it will produce enough heat to melt the solder when the motor is drawing normal full-load current.

Heaters are rated by NEMA according to the amount of time it will take them to melt the solder in the overload mechanism when the motor draws 600 percent of the rated full-load current. This means that the motor is drawing an overload equal to six times the rated full-load current. The heaters are grouped into three classes. A class 10 heater will melt the solder in the assembly in 10 seconds, a class 20 heater will melt the solder in 20 seconds, and a class 30 heater will melt the solder in 30 seconds.

If the amount of overload is less than 600 percent, it will take longer for the heaters to melt the solder. This is not a problem for the motor's protection since the motor can withstand smaller amounts of current for longer periods without causing damage to the windings.

Class 10 heaters should be used in applications for protecting hermetic motors such as air-conditioning compressors, submersible pumps, or other motors with short locked-rotor-time capabilities. Class 20 heaters are recommended for general-purpose applications, and class 30 heaters should be used in applications where motors require longer periods to come up to full speed, such as motors that are turning flywheels, grinding wheels, saw blades, and other high-inertia loads.

Figure 7-21 shows three tables for selecting class 10 heaters. Allen-Bradley identifies their class 10 heaters as type J, their class 20 heaters as type W, and their class 30 heaters as type WL. Table 164 shows the heater selection table for size 0 to 4 single-phase starters, Table 162 shows the selection table for size 0 to 4 three-phase starters, and Table 165 shows the selection table for size 5 to 9 three-phase starters. The heater type number is shown down the left side of the table. The maximum full-load current for a motor (as listed on the motor data plate) is listed in the columns. To select the proper-size heater, you need to find the column for the size of starter that you are using in your application and locate the amount of full-load amperage that is listed on the motor data plate. When you have found the motor's full-load current rating in the column for the starter that is being used, look for the part number of that particular heater in the left-hand column. For example, if the motor in your application is a three-phase motor and you were selecting the heater for a size 2 starter, you would use Table 162. If the motor's data plate indicates that the motor draws 6.43 A, you would select a J23 heater to protect the motor.

If the current listed on the motor's data plate did not exactly match the current listed in the table, you would select the heater with the amount of current listed that was nearest the motor's rating. If the temperature at the motor starter is higher than the temperature at the motor, the heater with the next-higher rating should be selected. If the temperature at the controller is lower than the temperature at the motor, the heater with the next-lower rating should be selected.

Sizing Magnetic Motor Starters

Magnetic motor starters are sized according to their ability to carry motor current. This ability is based on the size of the load contacts. Figure 7-22 shows the NEMA ratings for magnetic starters. The NEMA sizes for magnetic starters are the same as those used for manual starters. This table shows continuous amperage rating for the motor as well as the horsepower rating. Other ratings are shown in this table for tungsten and infrared lamp loads, resistance heating loads rated in kW, switching transformer primaries in kWA, and kVAR ratings for switching capacitive loads. From this table you can see that a NEMA size 2 starter can handle 3 hp for a single-phase motor at 115 V and 7.5 hp for a single-phase motor at 230 V. If the starter is used in a three-phase application, it can handle 10 hp at 200 V, 15 hp at 230 V, and 25 hp at 380, 460, or 575 V.

You can also see from this table that the size of the load must be derated if the motor is used for plugging or jogging. The starter must be derated for plugging and jogging applications because they cause the motor to draw locked-rotor current frequently, which causes the starter to overheat.

Types of Auxiliary and Interlock Contacts

Several types of auxiliary contacts are available for use on magnetic starters. These contacts are available as normally open and normally closed contacts rated for pilot duty. This means that the contacts are sized for small control currents. They can be added to the motor starter after it has been installed for some time or during initial installation. The auxiliary contacts are mounted on the side of the starter and are activated by the movable contact carrier when the coil pulls it into place. These auxiliary contacts are similar to the ones used on the contactors.

Several types of interlocks are also available. These are shown in Figure 7-23. The interlocks are a type of contact that can be added directly to the side of the magnetic motor starter. These interlock contacts are activated in the same way that auxiliary contacts are operated. In Figure 7-23, you can see the activating mechanism of these contacts sticking out of the bottom of the contact assembly.

From this figure you can also see the types of interlock contacts that are available. The top set of contacts provide a normally open and a normally closed set of contacts. The voltage that these contacts control must be from the same phase. The second set of contacts shown in this figure is exactly the same as the top set, except that they are designed to handle voltage from one phase through the normally open set of contacts and from another phase through the normally closed set of contacts. The three sets of contacts shown below the two circuit contacts each provide one set of either normally open or normally closed contacts.

It is also possible to stack contacts on top of each other. This means that you can add two or more sets of contacts on top of the first set that is mounted directly on the base of the magnetic motor starter. It is important to make sure that the activator for each set is correctly located against the activator on the set that is mounted below it.

Coil Ratings

Ac and dc coils are available for magnetic starters. Ac coils are available for 24-, 110/115-, 120-, 208/220-, 240-, 277-, 380-, 440-, 480-, 550-, and 600-V circuits. Dc coils are available for ac and dc starters. Typical coil voltages include 24, 48, 120, and 240 V. Dc coils are made to fit the smaller starters, such as NEMA sizes 0 to 5.

When you are selecting a starter for your application, you can specify a coil that will operate on ac or dc control voltage. Be sure that the coil voltage specification matches the exact amount of voltage supplied by the control circuit.

Coils for magnetic starters have typical pickup and seal-in currents similar to those listed for contactors. In most cases, all you need to do is match the coil for the proper voltage that is used in the control circuit and match the coil for the size of magnetic motor starter it will be expected to operate. All manufacturers provide extensive data for coil sizes and voltages for various applications. This information is useful for initial installations and as replacement parts.

Types of Magnetic Motor Starters

Full-Voltage Starters. Several different types of magnetic motor starters are available for different motor control applications. The most common type of control is called the full-voltage starter. The starter shown in Figure 7-15 is a full-voltage starter. This means that the starter applied full voltage to the motor for starting and control purposes. Figure 7-17 shows the wiring diagram for the full-voltage contactor. You should notice from the diagram of

Example type J heaters

FIGURE 7-21 Tables used to select class 10 heaters for single-phase and three-phase motors. (Courtesy of Allen-Bradley, Inc.)

Heaters mounted in overload assembly

Test module allows opening N.C. contact for testing purposes with no need to trip relay. Timesaving convenience for setup and test procedures.

Starter mounted relay has direct bus connection to load side of contactor—improves reliability and saves space.

Large reset button has "non-slip" surface.

Unique optical trip indicator uses reflected ambient light, to indicate when relay is tripped.

FIGURE 7-21 continued Tables used to select class 10 heaters for single-phase and three-phase motors. (Courtesy of Allen-Bradley, Inc.)

ELECTRICAL RATINGS
FOR AC MAGNETIC CONTACTORS AND STARTERS

NEMA Size	Volts	Maximum Horsepower Rating — Nonplugging and Nonjogging Duty		Maximum Horsepower Rating — Plugging and Jogging Duty †		Continuous Current Rating, Amperes — 600 Volt Max.	Service-Limit Current Rating, Amperes *	Tungsten and Infrared Lamp Load, Amperes — 250 Volts Max. ★	Resistance Heating Loads, KW — other than Infrared Lamp Loads ‡		KVA Rating for Switching Transformer Primaries at 50 or 60 Cycles ▲		3 Phase Rating for Switching Capacitors ◐
		Single Phase	Poly-Phase	Single Phase	Poly-Phase				Single Phase	Poly-Phase	Single Phase	Poly-Phase	Kvar
00	115	⅓	...	...	...	9	11	5	...	...	...	...	...
	200	...	1½	...	...	9	11	5	...	...	...	...	...
	230	1	1½	...	...	9	11	5	...	...	...	...	...
	380	...	1½	...	...	9	11	...	...	...	...	...	...
	460	...	2	...	...	9	11	...	...	...	...	...	...
	575	...	2	...	...	9	11	...	...	...	...	...	...
0	115	1	...	½	...	18	21	10	...	...	0.9	1.2	...
	200	...	3	...	1½	18	21	10	...	...	...	1.4	...
	230	2	3	1	1½	18	21	10	...	...	1.4	1.7	...
	380	...	5	...	1½	18	21	...	...	...	...	2.0	...
	460	...	5	...	2	18	21	...	...	...	1.9	2.5	...
	575	...	5	...	2	18	21	...	...	...	1.9	2.5	...
1	115	2	...	1	...	27	32	15	3	5	1.4	1.7	...
	200	...	7½	...	3	27	32	15	...	9.1	...	3.5	...
	230	3	7½	2	3	27	32	15	6	10	1.9	4.1	...
	380	...	10	...	5	27	32	...	...	16.5	...	4.3	...
	460	...	10	...	5	27	32	...	12	20	3	5.3	...
	575	...	10	...	5	27	32	...	15	25	3	5.3	...
1P	115	3	...	1½	...	36	42	24	...	...	...	...	...
	230	5	...	3	...	36	42	24	...	...	...	...	...
2	115	3	...	2	...	45	52	30	5	8.5	1.9	4.1	...
	200	...	10	...	7½	45	52	30	...	15.4	...	6.6	11.3
	230	7½	15	5	10	45	52	30	10	17	4.6	7.6	13
	380	...	25	...	15	45	52	...	...	28	...	9.9	21
	460	...	25	...	15	45	52	...	20	34	5.7	12	26
	575	...	25	...	15	45	52	...	25	43	5.7	12	33
3	115	7½	...	...	...	90	104	60	10	17	4.6	7.6	...
	200	...	25	...	15	90	104	60	...	31	...	13	23.4
	230	15	30	...	20	90	104	60	20	34	8.6	15	27
	380	...	50	...	30	90	104	...	...	56	...	19	43.7
	460	...	50	...	30	90	104	...	40	68	14	23	53
	575	...	50	...	30	90	104	...	50	86	14	23	67
4	200	...	40	...	25	135	156	120	...	45	...	20	34
	230	...	50	...	30	135	156	120	30	52	11	23	40
	380	...	75	...	50	135	156	...	...	86.7	...	38	66
	460	...	100	...	60	135	156	...	60	105	22	46	80
	575	...	100	...	60	135	156	...	75	130	22	46	100
5	200	...	75	...	60	270	311	240	...	91	...	40	69
	230	...	100	...	75	270	311	240	60	105	28	46	80
	380	...	150	...	125	270	311	...	...	173	...	75	132
	460	...	200	...	150	270	311	...	120	210	40	91	160
	575	...	200	...	150	270	311	...	150	260	40	91	200
6	200	...	150	...	125	540	621	480	...	182	...	79	139
	230	...	200	...	150	540	621	480	120	210	57	91	160
	380	...	300	...	250	540	621	...	...	342	...	148	264
	460	...	400	...	300	540	621	...	240	415	86	180	320
	575	...	400	...	300	540	621	...	300	515	86	180	400
7	230	...	300	...	...	810	932	720	180	315	...	...	240
	460	...	600	...	...	810	932	...	360	625	...	...	480
	575	...	600	...	...	810	932	...	450	775	...	...	600
8	230	...	450	...	...	1215	1400	1080	...	...	...	...	360
	460	...	900	...	...	1215	1400	...	...	...	...	...	720
	575	...	900	...	...	1215	1400	...	...	...	...	...	900

Tables and footnotes are taken from NEMA Standards Publication No. IC 1-1965 Section 2, Part 11 for Magnetic Contactors and Section 3, Parts 21B, 21C, 21D and 21F for Magnetic Starters and includes 1971 revisions for 200 V. and 380 V. ratings.

†Ratings shown are for applications requiring repeated interruption of stalled motor current or repeated closing of high transient currents encountered in rapid motor reversal, involving more than five openings per minute such as plug-stop, plug-reverse or jogging duty. Ratings apply to single speed and multi-speed controllers.

*Per NEMA Standards paragraph IC 1-21A.20, the service-limit current represents the maximum rms current, in amperes, which the controller may be expected to carry for protracted periods in normal service. At service-limit current ratings, temperature rises may exceed those obtained by testing the controller at its continuous current rating. The ultimate trip current of overcurrent (overload) relays or other motor protective devices shall not exceed the service-limit current ratings of the controller.

★FLUORESCENT LAMP LOADS — 300 VOLTS AND LESS — The characteristics of fluorescent lamps are such that it is not necessary to derate Class 8502 contactors below their normal continuous current rating. Class 8903 contactors may also be used with fluorescent lamp loads. For controlling tungsten and infrared lamp loads, Class 8903 ac lighting contactors are recommended. These contactors are specifically designed for such loads and are applied at their full rating as listed in the Class 8903 Section. Do not use Class 8903 contactors with motor loads or resistance heating loads.

‡Ratings apply to contactors which are employed to switch the load at the utilization voltage of the heat producing element with a duty which requires continuous operation of not more than five openings per minute.

▲Applies to contactors used with transformers having an inrush of not more than 20 times their rated full load current, irrespective of the nature of the secondary load.

◐Kilovar ratings of contactors employed to switch power capacitor loads. When capacitors are connected directly across the terminals of an alternating current motor for power factor correction, the motor manufacturer should be consulted as to the maximum size of the capacitor and the proper rating of the motor overcurrent protective device.

"CAUTION: For three phase motors having locked-rotor KVA per horsepower in excess of that for the motor code letters in the right table, do not apply the controller at its maximum rating without consulting the factory. In most cases, the next higher horsepower rated controller should be used."

Controller HP Rating	Maximum Allowable Motor Code Letter
1½-2	L
3-5	K
7½ & above	H

FIGURE 7-22 Electrical ratings for ac magnetic starters and contactors. (Courtesy of Square D Company.)

FIGURE 7-23 Interlock contacts available for use on magnetic motor starters. (Courtesy of Eaton Corp./Cutler-Hammer Division.)

this starter that the coil is powered from voltage from phase A and phase B of the load circuit.

When the contacts of this magnetic starter are closed, full line voltage is applied to the motor and it will draw maximum locked-rotor current. This causes the motor to start with the maximum amount of starting torque on its shaft so that it can start heavy loads.

This type of starter is also called an across-the-line starter since it simply closes the line contacts that connect the motor across the incoming voltage wires (lines). This means that no control other than the contacts are placed in the circuit with the motor. In this type of configuration the motor is allowed to see full applied voltage as soon as the contacts are closed.

Combination Starters. A combination starter combines a magnetic motor starter in a cabinet with a disconnect switch. The disconnect switch is the same type of switch that was shown in Chapter 4. Figure 7-24 shows a photograph of a combination starter. The disconnect switch can be the fused type or the nonfused type, or may have a circuit breaker instead of a fuse for circuit protection. If a fused disconnect is used, the fuse must be a dual-element type (inverse-trip-time characteristic) so that enough locked-rotor current during starting is allowed to pass through the fuse without tripping it.

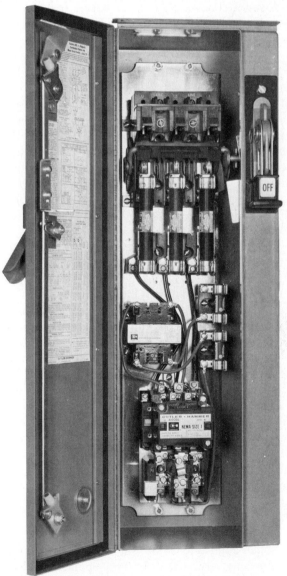

FIGURE 7-24 Combination starters. (Courtesy of Eaton Corp./Cutler-Hammer Division.)

Magnetic Motor Starters 181

The inverse time characteristic will trip or blow the fuse if the motor draws excessive current for an extended period during its running cycle. If a circuit breaker is used, it should also have an inverse time characteristic so that the motor is allowed enough time to come up to speed when it is pulling locked-rotor current during the time that the motor is trying to start.

The combination starter allows a compact installation by including the motor starter in the same cabinet as the disconnect. The start/stop button for the starter is mounted on the outside of the cabinet door. This allows the starter and disconnect to meet the requirements of National Electrical Code, which requires the motor to have a disconnecting means when it is used to power machinery.

This type of control also allows the motor starter to be mounted without requiring a separate enclosure. Time is also saved during installation since the starter is wired directly to the disconnect by the manufacturer.

Motor Control Centers

In some applications, such as with multiple conveyors that pass through several floors in a factory, it is not practical to mount the motor starter near the motors. Instead, all the motor starters are located in a panel called a motor control center. This type of installation groups all the motor starters together for easy installation and troubleshooting. Figure 7-25 shows several examples of motor control centers.

The motor control center is also used to control remotely located equipment such as ventillating fans and equipment mounted in the ceiling of a building. A reset button is mounted on the door of each motor starter. If a motor starter's overload trips, the reset button can be pressed without opening the door.

The motor starters in the motor control center are available with circuit breakers or fuses for protection with their heaters and overloads. Starters are also available as reversing starters or two-speed starters. The motor control center provides an option that allows you to mount motor starters together, or you can mount them in individual enclosures near the motor, as explained previously.

Control Transformer with the Starter

In some applications the supply voltage to the disconnect is over 400 V. In these applications it is not safe to use the high voltage as the control voltage since it is a safety hazard to apply high voltage to start/stop buttons. For these applications, a step-down control transformer is provided in the combination starter. The control transformer drops the voltage to 120 or 208 V, which can be used safely in the control circuits and pilot devices. It can be see above the starter in Figure 7-24.

Reversing Starters

Another specialty type of starter is the *reversing magnetic starter*. The reversing starter is actually a starter and contactor mounted in the same cabinet. The starter is used for forward operation of the motor and the contactor is used for reverse operation. The starter and the contactor each have their coil. The overload assemblies for both of the controls are mounted on the starter.

Figure 7-26 shows a photograph of the reversing starter. The two controls are located side by side in the same enclosure. The wiring diagram in Figure 7-27 shows the location of the components for these controls. This is a good example of where a wiring diagram is not suited for explaining the operation of a circuit. The wiring diagram is important because it shows the location of all the components and the terminal points for all field wiring connects. The ladder diagram does not show locations where the field wires should be connected, but it does show the sequence of operation of the reversing circuit.

From the ladder diagram you can see that the coil for the forward starter is identified with the letter F and the coil for the reversing contactor is identified with the letter R. All the contacts in this diagram that are controlled by the forward coil are also identified by the letter E, and all the contacts controlled by the reverse coil are identified with the letter R.

When the forward start button is depressed, the forward coil is energized, which activates all the sets of contacts marked with an F. Notice that the forward coil closes the normally open seal in contacts that are wired in parallel with the forward start button, and open the normally closed set of contacts that are wired in series with the reverse coil. The normally closed contacts act as an electrical lockout for the reverse coil circuit. The forward load contacts are also closed, so that the motor begins to operate in the forward direction.

Some reversing starters provide a mechanical interlock in addition to the electrical interlock. The mechanical interlock provides a mechanism that physically prevents the reversing contacts from coming closed any time the forward contacts have been energized. The interlock also prevents the forward contacts from closing if the reversing con-

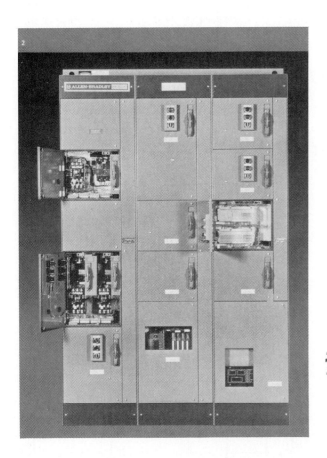
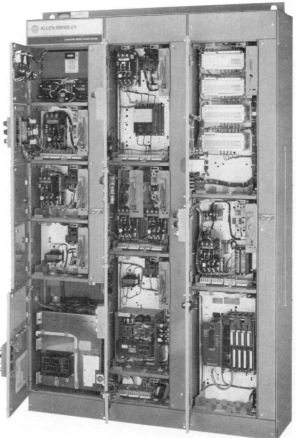

FIGURE 7-25 Motor control centers. (Courtesy of Eaton Corp./Cutler-Hammer Division and Allen-Bradley, Inc.)

Magnetic Motor Starters **183**

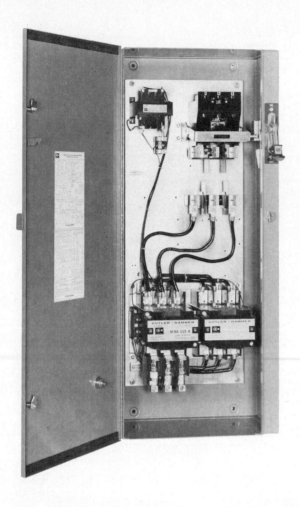

FIGURE 7-26 Picture and diagram of reversing starters. (Courtesy of Eaton Corp./Cutler-Hammer Division.)

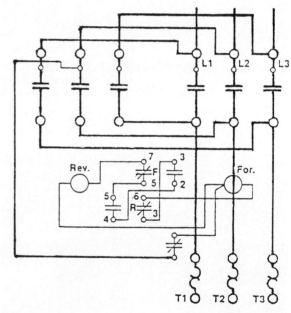

FIGURE 7-27 Diagram of a reversing starter that allows a motor to reverse without using the stop button. (Courtesy of Allen-Bradley, Inc.)

tacts have been closed first. The mechanical interlock is identified on the ladder diagram as a dashed line.

If you look at the load contacts of each control, you will notice that two legs of the three-phase voltage are exchanged at the reversing starter to make the motor run in the opposite direction. This also presents a potential problem if both sets of contacts would be closed at the same time, since the reverse contactor would provide a direct short circuit between L1 and L2. The electrical and mechanical interlocks are provided to prevent both contacts from being closed at the same time. *Neer try to press the reerse contacts closed by hand with a screwdrier or other tool when the forward contacts are closed, as this will cause a seere arc to occur from a short circuit.*

When the motor is running in the forward direction, the stop button must be depressed to deenergize the forward contactor before the reverse contactor can be energized. When the forward contactor is deenergized, the normally closed contacts that make up the forward electrical interlock are allowed to return to their normally closed condition. This allows the reverse start button to be depressed so that the reverse coil can be energized. When the reverse coil becomes energized, all the contacts identified by the letter R are activated. The normally open set of reverse contacts that are connected in parallel across the reverse pushbutton act as a seal for it in the circuit. The normally closed set of reverse contacts that are connected in series with the forward pushbutton will be opened so that they act as an electrical interlock for the reverse contactor. The dashed line on this contactor indicates that a mechanical interlock will be engaged so that both sets of contacts cannot be energized. If the motor is required to run again in the forward direction, the stop button must be depressed first to disengage the electrical and mechanical interlocks.

The heaters for the overloads are placed in lines so that they are in the circuit when the motor is running forward and reverse. The contacts for the overloads are in the control circuit next to the stop button. This location ensures that they are wired in series with both the reverse and forward control circuits so that they can protect the motor at all times.

Another type of reversing starter is also available. This type of control allows the motor to be switched from the forward direction to the reverse direction without using the stop button. This type of control is used for jogging and similar applications that require the motor to be reversed quickly. Figure 7-27 shows a diagram of this type of control. This control used only electrical interlocks. The interlocks in this case are slightly different in that they allow control to be switched from the forward direction directly to the reverse direction without using the stop button. The interlock still prevents both sets of contacts from closing at the same time, but it does allow the reverse coil to be energized while the forward coil is energized. Operation of the control contacts allows the reversing coil to become energized while it is deenergizing the reverse contacts.

Two-Speed Magnetic Motor Starters

Two-speed motor starters use two magnetic starters to provide control for two-speed motors. Figure 7-28 shows a photograph and diagrams of a typical two-speed motor starter. This type of control uses two separate magnetic motor starters housed on the same mounting plate, which allows the two starters to be mounted side by side in the same enclosure. Each starter has its own set of overloads, which allows the motor to be protected when the motor is operated on either high speed or low speed.

The operation of the starters in this application is best explained by the ladder diagram. The wiring diagram shows the two starters in the location where you would find them. You should also notice that the dark lines in the diagram indicates the heavier load wires and the thin lines in the diagram indicates the lighter control voltage wires. The motor in this application is a two-speed two-winding three-phase motor. When the motor is wired for low speed, L1, L2, and L3 are connected to T1, T2, and T3, respectively. Terminals T11, T12, and T13 remain open. When the motor is wired for high speed, L1, L2, and L3 are connected to T11, T12, and T13, respectively. T1, T2, and T3 remain open. Additional information on motor speed applications is provided in Chapter 11. The terminals for the high-speed winding, T11, T12, and T13, are connected to the three load-side terminals in the first starter. Terminals for the low-speed winding, T1, T2, and T3, are connected to the load-side terminals of the second starter. It is very important that you connect the motor leads to the load terminals on the starters in the exact order as listed in the diagram. If any of the leads are interchanged, the motor will try to operate in the opposite direction when it changes speed.

Operation of the two-speed circuit is indicated in the ladder diagram. Each motor starter has its own coil. The high-speed coil is identified by the letter H and the low-speed coil is identified by the letter L. All the contacts in the diagram identified by the letter H are activated by the high-speed coil and all the contacts identified by the letter L are activated by the low-speed coil.

This circuit is controlled by a set of pushbuttons. The top pushbutton is for high speed, the middle pushbutton is for low speed, and the bottom pushbutton is the stop button. The pushbutton for high speed consists of a set of normally open contacts and a set of normally closed contacts. The stop button and the low-speed button consist of only one set of normally open contacts.

When the high-speed pushbutton is depressed, the normally open set of contacts in the switch energize the high-speed coil. The normally closed set of contacts in the high-speed switch are opened when the high-speed pushbutton is depressed. This ensures that the low-speed coil cannot energize at the same time as the high-speed coil. When the coil to the high-speed motor starter is energized, the main load contacts are closed and the motor begins to run at high speed. The normally open set of auxiliary contacts from the high-speed starter are wired in parallel across the normally open set contacts on the high-speed pushbutton. These auxiliary contacts close and seal the high-speed pushbutton when it is depressed momentarily. Any time the stop button is depressed, the high-speed coil is deenergized. If the overload contacts open, the coil will also be deenergized.

Any time the low-speed button is depressed, the coil for the low-speed motor starter is ener-

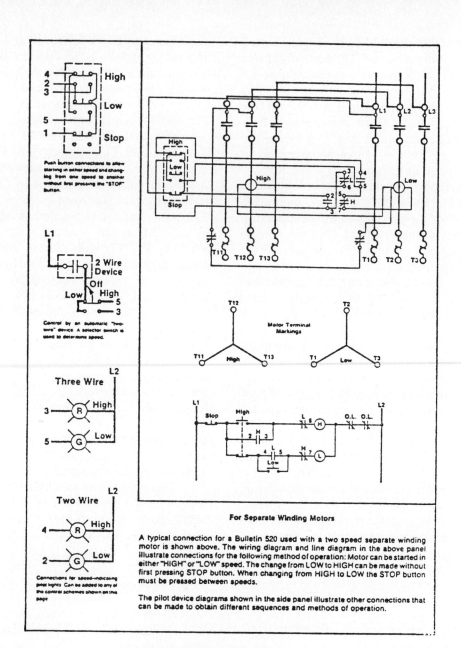

FIGURE 7-28 Picture with ladder and wiring diagram for two-speed motor starter. (Courtesy of Allen-Bradley, Inc.)

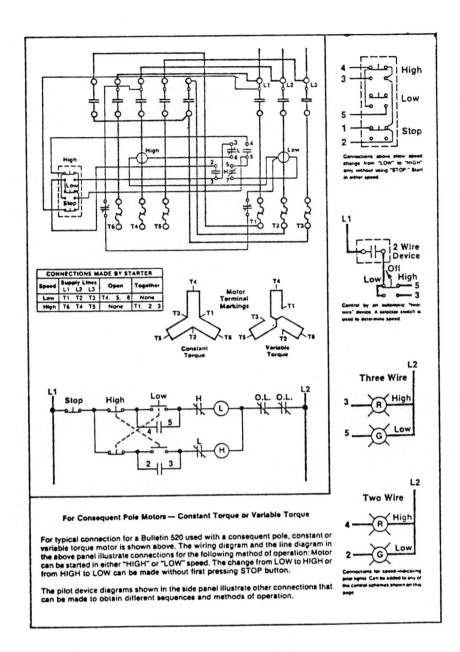

FIGURE 7-28 Two-speed motor starter (*Continued*)

Magnetic Motor Starters

gized. The low-speed coil can even be energized if the high-speed coil is energized. This can occur since the normally closed set of contacts on the high-speed pushbutton will be returned to their closed position whenever the high-speed pushbutton is released. Since these contacts are closed when the motor is running at high speed, the low-speed pushbutton can be depressed and the low-speed coil will become energized. The high-speed load contacts will be dropping out while the low-speed contacts are pulling in. The slight time difference will ensure that both sets of contacts will not be energized at the same time. The extra set of contacts in the high-speed pushbutton also ensures that only the high-speed pushbutton will be energized if both the high-speed and low-speed pushbuttons are depressed at the same time.

Two-Speed Motor Starters for Consequent Pole Motors

Figure 7-28 also shows the wiring diagram and ladder diagram for a two-speed consequent pole motor. The motor connection chart is also provided with this diagram so that you can see how the motor leads are connected to the two motor starters. The high-speed pushbutton and the low-speed pushbutton each have one set of normally open contacts and one set of normally closed contacts. A mechanical interlock on the pushbutton switches prevents the high- and low-speed coils from being energized at the same time. This wiring configuration of the high-speed and low-speed pushbuttons allows the motor to be switched from high to low speed or from low to high speed without using the stop button. Whenever the stop button or overload contacts are opened, both of the motor starter's coils will be deenergized.

Interfacing Magnetic Motor Starters with Programmable Controllers and Robots

Magnetic motor starters are often interfaced with programmable controllers and robots. In most cases the starter's coil is connected to the programmable output module. It is also possible to connect the auxiliary contacts of the motor starter to input modules on the P/C.

Another reason a programmable controller is used to control motor starters is that its program can duplicate complex motor control logic without involving intricate wiring. The P/C program can provide electrical interlocks for reversing and two-speed applications without using multicontact switches. The P/C output that controls the motor starter coil can also be connected to the output of a programmed timer, counter, or sequencer for specialized controls.

Another advantage of using a programmable controller to control a motor starter is that it eliminates the need for additional sets of auxiliary contacts to be mounted physically on the motor starter. This is possible because the programmable controller allows any input contacts to be used in the program as many times as is needed. This also allows additional safety circuits to be programmed into the system without using any additional hard wiring, which is expensive and time consuming to install.

The programmable controller also allows the motor starter to be troubleshooted easily when a fault is suspected. The status indicators on the output module can be used to determine if the motor starter's coil has been energized, and the input module's status indicator can be used to determine if the auxiliary contacts on the motor starter have closed, which indicates that the motor starter is energized.

Other advantages of using a programmable controller for controlling a motor starter is that the P/C's output modules are available to energize a variety of ac and dc voltages. This provides a very wide variety of choices of voltages when a system is being designed.

Figure 7-29 shows several motor starters with ac and dc coils connected to programmable controller output modules. The auxiliary contacts of each motor starter are shown connected to input modules. The program that would be entered into the programmable controller is also shown. Notice that the contacts in the programmable controller program are identified with the address to which the coils and contacts from the motor starter are connected. Since the program can be as large as the memory of the programmable controller, a very complex control system can be designed. The overload heaters are still connected in series with each of the three-phase lines connected to the motor, and the overload contacts are connected in series with the motor starter's coil. Most designers prefer to leave the overload contacts in the hard-wired circuit connected in series with the coil. This ensures that the overloads cannot be by passed in the P/C's program, which would allow the motor to be damaged. If the overload contacts must be monitored with the P/C, one of the extra sets can be used or a logic circuit can be designed in the P/C program to monitor the contacts' activity.

More programs for motor starter control are provided in Chapter 16. After you understand the wide variety of logic possibilities that are available in a P/C you can design more complex control circuits with a minimum of two wires for an output and two wires for an input. This means that the control circuit can be entered and later changed with a minimum of effort.

The motor starter can also be connected directly to most robot outputs. Since the voltage at a robot output bus may be 24 V dc or TTL voltage, you must match the coil to these voltages. If a coil is not available at the voltage that your robot uses on its output bus, an interface device such as a small control relay or a solid-state relay can be used. It is also important to check the amount of current that is available at the robot's output to make sure that it is sufficient to energize the motor starter's coil. If the starter is larger than a NEMA size 5, a small control relay may be required. The coil of the control relay is connected directly to the robot output, and the coil of the motor starter is connected to the control relay's contacts. This allows the robot to control virtually any size of motor starter that is available even though the robot output bus provides a small amount of signal voltage and current.

Figure 7-29 also shows the motor starter connected to a robot bus. Notice that the coil of the motor starter is rated for 24 V dc to match the output voltage of the robot bus. Since the motor starter is controlled directly from the robot, the language to turn it on or off will be written into the robot program.

Installing the Magnetic Motor Starter

Motor starters are installed and field wire connections are made according to the wiring diagram. The ladder diagram is not much help in the installation procedure. Remember that some manufacturers install a wire between the starter's coil and overload contacts and between the coil and the auxiliary contacts at the factory. This prewiring allows the auxiliary contact to be used to seal in a momentary start pushbutton, and it ensures that the overload contacts are used in series with the coil. It also saves time and money for field wiring that would need to take place to make these connections at the time of installation. If the starter is being used in a two-wire control circuit, the wire from the motor starter coil and the auxiliary contacts will need to be removed.

The field wiring should be connected to the motor starter in accordance with the wiring diagram. The control voltage may not use the same source as the load voltage. In some circuits the control voltage will be supplied from a control transformer. Be sure to test the amount of voltage that will be used for the control circuit prior to applying power to the control circuit.

If a reversing motor starter or two-speed motor starter is being installed, be sure that you determine the correct phase sequence of the supply voltage lines for each motor starter. If the sequence is reversed on either of the starters, a rotational reversal may occur when the motor changes speed, or the motor may not change rotation when you expect it to with the reversing starter.

A variety of NEMA enclosures are available for the different types of applications that the motor starter may be used for. Some of the enclosures provide an airtight seal for explosion-proof applications. The complete list of enclosures was provided in Chapter 5, where installation of manual motor starters was explained. If you need to review the types of enclosures that are available, refer to the information in that chapter.

The enclosure type is specified on the wiring diagram for each application. Be sure that all connecting hardware such as nipples and conduits matches the enclosure type. Also be sure that the enclosure cover fits properly when you have completed the field wiring. After the installation is complete, the application should be throughly tested. If any problems are detected, use the troubleshooting procedure listed in the next section.

Troubleshooting the Magnetic Motor Starter

The magnetic motor starter is troubleshooted with a combination of procedures that are used for manual motor starters and relays. The contacts and overloads should be tested using the procedures from Chapter 5, and the coil and auxiliary contacts should be tested using procedures from Chapter 6. You should always start the troubleshooting procedure with a test of the coil circuit. *The most important fact to remember about the magnetic motor starter is that the coil must be energized before the contacts will close.*

A diagram is provided for use during the installation procedure. From this diagram you can see that the line wires are connected to the terminals marked L1, L2, and L3. The wires to the motor will be connected to the terminals marked T1, T2, and T3. The control wires should be connected at the coil terminal and at the last overload. In some motor starters the interconnecting wires from the coil through the overloads and auxiliary contacts may be installed by the manufacturers to save you time. If this has not been completed, or if you must make minor changes in your circuit, be sure to

Magnetic Motor Starters

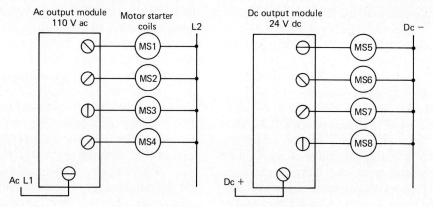

(a) Motor starters connected to output modules

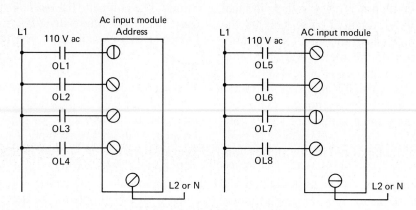

(b) Over loads connected to input module

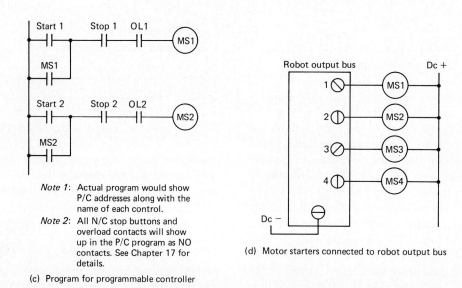

Note 1: Actual program would show P/C addresses along with the name of each control.
Note 2: All N/C stop buttons and overload contacts will show up in the P/C program as NO contacts. See Chapter 17 for details.

(c) Program for programmable controller

(d) Motor starters connected to robot output bus

FIGURE 7-29 Inputs and outputs for programmable controllers and robots.

check these connections carefully as they are completed.

Since the coil must energize before the contacts will close, it is important to determine if the coil is receiving the proper amount of voltage to energize. Most manufacturers provide some type of visual indicator on the face of the contactor's coil to indicate that the coil is energized. Figure 7-15 shows a typical indicator. The indicator is usually quite visible so that troubleshooting technicians can see it from a distance. Other starters also provide indicator lamps or pilot lamps to show when the coil is energized. A diagram of a pilot lamp connected to a motor starter is also shown in this figure. If the

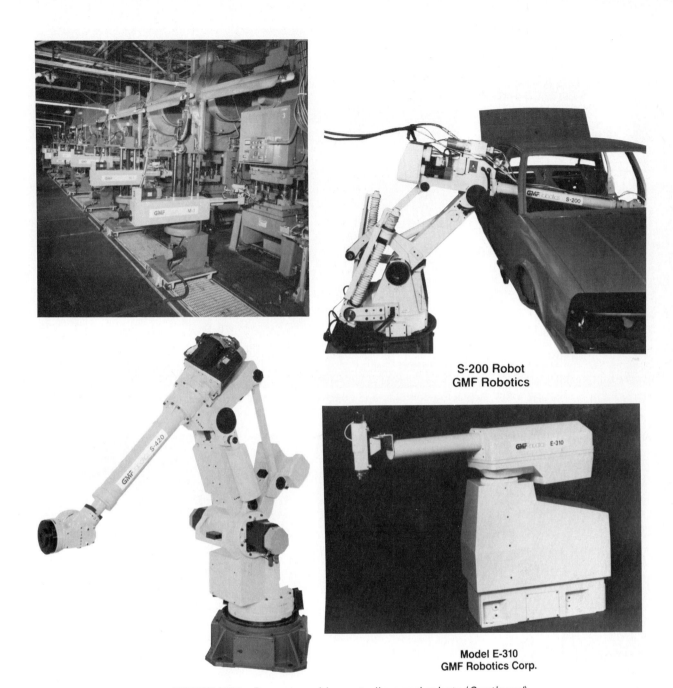

FIGURE 7-29 Programmable controllers and robots (*Continued*)

starter is housed in an enclosure, the pilot lamp or indicator may be mounted on the cover of the enclosure.

Once you have determined the status of the coil, you can begin to form a troubleshooting strategy. If the coil is not energized, your test should focus on the control circuit. If the coil is energized, the tests should focus on the load circuit. Be sure to test that the incoming voltage is the proper amount. The next tests should be to see that voltage is passing through each set of contacts. It is possible that one or more of the load contacts is defective from arcing. Another problem in the circuit is an improperly sized overload heater. If the heater is too small, the motor will continually trip the overload contacts when the motor is operating correctly.

If the contacts are passing voltage correctly, the next test should be performed on the motor. You may need to refer to Chapters 10 and 11 for troubleshooting procedures for dc and ac motors. If the motor is receiving the proper amount of voltage but does not operate correctly, it is defective and must be tested further or replaced.

Magnetic Motor Starters

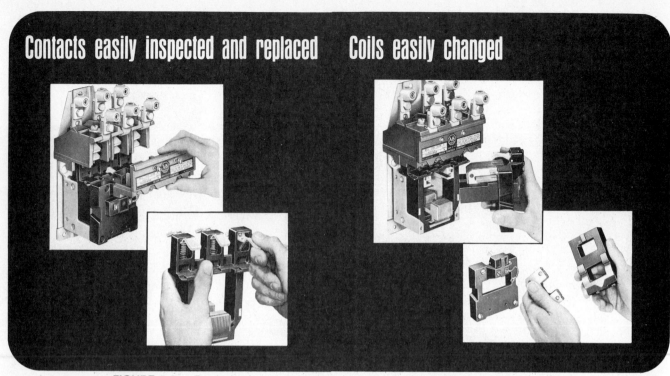

FIGURE 7-30 Replacing contacts and coils in the field. (Courtesy of Allen-Bradley, Inc.)

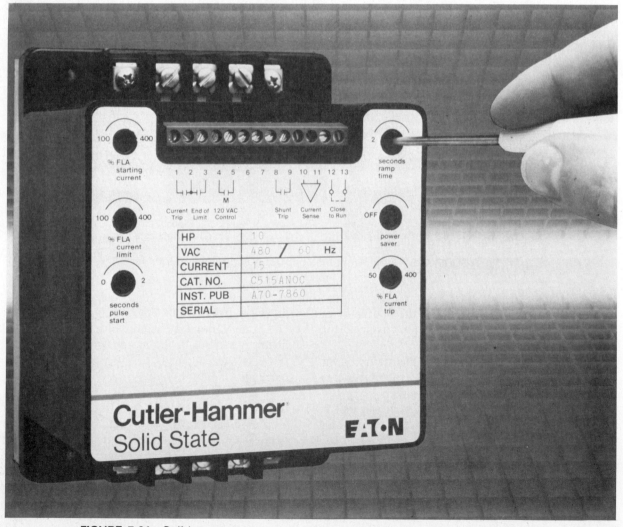

FIGURE 7-31 Solid-state motor starter. (Courtesy Eaton Corp./Cutler-Hammer Division.)

Replacing Contacts and Coils

If the motor starter line contacts or auxiliary contacts are found to be defective, they can be changed without replacing the complete starter. Most manufacturers provide contact kits that contain new contacts and springs. The contacts can be installed when the motor starter is mounted in place. The front cover of the starter can be removed and the contacts can easily be replaced. Figure 7-30 shows motor starter contacts being replaced in the field. You may need to refer to the exploded view diagram in the beginning of this section to get a better idea of where the contacts are located.

The coil of a motor starter can also be replace while the it is mounted in place. A picture of a coil being replaced is also shown in Figure 7-30. It is important to ensure that the coil voltage specification matches the coil that is mounted in the starter. For some coils color codes are used to indicate the amount of coil voltage, while for other coils the proper voltage and frequency for the coil are shown stamped on the front of the coil. The motor starter must be disassembled so that the coil can be removed and replaced. After the coil has been replaced it can be tested and the circuit can be put back into commission. Today solid state motor starters are also available. These starters are used in any application where a normal motor starter could be used. Figure 7-31 shows a solid state motor starter.

QUESTIONS

7-1. Explain how a contactor and motor starter are similar.

7-2. Draw a diagram of a motor starter and identify all its parts.

7-3. Explain how a motor starter is different from a contactor.

7-4. List two applications for a contactor.

7-5. Draw a diagram of a contactor and identify all its parts.

7-6. Explain what the contact carrier does on a contactor.

7-7. Of what material are the contacts on a contactor made?

7-8. Explain what arc shutes and arc traps do on a contactor.

7-9. List all the sizes of contactors that you would find on a machine. Indicate which is the largest.

7-10. What does "NEMA" stand for?

7-11. Explain how auxiliary contacts can be mounted on a contactor and what they could be used for.

7-12. Explain how the mechanical interlock operates on a contactor. How is it released?

7-13. Does the coil voltage for a contactor have to be the same as its contact voltage? Why?

7-14. Explain the methods that manufacturers use to identify different voltage ratings for their coils.

7-15. Explain the differences between an ac and a dc coil.

7-16. Explain the function of a thermal cutout on a coil. How does it work?

7-17. Draw a diagram that shows a contactor used to jog a motor.

7-18. Draw a diagram that shows a set of contactors used to reverse the rotation of a motor.

7-19. What does "IEC" stand for?

7-20. When was NEMA established?

7-21. When was IEC established?

7-22. Give several differences between NEMA and IEC standards.

7-23. Where would you expect to find IEC rate controls and equipment?

7-24. Explain how you would test a contactor that was controlling a motor if the motor would not run when you pressed the start button.

7-25. Explain the operation of the overloads in a motor starter.

7-26. What function do heaters provide?

7-27. Explain where you will find the identifications L1 and T1 on a motor starter.

7-28. Draw a ladder diagram of start/stop buttons controlling a motor starter. Be sure to use a ladder diagram and show a three-phase motor connected as the load.

7-29. Explain how the ratchet and pawl and pivot operate as an overload device. Draw a brief sketch of this device.

7-30. What is the maximum amperage that a size 2 motor starter can carry if the motor runs on 480 V?

7-31. Why would frequent starts affect the time it takes an overload to kick out?

7-32. Explain what a combination starter is.

7-33. Why shouldn't you try to "plug" a forward or reversing motor starter?

7-34. Draw a diagram of a motor starter connected to the output module of a programmable controller.

7-35. Explain how you would troubleshoot a motor starter. Do not forget to include checking the overload.

Pilot devices   8

INTRODUCTION

Pilot devices are switches used to control small amperages in motor control circuits. These switch contacts can be activated by motion, temperature, flow, liquid level, or other industrial actions. The main feature of a pilot device is the amount of current (amperage) it can handle. Since these switches are typically used in the control portion of a circuit, they are also called control devices. In some cases these devices can be connected directly to small motors, but generally they are used to control current for the coils of relays, motor starters, and solenoids. Small loads such as indicator lamps are also considered pilot devices. Information concerning indicator lamps used as pilot loads is provided later in this chapter. Figure 8-1 shows the typical current ratings of pilot switches. You should notice that the type of current and the amount of current are both listed. Additional columns indicate the amount of current the switch will control on a continuous basis and during make or break. The term *make* indicates the time when the contacts close and allow current to begin flowing to the load. The term *break* indicates the time when contacts open and cause the interruption of current flow to the load. The characteristics of inductive and resistive loads vary considerably when current begins to flow and again when current is interrupted. For instance, an inductive load such as a coil will draw very large currents when current first begins to flow. This current is called inrush current. It will be two to five times larger then the continuous current the load will draw. The amount of inrush current will also vary between loads that use dc voltage and loads that use ac voltage. Inrush current for ac loads is typically much higher than for loads in dc circuits. Since the inrush current exists for only a few ac cycles, the contacts may have a continuous current rating of 15 A and be able to withstand inrush currents up to 40 A. When current to an inductive load is interrupted by switch contacts opening, the inductive load generates a moderate amount of back EMF (voltage), which will cause an increase in the current across the switch contacts. This increase of current looks like a spark or arc across the contacts and tends to continue for only a few ac cycles. If the load is pure resistive, such as an indicator lamp, in rush current will not occur. In this type of circuit the contact rating is based on the amount of continuous current to the load. If the contacts are used as part of multiple sets of contacts, such as double pole, double throw, the current rating must be derated.

MAXIMUM CURRENT RATINGS FOR CONTROL CIRCUIT CONTACTS — TYPES C, XA, T, FT, and A

Switch Type	Contacts	Volts	AC — 50 or 60 Hz					Volts	DC			AC or DC
			Inductive 35% Power Factor				Resistive 75% Power Factor		Inductive and Resistive			Continuous Carrying Amperes
			Make		Break		Make and Break Amperes		Make and Break Amperes			
			Amps.	VA	Amps.	VA			Single Throw	Double Throw		
C	SPDT	120 240 480 600	60 30 15 12	7200 7200 7200 7200	6 3 1.5 1.2	720 720 720 720	6 3 1.5 1.2	125 250 600	0.55 0.27 0.1	0.22 0.11	10 10 10 ....
C	DPDT	120 240 480 600	60 30 15 12	7200 7200 7200 7200	6 3 1.5 1.2	720 720 720 720	6 3 1.5 1.2	125 250 600	0.22 0.11	0.22 0.11	10 10 10 ....
C	Reed✱ SPST	120 240 480 600	15 7.5 3.75 3	1800 1800 1800 1800	1.5 0.75 0.375 0.3	180 180 180 180	1.5 0.75 0.375 0.300	125 250	0.55 0.27	....	5 5 5 5
XA	Reed✱ SPST	120 240	2.0 1.0	240 240	0.2 0.1	24 24	0.2 0.1	125	0.2	....	0.5 0.5
AW, AO-2 and AO-6 AB, AP	SPDT SPST●	120 240 480 600	40 20 10 8	4800 4800 4800 4800	15 10 6 5	1800 2400 2880 3000	15 10 6 5	125 250 600	2.0 0.5 0.1	0.5 0.2 0.02	15 15 15 15
AW, CO-3 and CO-6 CB, CC, CP	DPDT DPST	120 240 480 600	30 15 7.5 6	3600 3600 3600 3600	3 1.5 0.75 0.6	360 360 360 360	3 1.5 0.75 0.6	125 250 600	1.0 0.3 0.1	0.2 0.1	10 10 10 10
AO-1, AC	SPDT SPST●	120 240 480 600	40 20 10 8	4800 4800 4800 4800	15 10 6 5	1800 2400 2880 3000	15 10 6 5	125 250 600	0.5 0.25 0.05	0.25 0.1	15 15 15 15
T and FT	SPDT Quick Make and Break	120 240 480 600	150 75 37.5 30	18,000 18,000 18,000 18,000	20 12.5 6.25 5	2400 3000 3000 3000	20 12.5 6.25 5.0	125 250 600	5.0 1.0 0.2	....	20 20 20 20
T and FT	All Slow Make and Break	120 240 480 600	60 30 15 12	7200 7200 7200 7200	6 3 1.5 1.2	720 720 720 720	6 3 1.5 1.2		....	....	20 20 20 20

● SPST versions are rated ½ HP at 110 and 200 VAC.
✱ Use of a transient suppressor will extend life of the switch when using on heavy electrical loads.   + VITON is a Registered Trademark of DuPont.

FIGURE 8-1 Table for electrical contact ratings for pilot devices. (Courtesy of Square D Company.)

The derating is required because larger currents may arc between the different sets of contacts. The voltage rating of contacts will also affect the amount of current the contacts can safely handle. As the voltage rating increases, the current rating decreases. This derating occurs because the VA rating of the contacts remains essentially the same. If the contacts are used in double-pole, double-throw arrangements, the VA rating remains exactly the same. This chart represents contact ratings for one manufacturer, but they will be nearly the same for pilot devices made by other companies.

The contacts used in pilot devices can be either normally open, normally closed, or a combination of normally open and normally closed. They will generally be made of a material that will withstand arcs without becoming damaged.

Other types of contacts are also frequently used in pilot controls. One of these is the *reed switch,* which contains a set of contacts sealed in a glass or plastic capsule. Terminal leads that are connected to the contacts do not break this seal as they leave the capsule. This keeps the contacts hermetically sealed so that the environment will not cause them to corrode. The capsule also allows the reed switch to be used in explosive atmospheres without danger of starting fires.

Since the reed switch is fully encapsulated, the pilot operator cannot touch it to cause the switch contacts to activate. Instead, a magnet or magnetic strip is provided on the movable part of reed switch. A stronger magnet is provided on the pilot switch's operator and is placed in close proximity to the reed switch. When the pilot switch's operator moves close to the reed, the contacts come directly under the magnetic force and are pulled closed. When the pilot switch's operator moves away from the reed switch, it moves the strong magnet far enough from the magnet on the reed switch to limit the magnetic force so that spring tension can return the contacts to their normal location. This type of operation allows the reed contacts to be used extensively with any type of pilot device. Figure 8-2 shows specialized contacts provided for a variety of applications, such as fast response, microprecision movement, and in a logic circuit. For quick response, springs are added to the contacts to provide rapid snapover center action to make the contacts close and open faster. Other contacts are built so that a very small amount of

Introduction

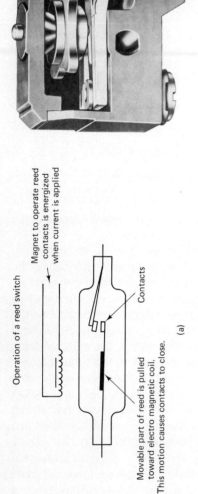

Operation of a reed switch

Magnet to operate reed contacts is energized when current is applied

Contacts

Movable part of reed is pulled toward electro magnetic coil. This motion causes contacts to close.

(a)

Sealed Contacts — Both of the relay configurations utilize sealed contact cartridges which contain a sealed switch. These contacts are hermetically sealed in a glass envelope which contains a controlled gas atmosphere.

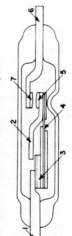

The sealed contact switch is comprised of a stationary element and a movable element sealed in a glass envelope. The movable element consists of a terminal (1), an armature (2), a pole piece (3), a spring member (4), and a tungsten-tipped contact (5). The stationary element consists of a terminal (6) and a stationary tungsten-tipped contact (7).

A unique magnetic design provides snap-action contact force that establishes a current handling capability yet provides for consistent low contact resistance throughout life for reliable switching of low level circuits.

(b)

FIGURE 8-2 (a) Types of miniature and sealed contacts found in pilot switches; (b) sealed contacts used in relays. [(c) Courtesy of Honeywell/Microswitch Division; (d) courtesy of Allen-Bradley, Inc.]

travel from the pilot device actuator will cause the contacts to change from open to close.

In this chapter we explain the function of pilot devices used in modern motor controls. Each family of pilot devices is introduced with pictures and electrical symbols. Operation, installation, and troubleshooting are also explained. Typical applications and electrical diagrams are used to provide an understanding of their use. The contacts in each of these devices are very similar, if not exactly the same. The major difference between pilot devices is the type of operation used to make these contacts open and close. For example, in a pushbutton switch, the contacts are closed by manual motion (someone pressing on the switch). In a limit switch the contacts are closed or opened by machine motion, and in a pressure switch the contacts change position by an increase or decrease of pressure. The electrical symbol for each pilot device will be similar. The contact portion of the symbol will show if the contacts are normally open or normally closed. They will also show if the pilot device controls single or multiple sets of contacts. The lower portion of each symbol will show what type of operation is required to activate the switch contacts. These fundamentals will be explained in depth in the remainder of this chapter.

PUSHBUTTON SWITCHES

Pushbutton switches are very widely used in motor control circuits for modern industry. Figure 8-3 shows pictures and electrical symbols of various pushbuttons. From this figure you can see that the pushbutton can operate normally open or normally closed contacts. It can also operate multiple sets of contacts, such as single-pole, double-throw and double-pole, double-throw. This figure also shows other variations of the pushbutton switch, including momentary type, maintained type, time-delay type, and pushbuttons with pilot lights.

In the event that multiple sets of contacts are operated by the same pushbutton, a dashed line interconnects the sets of contacts. This is a standard method of identifying multiple sets of contacts. In some diagrams the abbreviation "PB" is included with the pushbutton switch symbol. If multiple pushbuttons are used, they are identified as PB1, PB2, PB3, and so on. An example of circuits using multiple pushbuttons is shown in Figure 8-4.

The time-delay function is provided with a pneumatic operator. The movement of the contacts is delayed by air escaping from a chamber. The amount of time delay can be changed by adjusting the rate at which the air leaves the chamber. This type of switch provides simple time-delay operation without the cost of an expensive motor-driven time-delay device.

Other types of pushbutton applications include multiple pushbuttons. A number of switches may be mounted together for similar applications, such as start–stop, run–jog, forward–reverse, or up–down. All the pushbuttons for a machine's operation may be mounted together in one panel, called the operator's panel.

Another type of pushbutton switch available to make a complex automated system operation safe is the security pushbutton. This type of pushbutton can be locked with a key, and when locked, cannot be operated. This ensures that only qualified personnel operate the system. Another variation of this type of pushbutton has a lockable cover. Such a cover can be added to any pushbutton switch to ensure secure operation.

Another way to classify pushbutton switches is by the type of operator (button) that is used. For example, the operator can be flush mounted with the panel or it can be extended. It can also be fully shrouded or half-shrouded. The shroud protects the switch from being depressed accidentally by activity around the operator panel. Another type of button used on pushbutton switches is the *mushroom head*. The head is enlarged in the shape of a mushroom so that it is easy to depress. The mushroom switch is generally used as an emergency stop switch and may also be used as a start cycle switch. In these applications it is called a *palm button*. The size of the operator button is also used to classify pushbutton switches. In some applications the switch button is enlarged to a jumbo size so that it is easy to locate on the operator panel. This type of switch is generally used for emergency stop applications. The size of the pushbutton switch can also be reduced so that more switches can be fitted into the existing panel area. The reduced-size pushbutton switch is called a *miniature* or *compact switch*. Even though the size of the switch has been reduced, the amperage and voltage ratings are the same as for full-sized switches.

The shape of the pushbutton head is also used for classification. Pushbutton switches are available in round-head, rectangular-head, and square-head types. The heads are also available in different colors, such as red, black, green, yellow, white, blue, and orange. The pushbutton head may also contain an indicator lamp. The color of the indicator lamp may be red, amber (yellow), green, or white.

Another classification of pushbutton switches is by the environment or location where they are installed. These switches are available in oiltight, watertight, weatherproof, corrosion-resistant, and

FIGURE 8-3 Picture and electrical symbols of pushbutton pilot devices. (Courtesy of Eaton Corp./Cutler-Hammer Division.)

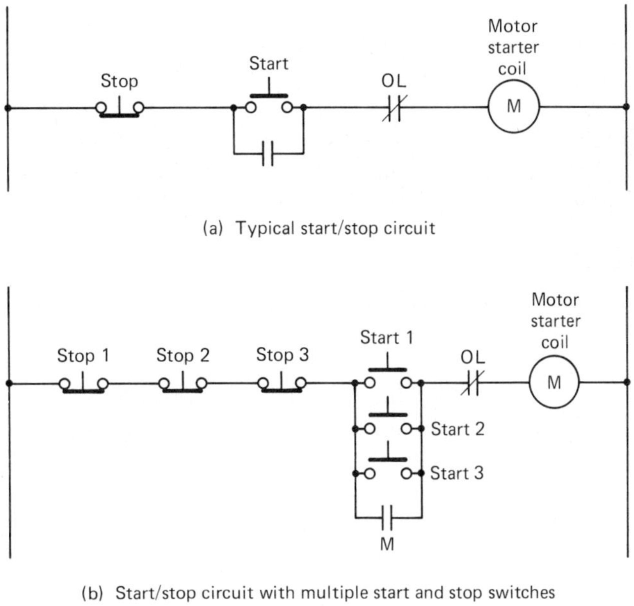

FIGURE 8-4 Pushbutton used in circuit application.

(a) Typical start/stop circuit

(b) Start/stop circuit with multiple start and stop switches

explosion-proof assemblies. Another way to provide protection for the pushbutton switch is to add a protective boot. This rubber boot will keep the switch protected against dirt and larger particulates but will not make the switch moisture-proof or explosion-proof.

Installing the Pushbutton Switch

You may be required to install or replace a pushbutton switch. To complete these tasks, you will need to understand both the ladder diagram and the wiring diagram. The electrical symbols provided earlier in this chapter were also used in the diagrams in Figure 8-3. These diagrams will be useful in the installation procedure because they show the wiring from component to component. The wiring diagram is also used to show the location of terminals on the switch and the location of switches on a panel board. Terminal diagrams are usually available in the manufacturer's technical data or you can determine the contact sets with an ohmmeter. This test can be accomplished by placing the ohmmeter across a set of terminals and depressing the pushbutton. If the meter shows continuity when the button is depressed and infinity when it is released, the two terminals that the meter leads are touching are part of a set of normally open contacts. If the meter reads continuity when the button is not depressed and reads infinity when the button is depressed, the terminals belong to a set of normally closed contacts. If the meter shows infinity when the button is depressed and when it is released, the two terminals are not part of a set. If the meter shows continuity when the button is pressed and when it is released, the two terminals are connected together as a common point.

The second step of installation for the pushbutton switch is to disassemble the switch so that it will fit through the face of the panel. The wires are connected directly to the part of the switch called the *contact block*. The contact block is connected to the switch operator with screws. This allows the switch to be changed out later without removing any wiring. The front side of the switch includes a locking nut and legend plate. The locking nut is also called the *mounting ring*. When it is screwed down, it will pull the switch assembly tight to the panel. Grounding nibs are provided on the mounting surface of the switch to provide a positive ground circuit. The nibs will cut through the paint and into the metal of the panel surface as the mounting ring is tightened.

Legend plates are used to identify each switch. Since the pushbutton switch can be used for a variety of applications, it is important to identify each switch on the panel for safe operation. Typical legend plates available for use with pushbuttons include *start/stop, on/off, up/down, jog/run, forward/reverse, hand/auto, high/low, manual/auto,* and *open/closed*. Other legend plates are also available, but they may have to be special ordered.

Troubleshooting the Pushbutton Switch

The pushbutton switch is easy to test if you suspect is to be faulty. You should remember that if the circuit will not operate correctly, the problem could be any one of several things, such as the loss of the power supply, a loose or broken wire, a faulty pilot switch, or a faulty load. This means that you must troubleshoot the complete circuit instead of picking on individual components and testing them at random. Start the procedure by testing for voltage across the load terminals. If voltage is present at the power supply and through all the pilot switches and interconnecting wires, the remainder of the test should focus on the load. If no voltage is present at the load terminals, the test should focus on the source voltage and the loss of it through a pilot switch. If voltage is not present at the power supply, be sure to check for a blown fuse, open disconnect, or other power supply problems. If voltage is present, it may be 24, 120, 240V or higher, and it may also be ac or dc. At this point in the test, all that you are concerned with is that the supply voltage is the same voltage that is specified on the electric control diagram.

After you have determined that the proper amount of supply voltage is available, the remainder of the tests should focus on the pilot switches and interconnecting wires. The fastest and most accurate test for locating an open wire, loose terminal, or faulty pilot switch is to test for a voltage drop (loss). This test requires that power remain applied to the circuits while you touch the meter probes to each test point. *You must be aware of where you are placing your hands, tools, and meter probes at all times because severe electrical shock could result if you come in contact with the electrical circuit.* Even though leaving power applied to a circuit while you test it presents an electrical safety condition, it is necessary because control circuits tend to be complex and you may not be able to find the problem with the power off.

Figure 8-5 shows the test points where you should place the voltmeter probes to execute the test. Remember that this test can be used to locate problems in any circuit, regardless of the number and types of pilot devices. The test points are identified as A through E. The first test should be made with the voltmeter probes touching points A and E, which is across the power supply. Using this test to make sure that your meter is on the right range. From this point, leave one of the voltmeter probes on point E and move the other probe to point B. If no voltage is available at point B, check to be sure that the stop button is not depressed. Remember that it may be a maintained type of switch and require someone to return it to its normal condition. If the switch is in its normal condition and no power is present at point B, the switch is faulty and must be replaced. If you are testing for a stop switch that will not deenergize the relay coil, be sure to test point B and ensure that voltage is not present there when the stop switch is depressed. If voltage is still present at point B even when the stop button is depressed, you can assume that its contacts are welded or stuck together and that it must be replaced. If voltage is present at point B, continue the test by moving the probe on to point C.

If the voltmeter indicates that voltage is not present at point C, depress the start switch and check for voltage. If voltage appears at point C when the start switch is depressed but does not remain when the switch is allowed to return to its open state, the problem is with the auxiliary contacts of the motor starter. Check to see if the motor start coil can pull its contacts in. If the contacts are pulled in but they drop out again, the problem is in the wiring or auxiliary contacts. If the coil does not energize its contacts, continue with this procedure.

If voltage appeared at point C when the start button was depressed, it indicates that the start pushbutton is operating correctly. Continue the test by checking for voltage at point D. If no voltage is present at point D, the overload contacts are open. Try resetting the overload contacts on the motor starter. Remember to check for an overcurrent condition when the motor is running again. If the overload contacts cannot be reset, the motor starter heaters may still be too hot from previous motor overload conditions. If the contacts will not reset after the heater has had time to cool down, you can assume that the overload contacts are faulty and must be changed.

If voltage is present at point D, the circuit has supplied voltage to the motor starter coil. If the motor starter coil does not pull the motor starter contacts closed, you can assume that the motor starter coil is inoperative and must be replaced.

Some electricians or technicians would rather start this test at points D and E and complete the tests in reverse order. If voltage is present, the circuit is operating correctly and the problem is with the motor starter coil. If voltage is not present, the problem is with the switches or supply voltage. Both of these test methods are very reliable, but it is important to remember to leave one of the meter probes at point A or point E as a reference.

It is also possible to test the pushbutton switch when it is removed from a circuit. Use an ohmmeter for this test. Place the meter probes across terminals of each set of contacts and depress the pushbutton. The contacts should show continuity when the button is depressed if they are normally open, and they should show infinity if they are normally closed. Be sure to test the switch when the pushbutton is depressed and when it has been released.

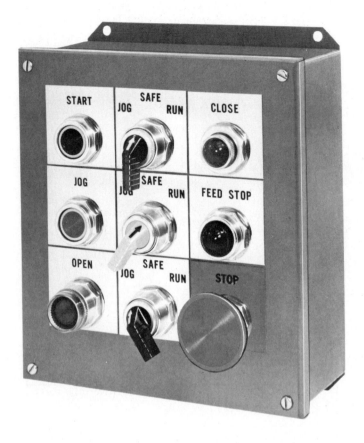

(a)

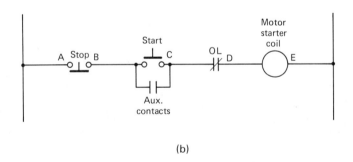

(b)

FIGURE 8-5a Multi-pushbutton. (Courtesy of Square D Company.) (b) Electrical diagram used for troubleshooting pilot devices in a motor control circuit.

LIMIT SWITCHES

Limit switches are one of the most widely used pilot device in modern motor control circuits. They are quite useful since they can convert machine motion into an electrical signal. Limit switches provide a variety of functions, such as limiting machine travel and sequence systems when machine motion has traveled a predetermined distance. Figure 8-6 shows an example application of a limit switch used to determine the depth to which a hole should be drilled. The drill will move down into the hole until the cam trips the limit switch and turns the motor off and disengages the drill clutch. An electrical diagram for this sytem is also shown in this figure. You can see that the limit switch cycles power to

Limit Switches **201**

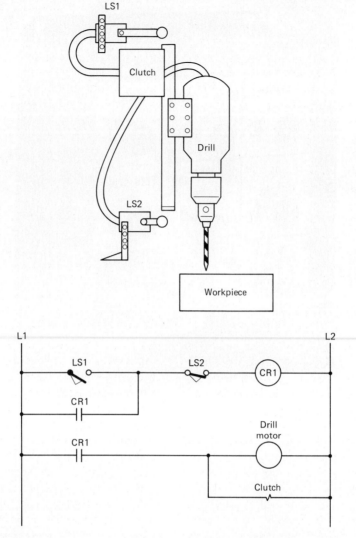

FIGURE 8-6 Electrical diagram and sketch of a limit switch controlling the depth of a drilled hole.

the drill and clutch when the drill travels to the predetermined depth. If a new hole depth is required, the cam on the arm can be adjusted to the new depth. Whenever the cam travels to a point where it activates the limit switch, it will deenergize the clutch and drill and a spring will return the drill to the top of its stroke.

Figure 8-7 shows the electrical symbol for various types of limit switches. There are basically four types of limit switches: normally open held open, normally open held closed, normally closed held open, and normally closed held closed. You should notice that the electrical switch symbol is essentially the same for all types of switches, while the type of operator varies slightly. The operator part of the symbol is shown as part of a triangle that is attached under the switch portion. The triangle signifies the limit switch cam or arm, which causes the limit switch contacts to activate. You can see from the electrical symbols that the location of the switch portion will be above the line for a limit switch that is held open, and below the line for a limit switch that is held closed. If the switch is shown in the closed position in the symbol, it is considered normally closed. If the switch is shown in the open position, it is considered normally

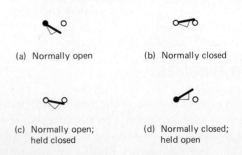

FIGURE 8-7 Electrical symbols for limit switches.

202 Chap. 8 / Pilot Devices

closed. Remember that the "normal" position is the position the switch would be in had no machine motion occurred. The "normal" position is also the position the contacts would be in had you removed the switch from the machine and bench tested it. The abbreviation for the limit switch is LS. If multiple limit switches are used in a circuit, they will be identified as LS1, LS2, LS3, and so on.

The limit switch can also activate multiple contacts. These contacts can be groupings of normally open and normally closed contacts. They can be activated as multiple set of single-pole contacts or as combinations such as single-pole, double-throw or double-pole, double-throw. The switch contacts are generally mounted in the switch body so that they can be changed without having to unmount the switch. Some switch bodies also allow for plug-in contacts, which allow for quick replacement.

The basic difference between limit switches relates to the type of operator used to activate the contacts. Figure 8-8 shows the different types of limit switch operators. The operators are grouped as roller arm actuators, plunger actuators, fork lever or yoke actuators and wobble head and cat whisker actuators. Each of these types of switches is shown in Figure 8-8.

The *roller arm actuator* is activated by the motion of the arm as it is moved through an arc (see Figure 8-8a). The shaft of the switch is connected directly to the shaft of the switch activator cam. When this shaft rotates, its cam will depress the switch contacts and cause the normally open set to close and the normally closed set to open. A roller is connected to the arm to provide smooth movement and easy travel as machine motion moves the arm. The roller also prevents the machine from causing wear on the switch. In some roller arm actuators, the roller is adjustable on the arm to provide precision limit control. The roller can be set for preliminary trials and then accurately adjusted to cause precise machine travel.

Another type of limit switch operator is the *plunger actuator*. This type of operator requires linear motion to depress the plunger. The top roller plunger (Figure 8-8c) is mounted so that an inclined cam can move past the roller, causing the plunger to depress. This type of switch must be accurately mounted since overtravel on the plunger will damage the switch. The side roller plunger is mounted in a similar manner. The cam that moves across the roller is tapered so that at one end of the travel it will cause the plunger to depress, and at the other end of the travel it will not touch the roller at all. In some cases the cam may have a double taper or multiple tapers. Each time the high point of one of the tapes passes directly over the roller, the plunger will be depressed and activate the switch. This type of switch is very useful in applications where mid-travel points are being sensed. An example of this application would be sensing the near-zero or near-home position on liner-actuated robots, where the robot will travel at high speed until it reaches the near-home point. At this point the limit switch would be activated and cause the robot motor to change to low speed as it continued the short distance to the actual home position.

The *cross roller plunger* is a variation of the side and top roller plunger switches. It is very useful in locations where the top or side plunger types cannot be mounted to detect machine travel accurately. The top round plunger (Figure 8-8f) is activated by motion that presses directly on the top of the plunger, causing it to depress. This type of switch is very useful to indicate that a sliding door has moved to its proper location. It can also be used to indicate that a moving part of a machine has traveled the predetermined distance and activated the switch. The adjustable top round plunger provides a screw mechanism on the plunger for precision adjustment. The screw mechanism allows the length of the plunger to be increased or decreased slightly by rotating the screw mechanism. Since the thread on this mechanism is very fine, the amount of machine travel can be adjusted for precise repeatability. The side round plunger provides the same operation for horizontal machine travel. These two types of limit switches allow both horizontal and vertical machine travel to be detected without having to design special brackets to make one type of switch design fit every application.

The *fork lever actuator* is used where machine travel in two directions must be detected. This type of actuator is also called a *yoke*. When the machine travel passes the fork in one direction, the yoke is rotated forward so that the front actuator is pressed down, which in turn leaves the rear wheel up where it will be activated when the machine travel reverses. When the machine travel is reversed, it strikes the rear yoke and moves it down. When the rear yoke moves down, the front yoke moves up where it will be in the path of forward machine travel. Since the switch is a maintained type, each time the machine travel moves one side of the yoke down and the other side up, it will remain in that position until the machine travel is reversed and moves the yoke in the opposite direction. This type of switch is very useful in surface grinding applications, where the grinding wheel is moved over the surface in one direction and then reversed. When the machine bed travels forward to the end of its stroke, it will move the yoke and activate the limit

Cross roller types

Fork level roller "yoke"

Roller arm

Side plunger

Side plunger

204 Chap. 8 / *Pilot Devices*

FIGURE 8-8 Types of limit switch actuators. (Courtesy of Honeywell/Micro Switch Division.)

switch contacts, which will reverse the motor that drives the bed travel. When the motor reverses, the bed will travel full stroke in the opposite direction until the cam on the other end of the bed trips the yoke in the opposite direction. This action causes the motor to switch to the forward rotation again. As this action continues, the bed cross travel is indexed by screw action. When the cross travel has moved the bed the full distance, it will strike another limit switch, which will turn the motor off.

The *wobble head* and *cat whisker actuators* operate on a similar principle. These limit switches use an extended wobble stick or cat whisker to detect machine movement. The extended wobble stick allows the limit switch to be mounted in a safe location where machine travel will not damage it. In applications such as where a robotic arm is used to remove parts from a press, the arm must be clear of the press before it begins its travel. The wobble stick allows the arm to test to see if it is coming close to any fixed portion of the machinery where it could be damaged. Since the wobble stick can be extended, they are available in several lengths. It is also important to remember that the cat whisker actuator is much more sensitive than the wobble stick type. The cat whisker actuator uses a set of microprecision contacts which are activated by a very small amount of whisker travel.

Another aspect of limit switches is their method of returning to their original positions. Some limit switches are returned by spring action. When the switch lever is activated, the return spring is wound up so that when the machine travel is not pressing on the limit switch lever, the spring will return it to its original position. Other types of return action include gravity and yoke or fork lever return. The gravity return requires the switch to be mounted in such a way that gravity will return the switch lever to its original position when machine travel is complete. The fork level actuator uses the return action of the machine travel to return the switch lever to its original position. It is important to mount this type of switch in such a way that the machine travel in one direction activates the switch by moving the forward lever of the yoke down. This motion causes the rear lever part of the yoke to move up so that it is in the path of machine travel in the opposite direction. On the return travel, the machine presses the rear yoke down, which returns the yoke to its original position.

Installing and Adjusting the Limit Switch

When you are planning the installation of a limit switch, you must consider the following environmental conditions that may affect the operation of the switch. Be sure to watch for mounting locations where heat may build up. These locations may include surfaces where heaters are used, such as injection molding machines or foundry application. This also means that you should not mount limit switches near the opening of furnace doors. When the door of the furnace opens, excessive heat may alter the plastic parts of the cam rollers or parts to the limit switch lever. If you must mount a limit switch in areas where high temperatures occur, be sure to mount it away from direct heat or use reflective shields. Sometimes a rod can be attached to the part of the machine that is being sensed. This rod will act as an extension of the machine motion and activate the limit switch even though it is mounted some distance from the heat source. A reflector can be used to deflect heat away from the switch. A rod can also be used in this application. In this case the rod could be mounted so that it can travel through the reflector shield. In this way the switch is mounted behind the shield and the rod passes through the shield to activate the limit switch.

You should be careful not to mount limit switches in the lower parts of metal-cutting machines such as mills and lathes. Be aware that in these applications, machine-cutting oils or moisture from coolants may drip down into the switch and cause it to corrode and malfunction. In these applications you should look for locations that are above the level where the cutting oil or coolant may fall. It is usually possible to design a cam or operator that can trip the limit switch when it is mounted away from these conditions. Another possible solution to this type of application is an oiltight or watertight switch.

Some locations on stamping presses may be unsuitable for mounting limit switches because the excessive vibrations from the stamping operation may cause the switch to become loose in its mounting. In this type of application look for locations where you can mount the limit switch so that it is not attached to the press. This means that the machine motion that is being detected is transferred from the press by means of a rod or cam to the lever of the limit switch. The limit switch can be mounted nearby on a separate mounting stand or pole so that it is isolated from direct machine contact and vibration.

In all limit switch installations you must test the machine travel so that you are sure it will activate the switch each time motion occurs. In some installations the motion at the very end of a machine part may not be uniform during each machine cycle. In these applications you may need to mount the limit switch closer to the center point of the part, where the machine motion will be more uniform. Remember: there are a variety of cams, levers, and rods commercially available that can extend the range of motion that a limit switch can detect. Other applications may require that you fabricate special trip mechanisms that will activate the limit switch.

If the amount of machine travel that must be detected is minimal, you may select a precision limit switch. A precision limit switch is specially manufactured to activate on a very small amount of travel. Essentially, the part of the limit switch that is different is the amount of travel that is required to activate the switch. Most manufacturers will list the degree of travel or the minimum amount of movement that is required to activate the switch. These distances must be taken into consideration when selecting a switch for sensitive applications. Keep in mind that a proximity switch or photoelectric switch will also provide the function of detecting machine motion. In some applications the amount of machine travel is so small that a limit switch cannot accurately sense the motion repeatably and a proximity switch may be a suitable substitute. The operation of the proximity switch is explained in more detail in Chapter 13.

When you are surveying the location to mount a limit switch, you should keep in mind that limit switches can be mounted as a surface-mounted device, a flush-mounted device, or a duplex-mounted device. The location that you select should be based on the amount of wear the limit switch will encounter, the assessibility of the switch for troubleshooting and replacement, and the amount of adjustment area the switch can utilize during calibration or adjustment.

Other considerations must be given to the amount of pretravel, total travel, and differential travel that the limit switch lever must make to activate the switch contacts. In some cases the amount of torque or force required to move the limit switch lever must also be calculated. If the amount of torque the machine movement can exert is limited, a precision limit switch should be considered.

The actual installation and wiring of a limit switch can be very easy. Figure 8-9a shows the basic parts of a limit switch. Notice that the base of the switch is mounted securely to the location where the machine travel will be sensed. The mounting base has adjustable holes so that the switch can be aligned accurately. The mounting base also provides a threaded hole for conduit fittings where the field wiring will enter the switch. The field wiring terminals will be connected to the terminal screws in the base. The second part of the assembly contains the normally open and normally closed switches and a set of pins that will make contact with the field wiring terminals when the switch assembly is screwed to the base assembly. Notice that this part of the assembly can easily be removed for quick change out of worn or faulty switch contacts. The switch assembly is available as spring-return or maintained contacts. After the field wiring is attached to the base unit, the switch assembly is screwed onto the front of the base assembly. The third part of the assembly is the actuator head. The acuator head can be any one of the types of actuators shown in Figure 8-8. After the head has been installed, the actuator arm or lever can be adjusted lightly for proper operation. Remember that this switch assembly is very similar to all other pilot devices. It is designed to provide simple installation and it is easy to troubleshoot and replace if problems develop later. Figure 8-9b shows correct methods for installing limit switches.

Troubleshooting the Limit Switch

The contacts of the limit switch can be tested in a manner similar to the contacts of other switches. If the switch is out of the circuit, an ohmmeter may be used to test for normally open or normally closed contacts. If the switch is in the circuit and under power, a voltmeter should be used to test the complete circuit prior to testing the limit switch at random. Keep in mind that the reason you will be called to troubleshoot the circuit is because the total machine operation is malfunctioning. By checking the machine operation, systematically you will be able to narrow the troubleshooting tests to one circuit and later to an individual component. When you are ready to troubleshoot the faulty circuit, be sure to leave one of the meter probes at a reference point and test both the input and output terminals of the contacts of each pilot device in the circuit. If you suspect that the limit switch is faulty, the most important part of testing it is to be sure to duplicate the type of travel that causes the switch to activate. If the switch is mounted on a machine, be sure to jog the machine back and forth across the limit

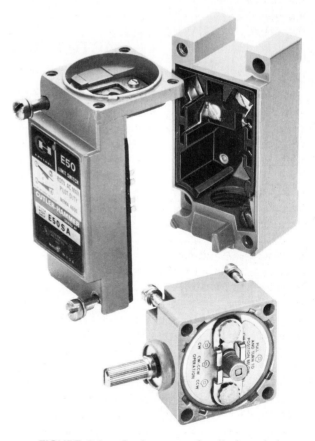

FIGURE 8-9a Basic parts of a limit switch shown ready for installation. (Courtesy of Eaton Corp./Cutler-Hammer Division.)

Limit Switches 207

LIMIT SWITCHES
Application and Installation Data

Limit switches are used in electrical control systems to sense position. They are actuated by the predetermined motion of a cam, machine component or piece part. This mechanical motion is then converted to an electrical signal through the actuation of a set of contacts. These signals can be used in the control circuits of solenoids, control relays and motor starters to control the operation of conveyors, hoists, elevators, machine tools, etc. (typically motor driven machines, processes and systems).

Experience has shown that the mechanical and electrical operating life of a limit switch is influenced to a large degree by proper installation and application procedures. This publication is intended to be used as a guide.

WIRING

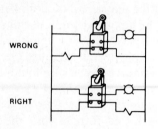

Opposite polarities should not be connected to the contacts of one limit switch unless the limit switch is specifically designed for such service.

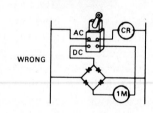

Power from different sources should not be connected to the contacts of one limit switch unless specifically designed for such service.

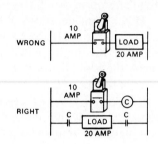

Limit switches should be used within their contact ratings.

ACTUATOR CONSIDERATION

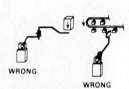

Limit switches are designed for proper performance with the actuators with which they are supplied. Supplementary actuators should not be used unless the limit switches are specifically designed for them.

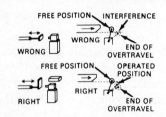

Operating mechanisms for limit switches should be so designed that, under any operating or emergency conditions, the limit switch is not operated beyond its over-travel limit position. A limit switch should not be used as a mechanical stop.

For limit switches with lever actuators, the actuating force should be applied as nearly perpendicular to the lever as practical and perpendicular to the shaft axis about which the lever rotates.

Where relatively fast motions are involved, cam arrangements should be such that the actuator does not receive a severe impact. Cams should be designed such that the limit switch will be held operated long enough to operate relays, valves, etc.

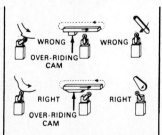

Cam or dog arrangements should be such that the actuator is not suddenly released to snap back freely.

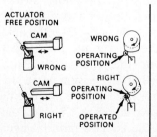

A limit switch actuator must be allowed to move far enough for positive operation of the contacts.

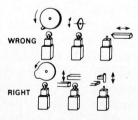

For limit switches with pushrod actuators, the actuating force should be applied as nearly as possible in line with the pushrod axis.

FIGURE 8-9b Correct methods for mounting limit switches. (Courtesy of Allen-Bradley).

LOCATION AND INSTALLATION

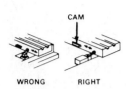

Limit switches should be mounted rigidly and in readily accessible locations, with suitable clearances to permit easy service and replacement when necessary. Cover plates should face the maintenance access point.

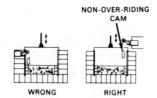

Limit switches should not be used in locations where temperature or atmospheric conditions are beyond those for which they have been specifically designed.

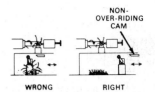

Limit switches should be placed in locations where machining chips do not accumulate under normal operating conditions.

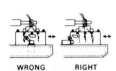

Limit switches should not be submerged in or splashed with oils, coolants, or other liquids.

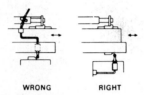

The location of oiltight limit switches and the method of connecting them should be such that condensation in the conduit cannot enter the switch enclosure.

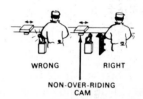

Limit switches should be mounted in locations which will prevent false operation by normal movements of operator or machine components.

FIGURE 8-9b (continued)

Limit Switches

switch so that the contacts are activated during the time you are checking for the change of voltage. If the contacts are not activating, you should suspect the machine travel as well as the electrical part of the switch. Make adjustments to the cams or levers if necessary to provide adequate machine travel to trip the switch.

If the switch is causing intermittent problems, be sure to check the speed of the machine travel across the limit switch. If the travel is exceedingly rapid, the contact closure may not be long enough in duration to allow proper operation. In this situation you may be able to prolong the contact closure by lengthening the cam on which the limit switch is activating. This problem is especially critical when the limit switch is connected to a programmable controller or robot. In these applications the programmable controller may scan its input modules at a rate of 50 ms. If the switch contacts activate faster than 50 ms, the programmable controller will not sense this switch activation as a legitimate input signal.

When you have located the faulty component, it should be removed and replaced as quickly as possible. In some cases this means that only the contact portion or the actuator portion of the switch needs to be changed. Always be aware of electrical shock hazards when you are making the tests, and be sure to disconnect the power supply when the components are removed and replaced.

SELECTOR SWITCHES

Selector switches are similar to pushbutton switches in that they require manual movement to cause the switch contacts to change position. Instead of pressing the switch to cause operation, the selector switch is turned with its handle to cause the contacts to change. Figure 8-10 shows several pictures of selector switches. From this figure you can see that the selector switch can have more than the on and off positions of a pushbutton switch. Typical selector switches have two, three, or four positions.

The two-position selector switch is normally used for such applications as on/off, up/down, and other dual conditions. Figure 8-11 shows the electrical diagram for these multiple-position selector switches. These diagrams are used as the switch symbol in all electrical diagrams. You should also notice that the selector switch abbreviation is SS. If multiple selector switches are used in a circuit, they will be identified as SS1, SS2, SS3, and so on. This figure also shows the switches in typical circuit applications. The input side of these switches is considered to be the left side of the switch or the

FIGURE 8-10 Picture of typical selector switches. (Courtesy of Honeywell/Micro Switch Division.)

point where the single terminal is located. If the switch is a two-position selector switch, either terminal may be used as the input of the switch. The three-position selector switch has one input terminal and two or three output terminals. If the middle position is used as the off position, only two output terminals are required. If the switch has three separate functions, three output terminals are required. The application circuit in Figure 8-11 shows the top position of the switch connected to a relay coil marked "hand." The middle position is set for the off position, with nothing connected to the middle output terminal. The lower terminal of the switch is connected to a relay marked "auto." In this application the switch must be in one of the three positions.

The selector switch's operation is similar to that of the maintained pushbutton switch, in that it will stay in the position to which it is changed. This function is quite useful in motor control circuits because it allows an operator to select an application and the switch will remain in that position. You must remember that the selector switch requires manual motion from an operator to cause the switch to change position.

The selector switch is installed in the operator's panel in a manner similar to the pushbutton switch. This means that all electrical connections are made on the contact block at the rear of the panel. This also means that various legend plates may be installed to explain or define the function of the selector switch.

Several types of lever or knobs are also available. One of these is the keyed knob, which allows the switch to be locked into position. If a change is required, the switch must first be unlocked. This type of switch is useful in applications where security is needed to guard against unauthorized changes of machine operation.

JOYSTICK CONTROLS

Joystick control is used in many applications for industrial control. It is used primarily for multiaxis motion control, such as overhead hoist operation, moving molten iron in ladles for pouring operations, and semiautomatic robot operation. A photograph and electrical diagram for joystick control are included in Figure 8-12.

The Joystick control actually consists of a multiple set of switch contacts located so that the stick's movement will activate one or more contacts at the same time. Figure 8-12 shows the location of switches for a two-position, four-position, and light-position joysticks. Typical electrical diagrams are also provided to show industrial applications. From this diagram you can see that when the two-position control is used, the operator can move the stick forward to activate the upper set of contacts. When the stick is pulled back, the lower set of contacts are activated. When the stick is returned to the neutral position, all contacts are deactivated and will return to their normal position. Most joysticks use a spring to return the stick to the neutral position.

The electrical diagrams are for an overhead hoist operation. The first diagram shows a simple control circuit for activating the lift portion up and down. From this electrical diagram you can see that when the upper set of contacts is activated, the up motor starter is energized and the motor will turn in a clockwise direction, moving the hoist hook upward. When the stick is returned to the neutral position, the up motor starter is deenergized and the motor is turned off. This type of motor includes an electrical brake that would be energized any time the motor is deenergized. The brake would prevent the motor from creeping, which would allow heavy loads from dropping when the motor is not running. When the jobstick is pulled back, the lower set of contacts is energized and the motor starter to allow the load to be lowered is activated. When the motor starter is energized, the motor will reverse and turn in the counterclockwise direction. In this sense the

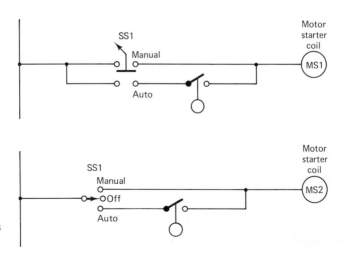

FIGURE 8-11 Electrical diagrams and symbols of selector switches.

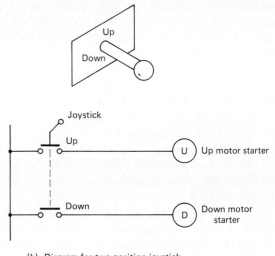

(b) Diagram for two-position joystick

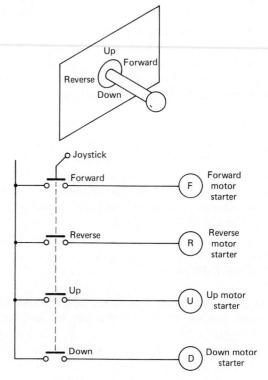

(c) Diagram for four-position joystick

FIGURE 8-12 Picture and electrical diagram of a joystick. (Courtesy of Eaton Corp./Cutler-Hammer Division.)

motor starter is wired for forward/reverse operation, which will cause the crane's cable to move up or down.

The second electrical diagram shows the operation of an overhead crane that can raise and lower heavy loads such as rolls of steel. The crane in this application can also move forward and reverse along a set of overhead rails. Another motor allows the crane to move back and forth across the width of the rails. These four motions allow the crane to be positioned anywhere on the tracks. They also allow the crane to move heavy loads to any machine that is located under the rails. This includes all of the area from one end of the rail to the other, and all of the area side to side between the rails.

The best way to explain the operation of this joystick is to use the directions of the compass. If the joystick is moved north or south (pushed ahead or pulled back), the crane will move forward or reverse along the length of the rail. If the joystick is moved east or west (left or right), the crane will move back and forth across the rail from left to right. This motion allows the crane to reach all the positions from side to side between the rails. These

212   Chap. 8 / Pilot Devices

four motions are controlled by two reversing-type motor starters that control the motor for forward and reverse travel, and the motor for left-to-right travel.

If the joystick were pushed in the northeast direction, the forward motor starter and the right motor starter would be energized. This would cause the crane to move down the track and to the right in one smooth simultaneous motion. If the joystick were pushed in the northwest direction, the forward motor starter and the left motor starter would be activated, causing the crane to move forward and to the left. When the joystick is pulled southeast the reverse and right motor starters are energized, causing the crane to move in reverse and to the right. Notice that each motor can be activated individually by moving the joystick to the north, south, east, or west location. An up/down pushbutton could also be added to raise or lower the load. Other safety features, such as key lock and indicator lamps, could be added to ensure safe operation. This type of switch allows total operation of the crane with minimum operator motion.

Troubleshooting the Joystick Control

If you think the joystick control is not operating correctly, it can be tested with power applied or with power removed. The first step in this procedure is to determine the number of contacts the switch has. After you have located each set of contacts, move the switch in each direction and test for proper switch operation. Sometimes the switches will have to be adjusted slightly for the joystick to activate them fully. After each individual motion has been tested, move the joystick so that the compound moves are active. If the stick is moved northwest, the upper and left-side circuits should be activated. Remember: the contact sets may be located in the opposite position on the switch from the function they provide. This means that when the stick is moved forward, the lower set of contacts is closed, and when the switch is moved to the right, the left set of contacts is activated. Your tests of the contacts will help you determine the location of each set. If any one of the sets of contacts is inoperative, adjust the contact actuator and retest the contacts. If the contacts remain inoperative, they must be replaced. After the single contact motions have been checked, continue the test of the multiple-contact moves. When the joystick is moved in the northwest direction, two sets of contacts will be activated. If either one of the two is activated, adjust the contacts to remedy this fault. If each of contact set passes the individual test, the fault must be in the adjustment of the contact actuator for the multiple-contact move. When you are certain that the contacts in the joystick are operating correctly, continue testing for faults in the motor starter circuits.

FOOT SWITCHES

Foot switches are pilot devices that are similar to a pushbutton switch, except that they are activated by pressing on them with your foot. The activator portion of the switch is manufactured with heavy-duty material to withstand repeated activation from someone stepping on it. The switch portion of the foot switch is very similar to that of other pilot contacts. The contacts are available as single-pole, single-throw or double-pole, double-throw combinations. If the switch contacts are shown above the terminal point in the diagram, the switch is a normally open switch and the contacts will be closed when the foot switch is activated. If the contacts are shown below the terminals, the switch is a normally closed type and the contacts will be opened when the switch is activated. The abbreviation for the foot pedal switch is FTS. If several foot pedal switches are used in a circuit, they are identified as FTS1, FTS2, and FTS3.

One typical application is for a foot switch to be used by a person unloading a conveyor. Parts are placed on the conveyor by robots that are unloading a plastics press. The conveyor is activated by the worker to bring parts to the end of the line, where they will be loaded into baskets. Since the switch can be activated by the operator's foot, the person's hands will be free to remove parts from the conveyor. Each time they are ready to remove a part from the conveyor, the pedal is depressed. The foot pedal activates the switch contacts, which energizes the conveyor motor starter. The motor starter energizes the conveyor motor, which moves the parts toward the person handling them. When the conveyor moves the parts close enough to be picked up, the operator releases the foot switch and the conveyor stops. After the operator places the part into the basket, the process is repeated. If the conveyor does not advance the parts to the end of the line, they will cause a jam-up at the point where the robot is placing the parts on the line. If a jam occurs, the robot senses that a part is in its way and will indicate a fault condition, and a technician must be called to test the circuit.

Other applications that utilize the foot switch are manual spot welding applications and manual stitching or stapling machines. In both of these applications it is important that operators have their hands free to move the parts under the machine head. In spot welding applications the metal parts are moved into position by hand. When the parts are in the proper location, the operator can depress the foot switch and activate the spot welder. A similar operation is utilized with manual stitching machines. A cardboard box can be stapled manually by moving the box from point to point under the stapler head. Each time the operator moves the box into position, the foot switch is depressed and the stapler head injects a staple into the box at that point.

Foot Switch Operation

The operation of a foot switch is similar to that of other pilot devices. It is important to remember that the switch operator must be depressed for the contacts to change position. The amount of travel the activation portion of the switch must travel can be adjusted. Since the pedal is returned to its original position with spring tension, the amount of clearance between the pedal and the swtich is very critical. If the clearance is too close, the switch will not return correctly to its normally open position. It will also have problems with false activation. When the switch is adjusted correctly, the pedal can be activated with the proper amount weight. If the switch requires too much weight to activate the pedal, the operator will become overly tired from repetitive operation.

FLOAT SWITCHES

Float switches are specialized pilot devices used to measure the level of liquids and granular materials used in industry. For this reason, the float switch is also called the *level switch*. Figure 8-13 shows photographs and symbols for the float switch. The symbol for the float switch looks like a ball connected directly under the contact symbol. The ball in the symbol represents the float device that is connected to the end of the actuator rod. The contacts for this symbol can show the switch opening under the terminal point or above the terminal point. The abbreviation of the float switch is FS. If multiple float switches are used in a circuit, they are identified as FS1, FS2, FS3, and so on.

If the switch contact is shown above the terminal point in the symbol it indicates that the switch contacts will open as the level increases and will close when the level decreases. This type of contact arrangement, called *normally closed,* would be used to keep a water tank from overflowing. When the water level increased to the proper level, the contacts would open and turn power to the

Symbol for normally closed float switch

Symbol for normally open float switch

FIGURE 8-13 Picture and electrical symbols for float switches. (Courtesy of Square D Company.)

motor off. When the tank level is lowered to a predetermined set point, the contacts would close again and the pump would turn on. Figure 8-14 shows an example of this application with the electrical diagram of the pump motor and level control.

If the symbol shows the contacts below the terminal point, it indicates that they will close when the level that is being measured increases and open when the level decreases. These contacts, called *normally open*, are used to activate a pump that is used to empty a sump. This application is also shown in Figure 8-14. When the level in the sump increases, the float will rise and cause the contacts to close and turn on the pump motor. When the pump has lowered the sump level to the predetermined level, the float will drop and cause the switch to open and turn off the pump motor.

Float Switch Operation

Several types of float switches are used in industrial control: the open switch, the closed switch, and rod and chain control. Figure 8-14 shows diagrams of the operation of these controls. The open float control (Figure 8-14a) has a set of contacts activated by a lever actuated by the float and rod. Several types of floats are used with these controls. The simplest type of float is made from plastic. This type of float is shaped like a ball and provides a threaded connector where the actuator rod can be connected.

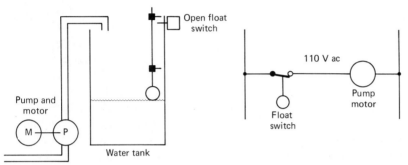

(a) Open float switch application and electrical diagram

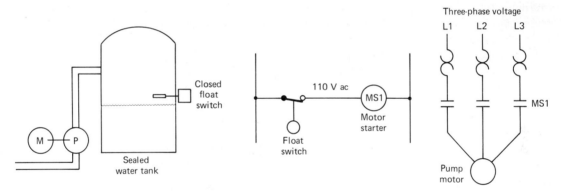

(b) Closed float switch application and electrical diagram

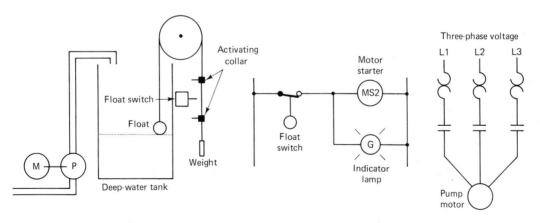

(c) Rod and chain float switch application and electrical diagram

FIGURE 8-14 Diagram of open, and closed, and rod and chain type float controls with application and field wiring diagrams for level controls.

Float Switches 215

The float is completely sealed so that liquid cannot enter it and cause it to sink. The top end of the rod has an adjustable collar that can be located to cause the switch to activate when the liquid level reaches a predetermined point. The length of the rod can be varied so that the switch can be located well above the liquid level. This allows the switch to be mounted in a dry location where it is easily accessible for installation, troubleshooting, and adjustment. After this type of control is installed, the float should be adjusted on a trial-and-error basis to ensure that the switch is activated when the liquid reaches the proper level. If the switch is not activating at the correct level, the collar on the activating rod can be adjusted up or down. Since most switches include both the normally open or normally closed contacts, this control can be used for both filling and emptying applications.

Open level control is named for the type of tank application. This means that the tank is not sealed or pressurized. The control is available with the switch portion open for dry locations or sealed for watertight, corrosion-resistant, and explosive-proof applications.

Another type of level control is called *closed tank control*. A diagram showing the operation of this control is shown in Figure 8-14b. The switch contacts are located in the same fixture as the level control. The switch contacts are isolated from the float with a seal. This allows the float control to be mounted directly onto a tank and still maintain the tank's seal. This is especially useful for applications where the level of a pressurized vessel is controlled. It is vitally important that the seal be maintained when the control is installed.

The *rod and chain float control* is shown in Figure 8-14c. This type of actuator allows the switch to be located remotely from the float mechanism. The switch is activated by collars attached to the chain. The chain has a float attached to one end and a weight attached to the other. This assembly is passed over a pulley so that the weight will counterbalance the float. When the tank level decreases, the float will move lower in the tank. This action will cause the chain to pull the weighted end higher. When the chain has moved far enough, the collar will activate the switch mechanism and engage the contacts. When the level of the tank increases, the float will rise with the liquid level and provide slack in the chain. The weight on the end of the chain will cause it to press the collar down against the switch mechanism and deactivate it. Since the chain can be varied in length, the switch can be mounted any distance from the tank to provide easy access for maintenance and troubleshooting. This type of assembly is also very useful for very deep tanks where an extended rod assemble may be too bulky or expensive.

Resistive-Sensitive Level Indicators

Another type of float switch or level indicator uses the difference of resistance between two probes to activate a relay when the level of a liquid in a tank changes. A solid-state circuit is used to sense the change of current that is caused by a change in resistance. In this type of control, current will flow between two sensors through a liquid as long as the liquid is covering both probes. If the liquid level falls below one of the probes, the current will not flow because the air gap between the lower liquid level and the probe causes a higher resistance. When the liquid level raises and covers the probe again, the resistance of the liquid is small enough to provide adequate current to allow the control to activate the relay again. This type of control can also be designed to operate with only one probe. In this application, the metal sides of the tank will carry current like the second probe would. The single probe would be placed at a point where it can sense the liquid level. A very small amount of current will flow through the liquid between the two sensors or between a single sensor and the side of the tank any time the liquid level is high enough. Even though the current flow is very small, the solid-state part of the control is able to distinguish between the small current flow and no current flow when liquid level is below the sensor. The electronic part of the control requires 12 V dc to operate correctly. The contacts of the relay are isolated and are rated for 5 A at 125 V ac, 1 A at 120 V dc, and 2.5 A at 230 V ac or dc for noninductive loads.

Installing and Troubleshooting the Float Switch

The contacts for float switches are very similar to the contacts in other pilot devices. Since normally open and normally closed contacts are available, this type of switch is easily adapted to a variety of control applications. The installation process involves two parts. The first part is the installation of the float apparatus. The float must be installed so that it can move with the liquid level of the tank or sump. You must be sure that the activator rod is long enough to reach the switch mechanism at both the high and low levels of the liquid. It is also important to determine the dead band for the application. The dead band is the distance between the point where the switch is activated at the high level and the point where it is activated at the low level. The dead band can also be measured as dwell time.

The *dwell time* is the amount of time the switch contacts are deactivated between the high-level point and the low-level point. If the dwell time is not sufficient, the pump motor may be required to start too many times per hour, which will cause it to overheat and wear out prematurely.

After the float mechanism has been installed and adjusted, the switch contacts can be wired. Use the wiring diagram to determine to which switch terminal the field wiring should be connected. Be sure to observe wire color codes and utilize wire markers. Complete field testing of the complete float control to ensure that it is operating correctly.

If the float control is not operating correctly, you must execute troubleshooting procedures to find the problem as quickly as possible. Test the complete circuit to determine the exact location of the fault. If you have determined that the float switch is at fault, the first step in troubleshooting it is to determine that the float mechanism is operating correctly. Remember that the switch contacts cannot be activated if the float mechanism does not move far enough. If the switch is covered, you may need to remove the cover so that the assemble can be clearly observed. If the float mechanism is not traveling correctly or if it has not moved far enough to activate the switch, correct this problem prior to continuing the troubleshooting procedure. After you have determined that the float mechanism is operating correctly, proceed to the electrical part of the test. If the switch is powered, use a voltmeter to test the contacts for proper operation. You should remember that the procedure for locating a problem in a control circuit will be the same regardless of the actuating device.

If the troubleshooting procedure determines that the problem is in the electrical contacts, be sure to test them again while you activate the switch manually. In this case you can manually activate the switch without harm to the pump motor. Remember that some pumps may be damaged if they are operated dry. If this is the case, be sure that the pump has an adequate supply of liquid during the testing procedures. If the motor control application involves a closed tank application, be sure to test the float control for leakage after all repairs have been completed.

PRESSURE SWITCHES

Pressure switches are used in a variety of motor control applications. These applications include detecting high pressures, low pressures, differential pressures, and vacuums. These devices can be used as safety control to protect systems from high pressures or the lack of pressure, or as operational devices to cycle an air compressor to maintain constant pressure in a system.

The electrical symbol for a pressure switch is shown in Figure 8-15. The symbol for a vacuum switch, differential switch, and pressure switch is the same. The symbol is very similar to those for other pilot devices, as it shows a contact and an operator. The operator in this case is shown as a sealed chamber. The symbol that shows the contacts above the terminal point indicate that this pressure switch opens when pressure increases and closes when pressure decreases. This indicates that the device is used as a high-pressure control. It could also be used to maintain pressure in a tank at a predetermined point. The symbol that shows the contact below the terminal point indicates that the switch will open when pressure decreases and close when pressure increases. This indicates that the device is used as a low-pressure control. A low-pressure switch may be used to indicate when a sealed pressurized system has lost pressure. This type of control is used extensively in air-conditioning and refrigeration systems to deener-

FIGURE 8-15 Diagram and electrical symbols for pressure switches.

gize the compressor motor if the system loses its refrigerant. Note that "PS" is the abbreviation for a pressure switch. If additional pressure switches are used in a circuit, they will be identified as PS1, PS2, PS3, and so on.

Diaphragm Pressure Switches

The diaphragm pressure control is used to sense relatively low pressures. It is manufactured with two halves separated by a rubber diaphragm. The bottom chamber is connected to the source of pressure. The top chamber is vented to the atmosphere (atmospheric pressure is 14.7 psi). The chamber that is connected to the pressure source will be manufactured with a threaded fitting to provide a means of connecting the operational part of the switch to the pressure part of the system. The pressure can be supplied from an air system or a fluid system such as water or hydraulic fluid. This fitting must be made to withstand the pressure the switch is sensing without leaking. When air pressure increases, the diaphragm moves upward against the switch mechanism and causes it to activate the switch contacts. The amount of pressure required to activate the switch contacts can be adjusted by spring tension. The contacts can be single-pole, double-throw, or double-pole, double-throw. They can be set against the switch mechanism to open when the pressure decreases or when the pressure increases beyond the switch's set point. You must be very careful when you adjust the pressure setting for this type of switch so as not to damage the spring mechanism.

After the switch contacts are activated by the movement of the diaphragm, they will remain in that state until the pressure returns to its preset value. The switch and diaphragm mechanism is manufactured so that it will not activate both off and on at the same pressure. For instance, if a high-pressure switch is used as an operational control to regulate the amount of pressure in an air compressor tank at 50 psi, the switch may have a range of 0 to 150 psi and a differential of 1 to 15 psi. The range is the highest value the switch will cut out and deactivate the switch contacts. In this switch, the pressure can be set at any value between 0 and 150 lb. To maintain the pressure at 50 psi, the range part of the switch should be set at 55.

The differential is the amount of pressure drop the system should have before the compressor is switched on again. If the differential is set at 0 or 1 psi, the pressure switch would deenergize the compressor motor when the tank pressure reached 55 psi, and reenergize the compressor motor when the pressure dropped to 54 psi. This would cause the compressor motor to cycle off and on continually any time air was used in the system. If the range was set to 55 psi and the differential was set at 5 psi, the switch would turn the compressor motor off at 55 psi and back on at 50 psi. This would allow a time delay between the turn-off and turn-on pressure, which would extend the life of the compressor motor. The differential pressure setting in the pressure switch will also extend the life of the switch's electrical contacts.

Vacuum and Extra-Low-Pressure Switches

The diaphragm switch can also be manufactured to sense very small changes in air pressure or vacuum. In this type of switch, the amount of movement the diaphragm provides is amplified through mechanical means. This allows a very small change of pressure of less than 2 psi to be detected. In a vacuum switch, the top side of the diaphragm is vented to atmospheric pressure and the lower side is connected to the air line that has the vacuum pressure. The springs and cams are used to set the amount of vacuum that will activate the switch contacts. Remember: vacuum is measured in feet of water column or inches of mercury. This means that the switch may be labeled in inches rather than in psi.

Bellows Pressure Switch

Diaphragm control is useful only for applications where the maximum pressure does not exceed 150 psi. Since the pressure presses against the rubber diaphragm, a pressure that is too high may rupture it. For application where pressures exceeding 150 psi must be sensed and controlled, a metal bellows is used. The pressure sensor may also be operated by a bellows. In this type of pressure switch, the pressure line is connected to the inside of the bellows. Atmospheric pressure will exert pressure on the outside of the bellows. When the pressure increases, the bellows expands upward until it activates the contact mechanism. The amount of pressure required to activate the contact mechanism can be adjusted by spring tension. An external adjustment knob or screw is provided to set the pressure that is required for activation. Since the bellows is made of metal, it can withstand pressures of up to several thousand pounds. This type of pressure switch is generally used for high-pressure air systems or high-pressure hydraulic systems.

The Bourdon tube actuator is very similar to the bellows operator except that it will operate more accurately with high repeatability. The Bourdon tube is sealed at one end and the air or liquid that is being sensed is allowed to enter the other end. Since the tube is wrapped in a spiral, it will try to straighten out as pressure increases. The amount of movement is directly related to the amount of pressure inside the tube. A set of springs, gears, and cams are used in conjunction with the tube to cause a set of contacts to activate. There will also be a setscrew and an offset spring that allows the set point and differential pressures to be adjusted. The contacts in these types of switches can be set by the manufacturer to activate on an increase or a decrease of pressure. The contacts can be single-pole, double-throw or double-pole, double-throw, which provides both normally open and normally closed contacts. The contact rating will be similar to the ratings listed in Figure 8-1.

Differential Pressure Switch

The differential pressure switch is activated by a diaphragm mechanism similar to a low-pressure switch, for the operation of this switch checks the difference between two pressures. The lower pressure of the two is connected to the top of the diaphragm chamber, where it will press down on the diaphragm. The higher pressure is then connected to the lower chamber, where it will press upward on the diaphragm. A set of springs and cams are used to set the exact amount of pressure differential that will activate the switch contacts.

This type of pressure switch is widely used to detect loss of oil pressure in an air-conditioning compressor. Since the compressor oil sump is part of the sealed system, it will have 70 to 80 psi in it when the compressor is not running. When the compressor begins to run, the oil pump in its sump will begin to pressurize the lubrication system and the compressor's pistons will increase pressure on the refrigerant. The oil pressure is typically 15 to 30 psi and will be in excess of the refrigeration pressure. A differential pressure switch is used to compare the refrigeration system pressure to the oil pressure. For application the differential pressure would be set to activate at a pressure difference of 15 psi. This means that the low-pressure side of the switch is activated by the refrigerant pressure. The high-pressure side of the switch is connected to the compressor oil sump, where it will "see" the same amount of refrigerant pressure plus the 15 to 30 psi of oil pressure. This causes enough pressure differential to activate the switch. If the oil pressure drops below 15 psi, the differential will become too low and the switch contacts will drop out and turn the compressor off. Since refrigerant oil may foam slightly or the oil pump may cavitate for a few seconds, a 30-second time delay is generally added to this switch. The time delay ensures that the pressure differential must exist for over 30 seconds before the switch activates and turns the compressor motor off.

You should remember that the oil pressure seems to float with the refrigerant pressure. This means that if the refrigerant pressures increases by 40 or 50 psi, the other side of the switch will also see the same increase in refrigerant pressure. The differential is still due to the addional pressure caused by the oil pump.

Installing the Pressure Switch

The installation procedure involves two distinct parts, the field wiring of the switch contacts and the connection and adjustment of the pressure side of the control. The field wiring should be completed after the switch has been connected to the pressure source. During the installation process, you will need to adjust the pressure side of the control with a gage. For this type of adjustment, connect an ohmmeter across any set of contacts. If the control is a high-pressure switch, increase the pressure until the switch contacts are activated. If the contacts do not activate at the correct pressure, release all pressure and adjust the range adjustment on the control. Retest the control by increasing the pressure. After the contacts activate at the correct pressure, the differential should be adjusted. The differential is adjusted by lowering the pressure slightly after the contacts have activated. Watch the ohmmeter to determine when the switch contacts close again. If the differential is too small or too large, adjust it by turning the differential screw adjustment. Remember that some pressure switches have a fixed differential; only the range can be adjusted.

If the control is a low-pressure control, test the contact activation as you release pressure from the control. Be sure that the pressure is higher than the activation pressure plus the differential. After you have determined the point where the switch contacts are activated, allow the pressure to rise slightly, to the point where the differential reactivates the contacts.

After you have adjusted the range and differential on the pressure switch, connect the field wiring to the switch contacts in accordance with the field wiring diagram. After the field wiring has been completed, apply power and test the switch for proper operation. Be sure to allow the pressure to rise and fall naturally with system operation while you observe the switch cycling. If the control is used as a safety device, such as for a low-pressure or high-pressure switch, manipulate the system so that these events will occur while you are observing the system to be sure that the safety switches operated properly.

Troubleshooting the Pressure Switch

The pressure switch can be tested while it is in or out of the circuit. During the troubleshooting procedure, begin troubleshooting the total machine operation and try to narrow the problem down to one circuit. Then concentrate your troubleshooting tests on that circuit. When you have determined the origin of the fault, you will need to perform diagnostic tests on that component. If the component is a pressure or vacuum switch, the most important point to remember when testing is to use the electrical symbol to determine whether the switch contacts open with an increase of pressure or whether the contacts will close when the pressure increases. After you have determined this, be sure that the pressure is changing enough to activate the contacts and return past the point where the differential will return the contacts to their normal state. You will need to install a pressure gage to make this determination. In some cases a pressure gage remains in the circuit at all times for this reason. If the pressure is not changing enough to activate the contacts, the problem may exist in the pressure part of the system rather than in the electrical part of the system.

If you have determined that the pressure part of the system is operating correctly, use a voltmeter to test the switch while it is connected to power. Leave one probe of the meter on the line 2 terminal of the circuit and move the other probe from point to point along the contacts in the circuit. Remember that the point where the voltage reading is zero indicates the point where the open circuit exists.

After you locate the faulty part, be sure to turn power off before you remove the field wires. Also remember to check the pressure side of the control so that you do not remove the pressure line with pressure on it. When you are ready to reinstall the pressure switch, use the information on installation procedures provided earlier in this section.

FLOW SWITCHES

Flow switches are used in a variety of applications to indicate the presence or absence of flow. This type of pilot device can be used to detect the flow of water, air, and other fluids. Figure 8-16 shows the electrical symbol and diagram of a flow switch. The symbol for the flow switch is a small flag or sail attached to the switch operator. The sail represents the device used to sense the flow. In some cases this device will be a small metal paddle that will move as fluid flows past it. The movement of the paddle activates the switch contacts. The abbreviation for the flow switch is FLS. In some diagrams a number is used in addition to the abbreviation to identify each flow switch (for examples, FLS1, FLS2, and FSL3).

From the electrical symbol you can see that the contacts can be shown as closing when flow is detected, or as opening when flow is detected. If the switch contact is shown above the terminal points,

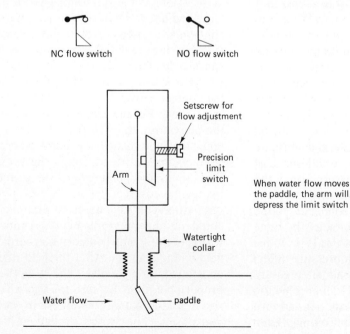

FIGURE 8-16 Diagram and electrical symbols for flow switches.

220   Chap. 8 / Pilot Devices

it indicates that the flow switch is normally closed and will open when fluid flow is established. If the switch contact is shown below the terminal point, the flow switch is normally open and will close when fluid flow is established.

Applications and Flow Switch Operation

In a simple application a flow switch is mounted in a cooling-water line to detect sufficient cooling water flow. The cooling water is used to keep the temperature of a plastics press at a predetermined level. The system is set so that the press cannot be placed in the automatic mode if cooling-water flow is not established. If the flow is interrupted at any point during press operation, a time-delay circuit will activate and deactivate the press after 30 seconds. The time-delay circuit allows the flow to cavitate for short periods without tripping the switch. Notice that the 30-second time delay is also used to allow the system to start. The flow switch will seal-in this circuit after flow is established.

The switch will have an adjustment screw to set the amount of flow required to activate the switch. Some switches may not have this type of adjustment; instead, the paddle must be cut or bent slightly to adjust the amount of flow required to activate the switch. If material is removed from the paddle, the switch become less sensitive. This means that more flow is required to cause the switch to activate.

In another application a flow switch is mounted where it can detect the presence of airflow in a chimney. For this application airflow is required through the chimney for 5 minutes prior to the ignition process. It is very important to be sure that the purge blower is operating correctly prior to ignition since most gas valves in industrial boilers and other gas-fired heaters have a possibility of leaking gas or fuel while they are turned off. The fuel tends to accumulate near the lower portion of the chimney close to the ignition source. If the ignition were turned on immediately before the chimney was purged, a tremendous explosion could result. The flow switch in this case ensures that a proper amount of fresh air has moved through the chimney to remove any residual fuel vapors prior to ignition. A flow switch is used instead of simply monitoring operation of the motor, because it is possible that the motor may be running and not enough air is moving. This could happen if the system had loose belts, a broken motor shaft or fan shaft, a blocked air passage, or a dirty fan. If the ignition system does not detect sufficient airflow for 5 minutes, the circuit will time out and lock out the ignition system. An indicator lamp will be energized to identify the system fault. The system must be reset manually to restart the fan motor. If the flow switch does not detect sufficient airflow for a full 5 minutes, the circuit will continue to lock out.

The flow switch operates a set of contacts when the paddle has moved a predetermined distance. The contacts can be single-pole, double-throw or double-pole, double-throw. These contacts can be any assortment of normally closed or normally open sets. The voltage and current ratings for these contacts are the same as for other pilot devices. In fact, some manufacturers use the same set of contacts in all their pilot devices. This allows users to stock one or two types of contacts that can be used for replacement in all their pilot devices.

Some flow switches detect very small amounts of airflow. A micro switch is used since it can be activated with a very small amount of travel from the switch operator. One application of this type of switch involves a small continuous flow of air direct from a nozzle at the flow switch operator. This switch assembly is used to detect the presence of moving plastic film. The airflow nozzle is placed on one side of moving film, and the paddle of the flow switch is located on the other side of the film. As long as the film is traveling correctly, the airflow cannot get to the nozzle. If the film breaks or runs out, the air from the nozzle will cause the flow switch to activate and signal that the plastic has broken.

It is very important with flow switches that they can activate without binding. It is also important that the flow not be reversed against the flow switch, as this will alter the activation point during normal operation.

The pilot device is basically an on/off switch. Flow switches are also available as analog devices, which can indicate the amount of flow from zero to maximum. The analog flow device used as a flow meter is discussed in a subsequent chapter.

Installing the Flow Switch

Flow switches are available for waterproof, explosion-proof, and other NEMA applications. This means that you can locate a flow switch in virtually any environment. The actual installation process should be completed in two distinct steps, similar to the pressure switch. The first step involves actual placement of the flow-detecting part of the switch. Be careful that the paddle is free to move and return as the flow is established and stopped. You should inspect the flow detection mechanism for proper direction of flow. The paddle is meant to detect flow in only one direction. If the flow is allowed to move against the paddle in the opposite direction, the switch will be damaged.

After the flow detection part of the switch has been installed, place an ohmmeter across the contact set you are planning to use, and activate the flow. Be sure to check the amount of flow with some other type of indicator or flow meter to ensure that the proper amount of flow is obtained before you try to adjust the flow switch. The flow switch may be adjusted by a spring and screw mechanism or by adjusting the amount of the paddle that is in the flow. This can be altered by raising or lowering the paddle into the flow or by removing part of the paddle. Remember that the farther the paddle is inserted into the flow, the more it will disturb the flow. Be aware that this turbulance may affect other flow-measuring devices in the system.

After you are assured that the flow switch is operating correctly, continue the installation by installing the field wiring. Use the field wiring diagram to complete the field wiring process. After the switch has been wired correctly, apply power and test the switch under actual conditions. Be sure to allow the flow to reach the low or high level of flow that you are trying to detect. Execute these tests several times to ensure that the switch installation is operating correctly.

Troubleshooting the Flow Switch

The flow switch can be tested like other pilot devices. It can be tested in or out of the circuit. In the case of this and other pilot devices, it is very important to do some preliminary testing to determine if the pilot device is actually the faulty part of the system. To execute this test, leave power connected to the circuit and use a voltmeter to determine if the load has power across its terminals. If power is present, continue testing the load. If power is not present, leave one meter probe on line 2 (L2) terminal of the circuit and continue moving the other probe from point to point, starting at line 1 (L1) side of the circuit. The point where you loose the voltage reading indicates the fault. If this test indicates that the flow switch is at fault, continue with this procedure to test it.

The most important part of the test procedure is to ensure that the switch is actually detecting flow. This means that the problem may be in the system, not in the switch itself. Remember: the flow switch is used to detect the presence or absence of flow. If the flow stops, the switch will actuate its contacts and indicate that flow has been lost. Electricians often assume that since the switch contacts are open, the switch is faulty, when in reality the switch is operating correctly and indicating that flow has been lost. There is also a strong possibility that the paddle that activates the flow switch contacts has become inoperative or is stuck in one position. This occurs because the fluid flow that the paddle is detecting can also corrode its mechanism. Sometimes the paddle will accumulate dirt or debris that is in the fluid and become immovable. When this occurs the system will actually have the proper amount of flow, but the paddle cannot move to indicate this. In these cases it is suggested that the paddle be removed from the fluid and inspected carefully for proper travel. Also inspect the sight where the paddle is installed. Sometimes a buildup of corrosion exists around the installation site, which will prevent the paddle from proper movement.

After you are satisfied that the operational part of the switch is functioning properly, continue the test to the electrical side of the switch. Test the contacts for proper operation. If they activate but do not pass power, they are faulty and must be replaced. Be sure to turn the power off prior to removing the switch contacts. It is also important to replace the contacts with a set that have the same voltage and current rating. Also make sure that they will activate at the same point where the previous set activated. To confirm this, execute the adjustment procedures listed in the earlier section covering installation.

In some cases it you may want to replace the entire flow switch. But sometimes this involves draining down the system or dealing with leakage of the fluid during the change. In these cases, the flow meter paddle is normally left in the system and only the switch contacts are changed. Prior to making your decision, you may need to check with the hydraulic or pneumatic specialist in your shop, who should be able to tell you which method can be accomplished most easily.

TEMPERATURE SWITCHES

Temperature controls are widely used in industrial applications. The electrical symbol for temperature pilot device is shown in Figure 8-17a. The abbreviation for the temperature switch is TS. If more than one temperature switch is used in a system, each switch is identified with a number in addition to the letters (for example, TS1, TS2, TS3).

From the symbols in this figure you can see that the switch contact can be shown above the terminal point or below the terminal point. If the switch contact is shown above the terminal contact, it is identifying a normally closed temperature switch that will open when the temperature in-

(a) Electrical symbols for temperature switches

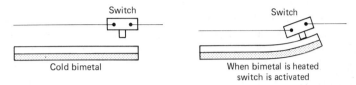

(b) Examples of changes in a bimetal strip caused by changes in temperature

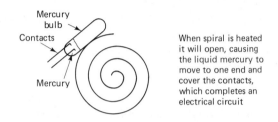

(c) Spiral element with mercury type switch

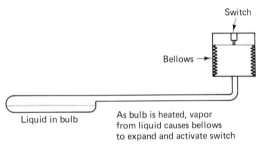

(d) Bulb control

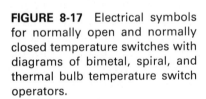

FIGURE 8-17 Electrical symbols for normally open and normally closed temperature switches with diagrams of bimetal, spiral, and thermal bulb temperature switch operators.

creases. This type of control is used as an overtemperature switch or as a heating thermostat. If the switch contact is shown below the terminals, it is considered to be a normally open temperature switch, which will close on temperature rise. This type of temperature switch is used to turn on a fan when the temperature gets too hot, or to turn on an air-conditioning system when the temperature increases.

Types of Temperature Operators

A variety of operators are used to cause the temperature switch contacts to activate. These operators are shown in Figure 8-17. The simplest of these operators to understand is the *bimetal strip* (Figure 8-17b). A bimetal operator is manufactured by bonding two strips of dissimilar metal together so that their surfaces touch. From this figure you can see that if one end of the bimetal switch is mounted to a fixture, the other end will move upward when it is heated. After the heat is removed, the free end will move back to its original position.

The simplest temperature switch is made by mounting a set of contacts where the movement of the bimetal strip can activate them. This type of switch can use springs to cause the contacts to snap over center when they open or close so that an arc cannot be drawn across them. Since the amount of movement can be predicted, the switch can be set for a single set-point temperature, or an adjustment screw and spring can be added to make the set point variable.

The motion of the bimetal can be amplified by bending the bimetal strip into a spiral. A spiral bimetal operator is shown in Figure 8-17c. From the diagram you should notice that the spiral will open as the temperature applied to it increases. When the temperature is returned to its original value or lowered, the spiral will tend to close. Since this movement is also predictable for any given change of temperature, hardware can be added to make an

Temperature Switches 223

adjustable set point. A set of cams and gears can also be added to activate a set of snap-over-center contacts, or a mercury switch can be used as contacts. Since this spiral operator is very sensitive to temperature and has to be mounted to take advantage of the maximum amount of travel as the temperature changes, you should not touch the spiral or attempt to adjust it. It is also important to keep the spiral clean and free of grease or oil, which might collect dirt and cause the operator to become out of adjustment.

Another type of temperature operator is the *sensing bulb control*. This type of control is also called *capillary tube control* (Figure 8-17d). You can see that this type of control uses a bulb filled with a liquid such as alcohol or refrigerant. The bulb is connected by a capillary tube to a pressure-sensitive switch activator. The switch activator can use a rubber diaphragm, a bellows operator, or a Bourdon tube, which are all similar to traditional pressure switches. When the temperature on the bulb increases, the liquid will experience a corresponding increase of pressure, which will cause the bellows or diaphragm to activate the switch contacts.

This bulb operator is considered a temperature switch even though the switch is activated by the pressure change that is caused by the temperature change. You should recall that if a liquid is confined in a bulb, the pressure that it exerts on the bulb and switch operator is directly proportional to the temperature. This means that for any given temperature, the amount of pressure can be determined and predicted. From this relationship it is easy to design a temperature switch that will be activated by the change of pressure on the fluid in the bulb.

The main advantage of bulb control is that the switch contacts can be mounted some distance from the bulb and away from the source of heat. The bulb can be mounted directly on the source of heat, or it can be mounted in a well where heat transfer can easily occur. This type of temperature switch is also useful where very cold temperatures must be measured. If the switch were mounted in very cold temperatures, any lubricants used on the switch mechanism may solidify and hamper the operation of the switch mechanism, which causes the switch to become inaccurate and unreliable. The thermal bulb allows the switch to be mounted in a cabinet at normal room temperature while the bulb is mounted remotely in a low-temperature environment.

Another type of temperature control uses a thermocouple to sense the temperature and activate a set of contacts. The termocouple is made of two dissimilar wires that will produce a small amount of voltage when they are heated. The two metals are connected together at a point called the *hot junction*. Thermocouples can use several different types of metals to make sensors for a variety of ranges. For example, a type J thermocouple is made from iron and constantan. Its temperature range is 32 to 1400°F. A type K thermocouple is made of chromel and alumel and can sense temperatures in the range 1000 to 2000°F.

Thermocouples are generally used as an analog device. This means that the output signal that they produce is proportional to the temperature that is being sensed. The thermocouple produces a millivolt signal. This signal must be amplified to be useful for control purposes.

The thermocouple can also be used as the sensor for an on/off controller even though the original signal is analog. In this type of controller, a thermocouple is connected to the control as a sensor 2. The set-point temperature is entered into the control by adjusting the set-point indicator on the face of the control. When the temperature reaches the set-point value, the signal from the thermocouple will be strong enough to activate the contacts. These terminals provide a set of contacts, and terminals 5 and 6 are normally open and normally closed, respectively. A heating contactor coil will usually be connected across these terminals, and the contacts of this contactor will be used to energize a heating element. In the normal operational mode, the normally open contacts will be closed any time the temperature is below the set-point value. This means that if the setpoint is adjusted for 350°F and the temperature that the thermocouple is sensing is 340°F, the normally closed contacts will be closed and the heater coil will be energized, which continues to cause the temperature to increase.

Solid-State Temperature Controllers Using RTDs and Thermistors

Another way to control a solid-state sensor is to use a resistive temperature detector (RTD) or thermister as a temperature-sensing input. The RTD uses a wire whose resistance will change when a change of temperature occurs. For support this wire is wound around a nonconductive core. The amount of resistance change is sensed by a Wheatstone bridge circuit. This means that the amount of milliampere current flow will change with a change in temperature. The RTD provides a fairly linear response to the temperature change. RTDs are made from three types of wire, to respond to a variety of temperature ranges. The platinum types has a sensing range of

−300 to 1300°F, the nickel type has a sensing range from −100 to 300°F, and the copper type has a sensing range of −60 to 300°F. The solid-state control for the RTD is essentially the same as thermocouple control. A set point can be entered into the control, and any time the temperature is below the set point, the control will close a set of normally open contacts. Since the contacts also have a set of normally closed contacts, the control can be used as a overtemperature control and deenergize a circuit when the temperature exceeds the set point.

The thermister sensor uses a temperature-sensitive resistor as the temperature detector. The thermister usually has a negative temperature coefficient. This means that as the temperature of thermister increases, the electrical resistance of the material decreases. This relationship is linear enough to provide accurate temperature indication.

Installing the Temperature Pilot Device

The temperature pilot device is installed in a manner similar to that for other pilot devices. The main concern in the installation process is to locate the temperature sensing element correctly. It is very important to locate the sensor close enough to the heat source so that it can monitor the temperature accurately. It is also important to keep the sensor located where it will not be damaged by excessive heat. Most sensors will provide instructions indicating the best location to mount the sensor. Remember that bulb (capillary tube) sensors must be mounted so that the liquid will not fall directly on the sensing diaphragm. This means that the sensing bulb should be mounted horizontally or in a vertical position so that the liquid will be in the bottom of bulb. In this location, only the vapor of the liquid will provide the pressure to activate the diaphragm. If the liquid reaches the diaphragm, it will cause the switch to become inaccurate.

After the temperature portion of the switch is installed, the temperature should be change to cause the contacts to activate. Allow the temperature to move above and below the set point several times while you have a meter on the contacts to observe their operation. This exercise will also allow you to identify the set of normally open or normally closed contacts that you intend to use.

Once you have identified the contacts and are sure that they are operating correctly, continue the installation procedure by installing the field wiring. Be sure to provide the proper source voltage to the control. If the device is a mechanical operator, the contacts can be wired in series as a pilot device. If the device is a solid-state control, follow the field wiring diagram that is provided on the back of the control and attach the thermocouple or RTD at the terminals indicated. Also be sure to provide line 1 and line 2 potential at the correct terminals. After the field wiring is completed, allow the temperature to change above and below the set point so that the control can be throughly tested.

Troubleshooting the Temperature Pilot Device

A temperature pilot device should be tested in the circuit when possible. It is important to determine if the temperature has changed enough to activate the switch contacts. In some cases it will be useful to identify where the problem exists. This means that you should use a voltmeter to test to see if voltage is present across the load terminals. If voltage is not present across the load terminals, continue to test the circuit to look for the open circuit. Use the procedure listed earlier in this chapter where one of the meter's probes is left on the line 2 terminal of the load as a reference. The other meter probe is moved from point to point in the circuit until the open is located.

If this procedure indicates that the temperature switch is causing the open, be sure to check the temperature where the sensor portion of the switch is located. Allow the temperature to increase and decrease to determine that the switch contacts are being activated. If the switch contacts are not activated, the switch operator is inoperative and must be replaced.

If the control is a solid-state control, the thermocouple or RTD should be tested separately to determine that they are producing the proper amount of voltage or current for the temperature. The control can be tested by supplying the proper amount of voltage or current with a separate power supply. A temperature–voltage curve should be used to determine the proper amount of voltage for any given temperature. If the voltage from the external supply cannot cause the solid-state control to activate its contacts, the control should be replaced. Be sure to turn the power off prior to replacing the control.

INDICATOR LAMPS

Indicator lamps are also considered pilot devices since they are located in the control portion of the circuit and because they draw a minimal amount of current. Figure 8-18 shows the electrical symbol for an indicator lamp. The abbreviation for the indicator lamp is LT. Multiple lamps are identified as

FIGURE 8-18 Electrical symbol for an indicator lamp.

LT1, LT2, or LT3. Some of these lamps are used as stand-alone devices, while others are incorporated into other pilot devices such as pushbuttons and selector switches. Indicator lamps are available as flush-mounted units, press-to-test units, flasher units, extended mount units, and master test units. Several types of lamps are used in the indicator assemblies, including the incandescent lamp, the neon bulb, and the LED indicator.

When the indicator lamp is shown in a electrical diagram, the electrical symbol is used in combination with a letter indicating the color of the lamp lens. The code for these colors are shown in Figure 8-19. The meaning of each of the colors has been standardized for motor control panels. The lens for the indicator lamp can be made of plastic or glass. Special lenses may be used to allow the lamp assembly to be used in hazardous environments where moisture or corrosive materials may be present.

The indicator lamp assembly used in most applications utilizes an incandescent lamp. The indicator lamp is actually an assembly. Figure 8-20 is an exploded-view photograph of a typical indicator lamp assembly. This assembly includes the terminal block where the wires are connected; the light unit, which is the receptacle where the bulb is secured; the mounting ring and mounting nut, which secure the light unit under the electrical panel; the locating ring, which prevents the lamp from twisting after it is mounted into the panel; the legend plate, which is used to identify the lamp; the

Pushbutton Color Code

Color	Typical Function	Example
Red	Stop, Emergency Stop	Stop of one or more motors; master stop.
Yellow	Return, Emergency Return	Return of machine elements to start position.
Black	Start Motors, Cycle, etc. Any operation for which no other color is specified	Start of one or more motors; start cycle or partial sequence.

Pilot Light Lens Color Code

Color	Typical Function	Example
Red	Danger, Abnormal Condition, Fault Condition	Voltage applied; cycle in automatic; faults in air, water, lubricating or filtering systems; ground detector circuits.
Amber (Yellow)	Attention	Motors running; machine in cycle; unit or head in forward position.
Green	Safe Condition (Security)	End of cycle; unit or head returned; motors stopped; motion stopped; contactors open.
White or Clear	Normal Condition	Normal pressure of air, water, lubrication.

Multiple Station Pilot Light Requirements

Color	Typical Function	Example
Red	Power On, Emergency On, Automatic Cycle	Ground detectors; lubrication failure; master relay on; pressure failure (water, air, gas).
Amber (Yellow)	Motors Running, Machine in Cycle, Full Depth	Machine elements in advanced position; manual cycle.
Green	End of Cycle, Heads in Returned Position	
White or Clear	Parts in Place, Lubrication Normal, Pressure Normal (Water, Air, Gas)	

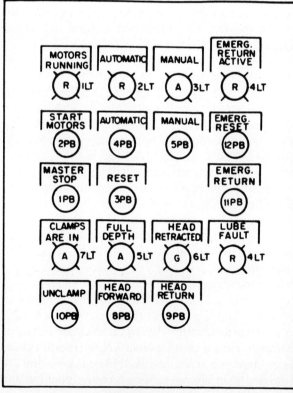

FIGURE 8-19 Color code for indicator lamp and pushbuttons with suggested locations of pushbuttons in operator panels. (Courtesy of Joint Industrial Council.)

FEATURES

- *Single seal for oiltight sealing of 1/16" to 3/16" panels*
- *Quick mounting plug-in installation*
- *Low profile, miniature, square appearance is compatible with other panel controls*
- *Wide variety of operator colors*
- *Mechanical, electronic duty, and solid state switch blocks*
- *UL and CSA listed*
- *Suitable for use in NEMA 4 and 13 enclosures*

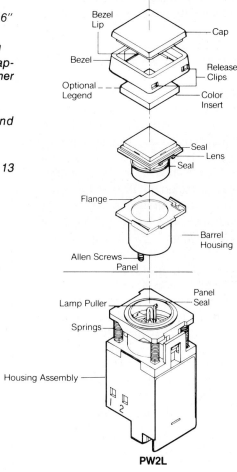

FIGURE 8-20 Exploded view for an indicator lamp assembly. [(a) Courtesy of Eaton Corp./Cutler-Hammer Division; (b) courtesy of Honeywell/Micro Switch Division.]

sealing washer; and the extended lens. This assembly allows the indicator lamp to be easily mounted into most electrical panels. This assembly also allows the device to be tested and serviced with a minimum of effort.

Indicator lamps are also classified by the voltage that the assembly requires. Typical voltages include, 6, 12, 24, 48, 120, 208, and 220 V ac and dc. Some assemblies require a transformer to be used to provide the lower voltages. The bulb must be matched to the voltage at the lamp terminals.

Installing the Indicator Lamp

Indicator lamps are usually installed into electrical control panels or operator panels. In many applications the panel layout is designed to fit the machine operation. For some machines, switches and indica-

Indicator Lamps **227**

tor lamps are mounted into the front or top of the electrical control panel, which serves as the operator's control unit. The first step for installing a pilot device into this type of panel is to lay out and cut the holes into the panel. After the holes have been cut into the panel, the indicator assembly can be mounted through the metal surface. All field wiring is attached to the terminal block. The terminal block is then connected to the rest of the lamp assembly. Be sure to follow the electrical diagram for the field wiring. This diagram will be slightly different for press-to-test lamps and master test lamps since they require three wires instead of the normal two.

Troubleshooting the Indicator Lamp

The indicator lamp assembly can be tested either with power applied or with power removed. The preferred testing method is with power applied to the assembly. If the lamp is a press-to-test or is part of a master test system, simply press the test lamp and see if the lamp illuminates. If the lamp illuminates during the test operation but does not light up during normal machine operation, the control circuit should be suspected and tested for an open circuit or malfunctioning part. Check the circuit diagram for a list of components in the lamp's circuit and use it for the test procedure.

If the lamp assembly does not illuminate during the press-to-test operation, change the light bulb or test for proper voltage. If you have an extra bulb, or if one is available in another indicator lamp on the panel, changing the bulb is the preferred test. If an extra lamp is not available and you have a voltmeter, test the circuit for proper voltage. Be sure to start at the lamp terminals and work your way back to the voltage supply. Remember that the lamp is in the system as an indicator. Voltage may be interrupted by a pilot switch indicating that a specific condition exists in the system. In this case the lamp circuit is not faulty; rather, it is operating correctly as it is designed.

CONNECTING PILOT DEVICES TO ROBOTS AND PROGRAMMABLE CONTROLLERS

It is likely that the modern pilot devices that you encounter will be interfaced to robots or programmable controllers as well as stand-alone hard-wired controls. Earlier in this chapter we explained the operation of pilot devices. In this section we explain the interface of pilot devices to robot and programmable controllers. It will explain the installation of the devices using wiring diagrams and the method

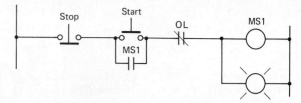

(a) Example of a hard-wired start–stop circuit

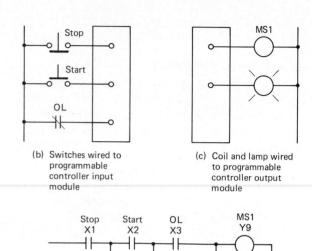

(b) Switches wired to programmable controller input module

(c) Coil and lamp wired to programmable controller output module

(d) Program for start-stop circuit in a programmable controller (note: X1 and X3 are wired as closed switch and programmed as open contacts)

FIGURE 8-21 Example of start/stop circuit connected to a programmable controller and as a hard-wired circuit.

of troubleshooting using the robot microprocessor and the programmable controller status indicators.

Interfacing Pilot Devices to Programmable Controllers

Programmable controllers (P/Cs) provide a variety of methods of interfacing pilot devices for complex control systems. In a traditional hard-wired motor control circuit, such as a start/stop circuit, the stop button is wired directly to the start button, and the start button is wired directly to the overloads and motor starter coil. Figure 8-21 shows this typical hard-wired start/stop circuit and compares it to the same circuit function as it appears in a programmable controller. You must first understand the operation of the programmable controller before you can understand the way pilot devices are wired to it. The programmable controller can be described in three distinct sections: input, output, and microprocessor sections. In the input section the input module converts the voltage of traditional pilot device into a computer signal. In the output section, the output module converts the computer signal to a voltage that is large enough to energize a motor control load such as a motor starter coil. A more detailed explanation of the input and output modules including component operation, is provided in Chapter 16.

The microprocessor part of the programmable controller also functions as several separate operations. These include the image register, the program that is in memory, and the microprocessor's scan function. Each of the input and output circuits is given an address by the microprocessor. The image register is a detailed map of these addresses. If a signal is present at a specific input address, a logic 1 is placed in that address in the image register. If the signal is not present at an address, a 0 is placed in the image register at that address. The addresses in the image register are constantly scanned by the microprocessor. The microprocessor then solves the logic of the circuit that is in the program memory. Figure 8-21 also shows an example of the pilot devices connected to the input module and pilot loads connected to the output module. The program stored in the P/C's memory is also shown in this figure. Notice that the program looks exactly like a electrical ladder diagram. The microprocessor solves this circuit exactly like a hard-wired circuit, in that if a pilot switch is activated, power will be passed to the switch next to it. If all the switches in a circuit are activated, power will be passed to the output and it will be energized. In this example the stop botton at address X1 is held close by a spring. This causes a 1 to be written into address X2 in the image register of the programmable controller. The image register is a register in the programmable controller that stores the results of switch operations and stores them on each scan cycle of the processor. If the switch is energized, a 1 is placed in the image register address for that switch. If the switch is deenergized, a 0 is written in the image register. When the processor solves the ladder logic during each scan, it reads the contents of each input image register address and writes the results of the logic operation in the output image register address assigned to each output. In this start/stop circuit application the start button is wired to address X2. When someone depresses the button, a 1 will be sent the address X2 in the image register. The microprocessor scans the program continually, and when it reads a 1 in the image register addresses of X1 and X2, it will solve the logic and write a 1 to the output image register address of Y9. If someone depresses the stop button, a 0 would be sent to address X2, and the next time the microprocessor scans the image register, it would find the 0 and use it when the logic is solved. Since the 0 represents an open switch, power could not be passed through contacts X1 and the output would be deenergized. After the microprocessor solved the logic and found that X1 was a 0, it would then write a 0 to the image register of output Y9. When the image register receives the 0 for output address Y9, the voltage to output module address Y10 would also be deenergized and the load connected to that terminal would be turned off. In most P/Cs the microprocessor scans the program about every 20 to 30 milliseconds.

Since the P/C's microprocessor reads the input image register, all the pilot switches are connected to the input module instead of to each other as in the hard-wired control circuit, and the motor starter coil is connected directly to the output module rather than being connected to the pilot switches as in the hard-wired control circuit. The program in the P/C's memory can be displayed on a cathode rate tube (CRT) monitor such as a television screen.

The outcome (operation) of the two circuits are very similar. In both circuits, when the start button is depressed, the motor start coil is energized and its auxiliary contacts seal the start switch. When the stop button is depressed, the motor starter coil is deenergized. In a hard-wired circuit, the electric current flows from the source at line 1 through the stop button contacts and the start button contacts the overload contacts and the motor starter coil. When the stop button is depressed, the contacts interrupt this current and the motor starter coil deenergizes. In the programmable controller, the signal from the start button sends a pulse to the P/C's input bus. The microprocessor continually scans the program and checks the input bus to see if a signal is present. When the signal is detected, the microprocessor solves the ladder diagram logic in a fashion similar to the hard-wired circuit. The contacts in the program that represent the start switch are closed and a signal is sent to the stop contacts in the program. Since the stop button is maintained closed with spring tension, it will also indicate to the P/C that it is energized. The P/C recognizes that both input signals are present and continues to solve the logic by energizing the output address where the motor starter coil is connected. This signal is converted to the correct voltage at the output module and sufficient power is sent to energize the motor starter coil.

Input and Output Modules. Since the input module can convert a voltage signal to the computer level signal, a variety of modules are available for interfacing nearly any input voltage to the computer level signal. For example, input modules are available to convert dc voltages from 12 to 28 V, 32 to 56 V and 86 to 120 V. Ac modules are available to convert ac pilot voltages from 12 to 24 V, 86 to 120 V, and 180 to 220 V. Figure 8-22 shows an example of the field wiring diagram for a typical ac module with several pilot devices connected to it. A typical output module is also shown.

If the module is the dc input module, it requires the negative side of the dc power source to be connected to the module to act as a common for each of the input circuits. The positive for the dc source is used as a supply for each pilot device. Each pilot device is connected to its own input terminal so that it can activate its circuit independently of the other pilot devices. It is important to remember to size the pilot switch contacts to exceed the voltage supply of the module. Another important point to remember is that multiple contact switches are not required for complex circuit logic in programmable controllers since the input switch can be programmed in the P/C as a normally open and normally closed switch as many times as required by the logic. The only restriction to the number of times an input switch may be entered in the program is the size of the P/C's program memory.

The wiring diagram for the ac input module is very similar to the dc module. The return line (neutral) is used as the common for the module. Each of the pilot devices is connected to the line 1 side of the ac voltage. This type of circuitry allows each pilot device to operate independent of the others. Again, as in the dc modules, the pilot device need only be a simple single-pole switch, while it can be programmed as a complex double-pole, double-throw switch with both normally open and normally closed contacts.

Troubleshooting a Pilot Device Connected to a P/C. The input signal from the pilot device is very easy to troubleshoot on a programmable controller input module. Each module provides a status indicator for each input circuit. If a pilot switch closes and passes power to the module, the status indicator for that circuit will light and show that the circuit is complete. The symbol on the CRT screen that represents the pilot device will also become highlighted to indicate that the circuit has been energized. Some programmable controllers will display the word "on" if the switch is made and the word "off" if the switch is open. Other programmable controllers may show a bit display of the input addresses. Any circuit address that is energized will display a 1, and any address that is deenergized will display a 0. The technician can troubleshoot the circuit using any of these status indicators. More information concerning the programmable controller is provided in Chapter 16. A voltmeter can also be used to troubleshoot for a missing signal from pilot switches to input modules. Since the signal from each individual pilot switch can be checked on the CRT, the actual voltage test can be limited to an individual switch that is suspected of being faulty.

Load Devices Connected to P/C Output Modules. The output module in a programmable controller can convert a computer signal to the proper level to power any pilot load. These loads include indicator lamps, solenoids, and relay and motor starter coils. The output module is usually selected to match the type of voltage the load will require. This means that if the load requires 120 V dc, a 120-V dc output module would be used.

When the ladder logic program in the programmable controller activates the output address, the circuit connected to that address will be ener-

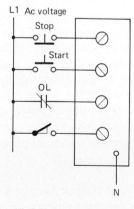

(a) Switches connected to an input module

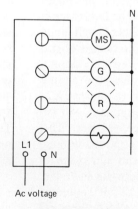

(b) Devices connected to an output module

FIGURE 8-22 Ac and input module with pilot devices connected and ac pilot loads connected to an output module.

gized. Figure 8-22 also shows a diagram of typical pilot loads connected to an ac output module. Notice that the output module requires both the line and neutral lines of the ac voltage to operate correctly. Each ac load should have one of its terminals connected to negative line and its other terminals connected to the terminal point on the module for the proper address. Each circuit in a ac output module switches the line side of the circuit. This wiring arrangement allows each of the loads to be energized or deenergized independently.

Some P/C manufacturers provide a fuse for protection in each circuit of the output module, while others prefer to fuse the power supply directly. If the fuse is located in the module, a blown-fuse indicator is usually provided to make troubleshooting of the module more convenient. Each circuit in the module also provides a status indicator, which will glow brightly to indicate the circuit is energized. The technician can troubleshoot the suspected circuit by using a ladder diagram and the status indicators. If a status indicator is not glowing, the technician should use the CRT to determine if the output address is being energized by the P/C. If the output is being energized and the module status indicator is not glowing, the problem will be between the P/C's output bus and the module itself. If the output address is not being energized by the P/C, the enabling logic should be checked to determine which set of contacts is holding the circuit in the denergized state. Use the test procedure for input switches described earlier to determine which inputs are not energized. Remember: the programmable controller is very powerful for troubleshooting. It provides several different methods to determine if a system fault has occurred. Learn to use its functions to troubleshoot both input and output faults.

Interfacing Pilot Devices to Robots

Pilot devices can easily be interface with robot input and output ports to help control the robot work cell or determine the status of the automated system. Pilot switches used as inputs to the robot and output pilot devices such as solenoids, coils, and lamps can be connected to the robot input and output bus. The robot's program is a series of steps that indicate the location of the end effector (robot hand). After the robot hand has moved to the location specified in the program, the robot's microprocessor can read the input bus to see if a signal is present, or it can write to the output bus to energize any of its outputs. For example, when the robot reaches the location over a conveyor, it can check its input bus address to see if the part-in-place limit switch is energized. If the part-in-place limit switch is energized, the robot knows that a part is ready to be picked up and it will proceed to the pickup part of the program. If the part-in-place limit switch is not energized, the robot knows that a part is not present and it can continue to another part of the program such as testing other conveyors for parts in place. In this way, the robot can service two or more conveyors and not waste time if parts are not ready on any one of the conveyors. If a part is in place, the robot may move to the location where the part can be picked up, and then send an output signal to an output address to energize a air solenoid to close its gripper.

The voltage for the robot input and output signals is provided by one power supply. For this application, the power supply is 24 V dc. The pilot switches must have a contact rating of 24 V or better. Since this is a fairly low voltage rating for the contacts, it will not present a problem of finding pilot devices with contact ratings of 24 V or better. The wiring configuration is very similar to the wiring of an input module for a programmable controller. The point of connection is in the work cell cable head. The cable head is used to connect input switches and output loads to the robot's microprocessor input and output bus. The microprocessor is usually mounted in the robot control cabinet. This wiring configuration allows the robot arm to move freely because the pilot devices are connected through a cable to the microprocessor. There are several extra pins that carry the positive and negative voltage in the cable head. The extra pins are provided because each input switch must be wired to the positive side of the power supply through one of the cable pins labeled as DC+, and each load needs to be connected to the return side of the power supply.

On the output side, each lamp, solenoid, relay coil, or motor starter coil must be rated for exactly 24 V dc. If the output loads require a higher voltage, a 24-V dc relay must be used to provide a larger voltage to the coil. For this type of application, a separate voltage supply is required since the voltage to the motor starter coil is ac. If the load connected to the output is inductive, such as a relay or motor starter coil, a diode should be connected in reverse bias across the output terminals. The diode is used to shunt inductive spikes when current to the load is deenergized. Since the diode is in reverse bias, a normal output signal will not be bothered. The spike that is created when the output current is interrupted will be of reverse polarity compared to the normal signal. The diode will provide a path from one side of the output load to the other when the spike of current occurs. This action allows the wire in the relay coil to dissipate the large surge of current before it can cause any damage.

Input and output signals may be routed through the same robot cable. The input signal are routed to the robot's microprocessor input bus, and the output signals are generated from the output bus.

Troubleshooting Robot Input and Output Signals. When pilot devices are connected to robots, the voltage that they control is called *input signals*. The voltage that is provided to output loads is called *output signals*. Input and output signals can be displayed on the robot controller's CRT screen. Some robots show these I/O addresses as diagnostic bits, while others use LEDs to display the status of input and output signals. The input and output signals can also be tested with a voltmeter. To test an input signal, one probe of the voltmeter should be placed on the negative voltage terminal, while the other probe is moved to the point where the input pilot device is connected. Remember: pilot devices used as inputs for a robot are wired as single-ended inputs. This means that each device is connected directly to the robot input bus through the work cell signal cable instead of being connected to each other like a hard-wired circuit. Each output from the robot is also connected as a single-ended load. This means that only one load is connected to each robot output address. The loads are also connected to the work cell signal cable. It is important to remember that as robots and programmable controllers become more complex, they will always be easily interfaced to pilot switches and loads and they will also provide an easy means of troubleshooting the signals coming and going to them.

QUESTIONS

8-1. Explain the operation of normally open and normally closed contacts.

8-2. Explain how the contacts of a reed switch are different from those of normal contacts.

8-3. Explain how reed switch contacts are moved from the open to the closed position.

8-4. Show the electrical symbol for normally open and normally closed contacts.

8-5. Show the symbol for normally open and normally closed pushbutton switches. Include the abbreviation for the switch.

8-6. Explain the difference between a normal momentary pushbutton and a maintained pushbutton.

8-7. Draw a diagram that shows how you would add additional start and stop pushbuttons to a circuit controlling a motor starter.

8-8. Explain why a keylock type of pushbutton might be used.

8-9. List the types of heads you may find on pushbuttons.

8-10. Explain what a legend plate is and why it is used with a pushbutton switch.

8-11. Explain how you would test the circuit you provided for Question 8-7 for a bad pushbutton switch. Remember that power is applied to the circuit.

8-12. Explain the operation of a limit switch. What is used to make the switch turn on or off?

8-13. Show the electrical symbol for a normally open and closed limit switch. Include the abbreviation for limit switch.

8-14. List six types of actuators that you would find on limit switches.

8-15. Draw a sketch of each of the actuators listed in Question 8-14.

8-16. Provide an example application where you would find each of the switch types listed in Question 8-4.

8-17. Name three ways that a limit switch can return to its original position.

8-18. List several conditions you must be aware of when you select a place to mount a limit switch.

8-19. Explain how you adjust a limit switch and test it during installation.

8-20. Explain how a precision limit switch is different from a normal limit switch.

8-21. Draw the electrical symbol for a normally open and a normally closed selector switch. Be sure to include the abbreviations for this switch.

8-22. Explain the difference between a two-, three-, and four-position limit switch.

8-23. Draw a symbol of a joystick switch.

8-24. Explain how a joystick operates.

8-25. List several applications where you would find a joystick.

8-26. Explain how you would troubleshoot a joystick control.

8-27. Draw the electrical symbol for a normally open and normally closed footswitch.

8-28. Explain the operation of a footswitch.

8-29. Draw the electrical symbol for a close-on-rise and close-on-fall float switch. Include the abbreviation for this switch.

8-30. List three types of float switches and explain their operations.

8-31. Explain the operation of a resistive-sensitive level indicator.

8-32. Draw the electrical symbol for a close-on increasing pressure and for an open-on increasing pressure switch. Include the abbreviations for this switch.

8-33. Explain the operation of a diaphragm pressure switch and a bellows pressure switch.

8-34. Explain the difference between a pressure switch and a vacuum switch.

8-35. Explain the operation of a differential pressure switch. Where would you be most likely to find this type of control?

8-36. Show the electrical symbol for a normally open and a normally closed flow switch.

Timers, counters, and sequencers 9

INTRODUCTION

Timers, counter, and sequencers provide complex control applications for today's automated systems. These devices can be used as add-on controls, stand-alone controls, or can be directly interfaced with programmable controllers and robots. In this chapter we cover the complete line of available timing devices. In the next part of the chapter we explain the operation and use of counters. In the final part of the chapter we discuss the use of counters. In the final part of the chapter we discuss the use of timer/counter devices and sequencers. The information provided in this chapter will help you identify the different types of controls, typical applications, theory of operation, diagrams and installation, and troubleshooting techniques.

TIMERS

Timers have been used to control industrial applications for many years. Timers provided time delay by using motors, gears, and cams to control sets of contacts. Other types of timers and time-delay devices uses air or liquid in a chamber that empties slowly to cause a delay in the action of contacts. As technology has advanced, new timing devices have incorporated modern synchronous motors, solid-state technology, and microprocessor technology. Some devices fill applications where the requirement is mainly low cost and simple operation. These devices are typically added on to standard relays, contactors, and motor starters to provide some degree of time delay.

There are four basic principles of operation used in modern timing devices: pneumatic, the electromechanical motor-driven, solid-state electronic and microprocessor, also called digital type. These devices come in a variety of models that are available with the following types of control for the outputs they activate: delay-on, delay-off, cycle times, and logic timers.

These types of timers can be used as reset timers where the timer is activated and then to time out and actuates a set of contacts, or they can be used as a repeat cycle timer that continually repeats its cycle and actuates a set of contacts at the end of each cycle. The final way to use a timer is as a totalizing timer, where the amount of time that a device operates is recorded. A comparison of the synchronous motor, solid-state, and the digital timers is presented in Figure 9-1.

● CLASSIFICATION BY OPERATING PRINCIPLE

Classification	Operating Principle	Merits	Demerits
Synchronous motor timer	The motor timer employs a synchronous motor, actuated by an electrical input signal, produces a mechanical motion, by which the contacts are caused to make and break after the lapse of a pre-determined time. In general, a desired time is obtained, through a clutch and reduction gear mechanism, by actuating a small synchronous motor by applying commercial 50 or 60Hz power as the input signal. The speed (r.p.m.) of the synchronous motor in principle is proportional to the input power frequency and is not affected by a change in the supply voltage or ambient temperature. The DC type motor timer is available in various types including the one employing a DC motor and the type with a similar mechanism to an AC timer, which actuates the synchronous motor by the built-in DC-AC converter.	1. Wide time setting range from short to long time. 2. Elapsed time indication by moving pointer. 3. Less affected by temperature and voltage fluctuations.	1. Precise, minute time setting impossible. 2. Short mechanical service life because of its construction with mechanical moving parts. 3. Manufacturing of DC operation type is difficult.
Solid-state timer	In the RC oscillation counting system, the RC oscillation circuit consisting of a resistor and a capacitor starts oscillating upon application of power. When the counter circuit counts up the set value by taking the oscillated signal as the reference signal, an output signal is generated. This output signal is then amplified to operate the output relay. The basic circuitry of the timer consists of a power circuit, a constant voltage circuit, an RC oscillation circuit, a counter circuit, an output amplifier circuit and a relay circuit. In the RC charging system, a desired time delay is obtained by utilizing the charge and discharge characteristics of the RC (resistor and capacitor) network. When the terminal voltage of the capacitor reaches the specified value, an output signal is generated from the relay or semi-conductor. Solid-state timers with less mechanical moving parts are highly resistant to vibration and shock, and are capable of highly frequent operation, thus boasting long service life.	1. Highly frequent operation with long mechanical service life. 2. Both DC and AC operations possible. 3. Solid-state output system and separate time setting system are possible.	1. Manufacturing of long-time versions is difficult. 2. Susceptible to temperature and voltage fluctuations. 3. Elapsed time indication impossible.
Digital timer	A constant frequency signal (usually, commercial power frequency) is employed as a time-limit element. This signal is counted by the counting circuit to obtain a time delay. The shorter and the more accurate the duration of the reference signal, the higher the timer accuracy. However, the duration of the reference signal must be such that permits its discrimination from externally occurring noises. The timer employs a thumbwheel switch in its time setting section to permit highly accurate time control, and also indicates the elapsed time digitally. The version employing an electromagnetic counter is also available. The counter advances mechanically at each input signal by the electromagnetic force of the electromagnetic coil. However, this version is not suitable for practical application because of its low counting speed and short mechanical service life.	1. High precision. 2. Error-free digital time setting. 3. Elapsed time indication also possible digitally.	1. Larger in size and more costly than other types. 2. Susceptible to electrical noise.

FIGURE 9-1 Comparison of types of timers. (Courtesy of Omron Electronic, Inc.)

Peneumatic Timers

Pneumatic time-delay devices are generally used as add-on or stand-alone controls. The basic operation of the pneumatic device is shown in Figure 9-2a. From this diagram you can see that the pneumatic control has a flexible diaphragm that seals an air chamber. This diaphragm has a metal plate attached to it that activates the electrical switch contacts when the diaphragm has moved far enough. A spring keeps constant pressure against the diaphragm during the timing cycle.

Air is released from the chamber through an orifice controlled by a needle valve. The needle valve controls the rate at which air can leave the chamber. When a sufficient amount of air has been released, the spring will cause the diaphragm to move far enough to activate the electrical switch element. When the pneumatic switch is reset, the chamber is refilled with air.

The amount of time delay is set by adjusting the needle valve and testing the response time. Several trials may need to be completed to get the time adjustment exactly right. The amount of time delay can be adjusted between 0 and 180 seconds. The accuracy can be controlled to about 0.5 second.

This type of device can be used to stand alone or an add-on timing control. Figure 9-2b shows a picture of a pneumatic add-on timer control. This type of control is added to a standard relay or contactor. It has an adjustment knob on the front to set the amount of time delay. The actual adjustment should be made using trial-and-error methods. This means that you should set the adjustment and then check the actual time with a watch. If the time delay is critical, minor corrections to the time delay can be made by adjusting the knob slightly.

Simple Time Delay-On and Time Delay-Off Examples. The time-delay devices can show a variety of applications though their wiring diagrams. The symbols for the timing functions are divided into two categories. The two categories refer to delay-on and delay-off. These may also be called on-delay and off-delay, delay-on make and delay-on break, or TD-on and TD-off. A switch turning on a motor is an example of a delay-on application. Instead of the switch turning on the motor directly, it would energize the time-delay element, which would allow its contacts to close some time later and energize the motor. This type of application could be used where three motors are started from the same circuit and a slight delay of 15 seconds is needed between the time each motor is started, to avoid the large power surge (in rush) that occurs when multiple motors are started. For this application, the first motor starter would energize directly from the control circuit. The second motor starter would be controlled by an on-delay timer that would have a preset time of 15 seconds. The third motor would be controlled by an on-delay timer that would have a preset time of 30 seconds.

When the control circuit was powered, it would energize the first motor starter immediately and begin the timing cycle for each of the on-delay timers. Fifteen seconds after the first motor started, the second motor's on-delay timer circuit would energize the second motor starter. Fifteen seconds later the third motor's timer circuit would time out and start the last motor. Each motor in this application would have time to start and come up to full speed, and the inrush current would return to normal limits, before the next motor would draw its large starting current.

In an off-delay circuit the timing circuit is

AC PNEUMATIC TIMING RELAYS

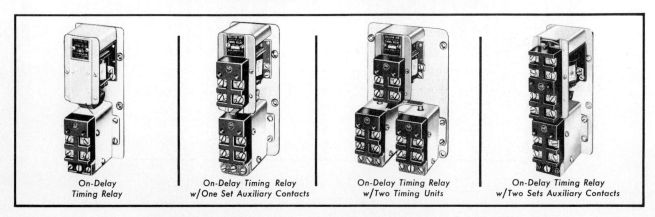

FIGURE 9-2a Diagram of pneumatic-type stand-alone time-delay switch and add-on timer. (Courtesy of Eaton Corp./Cutler-Hammer Division.)

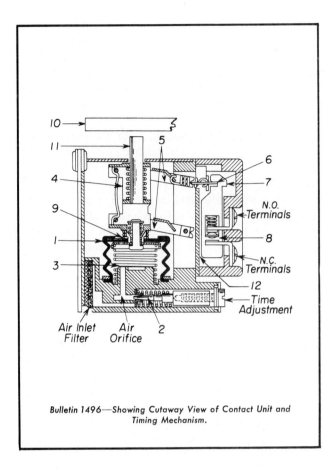

Bulletin 1496—Showing Cutaway View of Contact Unit and Timing Mechanism.

ACTION OF THE TIMING MECHANISM — When the external resetting device (10) is retracted from the push rod (11) it allows the spring (3) located inside the synthetic rubber bellows (1) to push the plunger (4) upward. As the plunger rises, it causes the over-center toggle mechanism (5) to move the snap action toggle blade (6) upward which in turn picks up the push plate (7) which carries the movable contacts (8).

The speed with which the bellows can expand is determined by the setting of the needle valve (2). If this needle valve is nearly closed, an appreciable length of time will be required for air to pass it and permit the bellows to expand. The setting of the needle valve determines the time interval which must elapse between operation of the actuator and expanding of the bellows to operate the contact unit.

When the push rod is again depressed, it forces the plunger to the lower position, exhausting the air through the release valve (9), and resetting the timer almost instantaneously.

★**REPLACING CONTACT UNIT** — Should the contact unit, for any reason, be removed from the front of the timing mechanism, care must be taken to see that the toggle blade (6) is in the down position when the unit is replaced. This enables the blade to fit into the notch in the push plate (7), as shown in the illustration. Hold down the push rod (11) and flip the toggle blade (6) down. Then put the contact unit in position, being sure that the "loose" black phenolic insulation (12), that isolates the contact cavity, is in place. Secure with the two mounting screws. After assembly, check for normal contact operation.

★Added or changed since previous issue.

PNEUMATIC TIMING UNITS

PNEUMATIC TIMER ATTACHMENT FOR SIZE 00 STARTER.

add on type timer

FIGURE 9-2b Comparison of types of timers. (Continued) (Courtesy of Eaton Corp./Cutler-Hammer Division, and Allen-Bradley.)

actuated when the control switch is turned off. In this type of circuit the timing device begins the timing cycle when power has been removed from it by the control switch. An example of an off-delay circuit could be shown with an overhead lighting application controlled by a switch with a time-delay device. When the switch is turned off, it does not turn the light off directly; instead, it activates the off-delay element. This allows the light to stay on for a short delay after the switch has been turned off.

This type of application allows personnel a few extra minutes to leave an area after turning off the switch before the lights go out. This is especially useful in large industrial areas. Many other motor control circuit that use time delay-off applications are discussed in this chapter.

Symbol and Electrical Diagrams for Time-Delay Circuits. The symbols for on-delay and off-delay timing circuits are shown in Figure 9-3. From this figure you can see that the symbol for the time-delay contacts is half an arrow. The time delay-on symbol uses the lower half or tail of the arrow. The time delay-off uses the upper half or arrowhead portion of the arrow. When the arrowhead is used for the time delay-off symbol, it is usually pointing down. In the case of the time delay-on symbol, the tail is usually pointing up. Time delay-on and time delay-off controls can both have normally open and normally closed contracts. Figure 9-3 shows the normally open and normally closed contacts.

Figure 9-4 shows the electrical diagram of an add-on pneumatic time-delay control which will delay turn-off time a motor. A limit switch in the line will open and deenergize the control relay's coil, and the contacts for that relay are controlled by a pneumatic timing device. This is indicated by the time delay-off symbol showing control of the relay contracts, and the contracts are identified by the same label as the relay coil.

When the electrical control to the relay coil is deenergized, the contacts push on the pneumatic actuator, causing air to begin to leave its chamber. When all the air has been pushed out of the chamber, the contacts are allowed to open. The net result is that the motor is shut off some time after the limit switch has deenergized the relay coil. The exact amount of time delay will be determined by the adjustment of the needle valve on the pneumatic add-on time-delay control.

From Figure 9-4 you can see that pneumatic add-on control can also be used as a time delay-on device. To change the device from time delay-off control to time delay-on control, the add-on pneumatic control is simply turned upside-down when it is installed onto the relay. The pneumatic add-on control has a marker with an arrow showing which way to mount the device for time delay-on and time delay-off. When the device is installed for time delay-on, the limit switch energizes the relay coil immediately after it closes. The magnetic pull from the relay coil causes the relay contracts to push on the bellows in the air chamber. Since the chamber is full of air, the contacts cannot close yet. As the air is removed from the chamber, the contacts are allowed to close and turn on the motor. The time it takes the air to leave the chamber provides the amount of time delay.

Electromechanical Timers

Another type of timer control uses a synchronous motor to turn a shaft on which cams are mounted. As the shaft rotates, the cams rotate to a point where they cause sets of contacts to open or close. Spring tension will cause the contacts will return to their original state (open or closed) when the timer is reset.

The synchronous motor turns the shaft at a predetermined speed. The amount of time delay will be determined by the speed of the motor and the distance the cams are from the switch contacts when the control is set. The amount of time delay can be extended by adding reduction gears between the motor shaft and the cam shaft.

Figure 9-5 shows a photograph of a synchronous motor-driven timer with a diagram of the internal electrical contacts and circuits. From this

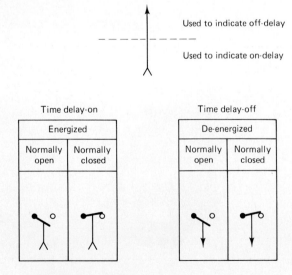

FIGURE 9-3 Symbols for time delay-on and time delay-off controls.

diagram you can see that the this type of timer has two separate controlling devices. The first one is shown as a solenoid, called a *clutch solenoid,* which controls an electromechanical clutch and two sets of isolated contacts. These contacts are called *instantaneous contacts*.

The second part of switch is controlled by the timing motor. The motor is identified by a "M" in the motor symbol. The timing motor provides the drive for the camshaft that actuates the timing contacts. There are two sets of isolated contacts controlled by the timer motor. These contact sets each has a normally open circuit and a normally closed circuit. There is also an indicator wired in parallel with the timer motor. This indicator is illuminated whenever that the timer motor is energized.

A dashed line extends from the solenoid symbol down through the two sets of contacts. This indicates that these contacts are under the control of this solenoid. A similar dashed line extends from the timer motor down through the remaining two

PNEUMATIC TIMING RELAYS

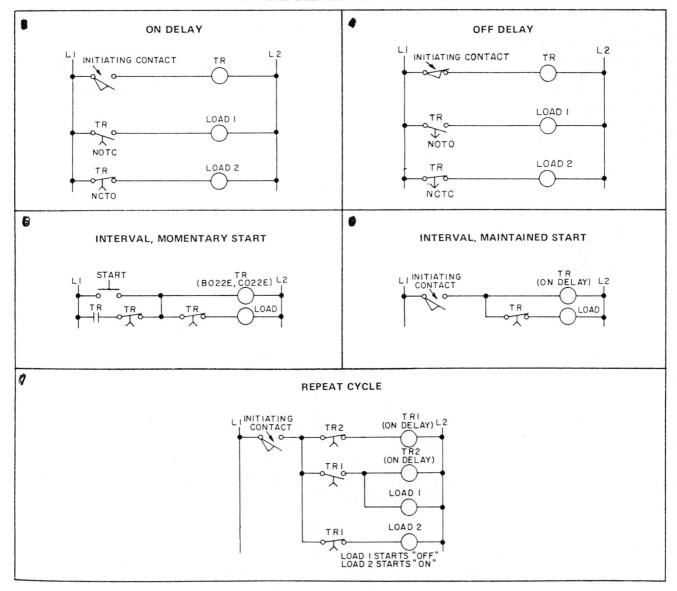

FIGURE 9-4 Diagram of off-delay control circuit using a pneumatic add-on timing device. (Courtesy of Allen-Bradley.)

Timers   239

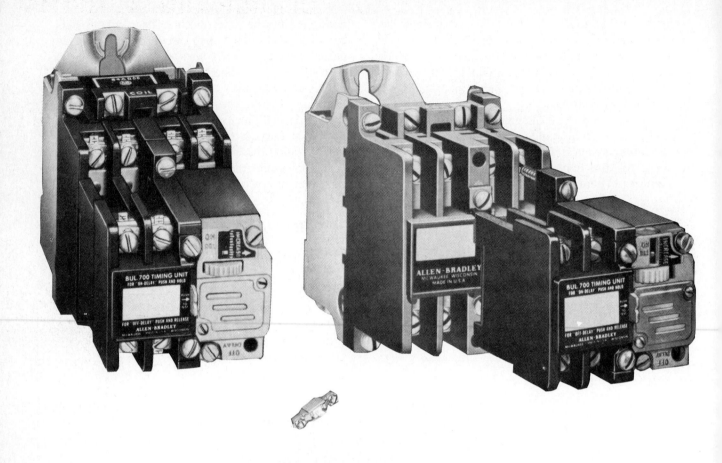

Pneumatic type add-on time delay relay

Solid state add-on time delay relay

FIGURE 9-4 (Continued)

SCHEMATIC DIAGRAM

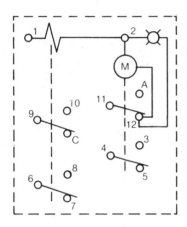

TERMINALS AND WIRING DIAGRAM ON REAR OF TIMER CASE

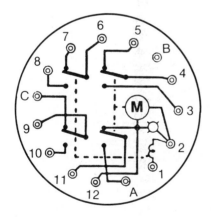

FIGURE 9-5 Synchronous motor-driven timer with a diagram of the internal electrical contacts and circuits. (Courtesy of Eagle Signal Controls.)

sets of contacts. This indicates that these contacts are under the control of the timer motor and will be affected by the time delay.

Motor-Driven On-Delay Timer Operation. The operation of the motor-driven timer will be explained in terms of the electrical diagram shown in Figure 9-6. This diagram shows the internal wiring of the timer in bold lines and the added field-installed wiring in lighter lines. The wiring setup makes this an on-delay timer.

The field wiring shows a limit switch connected in series with line 1 (L1) and terminal 1 on the timer. A field-installed jumper is connected between terminals 1 and 11 on the timer. This will provide L1 power to the timer motor through normally closed contacts 11 to 12. Line 2 (L2) is connected directly to terminal 2. Terminal 2 inside the timer is connected to the clutch solenoid, timer motor, and the indicator lamp. This provides the L2 potential side of the circuit to all of these loads inside the timer.

When the limit switch is closed, line 1 power is sent to the left side of the clutch solenoid and through the normally closed contacts at 11–12 to provide power to the timer motor and indicator. Since L2 power is connected directly to the other side of each of these loads from terminal 2, each load become energized.

When the clutch solenoid is energized, it pulls the gears into position to be driven by the timer motor shaft. The clutch solenoid also pulls in the two sets of instantaneous contacts. This causes normally closed contacts 9 to C to open and normally open contacts 9 to 12 to close. Similarly, normally closed contacts 6 and 7 are opened and normally open contacts 6 to 8 will close. These contacts can be used to control any load that you want to operate instantaneous with the limit switch.

At the same time, the timer motor is powered through normally closed contacts 11 and 12 and begins to turn its shaft at the synchronous speed. This in turns makes the timing gears (that have been engaged by the clutch) begin to turn. The timing gears drive the camshaft, which will cause the 11–12–A contacts and the 4–3–5 contacts to change when the timing cycle is complete.

The timer motor will continue turning the timing gears and camshaft until the camshaft touches the switch and causes it to open. The exact amount of time that the motor has run is called the *preset time*. The preset time is set by adjusting the hands on the face of the timer device with the adjustment knob.

The camshaft will activate both the 11–12–A and the 4–3–5 sets of contacts. Normally closed

Timers 241

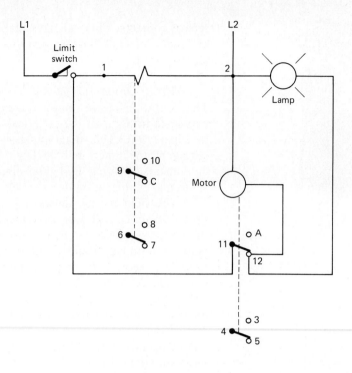

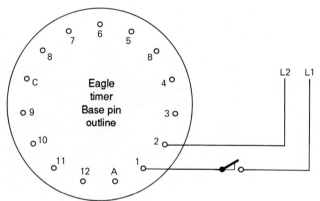

FIGURE 9-6 Electrical diagram for on-delay motor-driven timer.

contacts 11 and 12 will open and break the power circuit to the timer motor, which will stop its shaft from turning. This keeps the camshaft in position where it is activating the contacts.

This action would also turn on any electrical load such as a motor starter and three phase motor through contacts 4 and 3. Through this action, the limit switch would close, and the three-phase motor would turn on after the timer motor has completed its time delay cycle and closed contacts 4 and 3.

Whenever the limit switch is opened, the clutch solenoid and timer motor would deenergize. When the clutch solenoid is deenergized, it disengages the gears from the timer motor and returns the two sets of contacts under its control to their normal position. As the drive gears disengage, a rewind spring returns the timer cam to reset the original preset time. This prepares the switch for the next timing cycle.

Timing Diagrams

Two types of timing diagrams are used to show the operation of timing devices. Figure 9-7 shows both. The first diagram uses a matrix-type diagram to indicate when each part of the timer control is activated or deactivated. An "X" is used to indicate that part of the device is on or activated, and a "O" is used to indicate that a part of the device is deactivated or off. The parts are listed down the left side of the matrix. These include the clutch, timer motor, contacts 9–10, 9–C, 4–3, 4–5, 11–12 and 11–A.

The sections of the timing cycle (reset, timing, and timed out) are indicated across the top of the matrix. The first block at the top, labeled "on delay," indicates that the matrix diagram is for an on-delay timer. The term "reset" indicates that the timer clutch is disengaged and the rewind spring has

242   Chap. 9 / Timers, Counters, and Sequencers

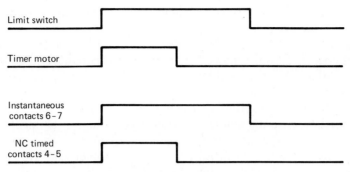

(a) Timing diagram for Eagle timer in Figure 9-6

X = on O = off	Standard Start	Reset	Timing	Timed Out
	CLUTCH	OFF	ON	ON
	MOTOR	OPTIONAL	ON	OFF
	9-10	O	X	X
	9-C	X	O	O
	4-3	O	O ǀ X	X
	4-5	X	X ǀ O	O
	11-12	X	X	O
	11-A	O	O	X

(b) On-delay time diagram

X = on O = off	Reverse Start	Reset	Timing	Timed Out
	CLUTCH	ON	OFF	OFF
	MOTOR	OPTIONAL	ON	OFF
	9-10	X	O	O
	9-C	O	X	X
	4-3	O	O ǀ X	X
	4-5	X	X ǀ O	O
	11-12	X	X	O
	11-A	O	O	X

(c) Off-delay time diagram

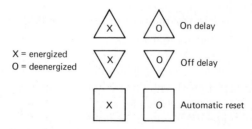

(d) Alternative method of marking timing diagrams

FIGURE 9-7 Timing diagrams. (Courtesy of Eagle Signal Controls.)

243

turned the camshaft back to its starting point. At this point the timer control is reset, and the full amount of time delay is ready to run down when the timer begins timing.

The term "timing" indicates that power has been applied directly to the timer motor and the clutch has engaged the motor shaft to the camshaft. The length of the timing period will be determined by the preset value that was set at the face of the control by adjusting the hands on the dial.

The term "timed out" indicates that the timer motor has moved the camshaft to a point where the cam activates the switch contacts. When this occurs, the timer motor has been deenergized by contacts 11–12 opening. Any other load that is connected to normally open contacts 4–3 will be energized and any load connected to contacts 4–5 will be deenergized. The timing device is ready for reset after this step.

If the device is an off-delay timer, the three categories at the top of the matrix will again be reset, timing and timed out. In the off-delay device, power must be removed from the clutch solenoid to cause the clutch to engage the timer motor shaft to the cam shaft. This means that during the time before timing, the timer motor is not energized and the clutch is not engaged. Once the input switch is opened, power is removed from the clutch and the timer motor is energized. This time is considered timer timing.

Figure 9-7a shows the timing sequence in a timing diagram. This diagram shows the on and off cycle of each part of the timer control in relationship to each other. This type of diagram is also called a logic diagram, as it shows the outputs relationship to the input signals. Both of these types of diagrams are used extensively in timing diagrams and in specification sheets for timing controls.

The "X" and "O" symbols used in the matrix diagram (Figure 9-7a and 9-7b) are also used to indicate the timing sequence of the load in a timing diagram. Another type of diagram, (Figure 9-7d) uses triangles located above the load to indicate that the load is controlled by a timing device. An X is placed inside the triangle to indicate that the load is energized, and a O is used to indicate that the load is deenergized.

If the triangle is pointing up, it means that the load is controlled by an on-delay device. If the triangle is pointing down, it means that the load is controlled by an off-delay device. The triangles will be displayed in groups of three, which indicate the three phases of the timing control sequence from left to right: reset, timing, and timed out. Figure 9-8 shows several types of time delay symbols used with application diagrams.

Thermal Time-Delay Devices

A simple and inexpensive time-delay device is the thermal time-delay element. This element is made of two dissimilar metals that have a predetermined movement when heated. The two strips are bonded together during manufacturing and are mounted in the switch mechanism directly over a heating element. One end of the bimetal is secured so that the other end will move a maximum distance when current is applied to the heating element. When the free end of the bimetal strip has traveled its maximum distance, it will activate a set of contacts. These contacts can be normally open or normally closed. When current is removed from the heating element, the bimetal strip begins to cool down. As the bimetal is cooling the end that is free to travel will begin to move back to its original location directly above the heating element. Figure 9-9 shows a diagram of the bimetal element with its element heated and again after it has cooled. The bimetal is mounted in the switch in such a way that it will snap open or closed as the contacts are opening and closing. This will allow the switch to take advantage of the slow travel that the bimetal provides, yet have quick enough action at the contacts to keep them from pitting and arcing.

The NO or NC contacts are insulated from the bimetal element, which allows the contacts to be connected to a control circuit using pilot voltages. The time it takes the heating element to move the contacts far enough to open or close the contacts is determined by the amount of wattage the heating element uses. In some thermal switches the wattage is fixed, which causes the time delay to be fixed (usually 5 to 10 seconds).

In other switches the heating element is a variable resistor that can be adjusted externally. If the resistor is variable, the amount of heat produced can be increased by adjusting the resistance to a lower value. From Ohm's law we know that if you keep voltage constant and reduce the resistance, the current will increase. When the current increases with the voltage staying constant, the wattage will increase. In this type of device the amount of time delay will be shortened as wattage is increased.

Solid-State and Digital Timing Devices

Since the advent of solid-state devices, solid-state time-delay controls have provided a variety of applications as add-on, stand-alone and programmable devices. These devices can utilize circuitry as simple as a variable resistor and a capacitor to provide the time delay or something as complex as a

WIRING DIAGRAMS — Bold Lines are Internal Wiring

ON DELAY HP5 CYCL-FLEX® TIMER

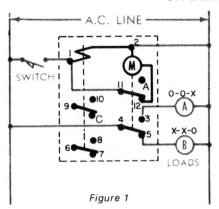

Figure 1

Sustained Control Switch. Close to Start, Open to Reset - Simple delayed closing and opening of load circuits.

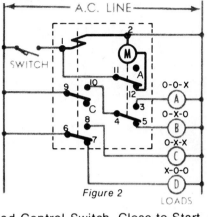

Figure 2

Sustained Control Switch. Close to Start, Open to Reset - Additional load circuit operations obtained by connecting contacts in series.

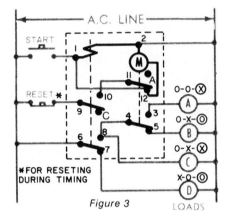

Figure 3

Momentary Control Switch. Close to Start, Resets Automatically - Pushbutton start. Remains in 3rd switch position indicated by Ⓧ or Ⓞ for ½ to 1½% of the dial range.

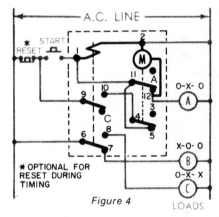

Figure 4

Momentary Control Switch. Close to Start, Resets Automatically - Use when the control switch may not always be opened before end of timing. This circuit insures shutting off timer motor at end of timing.

REVERSE START HP5-01 SERIES CYCL-FLEX TIMER

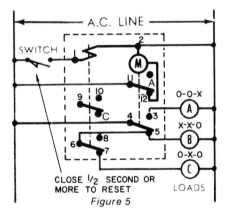

Figure 5

Open Control Switch to Start, Close to Reset - Simple delayed closing and opening of load circuits.

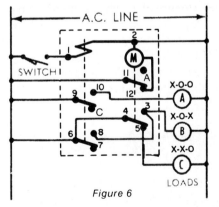

Figure 6

Open Control Switch to Start, Close to Reset - Additional load circuit operations obtained by connecting contacts in series or parallel.

FIGURE 9-8a Timing Application diagrams. (Courtesy of Eagle Signal Controls.)

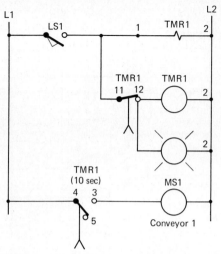

Conveyor 1 will start 10 seconds after limit switch 1 is energized (closed)

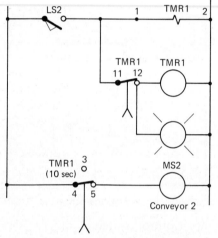

Conveyor 2 will stop 10 seconds after LS2 is energized

FIGURE 9-8b On- and off-delay application diagrams.

16-bit microprocessor chip. The timing controls used today provide high degree of accuracy and repeatability, in the range of 0.1 second, and can be reset as quickly as 20 ms for the next cycle.

The solid-state timing device can provide inexpensive control as add-on devices which work in conjunction with standard relays, or as stand-alone devices that provide pilot-type control. More advanced models provide digital readout and digital preset instead of a dial-type indicator. These devices also provide a variety of preset times, up to 120 minutes, and timing intervals (time base) of 0.1 second. Selecting time bases and preset time ranges can be accomplished by moving jumper wires within the device.

With the use of microprocessors, digital timers can now be fully programmable from a touch pad. The programmable features include time delay-on, time delay-off, time bases, preset times, and other conditions, such as auto resetting and number of cycles.

Solid-state time control devices use several types of circuits to produce exact time delays. The simplest type of circuit uses a resistor and capacitor with other solid-state devices to trigger the base of a triac, the gate of a silicon-controlled rectifier, or a relay coil. Figure 9-10 shows this type of circuit. The resistor and capacitor provide a time-constant circuit. When voltage is applied to the circuit, the resistor controls how fast the capacitor charges up. When the capacitor's voltage is large enough, it can be used to fire a variety of solid-state devices, such as a unijunction transistor. If a unijunction transistor is used, it will not fire until the voltage has hit the predetermined level that has been set by the size base 1 and base 2 resistors. Other electronic devices could be used in a similar manner. Generally, all of these devices are referred to as blackbox circuits, and all that is important is the output signal, which will turn on a triac or SCR.

A potentiometer (pot) is usually used as the resistor in this type of circuit. The size of the pot will determine the maximum amount of time delay. As the pot is adjusted, a smaller amount of resistance is put in the circuit with the capacitor, which causes the time delay to become shorter. The adjustment knob for the pot is usually accessible on the time-delay device so that the time delay can be adjusted while the device is still in the circuit.

Some devices allow the resistor for the time constant to be mounted remotely from the control in an operator's panel. In this way the operator can

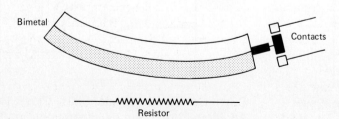

When power is applied to resistor it will heat the bimetal and it will move to close contacts; when power is removed from resistor the contacts will open some time later as the bimetal cools. Typical time delay is 1 to 5 seconds.

FIGURE 9-9 Diagram of thermal time-delay device using a bimetal element.

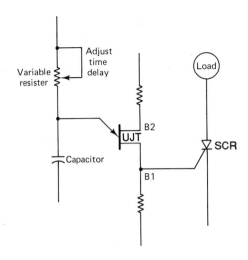

FIGURE 9-10 Resistor and capacitor time-delay circuits.

adjust the exact amount of time delay that is required for the operation. This is especially useful for batch mix timers or for plastic press operations. Figure 9-11 shows a picture and diagram of this type of device.

Cube Timers

The time-delay control shown in Figure 9-11 is called a *cube timer*. It is very accurate way of providing time delay to a load such as a small motor or to a motor starter coil for larger motor control. You should also notice that this type of device can be used in both an ac and a dc circuit without modification. This is possible because a triac is being used as the solid-state control device. The triac can control either ac loads or dc loads. If the timer control was rated for dc circuits only, it would be using a SCR as the solid-state device.

The load can be either 110 or 220 V. Again this is possible because the electronic solid-state circuit uses a voltage regulator circuit to set the incoming voltage to the time-delay part of the circuit at a constant level regardless of the incoming value.

Figure 9-12 shows several other diagrams for the cube timer. It can provide either two-wire or three-wire control. If two-wire control is used, it will get the supply voltage for its electronic circuit through the pilot device and the return part of the circuit will be completed through the load. The time delay will begin when the pilot switch is closed. This also means that if the load fails open or the pilot switch does not close, the timer will look like it is inoperative.

If the three-wire device is used, its timing circuit is connected across the line at terminals 2 and 3. The pilot devices that will initiate the timing circuit will be connected across terminals 6 and 7. The load is connected between terminal 1 and line 2. The operation of the timing begins when the circuit between terminal 6 and 7 has continuity. Notice that no external power is applied to these two terminals. The terminal where the load is connected will be capable of controlling loads up to 1 A. Remember that since a solid-state device is used as the switch, leak current of up to 2 mA is normal.

Adjusting the Time Delay of a Cube Timer. The time delay for some applications must be adjusted during installation and is not routinely readjusted during machine operation. In these applications the manufacturer will provide a formula to calculate the amount of resistance required for the specified time delay. The resistance can then be set and measured with an ohmmeter to give the preliminary setting. Trial-and-error tests will provide the fine adjustment. Figure 9-13 shows the formula for the preliminary setting.

A maximum and a minimum time is required as well as the approximate amount of time delay that you are trying to achieve. The maximum and minimum time range of the time delay is different for each model of device sold, and ranges fro 0.05 to 1 second through 0.5 to 10 hours.

An example calculation is worked out using a model that provides a range of 15 seconds as the minimum time and 300 seconds as the maximum time. The amount of time delay required is 100 seconds. From the calculation you can see that the potentiometer should be set at 298,250 ohms (Ω). This setting will get the time delay within 5 percent of the 100 seconds. Several trials may be needed to get the delay to exactly 100 seconds.

Eight-Pin (Octal)-Base Time-Delay Relays

Another type of solid-state timer is enclosed in a plastic case with eight pins protruding from the bottom. These eight pins allow the timer to be plugged directly into an eight-pin base. All field wiring is connected to the base, and if any problems are encountered with the timer circuit, the complete timer can be exchanged in several seconds. This is a very big advantage in industrial applications, where downtime must be kept to a minimum. Figure 9-14 shows a picture and with a wiring diagram of this type of solid-state timer.

The operation of this type of timer control is very similar to that of the cube timer. A resistor and capacitor circuit controls solid-state devices that switch on and off for the output control. The output can also use a single-pole, double-throw relay. The

Q1 Series

Operating Logic:
Upon application of input voltage, the time delay starts. At the end of the time delay the load is energized. Reset is accomplished by removing input power.

Note: 1) The load may be located on either side of the line. 2) R_t and terminals 4 and 5 are used for external time adjustment.

Logic Function Diagram

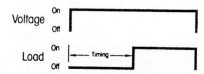

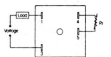

External Resistance/Time Delay Relationship
1 megohm external resistance is required to obtain the maximum time for all ranges. To determine the actual resistance needed to obtain the required time delay, use the following formula:

$$R_t = \frac{T\,req. - T\,min.}{T\,max. - T\,min.} \times 1{,}000{,}000 \text{ ohms}$$

Note: Due to component tolerances, the actual time obtained will normally be within 5% of desired time.

FIGURE 9-11 Cube timer with external resistor. (Courtesy of National Controls.)

time base can be set for divisions of 0.1 second and 1.0 second which provides time delay up to 10 hours. These controls can have an internal potentiometer for adjusting the amount of time delay. The knob and dial on the top of the control shows the approximate amount of time delay. Final adjustment of the time delay can be made by trial and error.

From the diagram you can see that the input to start this timer is a pushbutton switch. Any type of momentary pilot device would work well and should be wired across terminals 5 and 6. Several switches can be mounted in parallel or series for additional logic of necessary. You should notice that these terminals should be connected to voltage or ground circuity. Since they are looking for a reduction in resistance, which will be supplied when the momentary switch contacts close, the switch can be mounted remotely.

The output of the timer can control up to 10-A resistive loads or could control the coil of large motor starters which could start any size motor. A similar model of time-delay control has 11 pins that plug directly into a base. Again, as in eight-pin base, all field wiring is connected to the base so that the time-delay device can quickly be unplugged and replaced. The three extra pins in the 11-pin device provide connections to a second single-pole, double-throw set of contacts. Another model of this type of timer use slip-on clip connectors as terminals. These also provide easy removal and replacement of the timing control when it is suspected of operating incorrectly. Figure 9-15 shows examples of available logic outputs for their timers.

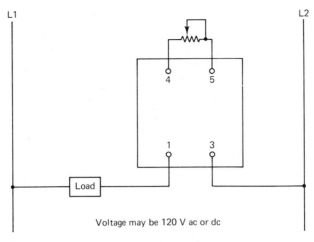

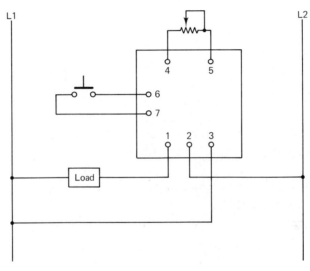

FIGURE 9-12 Two-wire and three-wire time-delay controls. (Courtesy of National Controls.)

CMOS Solid-State Controls

Newer models of time-delay controls use CMOS circuitry to provide accurate and repeatable time delays. The CMOS controls uses a crystal oscillator that is sent to sets of divider networks. The divider networks are used to divide the crystal frequency into increments that will be used to provide a variety of time bases for the timing control. Newer designs have incorporated this circuitry into existing timer control housings and packages so that they can easily be interchanged and mounted on existing bases for use with standard electromechanical timer controls.

The crystal and CMOS circuits are very reliable because of their low power consumption and their repeatability of millions of operations. This type of timing control is also reliable because of its construction on an integrated circuit (IC), and has been used in computers to provide clock and timing pulses for many years.

Integrated-circuit, or chip, construction has allowed high-quality controls to become available at a much lower price. This type of construction allows the timing device to be more compact at a time when space in the electrical control panel is a major concern. These devices have also proven to be easily interfaced to other electrical mechanical

$$R_T = \frac{T_{req} - T_{min}}{T_{max} - T_{min}} \times 1{,}000{,}000 \, \Omega$$

T_{req} = time required for time delay
T_{max} = maximum time delay of cube timer
T_{min} = minimum time delay of cube timer
R_T = total resistance required

FIGURE 9-13 Calculation for setting external resistor to provide a 100-second time delay. (Courtesy of National Controls.)

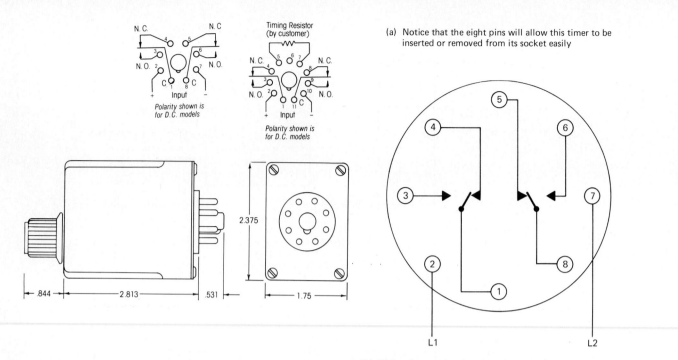

(a) Notice that the eight pins will allow this timer to be inserted or removed from its socket easily

(b) Wiring diagram showing pin assignments for the octal base timer

FIGURE 9-14 Picture and diagram of octal base timer delay devices. (Courtesy of Eagle Signal Controls.)

250 Chap. 9 / *Timers, Counters, and Sequencers*

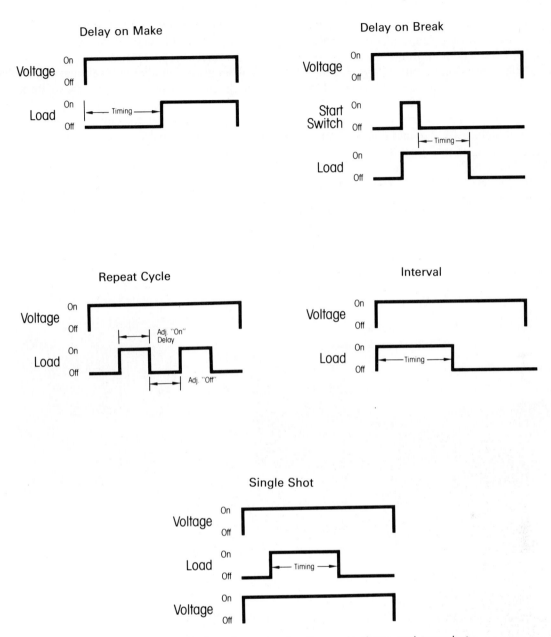

FIGURE 9-15 Timing diagrams for the cycle timer and one-shot outputs. (Courtesy of National Controls.)

control circuit devices such as relays, solenoids, and motor starters.

One problem of early CMOS circuits was their inability to exist in the industrial electricity environment where uncontrolled voltage spikes and strong magnetic fields caused undesirable operation or circuit failure. The CMOS circuits that are used in the newer industrial controls have been refined to a point where they are isolated from the voltage transients and magnetic fields. This allows them to be mounted in the same electrical cabinets as their electromechanical counterparts.

Examples of CMOS Solid-State Controls. Solid-state controls have been incorporated for use in some of traditional applications, such as add-on devices. The original add-on time-delay control used a pneumatic actuator. The solid-state add-on relay uses solid-state circuitry to control the time-delay portion of the control. The outputs for this control are sealed contacts that will allow the control of either ac and dc loads.

Since the timing circuitry is encapsulated it is immune to vibrations, humidity, and harsh environments with atmospheric contaminates. These devices use separate models to provide on-delay and off-delay control. A wide range of delay times is also available by selecting different models.

Solid-state time-delay controls are also available to fit into the same housing as that of the synchronous motor-driven timer. This allows the newer solid-state technology to be incorporated in existing installations with a minimum of changes. This control is pictured in Figure 9-16, from which you can see that the face of the control and its physical dimensions exactly fit the synchronous motor-drive time-delay device. The solid-state device uses CMOS circuitry to provide accuracy and repeatability to 0.1 second. Reset time is also decreased to several milliseconds for repeat operations. The preset time is set through the use of three thumbwheels, representing ones, tens, and hundreds units, which allow preset times up to 999. By moving internal jumper wires, these three values can represent several timing ranges. The timing ranges are 0.1 to 99.9 seconds, 1 to 999 seconds, and 0.1 to 99.9 minutes.

Three seven-segment light-emitting diodes (LEDs) are used to display the time remaining. This display will be synchronized with the time base that was selected with the jumpers. The digital readout on the display allows for the control to be observed from a greater distance than with a dial readout.

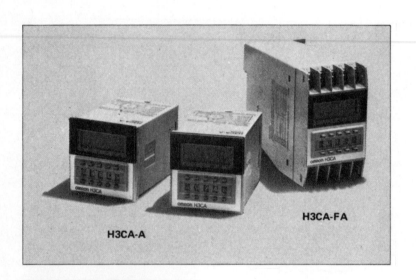

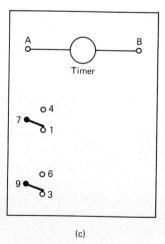

DG100 TERMINAL ASSIGNMENTS

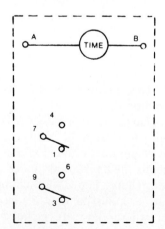

FIGURE 9-16 Picture of solid-state time control. (Courtesy of Eagle Signal Controls and Omron.)

This type of timer control is also available without the digital readout for applications where the timer control may be mounted inside an electrical cabinet.

Time delay-on and time delay-off functions are set with internal jumper wires. This versatility allows the user to stock only one timer control and get the functions of time delay-on or time delay-off with multiple time bases. Example wiring diagrams of outputs for these timers are provided in Figure 9-17.

Microprocessor-Controlled Timers

Microprocessor chips have become very usable in industrial control because of their size, versatility, and price. They have been incorporated into timer controls to provide a variety of timer ranges and

WIRING DIAGRAMS

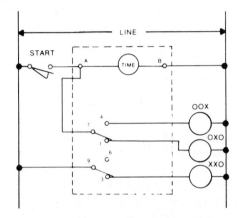

"00" FEATURE
Close switch to start timing; open to reset. Relay contacts provide delay output (OOX, XXO).

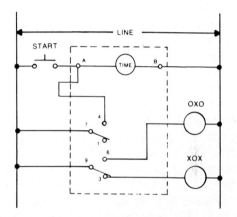

"02" FEATURE
Interval Timing - One Shot. Momentary or Maintained start switch. Close switch to start timing. Resets automatically using momentary start. Resets upon opening maintained switch. Relay contacts provide interval sequence (OXO, XOX). Contacts 4-7 must be wired as shown for timer control.

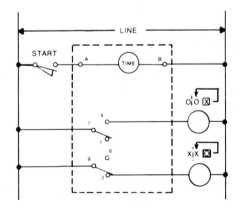

"03" FEATURE
Repeat Cycle Mode - Pulse Output. Output contacts transfer for approximately 200ms at the end of each timing period. Timing cycles continue as long as start switch is closed. A dashed line separates the output status during RESET from the status during TIMING.

FIGURE 9-17 Diagram of solid-state timer controls. (Courtesy of Eagle Signal Controls.)

operating logic that is selectable in one device. This control allows the user to select the timer range with a selector switch on the front of the control. The range selector allows the timer device to be set for ranges of 0.05 to 9.99 seconds, 0.1 to 99.9 seconds, 1 to 999 seconds, 0.1 to 99.9 minutes, and 1 to 999 minutes.

The timer also has five different logic modes that are also selectable from the front of the control: repeat cycle with 50 percent fixed duty cycle, single shot, delay-on break, interval, and delay-on make.

The preset time is dialed into the timer control through digit switches (thumbwheels) on the front of the control. A LED is also located on the front of the control to indicate when the timer is timing. This control has an eight-pin connector on the bottom that will plug directly into any octal base control. The diagram for this control is shown in Figure 9-17. From the diagram you can see that the timer operation will be initiated by the pilot device connected between terminals 5 and 6. The timer's operation will be identical to that of similar solid-state timer controls.

The microprocessor-controlled timer allows for a maximum number of functions in one control, which will allow industries to stock one control for all their timer functions. Since the timer range and logic control is user selected, this control can be used for all existing applications. It is also easily changed out during downtime because of its pin and socket mounting. All of these features are making the microprocessor timer the control of the future.

Programmable Timers

New advances in microprocessor technology have allowed them to be incorporated into programmable timer controls. The programmable timer provides timer control for continuous periods of time up to 365 days. This control is very similar to the 24-hour time clock controls used in business lighting and security systems.

The programmable timer can control up to 16 circuits and use 60 separate programs. Each program can control any or all of the 16 outputs. Typical applications for this type of control is for water treatment facilities to add water chemicals, small business control of lighting, security and fire protection, and energy management systems.

Each program can be set to repeat at differing intervals, such as hourly, daily, weekly, monthly, or annually. The programmable timer can also account for differences in the program cycle caused by holidays, changes for daylight savings time, and can even adjust for leap year.

The timer is programmed from a keypad and the program can be displayed on a printer for editing. Changes can be made from the keypad or from a modem (modulator/demodulator) over ordinary telephone lines. This allows for the timer to be installed at one location and serviced from another location. The program is backed up by batteries in case of power failure.

The outputs that are controlled by the timer can control resistive loads up to 15 A, and inductive loads up to 2 hp at 250 V ac. Standard pilot-duty loads can also be controlled, which allows the programmable timer to be used in conjunction with other control devices for a wide variety of applications.

Applications for Timers

Many applications for timers are in use in industry today. Some of these circuits allow the timing device to be interfaced with robots and other automated systems. One example includes a robot in a loading and unloading application. The robot will transfer parts from the end of the first conveyor and place the parts in a mill for machining. When the mill is finished with the parts, the robot is signaled to pick up the part and place it on the second conveyor. Each conveyor is controlled by a stand-alone cycle timer activated by the robot.

When the robot has picked up a part from the first conveyor, it will signal the timer to jog that conveyor for 2 seconds to bring the next part into position. After the robot places the part on the second conveyor it will send a signal to the timer that controls that conveyor to jog it for 2 seconds to move the parts ahead, which will provide adequate space for the next part to be placed.

Stand-alone timers are used so that maintenance personel can adjust the timers slightly without getting into the robot program. If the timers in the robot were used, the robot program would need to be edited each time an adjustment had to be made. This jogging technique causes the parts to be placed on the conveyor with the same space between each part. This accurately locates the parts for the next robot that must handle the parts to place them on a pallet.

Testing the Timer Control

After the timer control has been installed, it must be field tested and adjusted. The first step in this process is to check the load portion of the circuit to make sure that any unwanted signal will not harm personnel or damage equipment. This may be accomplished by turning off the main disconnect to the motor or hydraulics that the coils and solenoids may activate. If the timer is controlling a motor starter, you will still be able to see the motor starter coil pull in, even though the motor will not turn when the disconnect is open, and if the timer turns on the motor starter at the wrong time, it will not damage the motor or other equipment.

After the equipment safety check is made, power may be applied to the control circuitry. At this point, close the enabling switch to the timer and check to see that it is operating in the on-delay or off-delay mode as you have designed it. Also check the operation of the instantaneous contacts to make sure that the loads are connected to the proper set of normally open or normally closed as indicated in the circuit diagram. Do not be overly concerned about the amount of time delay during the first test.

After the first test has been completed and you are satisfied that the control is operating as the control circuit diagram indicates, trials can be made to adjust the time delay. You can use a watch that displays seconds to time the device. If a remote potentiometer is used to set the time delay, it would be helpful to calculate the amount of resistance required for the amount of time delay that you desire. Adjust the resistance in the potentiometer with an ohmmeter prior to the first trial. Additional trials can be use to make final adjustments.

After you are satisfied that the timer control is energizing all its load at the proper time, you can apply power to the motor disconnects and allow the complete application to run through a full cycle. During this test, be ready to disconnect power or hit the emergency stop buttons if any unwanted machine motion occurs. After you have completed these tests, be sure to add any notes to the electrical diagram, indicating any changes or adjustments that will be useful during troubleshooting. If you encounter any problems during these tests or at any time after the system has been operating for some time, use the troubleshooting instructions to aid in your diagnostics.

Troubleshooting the Timer Control

The most important tool in diagnosing problems with any automated controls or circuitry is the knowledge of how it is supposed to operate when it is working correctly. This means that at times you may need to review the theory of operation for the particular control that is malfunctioning. This will help you remember how it was designed to operate, and then through observation and electrical testing you can determine any differences. You should also review the electrical diagrams for the circuit prior to beginning the diagnostic procedure to see if the device is wired according to the field wiring diagram, and determine that the sequence of operation should be.

The diagnostic tests for timer controls will be grouped according to the type of control. Since there are some similarities, there will be some duplication in the tests. To use these tests effectively, you will need to determine which type of device is controlling the circuit. In addition to this, you must determine if the control is wired for a delay-on or delay-off application. Use only the tests that apply to your device and application.

It is very important to understand that some problems will be unique to a newly installed timer control, such as being wired improperly. If the device has failed after it has been operating correctly for some time, you may not need to worry about the device being wired incorrectly, and concentrate on a failed or broken component in the timer control. It is also important to remember that the timer control can operate correctly only when it receives the proper amount of voltage. If pilot switches are connected in series with the timer, they must be energized and supply the control with the proper amount of voltage for it to work correctly. The final point to remember is that the timer can only switch voltage to the load at the proper time. If the timer switches the proper amount of voltage to a load at the proper time it is considered to be operating correctly. If the electrical load (relay, solenoid, or motor starter) does not operate after the timer has switched power to it, it should be checked instead of the timer.

Troubleshooting the Solid-State Timer Control. Solid-state timer controls are very easy to troubleshoot. They essentially operate as a switch that is in line with the load being controlled. They require voltage to the operational part of their circuit and will act as a switch from L1 or the positive dc line to the load. These devices have a variety of field wiring assemblies, such as the octal base or push-on terminals. In some cases it will be faster to remove and replace the device from its base rather that try to make elaborate field tests. At one time the practice of removing and replacing components without field testing was considered a poor practice. In today's industrial applications, the time that equipment is malfunctioning and not operating may cost several thousands of dollars per hour. In these cases it is prudent to keep a variety of spare parts and simply try removal and replacement as the first order of operation.

You should also understand that many solid-state controls are encapsulated and cannot be repaired. They are relatively inexpensive and are easily replaced. This means that most of the tests in troubleshooting these controls involves testing for the presence of the proper amount of voltage at the supply terminals of the device and checking to see if the voltage is switched to the proper output terminals after the amount of time delay has occurred. Since most of the devices have a variable time adjustment, minor changes in the amount of time delay can be compensated for by small adjustments during trial-and-error timing tests.

Troubleshooting the Synchronous Motor-Driven Timer. The electromechanical motor-driven timer is also very easy to troubleshoot. You may need to review the theory of operation prior to making these tests. You will also need to check the wiring diagram to see if the control has been wired as a time delay-on or a time delay-off control. Remember that the off-delay and on-delay timer applications are achieved by using two different models of timer controls.

The timer motor and clutch for this type of control require power from lines 1 and 2 for ac control and from the positive and negative lines for DC control. Any number of pilot devices can be used to enable timer control. These switches may include photoelectric switches, proximity switches, and limit switches. To diagnose problems correctly with this type of control, you may also need to observe and test the pilot devices that enable the timer and the electrical loads that the timer output controls. Another troubleshooting technique involves starting at the electrical load end of the circuit and work backward to the time control. If you would rather use this method of troubleshooting, the tests in each procedure should be executed in reverse order for each timer control.

COUNTERS

Counters are used in industrial control systems to count (totalize) parts being made and to keep track of sequential steps in machine operations such as plastic presses and punch presses. Electromechanical counters have been in use for many years. In recent years, solid-state components and microprocessor technology have been integrated into counters. These advances have made counters easy to install, operate, troubleshoot, and replace. In this section we explain the operation of various types of counters such as cycle counters, used to control sets of contacts, and totalizing counters, used to keep count of parts being made. Typical applications, installation, and troubleshooting are also covered.

Types of Counters

Mechanical Counters. The simplest type of counter, which has been in use for many years, is the mechanical counter used to keep track or totalize the number of parts being made. The internal operation of this type of counter is still used on the electromechanical types of counter that you will find in industry today.

This type of counter consists of a set of wheels with the numbers 0 through 9 listed on them. As parts move past the counter arm, it is moved to trip the counter and provide an input to the counter clutch. When the arm moves to register an input, each wheel rotates and causes the number to show up in a window. As many as four of these wheels are placed side by side to display counts up to 9999.

Each wheel is connected to the one next to it by gears. The first wheel will display the ones column. The wheel that is next to the ones column will display the tens column. When 10 inputs signals are counted, the gears from the wheel displaying the ones column will cause the wheel displaying the tens column to move ahead one count. If you were looking at the counter display it would show a 10, indicating that a total of 10 counts had been registered. As each new count is received the counter will display 11, 12, 13, and so on. After the counter receives 10 additional inputs, the gears on the ones column wheel will move the tens column wheel ahead another count so that it is displaying a 2. The count will now be displaying 20. Each additional input will now display 21, 22, 23, and so on.

The wheel that is displaying the tens column will continue this operation until it reaches the value 9. After 99 counts are received by the counter, the gear on the wheel displaying the tens column will move the wheel next to it ahead one count. This will cause the counter to display the value 100. After 999 counts have been received, the wheel that displays the hundreds value will engage the gears of the wheel next to it and the counter would display a value of 1000. This will continue until a value of 9999 is counted. At this point, the next count that is received will cause the counter to reset to the value of 0000. There will also be a reset button on the counter that will allow it to be reset at any time. When the counter is reset, a spring returns all the numbers in the window to zero. This type of counter is used to provide the total number of parts that move past the arm. The total count will be displayed in the window, where personnel view the count and write it down.

Electromechanical Counters. Another type of counter, called a *reset* or *cycle counter,* operates on electromechanical principles. The signal that causes the counter to advance one count comes from an electrical signal instead of through mechanical movement of an arm. This allows devices such as proximity switches, limit switches, and photoelectric switches to be used to provide the input signal. This type of control is generally used to count a specified number of input signals (counts) and to activate its outputs. This type of counter control has both instantaneous and controlled contacts.

Figure 9-18 shows a picture and diagram of this type of counter. From the picture you can see that the counter is mounted in a housing that is identical to the timer controls. This allows the timer and counters to share a common field wiring base, which cuts the cost of stocking spare parts. The counter face is also similar to the timer control, which makes the control easy to read.

From the electrical diagram you can see that the counter has a clutch and motor. There are also several sets of single-pole, double-throw contacts. Two sets of the contacts are instantaneous and are controlled by the clutch. The other sets of contacts are single-pole, double-throw, which are controlled by the counter control. These contacts are actuated when the counter reaches the specified number of counts.

The clutch is energized by voltage applied to terminal 1. The count signal is input to terminal 12. When the clutch is energized it engages the gears so that the count signal will be acknowledged. This type of control uses a dial with numbers printed on its face instead of wheels with numbers 0 to 9. Two hands are used to indicate the preset count and the actual count.

A preset value is set into the counter by adjusting the preset hand on the face of the control. This value can be any value between 0 and 100. Each time a count is registered the counter subtracts one count from the preset value by energizing the counter motor and causing it to advance one step the hand that is indicating the actual count. This causes the counter to decrement the count (subtract one count from the preset value) each time an input signal is received. When the counter counts down to zero, the output contacts are activated. This type of counter is called a *down-counter*.

While the counter motor is decrementing the count, it is also winding the reset spring. This spring will cause the hand that is indicating the actual count to return to the preset value when the clutch is deenergized. The counter can be reset to the preset value at any time by deenergizing the counter clutch. This type of counter is also available as an *up-counter*. In this type of counter the actual count starts at zero and increments (adds 1) when an input signal is received. When the actual count equals the preset count, the contacts will be activated.

This type of counter control is generally used to keep track of input signals and control outputs. The counter is used in conjunction with a palletizing machine. A photoelectric switch detects boxes as they pass by the end of the conveyor to be stacked on each pallet. This switch inputs a signal to the counter, which will energize the wrapper solenoid after four boxes have been stacked on the pallet. The wrapper solenoid energizes a plastic wrapping machine that wraps in plastic the pallet containing the four boxes.

When the palletizing machine is changed to place six boxes on each pallet, the counter can be adjusted to accommodate this change by setting the preset count to 6. Each change to the palletizing machine can be made by changing the counter preset.

Solid-State Counters. Several types of solid-state counters are available for use in modern industrial control systems. These include up-counters and down-counters that will energize outputs when the preset count is reached, and totalizing counters that keep track of the total number of parts that have been made. These types of counters utilize CMOS solid-state circuitry to execute the counter function. Figure 9-19 shows a picture and diagram of this type of control.

From Figure 9-19a you can see that the counter has a thumbwheel on the face of the counter that is used to input the preset count value. Machine operators or production engineers can dial in the preset count by using the thumbwheels. This type of counter will be used as an up-counter. When the counter has recorded the number of input signals (counts) that is equal to the preset value, it will energize its outputs.

From the diagram (Figure 9-19c) you can see that the counter control uses four lines that operate the counter. One of these lines is connected to terminal B, which is the electrical common for the control. The other lines are called *counter start, counter input, and inhibit input*. The counter start line is used to enable the counter control similar to the action of the clutch solenoid in the electromechanical counter. When this line receives voltage, it will enable the counter control to record any count signals that it receives at the counter input line. Whenever this line is deenergized, the counter is reset to zero.

The counter input line is sensitive to the on-to-off transition of the input signal. This means

COUNTER OPERATION

INSTANTANEOUS CONTACTS 6-7 A-C are open and 6-8 A-B closed when the counter is in reset. Contacts 6-8 A-B are open and 6-7 A-C closed when the counter is in the counting or counted out position.

DELAYED CONTACTS 4-3 and 9-10 close, 4-5 and 9-11 open when the red progress pointer reaches zero. Contacts 4-5 and 9-11 close, 4-3 and 9-10 open when the counter is reset.

NEON PILOT LIGHT is built into dial to indicate counter clutch coil is energized.

SWITCH	RESET	COUNTING	COUNT OUT
6 - 7	O	X	X
6 - 8	X	O	O
A - C	O	X	X
A - B	X	O	O
4 - 5	X	X	O
4 - 3	O	O	X
9 - 11*	X	X	O
9 - 10*	O	O	X
HZ170A6 Clutch Coil	Deener.	Ener.	Ener
HZ170A601 Clutch Coil	Ener.	Deener.	Deener.
Switches trip to count-out position on last deenergized stroke of count solenoid.			

X — Switch Closed   O — Switch Open

SCHEMATIC DIAGRAM

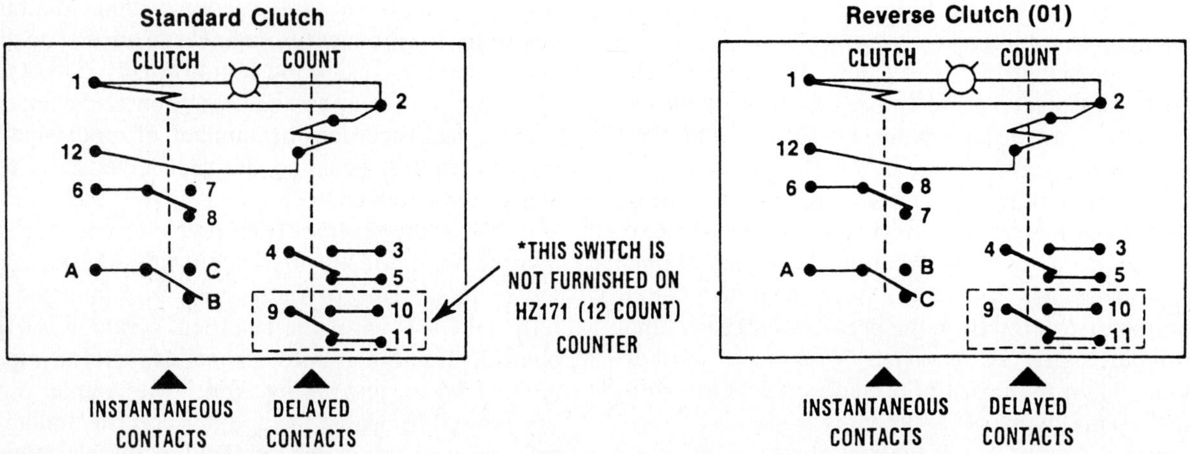

FIGURE 9-18 Picture and diagram of electromechanical counter. (Courtesy of Eagle Signal Controls.)

DZ100 TERMINAL ASSIGNMENTS

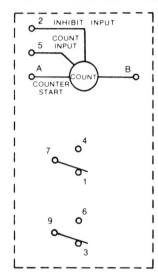

NOTE:
RELAY OUTPUTS AND TRIAC OUTPUT ARE NOT AVAILABLE SIMULTANEOUSLY.

RELAY OUTPUTS — STANDARD

START INPUT — CLOSE TO START, OPEN TO RESET
COUNT INPUT — COUNTS ON SWITCH OPENING
INHIBIT INPUT — CLOSE TO STOP COUNT PROGRESS, OPEN TO RESTART COUNT PROGRESS

D.C. APPLICATIONS REQUIRE A POSITIVE (+) POLARITY ON COUNT INPUT AND INHIBIT LINES.

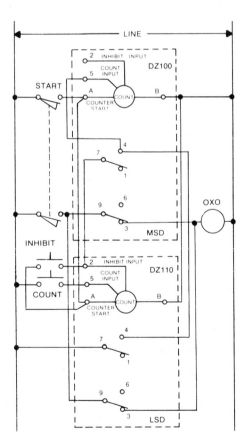

DZ100 OUTPUT SEQUENCES

The three output status conditions for a Reset Counter are RESET - COUNTING - and COUNTED OUT. Output status for each condition is indicated by an "X" or an "O", where "X" means the output is ON or energized, and "O" indicates the output is OFF or de-energized. Thus, an output which is OFF in RESET; ON during COUNTING; and OFF at COUNT-OUT is indicated by the notation OXO.

Four usable counter output sequences are available - OOX, XXO, OXO and XOX. The wiring diagrams shown in this product bulletin will reference the resultant output sequences in this manner. Certain operational modes cause the counter to RESET immediately after COUNTING. For these cases, the COUNT-OUT condition does not exist and is deleted from the output sequence designation.

FOUR DIGIT COUNTING

Two DZ100 Series Counters may be inter-wired to provide counting to 9999. The most significant digit pair is controlled by a DZ100 Counter. The least significant digit pair is controlled by a DZ110 Counter. (See Ordering Information)

For an OXO output sequence, wire the two units exactly as shown below. The start switch is closed to enable the counters to receive counts. Counts are entered into the DZ110 Counter. The standard inhibit function, if required, is also connected to the DZ110 unit. The counters reset when the start switch is opened.

FIGURE 9-19 Picture and diagram of CMOS counter. (Courtesy of Eagle Signal Controls.)

that the input signal will be registered as a count when it transitions from full voltage to turning off. When the actual count becomes equal to the preset count, the counter's outputs will be activated. This means that any set of open contacts will be closed and any set of closed contacts will be opened. This type of counter has two sets of single-pole, double-throw contacts.

The inhibit input line can be used to cause the counter to ignore signals that are received on the count input line. This line is especially useful to deenergize the counter from outside controllers such as robots, programmable controllers, or press controls. The internal operation of this counter utilizes existing CMOS chips that can increment or decrement a count. When the preset value is reached, the circuit activates its outputs. The outputs can be triacs or relay outputs.

Another type of solid-state counter utilizes a large-scale integrated circuit called LSI (large-scale integrated) technology. The LSI chip provides a counter with a thumbwheel input to set the preset count and LED seven-segment displays to show the current count. This type of counter is also available without the LED display, for applications where the control may be mounted inside a cabinet where the face cannot be seen. This type of control has preset ranges up to 9999 counts. It can read input signals of 40 millisecond on duration and 60 millisecond off duration as counts. This allows the counter to be interfaced with photoelectric and proximity controls for high-speed counting applications such as bottling lines and packaging lines. The input line can receive up to 1000 counts per minute.

Figure 9-19b shows this type of control. It is designed to fit the standard timer/counter mounting case, which makes it easy to install. This type of installation also makes the control easy for panel mounting. Figure 9-19d shows the field wiring terminals for this control.

From these diagrams you can see that control has two sets of single-pole, double-throw relay contacts that are activated when the actual count decrements to zero. These contacts are called *delay contacts*. There is also one set of single-pole, double-throw instantaneous contacts that are energized when power is applied to terminals 1 and 2. This counter is reset immediately after the actual count reaches zero, or any time that power is removed from the counter control relay (CR1).

Programmable Timer/Counters. Programmable timer/counters allow the user to purchase one control device that can be programmed to be a counter or timer. There are also several programmable options, such as up-counter or down-counter, counter ranges, and types of outputs. If the control is programmed as a timer, options are provided for timing ranges, time bases, time up or time down, and types of outputs.

As a programmable counter, the control provides a wide variety of options, which makes it very versatile. This allows industries to purchase only one model of control for all their timer and counter applications, which saves money and time when the control must be kept in stock as a replacement part.

The programmable counter uses a microprocessor chip similar to those used in computers. This chip provides the wide range of programmable options that makes the control so versatile. The programmable functions also provide the same options as those found in most programmable controllers, but much cheaper. This provides the greatest function at the lowest price.

Figure 9-20 shows an example of the programmable counter. The control has a seven-segment LED display that can show up to four digits. The front of the control also has a membrane keypad that allows set points to be programmed into the control right on the factory floor. This is accomplished by pressing the set key and then the up or down arrow to increase or decrease the value in the display. When the desired value is attained, the enter key is used to store this value in the control's memory. These controls can use a battery to provide memory failure protection during a power outage.

The control options are programmed or selected by setting dip switches on the main circuit board. The control must be removed from its housing to obtain access to these switches. These switches can also be changed on the factory floor to provide optimum use of the control.

Operation of the programmable counter is very similar to that of other counter controls dis-

FIGURE 9-20 Programmable counter/timer with diagrams. (Courtesy of Eagle Signal Controls.)

260   Chap. 9 / *Timers, Counters, and Sequencers*

OPERATION

Programming Procedure

A series of 7 miniature rocker switches located inside the unit housing are used to program the time/count range and the load sequence of the CX300 Timer/Counter. The following table defines the switch conditions to program the operating parameters. **CAUTION: The product label should always be marked to indicate the operating mode of the unit.**

Sym.	TIME OR COUNT INPUT RANGE	SWITCH NUMBER							Notes
		1	2	3	4	5	6	7	
1	COUNT—5000/Sec			X	X	X			a, c
2	COUNT—500/Min			O	X	X			b
3	COUNT—5000/Min			X	O	X			b
4	TIME—.01 Sec			O	O	X			
5	TIME—0.1 Sec			X	X	O			
6	TIME—0.001 Sec			O	X	O			
7	TIME—Min/Sec			X	O	O			
8	TIME—Hrs/Min			O	O	O			
	OUTPUT SEQUENCE								
1	OOX	X	X						
2	OXO	O	X						
3	OO X̄ Pulse Output	X	O						
4	OO X̄ Repeat Cycle	O	O						c
	BATTERY								
1	Battery ON						X		d
2	Battery OFF						O		d
	START MODE								
	Standard Start							O	
	Reverse Start							X	

X = Switch ON   O = Switch OFF   Blank = Don't Care

NOTES:

a. When programmed for count range 1, counts are registered upon application (leading edge) of voltage to the count line. Use range #1 with DC source only. In addition, count input line length must be considered and not exceed 50 feet of twisted pair, shielded cable. Shield must be attached to earth ground or machine frame at machine frame end of cable.

b. When programmed for count range 2 or 3, counts are registered when voltage is removed (trailing edge) from count line. Range 2 should be used whenever possible to provide the best protection from contact bounce of the count source.

c. When programmed for range 1, load sequence 4 (repeat cycle), the maximum count rate is 2500/second.

d. Battery must be switched off for storage or shipment. Battery must be switched on for proper operation of CX300 unit.

Entering and Displaying Setpoints

Whenever the CX300 unit is powered up and the previous setpoint has been lost, the digit display indicates four hyphens. The unit will not operate until it has been provided with a setpoint, clearing the display of hyphens.

To create or change a setpoint, press the **SET** key. The setpoint, if any, is displayed and the panel key pads become active. The operation of the timing or counting function and the output loads are not affected. For setpoint changes, the SET indicator appears on the graphics panel. The setpoint is changed by pressing appropriate △ or ▽ key pads. Pressing a △ key increments the setpoint digit located above the key; the ▽ key decrements the digit located above the key. If the pad is continually depressed the digit will change every .5 second until the pad is released. The display will carry to the digit on the left on the 9 to 0 transitions when using the △ pads. The display will borrow from the digits on the left on the 0 to 9 transitions when using the ▽ pads. On ranges 7 and 8, the display will carry on the 59 to 00 transition and borrow on the 00 to 59 transition of the two least significant digits.

When the desired setpoint is displayed, touch the **ENT** key. The new setpoint is entered, all △ and ▽ keys become inoperable and "SET" disappears from the graphics panel. If a new setpoint is entered when the unit is reset, the new setpoint will be in effect upon the next reset.

The setpoint may be displayed at any time without disturbing the timing or counting cycle by pressing **SET**. The actual value is returned by pressing **ENT**. If the unit is set at 0000, the load is always ON if programmed for OOX, and always OFF if programmed OXO.

A keypad "lock" is provided on the CX300 Timer/Counter to prevent unauthorized tampering. To initiate the keypad lock, press the word "SIGNAL" in the Eagle Signal logo for 8 seconds. To disable the lock to change setpoints, remove power from pin 11 and disconnect or turn the battery off. The unit will lose all setpoints and they must be re-entered for further operation.

CX300 TERMINAL ASSIGNMENTS

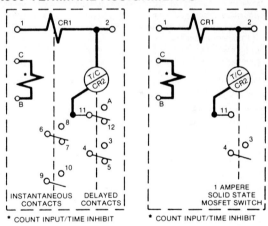

TERMINALS AND WIRING DIAGRAM ON REAR OF TIMER CASE

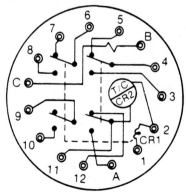

FIGURE 9-20 Programmable counter/timer with diagrams (Continued).

WIRING DIAGRAMS
Bold Lines are Internal Wiring

Power to unit is applied to terminals 11 and 2. The unit is started by applying power to terminals 1 and 2. Note that power should be applied to terminal 11 at least 300 msec before power is applied to terminal 1. An isolated input is provided on terminals C and B which functions as a count input when the unit is used as a counter, or as a time inhibit input when the unit is programmed to operate as a timer. Two sets of instantaneous contacts are provided which transfer whenever terminal 1 is energized. In addition, two sets of delayed contacts are provided. When programmed for Reverse Start, the unit will start when voltage is removed from terminal 1.

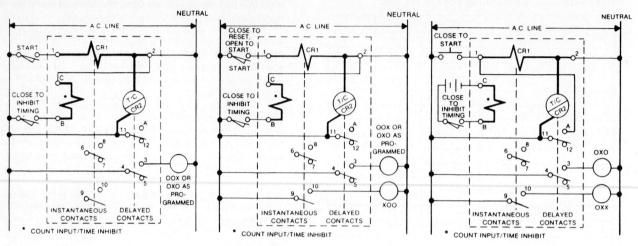

Standard Start — Close start switch to time; open to reset. (CX302 model count/inhibit input at service voltage level). Cycle progress is stopped without reset by closing inhibit switch. Unit resets on power failure. Setpoint is maintained by battery. Instantaneous contacts actuate with start switch. Delay contacts actuate as interval (OXO) or delay (OOX) sequence as programmed by rocker switches.

Reverse Start — Open switch to time; close to reset. (CX302 model count/inhibit input at service voltage level). Cycle progress is stopped without reset by closing inhibit switch. Unit does not reset on power failure. Battery retains cycle progress and setpoint.

Momentary or Sustained Start — Close to time. Automatic reset at end of timing period with start switch open or closed. (CX301 - Count/Inhibit not at service voltage level).

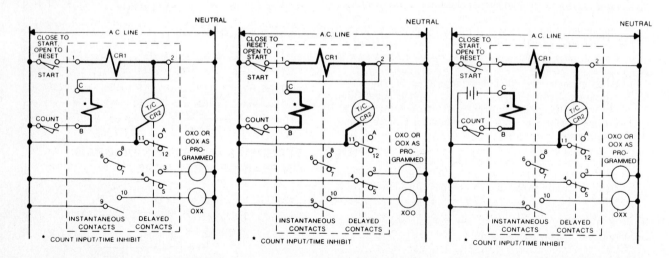

Standard Start — Close start switch to start, open to reset. (CX302 model count/inhibit input at service voltage level). Counts are registered when count switch opens. Unit resets on power failure. Setpoint is maintained.

Reverse Start — Close start switch to reset, open to start. (CX302 model count/inhibit at service voltage level). Unit does not reset on power failure. Battery retains cycle progress and setpoint.

Standard Start — Close start switch to start, open to reset. (CX301 model count/inhibit input not at service voltage level). Unit resets on power failure.

FIGURE 9-20 Programmable counter/timer with diagrams (Continued).

cussed in this chapter. Figure 9-20 also shows the diagram for this control. The control CR1 coil acts like a clutch coil, in that it controls the instantaneous contacts. The control also has an input line between terminals B and C that accepts the count signals. One of the terminals is connected to common and the other is connected to the device, such as a proximity or photoelectric switch that is providing the count pulse. The third part of the control is the CR 2 coil, which controls the delay outputs. CR2 energizes these output contacts after the counter has recorded the number of counts entered in the preset value. The controlled outputs are also called the *delay outputs*. These contacts are numbered 11, 12, and A for the first set and 4, 3, and 5 for the second set. These contacts can be used as outputs to control loads or can be wired to provide several variations of logic for the counter itself, such as automatic resetting. These contacts are rated to carry up to 10 A at 120 V ac. This means that the control can be used to drive small loads directly, or as a pilot for relay coils or motot starter coils for larger loads.

The counter will accept a preset value programmed into its keypad. This value can be retained in the counter's memory through battery backup. The battery is generally good for approximately two years of service even though it is rechargeable. This will ensure that the control will maintain its program even in the event of a power outage. The counter control will be enabled once power is applied to the CR1 relay. This enable condition also activates all sets of the instantaneous contacts. After the control is enabled it will register any signal on the counter input line that exhibits an on-to-off transition of 40 milliseconds or greater as a legitimate input signal (count). Every time the signal transitions from on to off, another count is registered in the counter control's memory.

After the number of counts in the control's memory equals the preset value, the delay contacts are energized. At this point the control will reset automatically. The timer required for resetting is approximately 30 miliseconds. The reset can also be executed any time by removing power from CR1. Output diagrams can be shown with X's and 0's to indicate when they are receiving power and when they are deactivated.

SEQUENCERS

Sequencers are used in a variety of automated control systems where events must occur in a predetermined sequence. This type of control is usually an electromechanical switch that has multiple cams and contacts that are opened and closed as the switch's shaft is rotated with an electric solenoid. Figure 9-21 shows an example of a sequencer switch. In this section we explain the operation of solid-state sequencers and how they are used in control systems.

The sequencer shown in Figure 9-21 is called a *step switch*. From the diagram you can see that it has 19 sets of contacts that can be open or closed during the 16 steps that the cam actuates during one rotation. The number of actual steps the cam uses can be adjusted by changing the programming pins in the cam. This switch can control 120- or 240-V devices at up to 10 A for the lower voltage and 5 A for the higher voltage.

The switch requires a $\frac{1}{10}$-second pulse as an input to advance the cam to the next step. The switch is capable of switching at a rate of 300 steps per minute from an external input provided by a pilot device or other control. The switch uses a solenoid stepping motor to rotate the camshaft and cams through 360 degrees. Each of the 19 sets of contacts can be activated to close on any particular step by inserting a cam pin at that location on the cam shaft for that step. The activator for each set of contacts rides on the cam shaft as it is rotated through its steps. If a pin is encountered on a particular step, the activator will move and close that set of contacts for that step. If a pin is not used on a particular step, the contacts will remain open for that step. In this manner, each set of contacts can be closed for one or any number of steps during the rotation. If 16 steps are not required, the switch can be adjusted to use fewer steps.

A timer control can be used to adjust the amount of time between steps on the sequencer. This type of device can be advanced by events in the logic circuits (contacts closing) or it can be advanced by the output from a timer. Some circuits use a combination of events and timer outputs to control the step increment of the sequencer device.

The step switch (sequencer) can be return to the first step, called the *home position*, at any time by energizing the reset terminal. This provides a method of returning the switch to its first step at any time with a minimum of effort.

Programmable Sequencers

The sequencer switches we have discussed thus far are all operated as electromechanical devices. Since the device has switch contacts and cams, it will tend to break down from extended operation. If the switch is making one complete revolution every 2 minutes, the contacts may be opened and closed up to 15 times a minute. When you calculate the

SYSTEM OPERATION	CONTROL DEVICE TO ADVANCE TO NEXT STEP	STEP	LOAD SWITCHES ON STEP SWITCH CLOSED = X OPEN = O						
			1	2	3	4	5	6	7
Off	Pushbutton	1	O	O	O	O	O	O	O
Solenoid valve no. 1 energized. Hydraulic Ram hits LS1 at end of movement.	Limit Switch No. 1	2	X	O	O	O	O	O	O
Solenoid valve no. 1 energized. Timer holds ram in position for adjustable time.	Timer No. 1	3	X	X	O	O	O	O	O
Solenoid valve no. 1 energized. pump and timer no. 2 starts.	Timer No. 2	4	X	O	X	O	O	O	O
Solenoid valve no. 1 energized. Solenoid valve no. 2 energized. 2nd ram hits limit switch 2.	Limit Switch No. 2	5	X	O	X	X	O	O	O
Design the control circuit by making a chart as above. List the system operation step-by-step, indicating which load circuits are closed in each step. Next, list the control device which is to terminate each step and to advance the switch to the next position. The chart above is easily translated into the electrical circuit diagram shown to the right. The input signal devices are connected to their respective tap switch terminals. The load circuits are connected to the switch terminals indicated by the chart.		LOADS	Solenoid Valve No. 1	Timer No. 1	Pump and Timer No. 2	Valve No. 2			

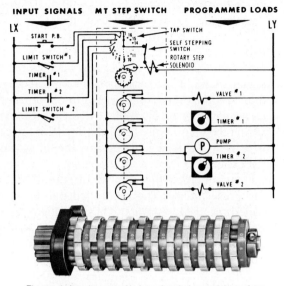

The cam shaft can be removed by loosening a lockscrew and removing a stud shaft in the right end plate. Another cam shaft with a different program can be inserted or the cams may be reprogrammed on the existing shaft. The cams are keyed so they must be correctly assembled on the hex cam shaft.

FIGURE 9-21 Sequence switch to control electrical loads in a sequential operation. A diagram is also provided with a matrix table to indicate the operation of the outputs. (Courtesy of Eagle Signal Controls.)

number of operations per year (over 7,000,000), it is easy to understand why the contacts tend to wear out. For this reason, microprocessors have been used for extended operations. Some programmable controllers (P/Cs) also provides sequencer operations through their program and input and output devices. Since the microprocessor controls the sequencer operation, and the solid-state components in the output modules control the on/off switching of the load, the P/C is better suited for sequencer applications. More information concerning the sequencer is provided in Chapter 16.

Installing and Troubleshooting the Sequencer

The installation of the sequencer switch should be accomplished with a detailed electrical diagram. This diagram will indicate what circuits (contacts) in the switch should be connected to each output, and it will also provide a matrix that indicates during which steps the contacts are closed. You will need to translate this information into locations on the cam so that you can insert the activator pins in the correct locations.

After the cam switch is wired and the pins are installed, the input signal should be activated, so that the switch can be stepped through a complete cycle. Pay particular attention to the sequence in which each load is energized. If any load is energized out of sequence, the activator pins should be checked for the correct locations.

If any other conditions occur that indicate that the sequencer is malfunctioning, the switch will need to be troubleshooted. The troubleshooting procedure can be broken into two distinct steps. First, you can check to see if the input signal is causing the sequencer to advance correctly. If the switch is advancing, you can begin to focus on the individual switch circuits activated by the cam. If the sequencer does not advance through its steps correctly, you should test the input circuit for the correct signal voltage and determine that the signal's duration is sufficient to cause the solenoid to advance the switch.

When you are testing the individual circuits of the switch, it is important to determine if the contacts close at any time and if the load operates during any of the steps. If the load operates correctly at some steps but is not energizing during other steps, the problem will be located in the area of the activator pins and cams.

If the loads do not energize on any step, you should advance the switch to a step where you are certain the contacts are closed and use the voltage-drop method of troubleshooting to look for the problem. If the circuit can provide the correct voltage to the load, be sure to test the load for

FIGURE 9-22 Twenty-four-hour programming time clock. (Courtesy of Eaton Corp./Cutler-Hammer Products.)

on the switch if they are not all used. Since all the contact sets are the same, the load can operate from any set as long as the activator pins have been adjusted to the same sequence. If all the sets of contacts are being used on the sequencer, you will need to change the faulty set of contacts when they are found. Whenever you make repairs or adjustments to the sequencer, you should be sure to test the complete control circuit before placing it back into operation.

Time Clocks

Time clocks are available in a variety of models that allow the user to program on times and off times by setting pins. These types of controls are used to set lighting and air-conditioning systems to turn on and off according to the time of day. These clocks will also allow the user to select a different turn-on and turn-off time for each day. Also, weekend hours need to be adjustable for times when the building is not in use. Figure 9-22 shows an example of this type of timer.

proper operation. If a set of contacts is faulty, it is possible to move the load to another set of contacts

QUESTIONS

9-1. List four types of outputs available on timers.

9-2. Explain the difference between a reset timer and a cycle timer.

9-3. Draw a diagram of a pneumatic diagram and explain its operation.

9-4. Explain the difference between an add-on and a stand-alone timer.

9-5. Explain how the change of time delay is accomplished with a pneumatic timer.

9-6. Show the electrical symbol for normally open and normally closed timed delay-on and timed delay-off contacts.

9-7. Show the electrical diagram for an Eagle timer and explain the operation of the instantaneous contacts and time-delay contacts.

9-8. Draw a timing diagram that shows a motor starter for conveyor 1 starting immediately when the circuit's start button is energized and conveyor 2 starting 20 seconds later.

9-9. Explain how "X" and "0" are energized and deenergized during the timing cycle. How is the reset condition shown in this type of diagram?

9-10. Draw a sketch of a thermal delay switch and explain what makes it activate.

9-11. Explain the operation of the solid-state timer.

9-12. Explain what occurs inside the solid-state timer to delay its operational time.

9-13. Draw a diagram of a cube timer controlling a motor in a time delay-on application.

9-14. Explain the difference between a delay-on and a delay-off application. When will the timer begin to time in each application?

9-15. Draw a diagram of an eight-pin (octal base) timer and identify each terminal.

9-16. What is an advantage of using an octal base timer?

9-17. Explain how a one-shot timer operates and how it would be used in industry.

9-18. What advantage does a microprocessor timer provide over standard timers?

9-19. Explain how you would troubleshoot and test a circuit that has a timer controlling a motor starter circuit.

9-20. Explain the operation of a counter.

9-21. Explain how a counter is used for totalizing applications and for resetting applications.

9-22. Explain the advantage of having a microprocessor device that can be programmed as either a timer or a counter.

9-23. Draw a diagram of a sequencer and explain how it can control four motors in sequential operation.

9-24. Explain how you would troublshoot a sequencer.

dc motors 10

INTRODUCTION

Dc motors are commonly used to operate machinery in a variety of applications on the factory floor. Dc motors were the first type of energy converter used in industry. It is important to remember that the earliest machines required speed control and dc motors could have their speed set by varying the voltage sent to them. The earliest speed controls for dc motors were nothing more than large resistors.

Dc motors required large amounts of dc voltage for operation. Because dc voltage cannot be distributed over a long distance, ac voltage became the industry standard. This meant that special generators were set up at the factory site to produce the amount of dc voltage that the dc motor would require. One system uses a large ac motor to drive a dc generator directly. The field current in the generator could be regulated to change the level of dc voltage, which in turn could be used to vary the speed of the dc motor. This system, called a Ward–Lennard system, was popular until solid-state diodes became available for rectifying large amounts of ac voltage to dc for use in motor-driven circuits.

During the 1950s and 1960s dc motors became more widely use in machinery control because their speed and torque was easy to control with simple SCR controllers. The SCR could rectify ac voltage to dc, provide current and voltage control at the same time, and were capable of being paralleled for larger loads up to 1000 A.

As solid-state controls became more reliable in the late 1960s and the 1970s a wide variety of low-cost ac motor speed controls became available. Transistors could handle larger loads, and microprocessors were relatively inexpensive so that they could be used to make variable-frequency ac motor controls. At this time, you had a choice of using good-quality ac or dc motors for all types of special speed and torque applications.

The most important points involve the concepts of controlling a dc motor's speed, its torque, and being able to reverse the direction of its rotation. It is also important to be able to recognize the features that make the series, shunt, and compound dc motors different from each other. You should also have a good understanding of the basic parts of the dc motor so that when you must troubleshoot a dc motor circuit, you will be able to recognize a malfunctioning component and make repairs or replace parts as quickly as possible.

It is also important to understand the methods of controlling the speed of a motor, its direction of rotation, and the amount of torque it can develop since these are the principles that are built into motor drive controls. Operation of the special drive controls is very easy to understand if you know what electrical principle they are trying to alter to provide control for the motor. If you do not understand the motor principle, it is doubtful that you will fully understand the motor control device, and this will make the system nearly impossible to troubleshoot. If you understand concepts, you will easily be able to understand the next generation of controls, those that will be produced during the next 10 years. If you do not understand principles, each new type of control device that you encounter will seem impossible to understand and repair.

This chapter is intended to be a review of the basic fundamentals that you learned in a class on dc motors. We provide enough detail for you to understand the operation of dc motors if you have not studied them previously. We also link the basic motor theories with the type of control that is available.

Magnetic theory are provided with a discussion of basic dc motor components. Additional information will explain the difference between series, shunt, and compound motors. Diagrams are provided to explain the methods of controlling speed, rotation, and torque. These diagrams are useful for making field wiring connections and for testing motors during troubleshooting procedures. You are also requested to review some of the theories of dc motor control found in this chapter when you are reading about more complex motor control circuits in other chapters.

MAGNETIC THEORY

Dc motors operate on the principles of basic magnetism. Some of these principles were introduced in Chapter 6 in the discussion of magnetism for use in relays and solenoids. The same principles are involved in dc motor operation. You should remember that a coil of wire can be magnetized when current is passed through it. When this principle was used in relay coils, the polarity of the current was not important. When the current is passed through a coil of wire to make a field coil for a motor, the polarity of the current will determine the direction of rotation for the motor.

The polarity of the current flowing through the coil of wire will determine the location of the north and south magnetic poles in the coil of wire. Another important principle involves the amount of current that is flowing through the coil. The amount of current was not important as long as enough current was present to move the armature of the relay or solenoid. In a dc motor, the amount of current in the windings will determine the speed (rpm) of the motor shaft and the amount of torque that it can produce.

A diagram of the left-hand rule of current flow through a coil of wire was presented in Chapter 6. This principle helped you understand that the direction of current flow will determine the magnetic polarity of the coil. Some students see this diagram and feel that they must try to identify the direction of flow of every electron in the motor winding. This diagram is not presented or intended to be used for this purpose. Instead, it is provided to show you a principle from which several facts can be determined. The first fact that you should understand is that the direction of current flow will determine which end of a coil of wire is negative or positive. This will determine which end of the coil will be the north pole of the magnet and which end will be the south pole. It is also easy to see from this diagram that by changing the direction of the current flow in the coil of wire, the magnetic poles will be reversed in the coil. This is important to understand, because the motor's rotation is caused by the changing magnetic field.

Another basic concept about magnets that you should remember is the relationship between two like poles and two unlike poles. When the north pole of two different magnets are placed close to each other, they will repel each other. When the north pole of one magnet is placed near the south pole of another magnet, the two poles will attract each other very strongly.

The third principle that is important to understand with the coil of wire is that the strength of the magnetic field can be varied by changing the amount of current flowing through the wire in the coil. If a small amount of current is flowing, a small number of flux lines will be created and the magnetic field will be relatively weak. If the amount of current is increased, the magnetic field will become stronger. The strength of the magnetic field can be increased to the point of saturation. A magnetic coil is said to be saturated when its magnetic strength cannot be increased by adding more current. Saturation is similar to filling a drinking glass with water. You cannot get the level of the glass any higher than full. Any additional water that is put into the glass when it is full will not increase the amount of water in the glass. The additional water will run over the side of the glass and be wasted. The same principle can be applied to a magnetic coil. When the strength of the magnetic field is at its strongest point, additional electric current will not cause the field to become any stronger. It will be easy to see these principles used to make a dc motor operate.

DC MOTOR THEORY

The basic operation of a dc motor is easy to understand. Figure 10-1 shows its operation in terms of magnetic theories. In Figure 10-1a you can see that a bar magnet has been mounted on a shaft so that it can spin. The field winding is one long coil of wire that has been separated into two sections. The top section is connected to the positive pole of the battery and the bottom section is connected to the negative pole of the battery. The current flow in this direction makes the top coil the north pole of the magnet and the bottom coil the south pole of the magnet.

The bar magnet is called the *armature* and the coil of wire is called the *field*. The arrow shows the direction of the armature's rotation. Notice that the arrow shows the armature starting to rotate in the clockwise direction. The north pole of the field is repelling the north pole of the armature, and the

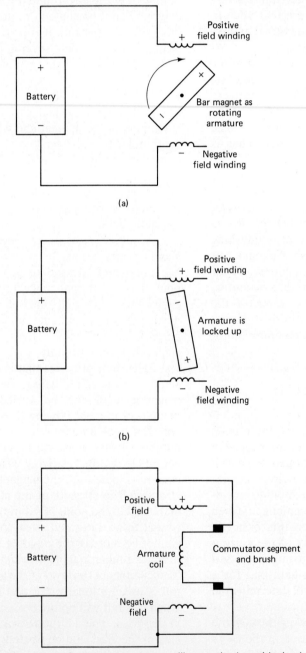

FIGURE 10-1 Elementary diagrams explaining the operation of simple dc motor.

south pole of the field coil is repelling the south pole of the armature.

As the armature begins to move, the north pole of the armature comes closer to the south pole of the field, and the south pole of the armature is coming closer to the north pole of the field. As the two unlike poles near each other, they begin to attract. This attraction becomes stronger until the armature's north pole moves directly in line with the field's south pole, and its south pole moves directly in line with the fields north pole (Figure 10-1b).

When the opposite poles are at their strongest attraction, the armature will be "locked up" and will resist further attempts to continue spinning. For the armature to continue its rotation, the armature's polarity must be switched. Since the armature in this diagram was a permanent magnet, you can see that it would lock up during the first rotation and not work. Therefore the armature must be changed to a coil (electro magnet) and a set of commutator segments must be added. One commutator segment is provided for each terminal of the magnetic coil. Since this armature has only one coil, it will have only two terminals, so the commutator has two segments.

Since the armature is now a coil of wire, it will need dc current flowing through it to become magnetized. This presents another problem; since the armature will be rotating, the dc voltage wires cannot be connected directly to the armature coil. A stationary set of carbon brushes is used to make contact to the rotating armature. The brushes ride on the commutator segments to make contact so that current will flow through the armature coil.

In Figure 10-1c you can see that the dc voltage is applied to the field and to the brushes. Since negative dc voltage is connected to one of the brushes, the commutator segment the negative brush rides on will also be negative. The armature's magnetic field causes the armature to begin to rotate. This time when the armature gets to the point where it becomes locked up with the magnetic field, the negative brush begins to touch the end of the armature coil that was previously positive and the positive brush begins to touch the end of the armature coil that was negative. This action switches the polarity of the armature coil's magnetic field at just the right time so that the repelling and attracting continues. The armature continues to switch its magnetic polarity twice during each rotation, which causes it to continually be attracted and repelled with the field poles.

This is a simple two-pole motor that is used primarily for instructional purposes. Since the motor has only two poles, the motor will operate rather roughly and not provide too much torque. Additional field poles and armature poles must be added to the motor for it to become useful for industry.

DC MOTOR COMPONENTS

The dc motor is made up of three basic components or assemblies. These components are designed slightly different for different types of dc motors. The three basic types of dc motors are the series motor, the shunt motor, and the compound motor. The *series motor* is designed to move loads with high starting torque, in applications such as a crane motor or lift hoist. The *shunt motor* is designed slightly different since it is made for applications such as pumping fluids, where constant-speed characteristics are important. The *compound motor* is designed with some of the series motor's characteristics and some of the shunt motor's characteristics. This allows the compound motor to be used in applications where high starting torque and controlled operating speed are both required.

It is important that you understand the function and operation of the basic components of the dc motor since motor controls will take advantage of these design characteristics to provide speed, torque, and direction of rotation control. These basic components include the armature assembly, which includes all rotating parts; the frame assembly, which houses the stationary field coils; and the end plates, which provide bearings for the motor shaft and a mounting point for the brush rigging. Each of these assemblies is explained in depth so that you will understand the design concepts used for motor control.

Armature

The armature is the part of a dc motor that rotates and provides energy at the end of the shaft. It is basically a electromagnet since it is a coil of wire that has to be specially designed to fit around core material on the shaft. The core of the armature is made of laminated steel and provides slots for the coils of wire to be pressed onto. Figure 10-2a shows a sketch of a typical dc motor armature. Figure 10-2b shows the laminated steel core of the armature without any coils of wire on it. This gives you a better look at the core.

The armature core is made of laminated steel to prevent the circulation of eddy currents. If the core were solid, magnetic currents would be produced that would circulate in the core material near the surface and cause the core metal to heat up.

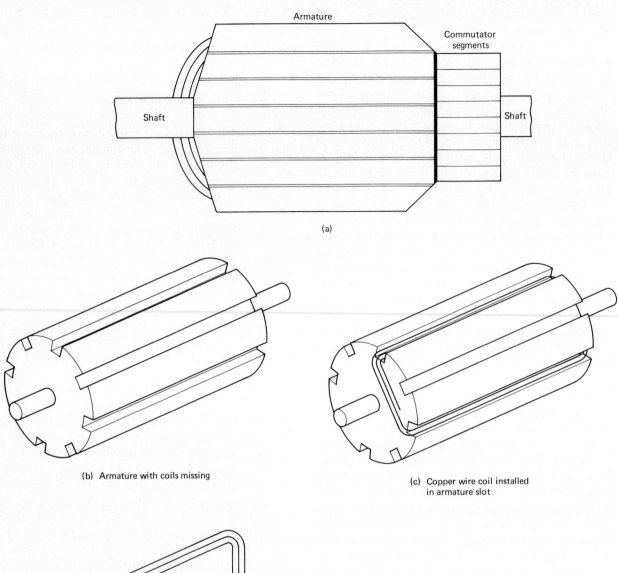

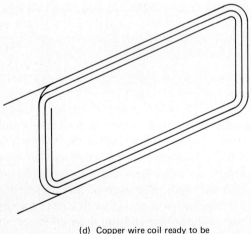

FIGURE 10-2 Sketches of the armature shows the commutator segments and laminated armature segments without the armature winding. The second sketch shows the armature coils mounted in place on the pole piece.

These magnetic currents are called *eddy currents*. When laminated steel sections are pressed together to make the core, the eddy current cannot flow from one laminated segment to another, so they are affectively canceled out. The laminated core also prevents other magnetic losses called *flux losses*.

These losses tend to make the magnetic field weaker so that more core material is required to obtain the same magnetic field strength. The flux losses and eddy current losses are grouped together by designers and called *core losses*. The laminated core is designed to allow the armature's magnetic field to be as strong as possible since the laminations prevent core losses.

Notice that one end of the core has commutator segments. There is one commutator segment for each end of each coil. This means that a armature with four coils will have eight commutator segments. The commutator segments are used as a contact point between the stationary brushes and the rotating armature. When each coil of wire is pressed onto the armature, the end of the coil is soldered to a specific commutator segment. This makes an electrical terminal point for the current that will flow from the brushes onto the commutator segment and finally through the coil of wire. Figure 10-2c shows the coil of wire mounted in the armature slot and soldered to the commutator segment.

The shaft is designed so that the laminated armature segments can be pressed onto it easily. It is also machined to provide a surface for a main bearing to be pressed on at each end. The bearing will ride in the end plates and support the armature when it begins to rotate. One end of the shaft is also longer than the other since it will provide the mounting shaft for the motor's load to be attached. Some shafts have a keyway or flat spot machined into them so that the load that is mounted on it can be secured. You must be careful when handling a motor that you do not damage the shaft since it must be smooth to accept the coupling mechanism. It is also possible to bend the shaft or cause damage to the bearings so that the motor will vibrate when it is operating at high speed. The commutator is made of copper. A thin section of insulation is placed between each commutator segment. This effectively isolates each commutator segment from all others.

Motor Frame

The armature is placed inside the frame of the motor where the field coils are mounted. When the field coils and the armature coils become magnetized the armature will begin to rotate. The field winding is made by coiling up a long piece of wire. The wire is mounted on laminated pole pieces called *field poles*. Similar to an armature, these poles are made of laminated steel or cast iron to prevent eddy current and other flux losses. Figure 10-3 shows a dc motor frame with field coils mounted in it.

The amount of wire that is used to make the field winding will depend on the type of motor that is being manufactured. A series motor uses heavy-gage wire for its field winding so that it can handle the very large field currents. Since the wire is a large gage, the number of turns of wire in the coil will be limited. If the field winding is designed for a

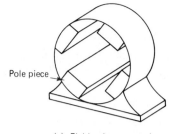

(a) Field poles mounted in a dc motor frame

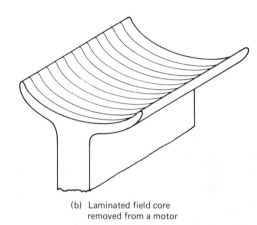

(b) Laminated field core removed from a motor

FIGURE 10-3 Frame with field coils.

shunt motor, it will be made of very small gage wire and many turns can be used.

After the coils are wound, they are coated for protection against moisture and other environmental elements. After they have been pressed onto the field poles, they must be secured with shims or bolts so that they are held rigidly in place. Remember: when current is passed through the coil, it will become strongly magnetized and attract and repel the armature magnetic poles. If the field poles are not rigidly secured, they will be pulled loose when they are attracted to the armature's magnetic field and then pressed back into place when they become repelled. This action will cause the field to vibrate and damage the outer protective insulation and cause a short circuit or a ground condition between the winding and the frame of the motor.

The ends of the frame are machined so that the end plates will mount firmly into place. An access hole is also provided in the side of the frame or in the end plates so that the field wires can be brought to the outside of the motor, where dc voltage can be connected.

The bottom of the frame has the mounting bracket attached. The bracket has a set of holes or slots provided so that the motor can be bolted down and securely mounted on the machine it is driving. The mounting holes will be designed to specifications by frame size. The dimensions for the frame sizes are provided in tables printed by motor manufacturers. Since these holes and slots are designed to a standard, you can predrill the mounting holes in the machinery before the motor is put in place. The slots are used to provide minor adjustments to the mounting alignment when the motor is used in belt-driven or chain-driven applications. It is also important to have a small amount of mounting adjustment when the motor is used in direct-drive applications. It is very important that the motor be mounted so that the armature shaft can turn freely and not bind with the load.

End Plates

The end plates of the motor are mounted on the ends of the motor frame. Figure 10-4a shows the location of the end plates in relation with the motor frame. The end plates are held in place by four bolts that pass through the motor frame. The bolts can be removed from the frame completely so that the end plates can easily be removed for maintenance. The end plates also house the bearings for the armature

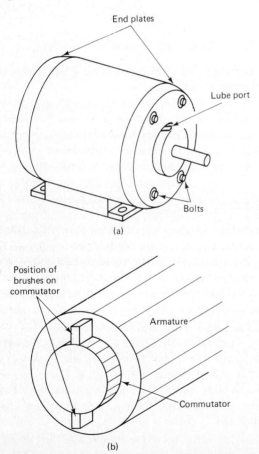

FIGURE 10-4 Diagram of the end plates for a dc motor. This diagram also shows the location of the lubrication system and brushes.

shaft. These bearings can be either sleeve or ball type. If the bearing is a *ball bearing* type, it is normally permanently lubricated, and if it is a *sleeve* type, it will require a light film of oil to operate properly. The end plates that house a sleeve type bearing will have a lubrication tube and wicking material. Several drops of lubricating oil is poured down the lubrication tube, where it will saturate the wicking material. The wicking is located in the bearing sleeve so that it can make contact with the armature shaft and transfer a light film of oil to it. It is important that the end plate for a sleeve bearing be mounted on the motor frame so that the lubricating tube is pointing up. This position will ensure that gravity will pull the oil to the wicking material. If the end plates are mounted so that the lubricating tube is pointing down, the oil will flow away from the wicking and it will become dry. When the wicking dries out, the armature shaft will rub directly on the metal in the sleeve bearing, it will quickly heat up, and the shaft will seize to the bearing. For this reason it is also important to follow lubrication instructions and oil the motor on a regular basis.

Brushes and Brush Rigging

The brush rigging is an assembly that securely holds the brushes in place so that they will be able to ride on the commutator. It is mounted on the rear end plate so that the brushes will be accessible by removing the end plate. An access hole is also provided in the motor frame so that the brushes can be adjusted slightly when the motor is initially set up. The brush rigging uses a spring to provide the proper amount of tension on the brushes so that they make proper contact with the commutator. If the tension is too light, the brushes will bounce and arc, and if the tension is too heavy, the brushes will wear down prematurely.

The brush rigging is shown in Figure 10-4b. Notice that it is mounted on the rear end plate. Since the rigging is made of metal, it must be insulated electrically when it is mounted on the end plate. The dc voltage that is used to energize the armature will pass through the brushes to the commutator segments and into the armature coils. Each brush has a wire connected to it. The wires will be connected to either the positive or negative terminal of the dc power supply. The motor will always have an even number of brushes. Half of the brushes will be connected to positive voltage and half will be connected to negative voltage. In most motors the number of brush sets will be equal to the number of field poles. It is important to remember that the voltage polarity will remain constant on each brush. This means that half of the brush will be connected to the positive power terminal, and the other half will be connected permanently to the negative terminal.

The polarity of each armature segment will alternate from positive to negative. When the armature is spinning, each commutator segment will come in contact with a positive brush for an instant and will be positive during that time. As the armature rotates slightly, that commutator segment will come in contact with a brush that is connected to the negative voltage supply and it will become negative during that time. As the armature continues to spin, each commutator segment will be alternately powered by positive and then negative voltage.

The brushes are made of carbon-composite material. Usually, the brushes have copper added to aid in conduction. Other material is also added to make them wear longer. The end of the brush that rides on the commutator is contoured to fit the commutator exactly so that current will transfer easily. The process of contouring the brush to the commutator is called *seating*. Whenever a set of new brushes are installed, they should be seated to fit the commutator. The brushes are the main part of the dc motor that will wear out. It is important that their wear be monitored closely so that they do not damage the commutator segments when they begin to wear out. Most brushes have a small mark on them called a *wear mark* or *wear bar*. When a brush wears down to the mark, it should be replaced. If the brushes begin to wear excessively or do not fit properly on the commutator, they will heat up and damage the brush rigging and spring mechanism. If the brushes have been overheated, they can cause burn marks or pitting on the commutator segments and also warp the spring mechanism so that it will no longer hold the brushes with the proper amount of tension.

If the spring mechanism has been overheated, it should be replace and the brushes should be checked for proper operation. If the commutator is pitted, it can be turned down on a lathe. After the commutator has been turned down, the brushes will need to be reseated.

After you have an understanding of the function of each of the parts or assemblies of the motor, you will better be able to understand the operation of a basic dc motor. Operation of the motor involves the interaction of all the motor parts. Some of the parts will be altered slightly for specific motor applications. These changes will become evident when the motor's basic operation is explained.

DC MOTOR OPERATION

An industrial dc motor operates very similarly to the simple dc motor described earlier in this chapter. The components described earlier in this chapter all operate together to make the motor function properly. Figure 10-5 shows an electrical diagram of a simple dc motor. Notice that the dc voltage is applied directly to the field winding and the brushes. The armature and the field are both shown as a coil of wire.

When voltage is applied to the motor, current begins to flow through the field coil from the negative terminal to the positive terminal. This sets up a strong magnetic field in the field winding. Current also begins to flow through the brushes into a commutator segment and then through an armature coil. The current continues to flow through the coil back to the brush that is attached to other end of the coil and returns to the dc power source. The current flowing in the armature coil sets up a strong magnetic field in the armature.

The magnetic field in the armature and field coil causes the armature to begin to rotate. This occurs by the unlike magnetic poles attracting each other and the like magnetic poles repelling each other. As the armature begins to rotate, the commutator segments will also begin to move under the brushes. As the commutator segments move under the brushes, they will continually change polarity from positive to negative. This action cause the magnetic field in the armature to change polarity continually from north pole to south pole. The commutator segments and brushes are aligned in such a way that the switch in polarity of the armature coincides with the location of the armature's magnetic field and the field winding's magnetic field. The switching action is timed so that the armature will not lock up magnetically with the field. Instead, the magnetic fields tend to build on each other and provide additional torque to keep the motor shaft rotating.

When the voltage is deenergized to the motor, the magnetic fields in the armature and the field winding will quickly diminish and the armature shaft's speed will begin to drop to zero. If voltage is

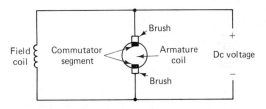

FIGURE 10-5 Diagram of simple dc motor.

applied to the motor again, the magnetic fields will strengthen and the armature will begin to rotate again.

Controlling the Speed of the Motor

The speed of the armature's rotation can be increased by increasing the amount of current passing through each coil. The amount of current in the armature and field coil can be changed by changing the amount of applied voltage. When the amount of voltage is increased, the amount of current in the coils will also increase. The increase in current through the coil will cause the strength of the magnetic field to increase up to the point where the coils reach saturation. Saturation usually occurs at a point where the applied voltage is increased slightly higher than the motor's voltage rating.

The voltage to a motor can be reduced by placing resistors in series with the voltage supply. SCRs are also used to control voltage to the motor. The SCR is particularly well suited for motor control because its gate can be controlled by a very low voltage phase shift circuit. The maximum current in the anode circuit can be up to several hundred amperes, which can be controlled down to near zero. This provides speed control from zero to maximum with little effort.

Reversing the Motor's Rotation

In applications such as lift and hoist motors, the dc motor must be able to operate in the forward and reverse directions. The rotation of a dc motor can be changed by changing the polarity of the field coil. This can be accomplished by exchanging the field coil wires at the power source. This means that the field coil wire that was connected to the positive terminal should be changed to the negative power supply terminal and the wire that was connected to the positive terminal should be changed to the negative power supply terminal. It is important to ensure that the armature's polarity stays the same. This means that the power supply wires connected to the brushes should remain the same. The motor would operate in the forward direction when the forward motor starter was energized and the field wires would be reversed when the reverse motor starter is energized.

If the dc motor provides only two wires for connection, they can be exchanged at the power supply and the motor will operate in the opposite direction. This provides a simple method of reversing the rotation of the motor, but it can cause a problem if the motor terminals are connected incorrectly during installation since the motor would rotate in the wrong direction. For this reason it is always important to test run the motor after it has been connected to ensure that it is operating in the proper direction.

Changing Torque Characteristics

Dc motors are used for a variety of applications in industry today. These applications vary from lifting very large loads such as hoist and crane operations to driving the axis on robots. Dc motors are also used in all types of motion control in addition to robots. These include intricate cutting applications such as with water jet cutters and laser or plasma cutters. Some of these cutters can travel at speeds up to 500 inches per minute and can hold tolerances to 0.003 inch.

These applications represent a wide variety of torque requirements. This means that some motors will need to lift very heavy loads for short periods. The largest amount of motor torque is required during the period when the motor is being started for these applications. Other applications require that a medium-sized load be controlled at specific speeds for long periods. These type of loads include machine tool and cutting applications. The speed of the load is more important than the size of the load. The load at startup for these applications will be very light since the cutter can be pulled away from the part while the motors come up to speed. These motors must be able of providing torque while the motor is at full speed. This would occur as the machine tool begins to make its cut on the part.

Some motors must be able to respond to a variety of torques through its entire range of speed. These motors will encounter full loads at any speed at which the motor is operating. This is also a machine tool application. Some materials must be cut at high speeds, while others must be cut at slower speeds. The load on the motor for each of these applications will be heavy when the tool begins to make contact and cut, and will become lighter if a smaller amount of material is being removed.

Other loads are very light and general purpose in nature, so that the speed of the motor is not critical. These types of loads may be used to operate conveyor motors or pumps. The load may be heavier at times but since the motor's speed is not critical, the small loss of rpm will not be critical.

It is not possible to change a motor for each of these loads. Rather it is better to manufacture a motor that fits the specific load and torque requirement. This means that you will not be able to use the same motor to provide heavy starting torque and then expect it to operate at a controlled rpm at high speed. These two characteristics are in opposition to each other and it would be difficult to provide one motor with these characteristics.

TYPES OF DC MOTORS

Three basic types of dc motors are used in industry today: the series motor, the shunt motor, and the compound motor. The series motor is capable of starting with a very large load attached, such as lifting applications. The shunt motor is able to operate with rpm control while it is at high speed. The compound motor, a combination of the series motor and the shunt motor, is able to start with fairly large loads and have some rpm control at higher speeds. In the remaining sections of this chapter we show a diagram for each of these motors and discuss their operational characteristics. A method of controlling their speed and direction of rotation is provided, together with procedures for installing and troubleshooting.

DC SERIES MOTORS

The series motor provides high starting torque and is able to move very large shaft loads when it is first energized. Figure 10-6 shows the wiring diagram of a series motor. From the diagram you can see that the field winding in this motor is wired in series with the armature winding. This is the attribute that gives the series motor its name.

Since the series winding carries the same amount of current that the armature will see, it must be made from wire that is large enough to carry the load. This means that the field coil will be made from very large gage wire. Since the wire gage is so large, the winding will have only a few turns of wire. The field winding in some series motors is made from copper bar stock rather than the conventional round wire that you use for power distribution. The square or rectangular shape of the copper bar stock makes it fit more easily around the field pole pieces. It can also radiate more easily the heat that has built up in the winding due to the large amount of current being carried.

The amount of current that passes through the winding determines the amount of torque the motor shaft can produce. Since the series field is made of very large conductors, it can carry very large amounts of current and produce very large torques. For example, the starter motor that is used to start an automobile's engine is a series motor may draw up to 500 A when it is turning the engine's crankshaft. Series motors used to power hoist or cranes may draw currents of thousands of amperes during operation.

The series motor can safely handle large currents since the motor does not operate for an extended period. In most applications the motor will operate for only a few seconds while this large current is present. Think about how long the starter motor on the automobile must operate to get the engine to start. This period is similar to that of industrial series motors.

Series Motor Operation

Operation of the series motor is rather easy to understand. From Figure 10-6 you can see that the field winding is connected in series with the armature winding. This means that power will be applied to one end of the series field winding and to one end of the armature winding (connected to the brush).

When voltage is applied, current begins to flow from negative power supply terminals through the series winding and armature winding and back to the positive power supply terminal. Since the armature is not rotating when voltage is first applied, the only resistance in this circuit will be provided by the large conductors used in the armature and field windings. Since these conductors are so large, they will have a very small amount of resistance. This causes the motor to draw a very large amount of current from the power supply. When the large current begins to flow through the field and armature windings, it causes a very strong magnetic field to be built. Since the current is so large, it will cause the coils to reach saturation, which will produce the strongest magnetic field possible.

Producing Back EMF. The strength of these magnetic fields provides the armature shafts with the greatest amount of torque possible. The large torque causes the armature to begin to spin with the maximum amount of power. When the armature begins to rotate, it begins to produce voltage. This concept is difficult for some students to understand

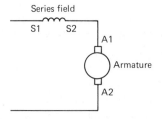

FIGURE 10-6 Electrical diagram of a dc series motor.

since the armature is part of the motor at this time. You should remember from the basic theories of magnetism that any time a magnetic field passes a coil of wire, a current will be produced. The stronger the magnetic field is or the faster the coil passes the flux lines, the more current will be generated.

When the armature begins to rotate, it will produce a voltage that is of opposite polarity to that of the power supply. This voltage is called *back voltage* or *back EMF* (electromotive force). The overall effect of this voltage is that it will be subtracted from the supply voltage so that the motor windings will see a smaller voltage potential. When Ohm's law is applied to this circuit, you will see that when the voltage is slightly reduced, the current will also be reduced slightly. This means that the series motor will see less current as its speed is increased. The reduced current will mean that the motor will continue to lose torque as the motor speed increase. Since the load is moving when the armature begins to pick up speed, the application will require less torque to keep the load moving. This works to the motor's advantage by automatically reducing the motor current as soon as the load begins to move. It also allows the motor to operate with less heat buildup.

This condition can cause problems if the series motor ever loses its load. The load could be lost when a shaft breaks or if a drive pin is sheared. When this occurs, the load current is allowed to fall to a minimum, which reduces the amount of back EMF that the armature is producing. Since the armature is not producing a sufficient amount of back EMF and the load is no longer causing a drag on the shaft, the armature will begin to rotate faster and faster. It will continue to increase rotational speed until it is operating at a very high speed. When the armature is operating at high speed the heavy armature windings will be pulled out of their slots by inertia. When the windings are pulled loose, they will catch on a field winding pole piece and the motor will be severely damaged.

This condition is called *runaway*. You can see why a dc series motor must have some type of runaway protection. A centrifugal switch can be set connected to the motor to deenergize the motor starter coil if the rpm exceeds the set amount. Other sensors can be use to deenergize the circuit if the motor's current drops while full voltage is applied to the motor. Additional safety circuits for the series motor are provided in Chapter 12. The rpm of the series motor is controlled by the size of the load.

Reversing the Rotation of the Motor. The direction of rotation of a series motor can be changed by changing the polarity of either armature or field winding. It is important to remember that if you simply changed the polarity of the applied voltage, you would be changing the polarity of both field and armature windings and the motor's rotation would remain the same.

Since only one of the windings need to be reversed, the armature winding is typically used because its terminals are readily accessible at the brush rigging. Remember that the armature receives its current through the brushes, so that if their polarity is changed, the armature's polarity will also be changed. A reversing motor starter is used to change the direction of the motor's rotation by changing the polarity of the armature windings. The armature's terminals are marked A1 and A2 and the field terminals are marked F1 and F2.

When the forward motor starter is energized, the A1 and F1 terminals are connected to the positive power supply and A2 and F2 are connected to the negative terminal of the power supply. When the reverse motor starter is energized, terminals A1 and A2 are reversed. A1 is now connected to the negative terminal of the power supply and A2 is connected to the positive terminal. Notice that F1 remains connected to the positive terminal and F2 remains connected to the negative power supply terminal. This ensures that only the armature's polarity has been changed and the motor will begin to rotate in the oposite direction.

Installing and Troubleshooting the Series Motor

Since the series motor has only two leads brought out of the motor for installation wiring, this wiring can be accomplished rather easily. If the motor is wired to operate in only one direction, the motor terminals can be connected to a manual or magnetic starter. If the motor's rotation is required to be reversed periodically, it should be connected to a reversing starter.

Most dc series motors are used in direct-drive applications. This means that the load is connected directly to the armature's shaft. This type of load is generally used to get the most torque converted. Belt-drive applications are not recommended since a broken belt would allow the motor to run away.

After the motor has been installed, a test run should be used to check it out. If any problems occur, the troubleshooting procedures should be used.

The most likely problem that will occur with the series motor is that it will develop an open in one of its windings or between the brushes and the commutator. Since the coils in a series motor are connected in series, each coil must be functioning properly or the motor will not draw any current. When this occurs, the motor cannot build a magnetic field and the armature will not turn. Another problem that is likely to occur with the motor circuit is that circuit voltage will be lost due to a blown fuse or circuit breaker. The motor will respond similarly in both of these conditions.

The best way to test a series motor is with a voltmeter. The first test should be for applied voltage at the motor terminals. Since the motor terminals are usually connected to a motor starter, the test leads can be placed on these terminals. If the meter shows that full voltage is applied, the problem will be in the motor, and if it shows that no voltage is present, you should test the supply voltage and the control circuit to ensure that the motor starter is closed. If the motor starter has a visual indicator, be sure to check to see that the starter's contacts are closed. If the overloads have tripped, you can assume that they have sensed a problem with the motor or its load. When you reset the overloads the motor will probably start again, but remember to test the motor thoroughly for problems that would cause an overcurrent situation.

If the voltage test indicates that the motor has full applied voltage to its terminals but the motor is not operating, you can assume that you have an open in one of the windings or between the brushes and the armature. Each of these sections should be disconnected from each other and voltage should be removed so that they can be tested with an ohmmeter for an open. The series field coils can be tested by putting the ohmmeter leads on terminal S1 and S2. If the meter indicates that an open exists, the motor will need to be removed and sent to be rewound or replaced. If the meter indicates that the field coil has continuity, you should continue the procedure by testing the armature.

The armature can also be tested with an ohmmeter by placing the leads on the terminals marked A1 and A2. If the meter shows continuity, rotate the armature shaft slightly to look for bad spots where the commutator may have an open or the brushes may not be seated properly. If the armature test indicates that an open exists, you should continue the test by visually inspecting the brushes and commutator. You may also have an open in the armature coils. The armature must be removed from the motor frame to be tested further. When you have located the problem you should remember that the armature can be turned down on a lathe and the brushes can be replaced while the motor remains in place.

It is possible that the motor will develop a problem but still run. This type of problem usually involves the motor overheating or not being able to pull its rated load. This type of problem is different from an open circuit because the motor is drawing current and trying to run. Since the motor is drawing current, you must assume that there is not an open circuit. It is still possible to have brush problems that would require the brushes to be reseated or replaced. Other conditions that will cause the motor to overheat include loose or damaged field and armature coils. The motor will also overheat if the armature shaft bearing is in need of lubrication or is damaged. The bearing will seize on the shaft and cause the motor to build up friction and overheat.

If either of these conditions occur, the motor may be fixed on site or be removed for extensive repairs. When the motor is restarted after repairs have been made, it is important to monitor the current usage and heat buildup. Remember that the motor will draw dc current so that an ac clamp on ammeter will not be useful for measuring the dc current. It is also important to remember that the motor can draw very high locked-rotor current when it is starting, so the ammeter should be capable of measuring currents up to 1000 A. After the motor has completed its test run successfully, it can be put back into operation for normal duty. Any time the motor is suspected of faulty operation, the troubleshooting procedure should be rechecked.

DC SHUNT MOTORS

The shunt motor is different from the series motor in that the field winding is connected in parallel with the armature instead of in series. You should remember from basic dc theory that a parallel circuit is often referred to as a shunt. Since the field winding is placed in parallel with the armature, it is called a shunt winding and the motor is called a shunt motor. Figure 10-7 shows a diagram of a shunt motor. Notice that the field terminals are marked F1 and F2, and the armature terminals are marked A1 and A2. You should notice in this

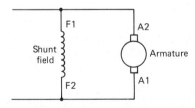

FIGURE 10-7 Diagram of a dc shunt motor. Notice the shunt coil is represented as a coil of fine wire with many turns that is connected in parallel (shunt) with the armature.

diagram that the shunt field is represented in the diagram with multiple turns using a thin line.

The shunt winding is made of very small gage wire with many turns on the coil. Since the wire is so small, the coil can have thousands of turns and still fit in the slots. The small-gage wire cannot handle as much current as the heavy-gage wire in the series field, but since this coil has many more turns of wire, it can still produce a very strong magnetic field.

Shunt Motor Operation

The shunt motor has slightly different operating characteristics than the series motor. Since the shunt field coil is made of fine wire, it cannot produce the large current flow required for starting like the series field. This means that the shunt motor has very low starting torque, which requires that the shaft load be rather small.

When voltage is applied to the motor, the high resistance of the shunt coil keeps the overall current flow low. The armature for the shunt motor is very similar to the series motor and it will draw enough current to produce a magnetic field strong enough to cause the armature shaft and load to start turning. Like the series motor, when the armature begins to turn, it will produce back EMF. The back EMF will cause the current in the armature to begin to diminish to a very small level. The amount of current the armature will draw is directly related to the size of the load when the motor reaches full speed. Since the load is generally small, the armature current will be small. When the motor reaches full rpm its speed will remain rather constant.

Controlling the Speed of the Motor. When the shunt motor will reach full rpm, its speed will remain fairly constant. The reason the speed remains constant is due to the load characteristics of the armature and shunt coil. You should remember that the speed of a series motor could not be controlled since it was totally dependent on the size of the load in comparison to the size of the motor. If the load was very large for the motor size, the speed of the armature would be very slow. If the load was light compared to the motor, the armature shaft speed would be much faster, and if no load was present on the shaft, the motor could run away.

The shunt motor's speed can be controlled. The ability of the motor to maintain a set rpm at high speed when the load changes is due to the characteristic of the shunt field and armature. Since the armature begins to produce back EMF as soon as it starts to rotate, it will use the back EMF to maintain its rpm at high speed. If the load increases slightly and causes the armature shaft to slow down, less back EMF will be produced. This will allow the difference between the back EMF and applied voltage to become larger, which will cause more current to flow. The extra current provides the motor with the extra torque required to regain its rpm when this load increased slightly.

The shunt motor's speed can be varied in two different ways. These include varying the amount of current supplied to the shunt field and controlling the amount of current supplied to the armature. Controlling the current to the shunt field allows the rpm to be changed 10 to 20 percent when the motor is at full rpm.

This type of speed control regulation is accomplished by slightly increasing or decreasing the voltage applied to the field. The armature continues to have full voltage applied to it while the voltage to the shunt field is regulated. When the shunt field's current is decreased, the motor's rpm will increase slightly. When the shunt field's current is reduced, the armature must rotate faster to produce the same amount of back EMF to keep the load turning, and if the shunt field current is increased slightly, the armature can rotate at a slower rpm and maintain the amount of back EMF to produce the armature current to drive the load. The field current can be adjusted with a field rheostat or an SCR current control.

The shunt motor's rpm can also be controlled by regulating the voltage that is applied to the motor armature. This means that if the motor is operated on less voltage than is shown on its data plate rating, it will run at less than full rpm. You must remember that the shunt motor's efficiency will drop off drastically when it is operated below its rated voltage. The motor will tend to overheat when it is operated below full voltage, so motor ventilation must be provided. You should also be aware that the motor's torque is reduced when it is operated below the full voltage level.

Since the armature draws more current than the shunt field, the control resistors were much larger than those used for the field rheostat. During the 1950s and 1960s SCRs were used for this type of current control. The SCR was able to control the armature current since it was capable of controlling several hundred amperes. The SCR was also well suited for this type of control because it could be controlled by a small variable resistor. The resistor was connected with a capacitor to provide phase shift control to the gate of the SCR. When the phase shift was change on the gate, the firing angle of the SCR was also changed and it would control the amount of current the SCR would pass to the motor's armature. In Chapter 14 we provide a more complex explanation of these types of speed controllers including photographs and diagrams. Figure 10-8 shows a diagram of an SCR motor speed control.

Torque Characteristics. The armature's torque increases as the motor gains speed due to the fact that the shunt motor's torque is directly proportional to the armature current. When the motor is starting and speed is very low, the motor has very little torque. After the motor reaches full rpm its torque is at its fullest potential. In fact, if the shunt field current is reduced slightly when the motor is at full rpm, the rpm will increase slightly and the motor's torque will also increase slightly.

This type of automatic control makes the shunt motor a good choice for applications where constant speed is required, even though the torque will vary slightly due to changes in the load.

Reversing the Rotation of the Motor. The direction of rotation of a dc shunt motor can be reversed by changing the polarity of either the armature coil or the field coil. In this application the armature coil is usually changed, as was the case with the series motor. Figure 10-9 shows the electrical ladder diagram of this circuit. You should notice that the F1 and F2 terminals of the shunt field are connected directly to the power supply, and the A1 and A2 terminals of the armature winding are connected to the reversing starter.

When the forward starter is energized its contacts connect the A1 lead to the positive power supply terminal and the A2 lead to the negative power supply terminal. The F1 motor lead is connected directly to the positive terminal of the power supply and the F2 lead is connected to the negative

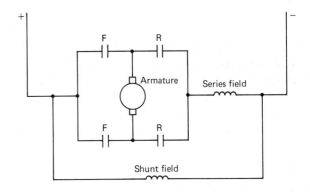

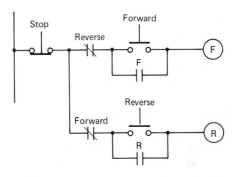

FIGURE 10-9 Circuit diagram for a reversing starter connected to a shunt motor.

terminal. When the motor is wired in this configuration it will begin to run in the forward direction.

When the reversing starter is energized, its contacts reverse the armature wires so that the A1 lead is connected to the negative power supply terminal and the A2 lead is connected to the positive power supply terminal. The field leads are connected directly to the power supply, so their polarity is not changed. Since the fields polarity has remained the same and the armature's polarity has reversed, the motor will begin to rotate in the reverse direction.

The shunt motor can be reversed with the reversing manual magnetic starter. These starters allow the motor to be safely reversed as often as required by the application. The reversing starter can also be used as a plugging circuit to stop the armature's rotation quickly.

Installing the Shunt Motor

The shunt motor can be installed rather easily. The motor is generally used in belt-drive applications. This means that the installation procedure should be broken into two sections, which include the mechanical installation of the motor and its load, and the installation of electrical wiring and controls.

When the mechanical part of the installation is completed, the alignment of the motor shaft and the

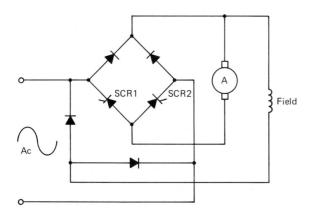

FIGURE 10-8 Diagram of a simple SCR speed control circuit used to control the speed of a dc shunt motor.

DC Shunt Motors **279**

load shaft should be checked. If the alignment is not true, the load will cause an undue stress on the armature bearing and there is the possibility of the load vibrating and causing damage to it and the motor. After the alignment is checked, the tension on the belt should also be tested. As a rule of thumb, you should have about $\frac{1}{2}$ to $\frac{1}{4}$ inch of play in the belt when it is properly tensioned. Several tension measurement devices are available to determine when a belt is tensioned properly. The belt tension can also be compared to the amount of current the motor draws. The motor must have its electrical installation completed to use this method.

The motor should be started, and if it is drawing too much current, the belt should be loosened slightly but not enough to allow the load to slip. If the belt is slipping, it can be tightened to the point where the motor is able to start successfully and not draw current over its rating.

The electrical installation can be completed before, after, or during the mechanical installation. The first step in this procedure is to locate the field and armature leads in the motor and prepare them for field connections. If the motor is connected to magnetic or manual across the line starter, the F1 field coil wire can be connected to the A1 armature lead and an interconnecting wire which will be used to connect these leads to the L1 terminal on the motor starter. The F2 lead can be connected to the A2 lead and a second wire which will connect these leads to the L2 motor starter terminal.

When these connections are completed, field and armature leads should be replaced back into the motor and the field wiring cover or motor access plate should be replaced. Next, the dc power supply's positive and negative leads should be connected to the motor starter's T1 and T2 terminals, respectively.

After all of the load wires are connected, any pilot devices or control circuitry should be installed and connected. The control circuit should be tested with the load voltage disconnected from the motor. If the control circuit uses the same power source as the motor, the load circuit can be isolated so the motor will not try to start by disconnecting the wire at terminal L2 on the motor starter. Operate the control circuit several times to ensure that it is wired correctly and operating properly. After you have tested the control circuit, the lead can be replaced to the L2 terminal of the motor starter and the motor can be started and tested for proper operation. Be sure to check the motor's voltage and current while it is under load to ensure that it is operating correctly. It is also important to check the motor's temperature periodically until you are satisfied the motor is operating correctly.

If the motor is connected to a reversing starter or reduced voltage starting circuit, their operation should also be tested. You may need to read the material in Chapter 12 to fully understand the operation of these methods of starting the motor using reduced-voltage methods. If the motor is not operating correctly or develops a fault, the troubleshooting procedure should be used to test the motor and locate the problem.

Troubleshooting the Shunt Motor

When the dc shunt motor develops a fault, you must be able to locate the problem quickly and return the motor to service or have it replace. The most likely problems to occur with the shunt motor include loss of supply voltage, or an open in either shunt winding or the armature winding. Other problems may arise which cause the motor to run abnormally hot even though it continues to drive the load. The motor will show different symptoms for each of these problems, which will make the troubleshooting procedure easier.

When you are called to troubleshoot the shunt motor it is important to try and determine if the problem occurs while the motor is running or when it is trying to start. If the motor will not start, you should try to listen to see if the motor is humming and trying to start. When the supply voltage has been interrupted due to a blown fuse or a deenergized control circuit, the motor will not be able to draw any current and it will be silent when you try to start it. You can also determine that the supply voltage has been lost by measuring it with a voltmeter at the starter's L1 and L2 terminals. If no voltage is present at the load terminals, you should check for voltage at the starter's T1 and T2 terminals. If voltage is present here but not at the load terminals, it indicates that the motor starter is deenergized or defective. If no voltage is present at the T1 and T2 terminals, it indicates that supply voltage has been lost prior to the motor starter. You will need to check the supply fuses and the rest of the supply circuit to locate the fault.

If the motor tries to start and hums loudly, it indicates that the supply voltage is present. The problem in this case is probably due to an open field winding or armature winding. It could also be caused by the supply voltage being too low.

The most likely problem will be an open in the field winding since it is made from very small gage wire. The open can occur if the field winding draws too much current or develops a short circuit between the insulation in the coils. The best way to test the field is to remove supply voltage to the motor by opening the disconnect or deenergizing the motor starter. *Be sure to use a lockout when you are working on the motor after the disconnect has been opened.*

After power has been removed, the field terminals should be isolated from the armature coil. This can be accomplished by disconnecting one set of leads where the field and armature are connected together. Remember that the field and armature are connected in parallel and if they are not isolated, your continuity test will show a completed circuit even if one of the two windings have an open.

When you have the field coil isolated from the armature coil, you can proceed with the continuity test. Be sure to use the R × 1k or R × 10 k setting on the ohmmeter because the resistance in the field coil will be very high since the field coil may be wound from several thousand feet of wire. If the field winding test indicates the field winding is good, you should continue the procedure and test the armature winding for continuity.

The armature winding test may show that an open has developed from the coil burning open or from a problem with the brushes. Since the brushes may be part of the fault, they should be visually inspected and replaced if they are worn or not seating properly. If the commutator is also damaged, the armature should be removed, so the commutator can be turned down on a lathe.

If either the field winding or the armature winding have developed an open circuit, the motor will have to be removed and replaced. In some larger motors it will be possible to change the armature by itself rather than remove and replace the entire motor.

If the motor operates but draws excessive current or heats up, the motor should be tested for loose or shorting coils in the field and armature. The field coils may tend to come loose and cause the motor to vibrate and overheat, or the armature coils may come loose from their slots and cause problems. If the motor continues to overheat or operate roughly, the motor should be removed and sent to a motor rebuilding shop so that a more in-depth test may be preformed to find the problem before the motor is permanently damaged by the heat.

DC COMPOUND MOTORS

The dc compound motor is a combination of the series motor and the shunt motor. It has a series field winding that is connected in series with the armature and a shunt field that is in parallel with the armature. The combination of series and shunt winding allows the motor to have torque characteristics of the series motor and regulated speed characteristics of the shunt motor. Figure 10-10 shows a diagram of the compound motor. Several versions of the compound motor are also shown in this diagram.

Cumulative Compound Motors

Figure 10-10a shows the cumulative compound motor. It is called *cumulative* because the shunt field is connected so that its coils are aiding the magnetic fields of the series field and armature. The shunt winding can be wired as a long shunt or as a short shunt. The shunt field is connected in parallel with both the series field and the armature which indicates that this is a long shunt. In Figure 10-10a you can see that the shunt field is connected only across the armature, which indicates that it is wired as a short shunt.

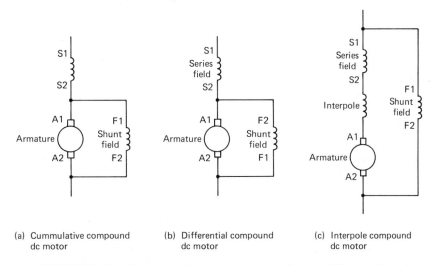

(a) Cummulative compound dc motor

(b) Differential compound dc motor

(c) Interpole compound dc motor

FIGURE 10-10 Electrical diagrams of cummulative, differential, and interpole compound motors.

The cumulative compound motor is one of the most common dc motors because it provides high starting torque and good speed regulation at high speeds. Since the shunt field is wired with similar polarity in parallel with the magnetic field aiding the series field and armature field, it is called cumulative. When the motor is connected this way, it can start even with a large load, and then operate smoothly when the load varies slightly at full rpm. You should recall that the shunt motor can provide smooth operation at full speed but it cannot start with a large load attached, and the series motor can start with a heavy load, but its speed cannot be controlled. The cumulative compound motor takes the best characteristics of both the series motor and shunt motor, which makes it acceptable to most applications.

Differential Compound Motors

Differential compound motors use the same motor and windings and connect them in a slightly different manner to provide slightly different operating speed and torque characteristics. Figure 10-10b shows the diagram for a differential compound motor with the shunt field connected reverse polarity in parallel with armature, which indicates that it is connected as a short shunt.

You should again notice the F1 and F2 are reversed indicating the direction of the current flow through each of the field. In this diagram the shunt field is connected so that its magnetic field opposes the magnetic fields in the armature and series field. When the shunt fields polarity is reversed like this, its field will oppose the other fields and the characteristics of the shunt motor are not as pronounced in this motor. This means that the motor will tend to overspeed when the load is reduced. Its speed will also droop more than the cumulative compound motor when the load increases at full rpm. These two characteristics make the differential motor less desirable than the cumulative motor for most applications.

It is important to remember that the only difference between cumulative and differential motors is the direction the shunt field is connected. The two motors have the same windings; only the direction of the windings is different. Since there is no difference in the physical parts between the two motor configurations, the motor is generally connected as a cumulative motor.

Compound Interpole Motors

The compound interpole motor is build slightly different from the cumulative and differential compound motor. This motor has interpoles added to the series field (Figure 10-10c). The interpole is connected in series between the armature and series winding. It is physically located behind the series coil. It is made of wire that is the same gage as the series winding and it is connected so that its polarity is the same as the series winding pole it is mounted behind. Remember that these motors may have any number of poles to make the field stronger.

The interpole prevents the armature and brushes from arcing due to the buildup magnetic forces. They are so effective that normally all dc compound motors that are larger than ½ hp will utilize them. Since the brushes do not arc, they will last longer and the armature will not need to be cut down as often. The interpoles also allow the armature to draw heavier currents and carry larger shaft loads.

When the interpoles are connected, they must be tested carefully to determine their polarity so that it can be matched with the series winding. If the polarity of the interpoles does not match the series winding it is mounted behind, it will cause the motor to overheat and may damage the series winding.

Compound Motor Operation

Reversing the Rotation of The Motor. Each of the compound motors shown in Figure 10-10 can be reversed by changing the polarity of the armature winding. If the motor has interpoles, the polarity of the interpole must be changed when the armatures polarity is changed. Since the armature winding is always marked as A1 and A2, these terminals should be connected to the reversing motor starter.

Controlling the Speed of the Motor. The speed of a compound motor can be changed very easily by adjusting the amount of voltage applied to it. In fact, it can be generalized that prior to the late 1970s, any industrial application that required a motor to have a constant speed would be handled by an ac motor, and any application that required the load to be driven at variable speeds would automatically be handled by dc motors. This statement was true because the speed of a dc motor was easier to change than an ac motor. Since the advent of solid-state components and microprocessor controls in the late 1970s, this condition is no longer true. In fact, today a solid-state ac motor drive can vary the speed of an ac motor as easily as that of dc motors. This brings about a conditon where you must understand methods of controlling the speed of both ac and dc motors. Information about ac motor speed control is provided in Chapter 11.

You must understand that the motors and controls you find in industry today were installed because they were the most modern for the technology of their day. This means that before industrial-grade solid-state diodes and SCRs were developed in the 1950s, the only way that dc voltage could be varied was through resistors. This meant that the speed of dc compound motors was controlled by adding or removing resistors from the applied voltage circuit. The resistors had to have a current rating equal to the current that the motor would draw, and it would also have to provide a means of getting rid of the excess heat that would build up in the resistors.

In the late 1950s SCRs were being produced that could control larger currents. Since the SCR could be used to rectify ac voltage as well as provide control, it was widely integrated into dc motor speed controls. A variable resistor and capacitor could be used as a phase shift network to control the firing angle on the gate of the SCR. The applied voltage supplied to the motor is controlled by the SCR. A potentiometer is used to control the firing angle of the SCR. It is usually mounted on the cover of the speed control so that it can be adjusted by a machine operator. When the firing angle of the SCR is decreased, the SCR will turn on earlier in its cycle and more voltage would be allowed to flow to the motor. When the firing angle was increased the SCR would turn on later in the cycle and less voltage would be allowed to pass to the motor. This type of circuit could be connected to ac or pulsing dc supply voltage. Some controls used diodes to rectify the ac voltage to pulsing dc prior to providing it for the SCR circuit, while other circuits connected the SCR directly to the AC circuit to provide pulsing dc to the motor.

When voltage to the compound motor is decreased, it will cause the armature current to decrease, which causes its rotation to slow down. Since the motor has a series winding, it will be able to carry the load on the armature shaft even though the speed has been reduced. When the voltage is increased again, the current in the armature's coil causes the armature's rotation to increase.

QUESTIONS

10-1. Why were dc motors used so much in the early 1950s?

10-2. Explain the operation of a simple dc motor. Provide a diagram and identify all the parts.

10-3. Identify the three magnetic principles used to make a dc motor operate.

10-4. Explain the function of the armature in a dc motor.

10-5. Explain the function of the field in a dc motor.

10-6. Explain the function of the brushes and commutator in a dc motor.

10-7. Explain what eddy currents are and how they are controlled.

10-8. Explain the function of the end plates on a dc motor and how are they cared for.

10-9. Explain why brushes must be mated correctly to the commutator. What is this process called?

10-10. Explain how the speed of a dc motor can be increased or decreased.

10-11. Explain how the rotation of a dc motor can be reversed.

10-12. Explain how the torque of a dc motor can be increased.

10-13. Draw an electrical diagram of a dc series motor.

10-14. How is the speed of a series motor controlled?

10-15. What determines the amount of current a series motor will draw?

10-16. What will happen to a dc series motor if the shaft breaks between the motor and the load?

10-17. Explain how you would reverse the rotation of a dc series motor.

10-18. Explain how you would troubleshoot a dc series motor.

10-19. Draw an electrical diagram of a dc shunt motor.

10-20. Explain how you would increase the speed of a shunt motor.

10-21. Explain how you would reverse the rotation of a shunt motor.

10-22. Explain how you would troubleshoot a shunt motor.

10-23. Explain the differences between a shunt winding and a series winding.

10-24. Draw an electrical diagram of a cumulative, differential, and interpole compound motors.

10-25. Explain how you would increase or decrease the speed of a compound motor.

10-26. Explain how solid-state devices are used to control the speed of dc motors.

10-27. Explain the difference between the operating characteristics of the three types of compound motors.

10-28. Explain the difference between a permanent-magnet dc motor and series and shunt dc motors.

10-29. What is the main advantage of using a permanent-magnet motor?

ac motors

11

INTRODUCTION

Ac motors are more widely used than dc motors in industrial applications. They are available to operate on single-phase or three-phase supply voltage systems. This allows the motor control designer to choose the type of motor to fit the application. Most single-phase motors are less than 3 hp; although some larger ones are available, they are not as common. Three-phase motors are available up to several thousand horse power, although most that you will be working with will be less than 50 hp.

The ac motor provides several advantages over dc motors. One advantage the ac motor has is in its design, since it has eliminated the need for brushes and commutators by making the field rotate and keeping the armature stationary. This is possible in an ac motor because of the nature of the ac supply voltage, which moves through a sine wave during each cycle.

The sinusoidal voltage allows the magnetic field to be created in the rotating assembly by induction rather than the actual flow of current. This rotating field is called the *rotor* in ac motors, and the stationary field is called a *stator*.

The design of the rotor is different from the rotating armature in the dc motor. Since the rotor will receive its magnetic field by induction, it will be made completely of laminated steel rather than having copper coils pressed on the laminated steel. Since the rotor receives its magnetic field by induction rather than having current flow through a copper coil, it does not require brushes, brush riggings, and commutators.

This allows the ac motor to operate longer than the dc motor with less periodic maintenance. This means that more ac motors are used in industry, and you must be able to understand their operation so that you can make the proper selection for each application and be able to troubleshoot and repair any motor that you find is faulty.

In this chapter we introduce each of the different types of ac motors and explain their basic parts, theory of operation, methods of controlling their speed and torque, changing the direction of rotation, and procedures for installation and troubleshooting. This basic information will also allow you to better understand methods that motor controls used to take advantage of the motor's design to provide control. This means that if you understand how the motor operates, you will understand what the motor control is trying to control.

In this chapter we introduce the basic parts that are found in all ac motors and explain their operation and function. After the operation of a basic three-phase motor is explained, each type of ac motors is introduced and its special design features and applications for which it is best suited are discussed. You will be able to use this information in recognizing the type of ac motor you are working with; understand the theory of its operation, which will allow you to install and interface it to motor controls; and be able to troubleshoot the motor and quickly determine what faults it has. The three-phase motors is presented in the first part of the chapter since some of the parts of a single-phase motor are designed specifically to compensate for the differences between three-phase and single-phase voltage. If you fully understand the characteristics of three-phase voltage and how three-phase motors take advantage of them, you will easily understand single-phase motors.

CHARACTERISTICS OF THREE-PHASE VOLTAGE

The three-phase ac motor is an induction motor. Its is designed specifically to take advantage of the characteristics of the three-phase voltage that it uses for power. For this reason it is important to review the characteristics of three-phase voltage. Figure 11-1 shows a diagram of three-phase voltage. From this diagram you should notice that each of the three phases represent a separately generated voltage. You should recall that the three-phase generator has three separate windings that produce the three voltages slightly out of phase with each other. The units of measure for this voltage are electrical degrees. An electrical degree represents one sine wave in 360 degrees. The sine wave may be produced once during each rotation of the generator's shaft or twice during each generator's shaft rotation. If the sine wave is produced by one rotation of the generator's shaft, 360 electrical degrees are equal to 360 mechanical degrees. If the sine wave is produced twice during each shaft's rotation, 360 electrical degrees are equal to 180 mechanical degrees. Since this can tend to be confusing, all electrical diagrams are presented in terms of 360 electrical degrees being equal to 360 mechanical degrees. In this way the degrees will be the same and you will not have to try and figure out if electrical or mechanical degrees are being used in the example.

The first voltage shown in the diagram is called *A phase* and is shown starting at 0 degrees and peaking positively at the 90-degree mark. It passes through zero volts again at the 180-degree mark and peaks negatively at 270-degree mark. After it peaks negatively it return to zero volts at the 360-degree mark, which is also the zero-degree point. The second voltage is called *B phase* and starts its zero voltage point 120 degrees later than A phase. B phase peaks positive, passes through zero voltage, and passes through negative peak voltage as A phase does, except that it is alway 120 degrees later than A phase. This means that B phase is increasing in the positive direction when A phase is passing through its zero voltage at the 180-degree mark.

The third voltage shown on this diagram is called *C phase*. It starts at its zero voltage point 240 degrees after A phase starts at its zero voltage

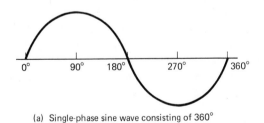

(a) Single-phase sine wave consisting of 360°

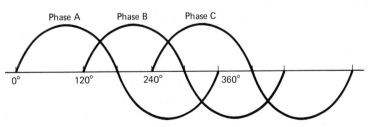

FIGURE 11-1 Example of three-phase and single-phase voltage.

(b) Three-phase sine wave; each sine wave is 120° out of phase with the next

point. This puts B phase 120 degrees out of phase with A phase and C phase 120 degrees out of phase with B phase.

The ac motor takes advantage of this characteristic to provide a rotating magnetic field in its stator and rotor that is very strong because three separate fields rotate 120 degrees out of phase with each other. Since the magnetic fields are induced from the applied voltage, they will always be 120 degrees out of phase with each other. Do not worry about the induced magnetic field being 180 degrees out of phase with the voltage that induced it. At this time this phase difference is not as important as the 120-degree phase difference between the rotating magnetic fields.

Since the magnetic fields are 120 degrees out of phase with each other and are rotating, one will always be increasing its strength when one of the other phases is losing its strength by passing through the zero voltage point on its sine wave. This means that the magnetic field produced by all three phases never fully collapses and its average is much stronger than that of a field produced by single-phase voltage.

THREE-PHASE MOTOR COMPONENTS

The ac induction motor has three basic parts: the stator, which is the stationary part of the motor; the rotor, which is the rotating part of the motor; and the end plates, which house the bearings that allow the rotor to rotate freely.

Stator

The stator is the stationary part of the motor and is made of several parts. Figure 11-2 shows the stator of a typical three-phase motor. The stator, the frame for the motor, houses the stationary winding and provides mounting holes for installation. The mounting holes for the motor are sized according to NEMA standards for the motor's frame type. Some motors will also have a lift ring in the stator to provide a means for handling larger motors. The lifting ring and mounting holes are actually built into the frame or housing part of the stator.

An insert is set inside the stator that provides slots for the stator coils to be inserted into. This insert is made of laminated steel to prevent eddy current and flux losses in the coils.

The stator windings are made by wrapping a predetermined length of wire on preformed brackets in the shape of the coil. These windings are then wrapped with insulation and installed in the stator slots. A typical four-pole three-phase motor will have three coils mounted consecutively in the slots to form a group. The three coils will be wired so that they each receives power from a separate phase of three-phase power supply. Three groups are connected together to form one of the four poles of the motor. This grouping is repeated for each of the other three poles so that the motor has a total of 36 coils to form the complete four-pole stator. It is not essential that you understand how to wind the coils or put them into the stator slots; rather, you should understand that these coils are connected in the stator, and 3, 6, 9, 12, or 15 wires from the coil connections will be brought out of the frame for external connections. These wires can be connected in the field to allow the motor to be powered by 208/230 or 480 V, or they allow the motor to be connected to provide the correct torque response for the load. Other changes can also be made to these connections to allow the motor to start up using less locked-rotor current.

After the coils are placed in the stator, their ends (leads) will be identified by a number that will be used to make connections during the installation procedure. The coils are locked into the stator with wedges. The wedges keep the coils securely mounted in the slots and allow them to be removed and replaced easily if they are damaged or become defective due to overheating.

Rotor

The rotor in an ac motor can be constructed from coils of wire wound on laminated steel or it can be made entirely from laminated steel without any wire coils. A rotor with wire coils is called a *wound rotor motor* and requires the use of brushes and slip rings to transfer current into the rotating coils. Since the wound rotor motor requires brushes, it is not very popular and in used only in limited applications. In fact, a wound rotor motor is rarely installed today. You may locate ones that have been installed in previous years and will need to treat them similar to the armature of a dc motor with regards to operation, troubleshooting, and repairs.

Motors that use a laminated steel rotor are called *induction motors* or *squirrel-cage induction motors*. The core of the rotor is made of die-cast aluminum in the shape of a squirrel cage. Laminated sections are pressed onto this core to make the rotating assembly. The squirrel cage received its name because the shape of the aluminum core looks like the exercise cage for a pet hampster or rat, or the frame of a cage to keep pets like a squirrel. Figure 11-3 shows a diagram of a squirrel-cage rotor. This diagram shows the squirrel-cage core without laminated sections. The fins or blades are

GREASE PLUGS for bearing lubrication standard on all motors 180-Frame and larger. Frames 210T and larger have additional relief plug for lubricant overflow.

ROTOR BARS are high density aluminum for uniformity and high electrical conductivity. End rings continuous with rotor bars. Rotor assembly dynamically balanced. Large 440 Frames have rotors fabricated of copper alloy bars and end rings.

STATOR CORE accurately positioned in the frame with press fit, then pinned to hold it securely.

VENTILATION on Frames 245T and larger is bi-directional. Cool air is drawn in through both ends of motor by fans integrally cast on rotor end rings. On steel frames, ventilation is uni-directional with air drawn in the shaft end bracket by a shaft-mounted fan.

BALL BEARINGS— Double shielded Frames 143T and 145T, single-shielded on Frames 182T through 449T.

PRE-LOAD SPRING restricts shaft end play. Provides quieter operation.

INSULATING MATERIALS are NEMA Class B, 40°C ambient or better. Classes F and H also available for applications where total motor temperatures exceed allowable temperature for Class B insulation, in high ambients for example. Special windings available including Moderate Moisture (MM), Excessive Moisture (EM), Fungus Proofing (FP), and Abrasive Dust (AD).

SHAFT DIMENSIONS are NEMA Standard. Short shafts available on Frames 284T and larger for direct connected applications.

END BRACKETS are cast iron for Frames 254T through 449T. Steel frame motors have die cast aluminum end brackets reinforced with steel bearing inserts in the bearing chambers.

ROLLED STEEL FRAMES with welded feet on 140T through 210T series motors.
CAST IRON FRAMES with integral feet standard on Frames 254T through 449T.

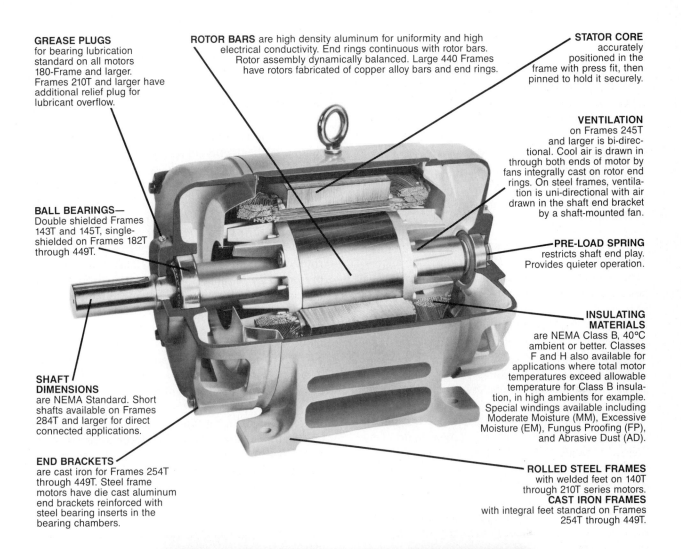

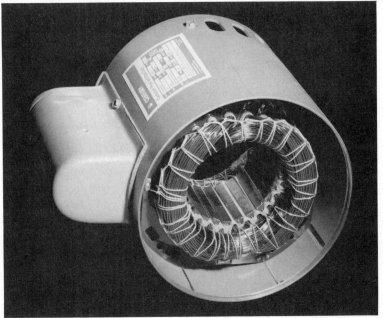

FIGURE 11-2 Parts of a typical three-phase induction motor, including the frame, the laminated stator insert, and the stator coil of single phase motor. (Courtesy of MagneTek Century Electric.)

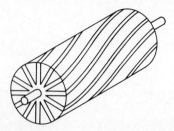

FIGURE 11-3 Sketch of the aluminum core of a squirrel-cage rotor with laminated sections pressed onto the core.

built into the rotor for cooling the motor. It is important that these blades not be damaged or broken since they are balanced so that the rotor will spin evenly without vibrations.

Motor End Plates

The end plates house the bearings for the motor. If the motor is a fractional-horsepower motor, it will generally use sleeve-type bearings and if the motor is one of the larger types, it will use ball bearings. Some ball bearings on smaller motors will be permanently lubricated, while the larger motor bearings will require periodic lubrication. All sleeve bearings will require a few drops of lubricating oil periodically.

The end plates are mounted on the ends of the motor and held in place by long bolts inserted through the stator frame. When nuts are placed on the bolts and tightened, the end plates will be secured in place.

If the motor is an open type, the end plates will have louvers to allow cooling air to circulate through the motor. An access plate may also be provided in the rear end plate to allow field wiring if one is not provided in the stator frame.

If the motor is not permanently lubricated, the oiler tube or grease fitting for lubrication should be mounted so that gravity will allow lubrication to reach the shaft. If you need to remove the end plates for any reason, they should be marked so that they will be replaced in the exact position from which where they were removed. This also helps to align the holes in the end plate with the holes in the stator so that the end plates can be reassembled easily.

AC INDUCTION MOTOR OPERATION

The basic principle of operation of an inductive motor is that the rotor receives its current by induction rather than with brushes and slip rings. Current can be induced into the rotor by being in close proximity to the stator. If a coil of wire is allowed to pass across magnetic flux lines, a current will be generated in the coil. This current will be 180 degrees out of phase with the current that produced it.

When the induced current begins to flow in the laminated segments of the squirrel-cage rotor, a magnetic field will be built in the rotor. This magnetic field will be very similar to the magnetic field produced by normal current flow rather than induced current flow.

Since the current is induced into the rotor, no brushes are required. The strength of the field in the rotor is not as strong as the field that could be produced by passing normal current through a coil of wire. This means that the stator's magnetic field must be much stronger than in the field windings used in a dc shunt motor. The stator in an ac induction motor is usually compared to the armature of a dc motor because of the size of its magnetic field, and the rotor is usually compared to the field winding of a dc motor.

When three-phase voltage is applied to the stator windings a rotating magnetic field is formed. The natural characteristic of three-phase voltage will cause the magnetic field to move from coil to coil in the stator, which appears as though the field is rotating. Figure 11-4 shows an experiment you can perform in your laboratory to prove that the magnetic field in the stator will actually rotate.

From this figure you can see that the end plates and rotor have been removed from the stator.

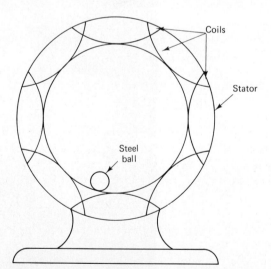

FIGURE 11-4 Experiment using a ball bearing to show the magnetic field in the stator is rotating. When power is applied, the rotating magnetic field will cause the ball bearing to rotate in the direction and at the speed of the magnetic field.

The three stator windings are connected to three-phase voltage through a switch. A large ball bearing approximately 1 inch in diameter is placed in the stator, and the switch is closed to energize the stator winding. When the magnetic field begins to form in the stator, it will begin to pull on the ball bearing. You may need to give the ball bearing a slight push with a plastic rod to get it to begin to rotate. Do not use a metal object such as a screwdriver to get the ball bearing to move because the tip of the screwdriver will be attracted to the stator by its magnetic field.

The ball bearing will continue to rotate at the speed of the rotating magnetic field until the switch is opened and the stator's magnetic field has collapsed. If you reverse two of the three-phase supply voltage wires, the magnetic field will begin to rotate in the opposite direction, and when you give the ball bearing a push to start it, it will also rotate in the opposite direction.

Induced Current in the Rotor. When the magnetic field in the stator cuts across the poles of the squirrel-cage rotor, a current is induced in the rotor. This current is out of phase with the applied current, but it is strong enough to cause the rotor to start to turn. The speed of the rotor is determined by the number of poles in the stator and the frequency of the incoming ac voltage. A formula is provided to determine the operating speed of the motor:

$$\text{operating speed of motor} = \frac{F \times 120}{P}$$

where F is the frequency of the applied voltage, 120 is a magnetic constant, and P is the number of poles. The full rpm is called *synchronous speed*.

From this formula we calculate that a two-pole motor will operate at 3600 rpm, a four-pole motor will operate at 1800 rpm, a six-pole motor will operate at 1200 rpm, and an eight-pole motor will operate at 900 rpm. These speeds do not include any slip or losses due to loads. From this example you can see that the only way an ac induction motor can have its speed changed is to change the number of poles it has or the frequency of the voltage on which it runs.

When power is first applied, the stator field will draw very high current since the rotor is not turning. This current is called *locked-rotor amperage* (LRA) and is sometimes referred to as *inrush current*. When LRA moves through the stator, its magnetic field is strong enough to cause the rotor to begin to rotate. As the rotor starts moving, it will begin to induce current into its laminated coils and build up torque. This causes the rotor to spin faster, until it begins to catch up with the rotating magnetic field.

As the rotor turns faster, it will begin to produce voltage of its own. This voltage is called *back EMF* or *counter EMF*. The counter EMF opposes the applied voltage, which has the affect of lowering the difference of potential across the stator coils. The lower potential causes current to become lower when the motor is a full load. The full-load amperage is referred to as FLA and will be as much as 6 to 10 times smaller than the inrush current (LRA). The stator will draw just enough current to keep the rotor spinning.

When the load on the rotor increases, it will begin to slow down slightly. This causes the counter EMF to drop slightly, which makes the difference in potential greater and allows more current to flow. The extra current provides the necessary torque to move the increased load and the rotor's speed catches up to its rated level. In this way, the squirrel-cage induction motor is allowed automatically to regulate the amount of current it requires to pull a load under varying conditions. The rotor will develop maximum torque when the rotor has reached 70 to 80 percent of synchronous speed.

The motor can make adjustments anywhere along its torque range. If the load becomes too large, the motor shaft will slow to the point of stalling and the motor will overheat from excess current draw. In this case the motor must be wired for increased torque, or a larger-horsepower motor should be used.

CONNECTING MOTORS FOR TORQUE, SPEED, AND HORSEPOWER CONDITIONS

The squirrel-cage induction motor can be connected in several different ways to produce constant torque, speed, or horsepower ratings. They can also be connected for variable torque. The load conditions or applications will dictate the method you select to connect the motor. These connections can be made after the motor is installed on the factory floor or when the motor is being installed in the machinery it is driving.

Wye-Connected Motors

The squirrel-cage induction motor may have 6, 9, or 12 terminal leads from the ends of its coils brought out of the motor frame for field wiring connections. Figure 11-5 shows the diagram of a nine-lead motor.

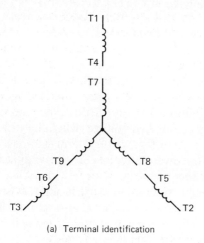

3φ Applied voltage

L1	L2	L3	Together		
T1	T2	T3	T4 T7	T5 T8	T6 T9

(b) Field wiring connections

FIGURE 11-5 Electrical diagram of a nine-lead wye connected motor. These diagrams include a table of terminals connections that are used in connecting the motor in a wye configuration.

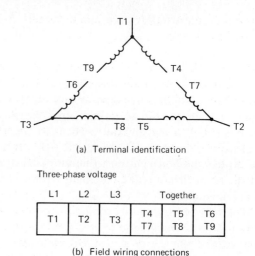

Three-phase voltage

L1	L2	L3	Together		
T1	T2	T3	T4 T7	T5 T8	T6 T9

(b) Field wiring connections

FIGURE 11-6 Diagrams of a nine-lead three-phase squirrel-cage induction motor. Tables are provided to show lead connections for making field wiring connections.

Notice that the coils are positioned in the shape of a the letter Y. This motor is called a *wye-connected motor* or *star-connected motor*.

The wye motor terminals are numbered in the clockwise direction. The two ends of the first coil are numbered 1 and 4, the second coil is numbered 2 and 5, and the third coil is numbered 3 and 6. The outside coils are isolated from each other while the inside coils are connected together at the wye point. The second diagram shows the motor connected to three-phase line voltage at T1, T2, and T3. A table is also provided to show which terminals are connected together and which ones are connected to line voltage for the motor to operate correctly. The information provided in this table is similar to data provided on the data plate of the motor.

You will need to make these connections in the field when you are installing the motor. Wire nuts or lugs should be used to make the connections for the coil terminals and supply voltage wires.

Delta-Connected Motors

Another method of connecting the nine leads of the squirrel-cage induction motor is in a series circuit called a *delta configuration*. Figure 11-6 shows the nine-lead motor connected in a delta configuration. The term *delta* is used because the formation of the coils in this diagram resembles the Greek capital letter delta. These connections are also made when the motor is being installed on the factory floor.

A table is also shown in this figure that provides you with the proper terminal connections as they would be shown on the motor's data plate. The motor is also shown connected to three-phase line voltage at terminals T1, T2, and T3. You should note that the terminals are numbered in a clockwise direction starting at the top of the delta. Each tip of the delta is number 1–3. The other ends of each of the first coils are numbered 4 and 9, and the ends of the second coils are marked 5 and 7, while the ends of the third coil are marked 6 and 8. Terminals 1, 2, and 3 are the midpoints in these coils.

When the motor terminals are wired for operation, terminals 1, 2, and 3 are connected to the power supply at terminals T1, T2, and T3. Terminal 4 is connected to 7, 8 to 5, and 6 to 9 to complete the series circuit. Since these coils are in a series circuit, when three-phase power is supplied, it will come in T1 through the winding and go out terminal T2. Another of the phases will come in T3 and go out T2, and the final phase will come in T2 and go out T3. When the sine wave reverses itself, the currents will reverse and come in from the opposite terminals. This means that at any one instant in time, two of the three wires from the power supply will be used to make a complete circuit. Since the ac three-phase voltage is 120 degrees out of phase and the windings in the motor are also 120 degrees out of phase, the three-phase current will energize each coil in such a way as to cause the magnetic field to rotate.

Reversing the Rotation of a Three-Phase Induction Motor

The rotation of a wye- or delta-connected motor can be changed by exchanging any two of the three phases of the incoming voltage. Figure 11-7 shows diagrams for wye- and delta-connected motors for forward and reverse rotation. From this diagram you can see that T1 and T2 supply voltage terminals have been exchanged in the diagram for motor reversal. The second diagram in each set shows the wye and delta motor connected to a reversing motor starter. Notice that the T1 and T2 terminals are switched by the contacts of the reversing motor starter. The forward starter is identified by the coil marked F and the reversing starter is identified by the coil marked R. The wiring diagrams for the reversing starters show the location of the parts rather than the sequence of operation. These diagrams are very useful for installation connections and troubleshooting.

This wiring configuration is also used when the motor is connected for plugging use. When the motor is used for plugging, it is reversed while running at full rpm. When the motor's stator is quickly reconnected for reverse rotation by switching two of the three input voltage lines, it will quickly build up a reverse magnetic field that will begin to rotate in the opposite direction. The rotor will follow this quick change in rotation and begin to rotate in the opposite direction. This will cause the load to decrease its rpm from full in the clockwise direction to zero, and begin to rotate counterclockwise. The moment the rotor begins to rotate in the opposite direction, the power is deenergized and the rotor shaft is stopped from rapid reverse torque.

Connecting Motors for a Change of Voltage

Delta- and wye-connected motors may have numerous sets of coils so that they can be wired to operate at two separate voltages. The extra windings are connected in series for the motor to operate at the higher voltage, and they are connected in parallel for the lower voltage. The higher voltage in these applications is usually 440 or 480 V, while the low

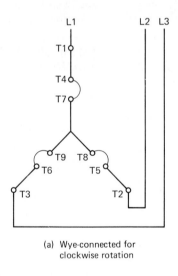

(a) Wye-connected for clockwise rotation

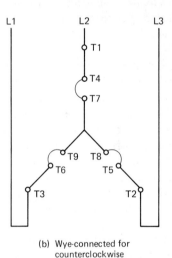

(b) Wye-connected for counterclockwise rotation

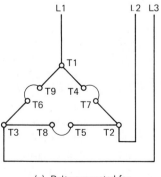

(c) Delta-connected for clockwise rotation

(d) Delta-connected for counterclockwise rotation

FIGURE 11-7 Diagrams for reversing wye- and delta-connected motors. Diagrams of these motors connected to reversing magnetic motor starters are also provided.

voltage is usually 208 or 220 V. The actual voltage will be specified on the motor's nameplate.

The dual-voltage option provides a larger variety of choices when the motor must be connected to power distribution systems. You should remember from Chapter 3 that it is very important to balance the loads on the power distribution system and not overload any of the transformers. When the motors are connected to operate at higher voltages, their current draw is reduced by half. The motor will use the same amount of wattage in both configurations, but smaller-gage wire and smaller contacts and switchgear can be used throughout the circuit when the motor is connected to the higher voltage.

The lower voltage is also useful when no high voltage is available in the area of the factory where the motor must be installed. The dual-voltage motor can be connected for the lower voltage and a savings can be realized by not having to install extra-long power cables to reach the remote source.

Figure 11-8 shows a set of diagrams and tables that indicate the proper connections for the motor to operate at both high and low voltage. The diagrams are presented as wiring diagrams with the numbers on the terminal leads. You should remember that the motor can be connected as either a delta motor or a wye motor, which will affect its starting torque and LRA characteristics, so high- and low-voltage diagrams are presented for each of these types of motors.

Figure 11-8a show a wye-connected motor wired for high voltage and for low voltage. Notice that the six coils are configured as three sets of parallel coils when the motor is connected for low voltage and as a series circuit when they are connected for high voltage. Since all the coils are used in the configurations for high and low voltage, the motor will have the same amount of torque and horsepower in both cases.

The diagrams that show the coils is presented to give you a picture of how the coils look in series and parallel, but this diagram is sometimes confusing when you must make changes in the field. For this reason a second diagram is present that shows only terminals. Each of the terminals are numbered as you would find them on a motor, and the heavy line shows the connections that must be made to complete the connections.

A table is provided in this figure that indicates the terminals that are connected to each other, to the power supply, and that are left open. When a terminal is left open, it means that a wire nut should be placed over the terminal end and wrapped with electrical tape to secure it. Do not cut this wire short, as it may be needed when the motor is reconnected later. Remember, machinery is moved around the factory rather frequently. This means that the motor may need to be connected for a different voltage when it is moved in a few years, or a new service entrance and power distribution system may be installed and several motors may be changed to operate on the new voltage.

Figure 11-8b shows the coils of a delta-connected motor wired for high and low voltage. This diagram shows the coils for the motor connected for low voltage wired in parallel and the coils for a high voltage connection wired in series. A terminal diagram and table are presented for the delta motor. Remember that these diagrams are presented to show you the variety of methods that manufacturers use to indicate the terminal connections for wiring their motor for high or low voltage.

It should also be pointed out at this time that all squirrel-cage motors that have nine leads brought out of their frame can be connected for high or low voltage. The diagrams presented in this figure are usable on all name brands of motors. This is very important, because the motor data plate is usually painted over, damaged, or removed when you need it to make field wiring connections. This means that you will be able to use the diagrams found in this figure to make the connections.

Connecting a Motor for a Change of Speed

Some delta-connected and wye-connected motor can be wired to operate at two different speeds. Unlike the nine-lead motor, which allow all motors to be wired for high or low voltage, not all motors are manufactured to be reconnected for a change of speeds. The motor must be specially manufactured with enough leads brought out of the frame to make the changes required to allow the motor to operate on the different speeds.

Some motors have enough leads brought out to operate at two different speeds, while other motors can be reconnected to operate at up to four speeds. It is very important that you understand that a motor's speed is changed by changing the number of poles that are used. You should remember that when the motor uses eight poles it will operate at 900 rpm, with six poles it will operate at 1200 rpm; with four poles it will operate at 1800 rpm, and the two-pole motor will operate at 3600 rpm. This means that the motor would provide less horsepower when poles are removed from the circuit completely to allow the motor to operate at a higher speed.

Some motors provide a means to reconnect the extra poles back into the circuit to keep the overall horsepower rating of the motor constant.

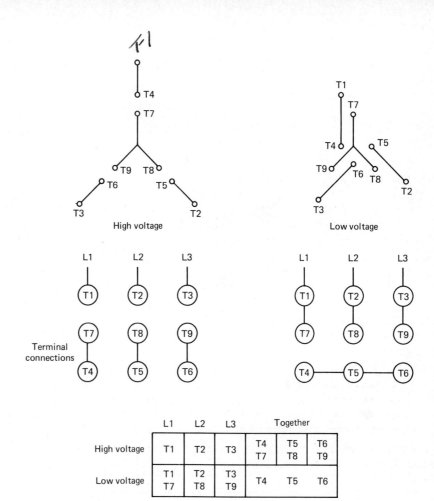
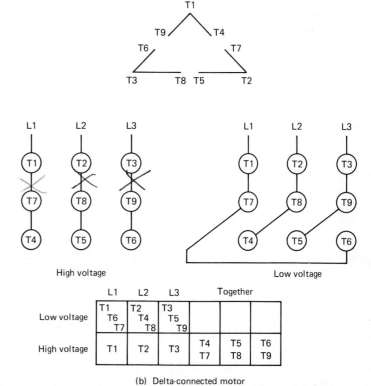

FIGURE 11-8 Diagrams of wye and delta motors connected for low and high voltages. The diagrams are presented showing coil locations, terminal identification, and as tables that indicate which terminals to connect to each other, to power, or left open.

Connecting Motors for Torque, Speed, and Horsepower Conditions   **293**

MULTISPEED MOTOR CONNECTIONS

SINGLE PHASE MOTORS

NOTE: THERE ARE NO GENERAL NEMA STANDARDS DEVELOPED FOR TERMINAL MARKINGS OF SINGLE PHASE MULTISPEED MOTORS BECAUSE OF THE GREAT VARIETY OF METHODS EMPLOYED TO OBTAIN MULTIPLE SPEEDS. THE FOLLOWING CONNECTION SCHEMES SHOW SOME ARRANGEMENTS. NOT ALL POSSIBLE ARRANGEMENTS ARE SHOWN.

❶ 1 PHASE 2 SPEED 2 WINDING
T1 T2 T3 T4

SPEED	L1	L2	OPEN	TOGETHER
LOW	T1	T2	T3, T4	—
HIGH	T3	T4	T1, T2	—

❷ 1 PHASE 2 SPEED 2 WINDING
T1 T2 T3 T4

SPEED	L1	L2	OPEN	TOGETHER
LOW	T3	T4	T1, T2	—
HIGH	T1	T2	T3, T4	—

❸ 1 PHASE 2 SPEED SINGLE WINDING
COM A B

SPEED	L1	L2	OPEN	TOGETHER
LOW	COM.	A	B	—
HIGH	COM.	B	A	—

❹ 1 PHASE 2 SPEED SINGLE WINDING
T1 COM T4

SPEED	L1	L2	OPEN	TOGETHER
LOW	COM.	T1	T4	—
HIGH	COM.	T4	T1	—

❺ 1 PHASE 2 SPEED SINGLE WINDING
T1 COM T4

SPEED	L1	L2	OPEN	TOGETHER
LOW	T1	T4	COM.	—
HIGH	T1	COM.	—	T1, T4

❻ 1 PHASE 2 SPEED SINGLE WINDING
T1 COM T4

SPEED	L1	L2	OPEN	TOGETHER
LOW	T1	COM.	—	T1, T4
HIGH	T1	T4	COM.	—

THREE PHASE MOTORS

NOTE: THE FOLLOWING DIAGRAMS ARE TYPICAL MOTOR CONNECTION ARRANGEMENTS, CONFORMING TO NEMA STANDARDS. NOT ALL POSSIBLE ARRANGEMENTS ARE SHOWN.

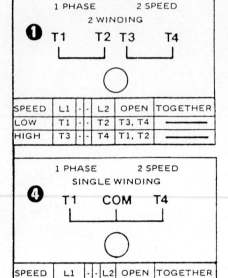

❶ 3 PHASE 2 SPEED 1 WINDING CONSTANT HORSEPOWER

SPEED	L1	L2	L3	OPEN	TOGETHER
LOW	T1	T2	T3	—	T4, T5, T6
HIGH	T6	T4	T5	ALL OTHERS	—

❷ 3 PHASE 2 SPEED 1 WINDING CONSTANT TORQUE

SPEED	L1	L2	L3	OPEN	TOGETHER
LOW	T1	T2	T3	ALL OTHERS	—
HIGH	T6	T4	T5	—	T1, T2, T3

❸ 3 PHASE 2 SPEED 1 WINDING VARIABLE TORQUE

SPEED	L1	L2	L3	OPEN	TOGETHER
LOW	T1	T2	T3	ALL OTHERS	—
HIGH	T6	T4	T5	—	T1, T2, T3

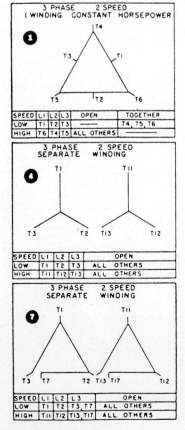

❹ 3 PHASE 2 SPEED SEPARATE WINDING

SPEED	L1	L2	L3	OPEN
LOW	T1	T2	T3	ALL OTHERS
HIGH	T11	T12	T13	ALL OTHERS

❼ 3 PHASE 2 SPEED SEPARATE WINDING

SPEED	L1	L2	L3	OPEN
LOW	T1	T2	T3, T7	ALL OTHERS
HIGH	T11	T12	T13, T17	ALL OTHERS

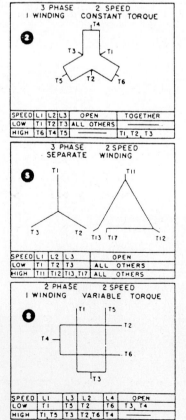

❺ 3 PHASE 2 SPEED SEPARATE WINDING

SPEED	L1	L2	L3	OPEN
LOW	T1	T2	T3	ALL OTHERS
HIGH	T11	T12	T13, T17	ALL OTHERS

❽ 2 PHASE 2 SPEED 1 WINDING VARIABLE TORQUE

SPEED	L1	L3	L2	L4	OPEN
LOW	T1	T5	T2	T6	T3, T4
HIGH	T1, T5	T3	T2, T6	T4	—

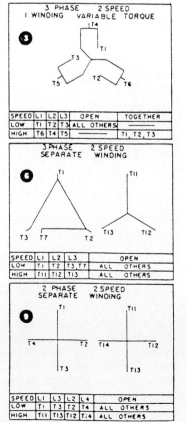

❻ 3 PHASE 2 SPEED SEPARATE WINDING

SPEED	L1	L2	L3	OPEN
LOW	T1	T2	T3, T7	ALL OTHERS
HIGH	T11	T12	T13	ALL OTHERS

❾ 2 PHASE 2 SPEED SEPARATE WINDING

SPEED	L1	L3	L2	L4	OPEN
LOW	T1	T3	T2	T4	ALL OTHERS
HIGH	T11	T13	T12	T14	ALL OTHERS

Three and Four Speed Connections on Next Page

MULTISPEED MOTOR CONNECTIONS

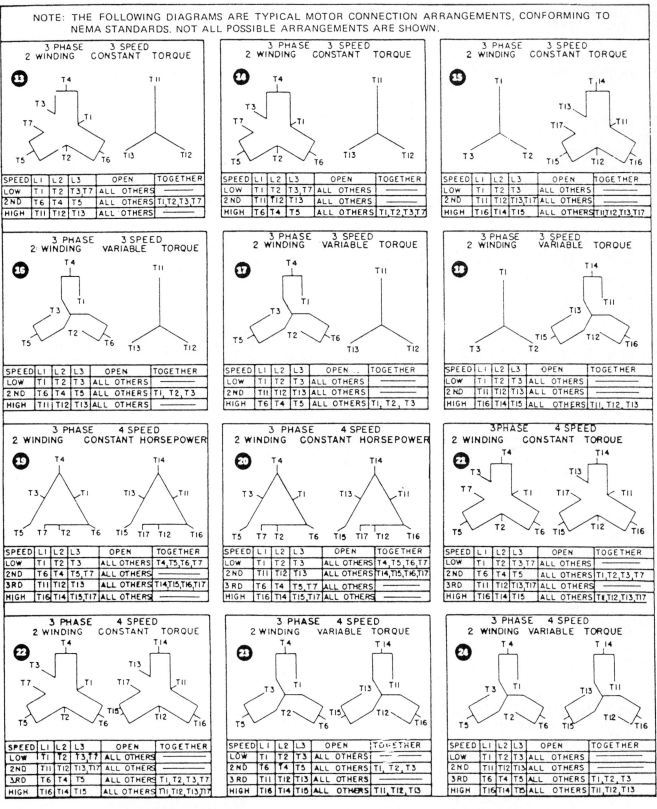

FIGURE 11-9 Diagram for changing motor speeds that provide selection of torque and horsepower requirements. (Courtesy of Square D Company.)

Other changes in the connections can be made to allow the motor to provide constant torque regardless of the speed at which it is operating.

Figure 11-9 shows a series of diagrams that provide methods of connecting these multilead motors for a change of speed. These diagrams allow you to select the torque and horsepower requirements for your application. From this figure you can see that the diagrams are listed in terms of the number of windings the motor has. The first group of diagrams shows diagrams for two-speed motors with one winding. These motors will have six leads brought out of the motor for these field connections. Diagrams are provided to allow the motor to operate at constant horsepower, constant torque, and variable torque.

The second group of diagrams is presented for two-speed two-winding motors. Diagrams are provided to allow the motor to be connected for constant torque, variable torque, and constant horsepower. The third group of diagrams are presented for connecting three-speed motors with two windings. These diagrams allow the motor to operate at constant horsepower.

Interpreting the Wiring Diagrams and Tables. The diagrams and tables in this figure present two ways of representing the connection that must be made to the motor terminals. The terminals in the motor are marked with numbers. Sometimes the numbers are stamped into the wire material, and other times they are metal tags that are crimped on the wire near the terminal end. After you have located all the terminals and have identified them by their numbers, you are ready to make the connections shown in the diagram for your application.

The table lists the terminals that should be connected to the three-input voltage wires. These are identified as L1, L2, and L3, and any terminal number listed in the category under the line number should be connected to that line. Be sure to check the row indicating the speed for which the motor will be wired.

The second column lists the wires that are left open. This means that the wires listed in this column should not be connected to anything. They are supposed to remain unconnected and they should have an insulated wire nut or cap placed over the end of the wire securely so that it does not come in contact with any metal parts of the motor or other energized wires. In some tables any lead that is not listed in another column should be left open.

The third column lists the leads or wires that should be connected together. This means that the wires listed in this column should be connected together and no power should be connected to these leads. The leads must be secured together with a wire nut and wrapped with insulating tape because they will be energized. In some diagrams no terminals will be listed in this column, which means that all the leads are used in one of the other columns. Be sure that you account for every lead before you return all the leads back into the motor and replace the field wiring access cover.

These diagrams are extremely useful when the application you are working with requires the motor to be reconnected on the factor floor. Many times these diagrams are not readily available when you need them, so this provides that much needed reference. These motors can also be reversed by exchanging two of the three supply voltage lines. This allows the motors to be used in the widest possible number of applications.

It is important to understand that you will be required to make these changes yourself or direct someone to make them for you. This means that you must understand the concept of changing the connections of motor leads to make the motor fit the application.

MOTOR DATA PLATES

The motor's data plate list all the pertinent data concerning the motor's operational characteristics. It is sometimes called the *nameplate*. Figure 11-10

NEMA Type Code	Phase	Hz	Amb. Temp.
C	3	60	40°C
Horsepower	Amperes		Volts
7.5	11/22		460/230
Service Factor	Frame		LRA
1.15	56		63/127
Model		Serial	
TC42M		71-469-82	

FIGURE 11-10 Example of a data plate and where it is usually located on a typical motor.

shows an example of a data plate and where it is usually located on a typical motor. The data plate contains information about the motor design, motor type; frame type; model; horsepower rating; RPM, including slip; phase; cycle (frequency); voltage; amperage; code; temperature rise; and rating.

Motor Design

The motor design is listed on the data plate by a letter A, B, C, or D. This designation is determined by the type of wire, insulation, and rotor that is used in the motor and is not affected by the way the motor might be connected in the field.

Type A motors have low rotor circuit resistance and have approximate slip of 5 to 10 percent at full load. These motor have low starting torque with a very high locked-rotor amperage (LRA). This type of motor tends to reach full speed rather rapidly.

Type B motors have low to medium starting torque and usually have slip of less than 5 percent at full load. These motors are generally used in fans, blowers, and centrifugal pump applications.

Type C motors have a very high starting torque per ampere rating. This means that they are capable of starting when the full load is applied for applications such as conveyors, crushers, and reciprocating compressors such as air conditioning and refrigeration compressors. These motors are rated to have slip of less than 5 percent.

Type D motors have a high starting torque with a low LRA rating. This type of motor has a rotor made of brass rather than copper segments. It is rated for slip of 10 percent at full load. Normally, this type of motor will require a larger frame to produce the same amount of horsepower as a type A, B, or C motor. These motors are generally used for applications with a rapid decrease of shaft acceleration, such as loads with a large flywheel, such as punch presses.

These standards are set by NEMA, and a motor must meet all the requirements of the standard to be marked as a type A, B, C, or D. This allows motors made by several manufacturers to be compared on an equal basis according to application.

Motor Type

The motor type catagory on the data plate refers to the type of ventilation the motor uses. These types include the open type, which provide flow-through ventilation from the fan mounted on the end of the rotor. In some motors the fan may be built into the end of the rotor.

Another type of motor is the enclosed type. The enclosed motor is not air cooled. Instead, it is manufactured to allow heat to dissipate quickly to and from the inside of the motor outward to the frame. In most cases the frame has fins built into it on the outside to provide more area for cooling air to reach.

Motor Rating

The motor rating on the data plate refers to the type of duty the motor is rated for. The types of duty include continuous duty, intermittent duty, and heavy duty, which includes jogging and plugging duty. Continuous duty includes applications where the motor is started and allowed to operate for hours at a time. The intermittent duty includes operations where the motor is started and stopped frequently. This type of application allows the motor to heat up because it will draw LRA more often than will a motor rated for continuous duty.

Motors that are rated for jogging and plugging are built to withstand very large amounts of heat that will build up when the motor will draw large LRA during starting and stopping. Since the motor can be reversed when it is running in the forward direction for plugging applications, it will build up excessive amounts of heat. Motors with this ratings must be able to get rid of heat as much as possible to withstand the heavy-duty applications.

Celsius Rise

The *Celsius rise* is the amount of temperature rise the motor can withstand during normal operation. This value is listed in degrees Celsius. A typical open motor can withstand a rise of 40°C (104°F) and an enclosed motor can withstand 50°C (122°F) rise.

Another classification that will help determine the amount of temperature a motor will be able to withstand is the type of insulation the motor winding has. The classes of insulation that are used with motors and the amount of temperature these classes can handle are listed in Figure 11-11.

Class	Temperature (°C)
A	105
B	130
F	155
H	180

FIGURE 11-11

Motor Code

Another rating that is listed on the motor data plate is the motor code. The code refers to the locked-rotor kVA per horsepower the motor can withstand. Figure 11-12 provides the code letters for

NEMA Code Letter	Locked-Rotor kVA per hp
A	0–3.15
B	3.15–3.55
C	3.55–4.00
D	4.00–4.50
E	4.50–5.00
F	5.00–5.60
G	5.60–6.30
H	6.30–7.10
J	7.10–8.00
K	8.00–9.00
L	9.00–10.0
M	10.0–11.2
N	11.2–12.5
P	12.5–14.0
R	14.0–16.0
S	16.0–18.0
T	18.0–20.0
U	20.0–22.4
V	22.4 and up

FIGURE 11-12

NEMA-rated motors. The code letters are in alphabetical order, with A representing the smallest LRA and V representing the highest LRA.

Service Factor

Service factor is a rating of the amount of extra horsepower a motor can safely deliver. The service factor is a type of safety factor that the motor may have built into it so that it can be overloaded and not damage the motor. The service factor can range from 1.75, which allows the motor to deliver 75 percent more horsepower than it is rated for, to 1.0, which means that the motor should not be overloaded at all. The service factor is also linked to the motor horsepower. If the application requires a 1.1-hp motor, you may select a 1-hp motor with a 1.15 service factor and it will be able to handle the load. It should be pointed out that motors with higher service factors are a little more expensive than motors with a service factor of 1.0, but they will often allow you to use a smaller motor for the application.

THREE-PHASE SYNCHRONOUS MOTORS

A *synchronous motor* is an ac motor designed to run at synchronous speed without any slip. As you know, the induction motor must have slip of approximately 9 to 10 percent to operate a maximum torque. The slip is required to allow the rotor to draw enough current to carry its load.

The synchronous motor is designed to operate with no slip by exciting the rotor with dc current once the motor reaches operating rpm. The motor can have dc applied from a dc power source or it can be developed through the use of diodes from a separately generated ac current. This current is produced by a small generator located on the end of the synchronous motor shaft.

Figure 11-13 shows a diagram of a synchronous motor with the dc voltage developed from an outside source and from an internal source. The diagram, which shows diodes being used to rectify the ac from an internal source, is the most common type of synchronous motor used today. This motor also provides an additional function in that it is capable of correcting power factor. The amount of dc current used to excite the rotor also determines the amount of improvement in the power factor.

When it is started, operation of the synchronous motor is similar to that of an iduction motor. This means that the motor is started as an induction motor until the motor reaches full rpm. When the motor is at its highest inductive rpm, a switch is closed to provide exciter current to the rotor from the external or internal source.

When the rotor is energized, it will cause its magnetic poles to lock, in step with the rotating magnetic field of the motor's stator. Since the speed of the magnetic field is determined by the number of poles in the motor and the frequency of the applied voltage, the speed of the rotor will be locked into this magnetic field's speed and they will rotate in unison. The amount of current in the rotor can be adjusted to provide more torque for the rotor. For this reason, it is important that the motor be started with out a heavy load applied. The load can be increased as the motor is synchronized.

AC SINGLE-PHASE MOTORS

Single-phase motors are used quite frequently in large and small industries. They are especially useful in applications where motors of less than 1 horsepower are required and in locations where a motor is required, three-phase voltage is not available. In these applications the single-phase motor can be installed on the single-phase voltage source and produce the required horsepower.

The single-phase motor uses a theory of operation that is similar to that of the three-phase motor. There are several minor changes in the single-phase motor design to achieve the same function as a three-phase motor. These changes include the addition of several components and the modification of several others.

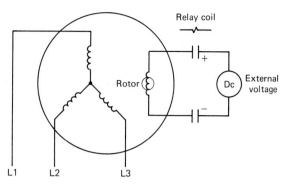

(a) Externally excited synchronous motor

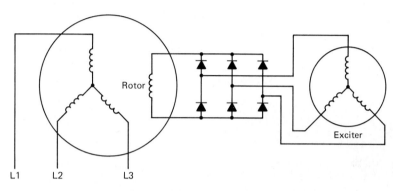

FIGURE 11-13 Electrical diagrams of a synchronous motor showing internally and externally produced exciter voltage.

(b) Internally excited synchronous motor

Single-Phase Motor Components

Like the three-phase motor: The single-phase motor has three basic components: the stator, rotor, and end plates.

Stator. The *stator* is the frame of the motor which houses the windings. The single-phase motor provides two separate stator windings to take advantage of the difference between three-phase voltage and single-phase voltage. These windings are called the start winding and the run winding.

The *start winding* is made of very fine gage wire which has many more turns than the run winding. The run winding is made from wire sized to carry the current for the motor at full-load amperage (FLA). This means the run winding wire will be much larger than the start winding and usually in the range of 12- to 16-gage wire.

The *start winding* is also placed in the stator offset from the run winding to give a phase difference of 90 degrees. This physical phase shift will cause a shift in the magnetic field produced by two windings. Since the start winding is made of very fine wire that has many turns, it can produce a very strong magnetic field for a short period of time. Figure 11-14a shows the locations of the start and run windings in the stator. You should notice that four run and four start windings are shown in the stator. The run windings are located to the outside of the stator and the start windings are shown located toward the inside of the stator, where they will be closer to the rotor. The start and run windings are connected together in parallel in the motor to provide the magnetic phase shift.

Several electrical diagrams are also presented in this figure to show you methods of representing the single-phase motor. In Figure 11-14b the windings are shown placed at right angles to each other. This is done to remind you that the windings are physically offset in the stator to produce more of a magnetic phase shift. You should also notice that the run winding is always represented by the larger coil. The rotor is shown in these diagrams as a circle in the middle of the windings. In Figure 11-14c the windings are shown connected in parallel with each other, which is how you would indicate their electrical relationship. You should also notice that in this diagram the run winding is identified with the letter R and the start winding is identified with the letter S. The point where the two windings are connected at the bottom end of the parallel circuit is called the *common point* and is identified by the letter C. In some motor theory, it is referred to as *terminal C*, even though it is the point where the two windings are connected together.

AC Single-Phase Motors

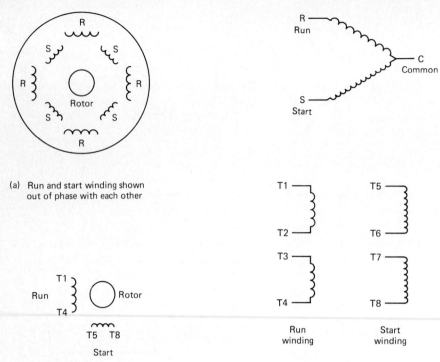

FIGURE 11-14 Diagram showing the location of the start and run windings in a single-phase stator. Typical electrical diagrams are also shown which will be the way the motor is represented on field wiring prints.

Terminal Identification. The terminals on a single-phase motor have a standard identification method. The ends of the start and run windings are numbered to help you identify and locate them when you must install or troubleshoot the motors.

Figure 11-15 shows the standard numbering method for a split-phase motor. In this figure two diagrams are presented to show an eight-lead and a four-lead motor. In Figure 11-15 the two windings are shown connected in parallel as they operate electrically. The run winding is shown in two sections. The numbering starts at the top of the diagram, with the terminals of the first section being numbered 1 and 2. The second section's terminals

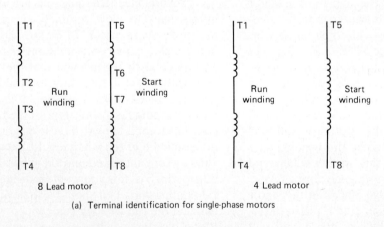

FIGURE 11-15 Numbering of terminals and a color code for terminal identification.

300   Chap. 11 / AC Motors

are numbered 3 and 4. The start winding is also shown in two sections, with the terminals of the top section numbered 5 and 6, while the bottom terminals are marked 7 and 8.

The run and start winding terminals can also be identified by the amount of resistance each has. Since the start winding is made of many turns of very fine wire, its resistance will be much higher than that of the run winding. If the terminal identification is missing, you can use a continuity test to group the terminal leads into coils, and then by measuring their resistance you can compare the readings. The highest readings will belong to the start windings, while the lower readings indicate that the winding is the run winding. Figure 11-15b shows a table with the color code for each of the motor terminal conductors for single-phase motors. The color codes may be used if the terminal identification is not used. You should remember that some manufacturers identify their leads with terminal markers, so the color of wires that are used in their motors have no meaning.

Rotor and End Plates. The rotor in the single-phase motor is very similar to the rotor in the three-phase induction motor in that it uses an aluminum frame to mount the copper rotor bars into. The rotor has the basic shape of a squirrel cage, so it is also called a *squirrel-cage rotor*. It has fan blades cast into the aluminum frame to provide cooling air for the motor.

The ends of the rotor provide the shaft for the load and the bearings. The rear part of the shaft is machined to mount inside the shaft bearing, and the front part of the shaft is extended 3 to 4 inches beyond the front bearing to provide a means of mounting pulleys or gears to drive the load.

The end plates are located on each end of the stator. They are secured in place by four bolts that are inserted completely through the stator. They house the bearings for the rotor shaft to ride on. The bearings can be the sleeve type or ball type. The ball bearing is usually lubricated for life and sealed, while the sleeve bearing must be lubricated frequently with several drops of high-grade electric motor oil. You should recall that the sleeve bearing uses felt wicking to hold the excess lubricating oil in contact to the shaft. This means that the end plates must be mounted with the lubricating port pointing upward so that the oil will be pulled to the wicking by gravity.

Centrifugal Switch for the Start Winding. Since the start winding can only stay in the circuit for a short time because its wire is too small and it will heat up rapidly, a switch is provided to disconnect it from the circuit as soon as the motor is started. This switch is a centrifugal switch that mechanically senses the speed of the shaft and opens when the shaft reaches approximately 90 to 95 percent of full rpm.

The centrifugal switch is mounted in the rear end plate. Since it is mounted on the end of the motor, it is commonly referred to as the *end switch.* The switch has two distinct parts, the switch and actuator. The switch is mounted in the end plate, and the actuator is mounted on the rotor shaft so that it will come in contact with the end switch when the rotor reaches full rpm.

The actuator has a weight built into its outer edges. These weights, called *fly weights,* are hinged on the inside near the rotor and allowed to move or swing at the outer edge. Since the outer edge is heavier, the centrifugal force caused by the shaft rotation will cause them to move away from the shaft. Since the actuator is hinged to the inside, this action will cause the actuator to move along the length of the shaft slightly in the direction of the switch. The movement is only $\frac{1}{4}$ to $\frac{1}{2}$ inch, but it is sufficient to actuate the end switch to the open position.

The end switch is made of spring steel, which provides tension to keep the switch contacts closed. Whenever the centrifugal actuator is not pressing on the switch, the contacts will remain in the closed position. When the actuator moves along the shaft slightly, it will provide enough force to cause the switch contacts to snap open.

When the motor is deenergized, the rotor will deaccelerate to a stop and the centrifugal actuator will return to its original position with the aid of return springs. When the fly weights return to their normal position, the actuator moves back away from the switch and allows it to return to its closed position so that it is ready for the next time the motor is started.

Since the fly weights snap over center to overcome the return spring's tension, you will hear a distinct snap when the motor is reaches approximately 95 percent full speed, which indicates that the end switch has opened, and again after the motor is deenergized and the rotor shaft is coasting to a stop. When you hear the snap as the motor is coasting to a stop, it indicates that the end switch has returned to its closed position.

Thermal Overload Protector. Single-phase motors provide a bimetal switch for use as a built-in overload device. This overload is mounted in the rear endplate near the centrifugal switch assembly and terminal board. The thermal overload consists of a heater and contacts.

When the motor is starting and running, all its current is pulled through the heating element. If the motor draws excessive current, the heating element will become warm enough to cause the bimetal contacts to snap open and deenergize the motor windings. When the bimetal cools down, the bimetal will cool again and snap closed, which will reenergize the motor. If the same fault still exists, the motor will overheat again and continue to cycle on the overload.

Some overloads do not reset automatically. Instead, they have a reset button that must be depressed manually to close the overload contacts. This necessitates someone going to the motor when it trips the overload and resetting it. At that time they should inspect the motor and its loads to ensure that it is operating correctly.

Changing Voltage and Speeds of Single-Phase Motors

The single-phase motor is available for connection on 115 or 230 V. The motor can be reconnected on the factory floor for 115 or 230 V. This allows the motor to be used in any voltage application.

Figure 11-16 shows the split-phase motor connected for high and low voltage. Figure 11-16a shows the two coils of the run winding connected in parallel for low voltage. The start winding is then connected in parallel with these coils. The table shown in this figure shows that leads T1, T3, and T5 are connected to L1 and that T2, T4, and T8 are connected to L2.

Figure 11-17b shows the connections for high voltage. From this diagram you can see that the two coils for the run winding are connected in series. The start winding is connected in parallel across the lower coil (T3–T4) of the run winding. When 230 V is applied to the two coils of the run winding, the voltage divides equally, 115 V across each coil. The start winding will also receive 115 V from the terminals to which it is connected. From these connections you can see that each coil section is rated for 115 V. If the motor is connected for low voltage, all the coil sections are connected in parallel so they will each have 115 V applied to them. When the motor is connected to 230 V, the run winding acts as a voltage divider so that each coil still receives 115 V.

The single-phase motor is also available for dual speeds. This type of motor must have the number poles reduced to increase the speed of the motor. In most cases the motor will lose half of its horsepower rating when it is operated on the faster speed, since fewer poles are used.

Changing Single-Phase Motor Torque

The torque of a single-phase motor can be changed by adding capacitors to the start or run winding of the motor. When the single-phase motor is used without capacitors, as has been shown in the diagrams present so far in this section, the motor is called a *split-phase motor*.

When a start capacitor is connected in series with the start winding and centrifugal switch, the motor is called a *capacitor start, induction run (CSIR) motor*. When the motor has a start capacitor in series with the start winding and a run capacitor is connected permanently across the run and start terminal, the motor is called a *two-capacitor motor,* or a *capacitor start, capacitor run (CSCR) motor*. If the motor has only a run capacitor connected permanently across the start and run winding, it is called a *permanent split capacitor (PSC) motor*. If the single-phase motor rotor is made of copper wire rather than a squirrel-cage rotor, it is called a *wound rotor motor* or *repulsion start motor*.

Each of these motors is introduced here and the theory of operation is explained. Methods and

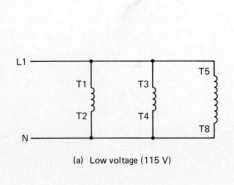

(a) Low voltage (115 V)

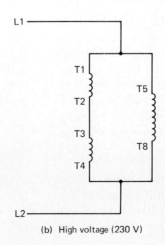

(b) High voltage (230 V)

FIGURE 11-16 Dual-voltage split-phase motors.

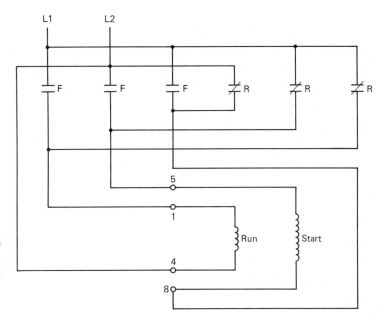

FIGURE 11-17 Diagrams for reversing the direction of rotation of a split-phase motor. A diagram is also presented that shows this motor connected to a reversing starter.

diagrams for reversing these motors are also presented. At the end of this section methods of troubleshooting each of these types of motor are provided. You should gain an understanding of how these motors operate and how their rotation is reversed. It will also help to understand the methods of reconnecting the motor to operate on dual voltage or as dual speeds. This information is important when you must connect motor control devices such as reversing starters or dual-voltage starters to them. It will also help you understand the motor characteristics that are being controlled with various motor control devices.

SPLIT-PHASE MOTORS — ISIR

The split-phase motor is a single-phase motor that does not have any capacitors or other devices in its circuit to alter its torque characteristics. Diagrams of this motor have been presented in Figures 11-15 to 11-18. This motor is also called the single-phase motor or the ISIR motor since it uses only induction to start and run.

This type of motor has the lowest starting toque of all single-phase motors. It uses the physical displacement of the run and start windings in the stator to provide the phase shift required to start the rotor moving. You should remember that the three-phase motor uses the 120-degree phase shift that naturally occurs in the three-phase voltage to cause starting torque. Since the single-phase motor does not have a natural phase shift, the split-phase motor uses the difference of the coil size to create a phase difference along with physically locating the start winding out of phase with the run winding to cause

a magnetic phase shift that is large enough to cause the rotor to start spinning.

When voltage is first applied to the motor's stator, the rotor is not turning and the windings will draw maximum current. This current is called *inrush current* or *locked-rotor amperage* (LRA). After the rotor starts to turn, it will induce current from the stator and produce its own magnetic field. This field will cause the rotor to increase speed until it reaches its rated speed. The rated speed is determined by the number of poles the motor uses and the frequency of the applied voltage. This means that a two-pole motor will operate at 3600 rpm, a four-pole motor will operate at 1800 rpm, a six-pole motor will operate at 1200 rpm, and an eight-pole motor will operate at 900 rpm.

Counter EMF

When the rotor increases its rpm it will begin to generate a back voltage or counter voltage. This voltage is also called *counter EMF* (electromotive force). The difference between the counter EMF and the applied voltage at full speed may be only 1 to 2 V. This voltage is called the *potential difference* between the two voltages and it is responsible for causing the current flow required to keep the rotor spinning at its rated speed. The current required to keep the rotor spinning at its rated speed when it is driving a load is called *full-load current*. It is also referred to as FLA.

If the load is increased, the rotor's rpm is slowed slightly, which will cause the counter EMF to be reduced slightly. Since the counter EMF has decreased and the applied voltage has remained the same, the difference of potential will increase and cause additional current to flow. The increased current will cause the motor to produce more torque, which will cause the rotor to come back up to speed. This feature makes the motor self-regulate its load. This characteristic of single-phase motors helps them operate a fairly constant speed throughout their load range. If the motor is overloaded too heavily the rotor will slow down to the point where it will stall. During this time it will continue to draw excessive current and overheat severely.

Methods of reversing the split-phase motor are presented in Figure 11-17. In this figure you can see that terminals 5 and 8 of the start winding are reversed to get the rotor spinning in the opposite direction. Once the rotor begins spinning in one direction, the magnetic phase shift that is created will cause the motor to continue to rotate in that direction.

If you find that the application requires more starting torque, a start capacitor should be added to the split-phase motor to make it a capacitor start, induction run motor.

Applications for Split-Phase Motors

The split-phase motor is used for general-purpose loads. The loads are generally belt-driven or small direct-drive loads like small drill presses, shop grinders, air-conditioning and heating belt-driven blowers and small belt-driven conveyors.

The main feature of the split-phase motor is that it can be used in areas of the factory where three-phase has not been distributed, or on small loads on the factory floor where a fractional-horsepower motor can handle the load. The motor does not provide a lot of starting torque, so the load must be rather small or belt-driven, where mechanical advantage can be utilized to help the motor start.

The motor is very inexpensive and it can be replaced when it wears out rather than trying to rewind it. It is also available in a variety of frame sizes, which allows it to be mounted easily in most machinery. If the application requires too much torque for the split-phase motor, one of the other motors, such as the capacitor start motor, can be used.

CAPACITOR START, INDUCTION RUN MOTORS

The capacitor start, induction run motor is a split-phase motor with a starting capacitor connected in series with its start winding. This modification to the split-phase motor's design is accomplished when the motor is manufactured, so the capacitor start motor is a separate choice in the manufacturer's catalog when the motor is being specified and selected for the application. Since the CSIR motor can provide more starting torque than the split-phase motor, it would be selected for direct-drive and other applications that require more power during startup.

Electrical Diagram

A start capacitor is connected in series with the start winding of the CSIR motor to provide a larger phase shift when voltage is first applied to the motor. The increased phase shift causes a stronger magnetic field to pull on the rotor to cause it to begin to spin. The start capacitor and electrical diagram for the CSIR motor are provided in Figure 11-18. The start capacitor is shaped like a small cylinder about the size of a small orange juice can, and it is usually rated for over 100 microfarads (μF) (up to 1000 μF) and is mounted in a plastic case.

The start capacitor can provide large amounts of capacitance for a very short period of time, such as during starting. After the motor is started, the capacitor must be removed from the circuit so that it will not overheat. This is accomplished with the end switch. The electrical diagram in this figure shows the capacitor connected in series with the start winding. When voltage is first applied to the motor, both the start winding and the run winding will be energized. When the rotor reaches nearly full rpm, the end switch will open and disconnect the start winding and capacitor from the circuit. This means that no current will flow through either start winding or capacitor and they can cool down and be ready for the next time the motor is started. When the motor is deenergized and the rotor slows to a stop, the end switch is closed again and the start winding and capacitor are reconnected to the circuit for the next start.

When the capacitor is physically mounted on top of the motor, a metal cover is placed over the capacitor to protect it from damage. Since the capacitor's cover is made of plastic, it can easily be cracked and damaged if the motor is used in a harsh industrial environment.

Connecting the CSIR Motor for Dual-Voltage or Dual-Speed Applications

The capacitor start motor is very similar to the split-phase motor since its terminals are identified in the same way and it can be connected to operate on either 115- or 230-V single-phase voltage. Like the split-phase motor, the CSIR motor will generally lose horsepower when it is operated at its higher speed since two poles are deleted from the motor to gain the speed. the diagrams shown in Figures 11-16 and 11-17 can be used to reconnect the motor on the factor floor for a change of voltage or a change of speeds. Some factories prefer to stock dual-voltage, two-speed motors which will cover nearly every application with their equipment. This means that they do not have to stock four separate types of motors for different applications, such as high speed, low voltage; high speed, high voltage; low speed, low voltage; and low speed, high voltage. By stocking the variable-speed variable-voltage motor, they only need to stock different horsepower sizes, such as $\frac{1}{2}$ and 1 hp; the other changes can be made by reconnecting the motor for the correct voltage or speed at the time the motor is installed.

Connecting the CSIR Motor for a Change of Rotation

All of these motors can easily have their rotation reversed by changing the start winding leads. Since the start capacitor is connected in this circuit, a diagram is provided in Figure 11-18b. The end switch is connected near the end where terminal 5 is located, and the capacitor is connected near the end marked terminal 8.

When the motor is operating in the clockwise direction, T5 is connected to T1 and T8 is connected to T4. When the motor is connected for counterclockwise rotation, T8 is connected to T1 and T5 is connected with T4. This reverses the current flow in the start winding with respect to the current flow in the run winding, and the rotor will begin to spin in the opposite direction. Remember: motor rotation is determined by looking at the motor shaft from the end opposite where the load is connected on the shaft.

Applications for CSIR Motors

The CSIR motor can be used for a wider variety of applications where motor starting torque is required than a split-phase motor can supply. These applications include direct-drive water pumps, air compressors, and larger conveyors. The capacitor start motor is also used to drive hermetic compressors for single-phase air-conditioning systems on rooftops and in window units. The hermetic compressor is the sealed motor that is used to drive the refrigeration pump. Since this is a direct-drive application, the CSIR motor is generally used. The split-phase motor may be used on the smaller hermetic motors and systems that use a capillary tube for its refrigerant metering valve.

Since the compressor is hermetically sealed, it is not practical to place the end switch inside the compressor housing with the motor because it is likely to wear out and require replacement. It is also dangerous to allow the end switch to be mounted inside the compressor housing because it may cause a spark that could ignite fumes from the compressor oil and cause an explosion. Since an end switch cannot be use, a current relay is used to disconnect the start winding after the motor starts. A sketch and electrical diagram of the current switch is provided in figure 11-19.

From this figure you can see that the current relay is generally connected directly on the motor terminals. You can see from the electrical diagram that the current relay consists of a coil and a set of contacts. The coil is connected between terminals 1 and M and the normally open contacts are connected between terminals 1 and S. You should also notice that the motor is shown as a run winding and a start winding connected at one end marked common.

The diagram shows the current relay connected to the motor. Notice that the coil of the

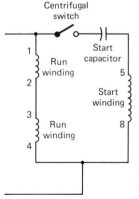

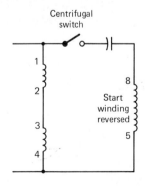

FIGURE 11-18 Diagrams for reversing the rotation of a CSIR motor.

(a) Capacitor start, induction run (CSIR) motor wired for forward direction

(b) CSIR motor wired for reverse rotation

Split-Phase Motors **305**

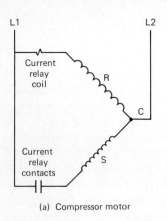
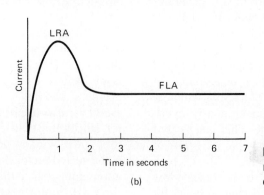

(a) Compressor motor (b)

FIGURE 11-19 Current relay connected to a split-phase sealed compressor motor.

current relay is connected in series with the run winding and will remain in this part of the circuit even when the motor is running. The contacts of the current relay are connected in series with the start winding. Since a start capacitor is used with this motor, the contacts (terminal S) of the relay are connected to the start capacitor and the other terminal of the start capacitor is connected to the start winding of the motor.

Current Relay Operation

From Figure 11-19a you can see that when voltage is applied to the compressor, the run winding will pull locked rotor amperage (LRA) of up to 40 A. Since the coil of the current relay is in series with the run winding, it will also see this large current, which will be strong enough to pull the current relay's contacts closed. When the contacts close, it will provide a path for voltage to reach the start winding. This voltage will also pass through the start capacitor, which causes a phase shift in the magnetic field in the start winding and the torque will be strong enough to start the compressor motor even though it is under 70 to 80 psi of pressure.

The current relay has a return spring connected to its contacts that tries to pull them open. After the motor starts and picks up speed, the rotor will produce enough counter EMF to allow the full-load current (FLA) to drop substantially. When the motor's current drops to the FLA level of 3 to 4 A the spring will pull the contacts open. This drop in current from LRA to FLA occurs in 1 to 2 seconds as the motor reaches full speed. When the contacts open, the start capacitor and start winding are remove from the circuit and their current drops to zero. Since the run winding continues to be energized, it will draw FLA and the motor will continue to run. A graph of the locked rotor current and full-load current is also shown in this figure. You can see that LRA only exists for several seconds while the rotor is coming up to speed. This current is strong enough to pull the current relay's contacts closed for the few seconds the motor requires to start.

Since the current relay is mounted on the outside of the compressor, it can easily be replaced if it becomes faulty. It can also be tested easily by checking voltage from L2 to the S terminal of the motor during the time the motor is starting. If no voltage is present at terminal S during start, the start capacitor should be removed from the circuit and an attempt made to restart the motor. Allow the motor to try and start for only a few seconds during this test because continued LRA current will damage the motor.

If the terminal S on the motor receives voltage when the capacitor is removed, you can assume that the capacitor is open. If the terminal still does not receive voltage, the current relay is faulty and should be replaced.

CAPACITOR START, CAPACITOR RUN MOTORS

The capacitor start, capacitor run (CSCR) motor adds a run capacitor to the start capacitor, which provides the motor with better torque characteristics when the motor is operating at full speed.

The run capacitor is usually oval or square shaped and has a metal housing rather than plastic. The metal housing allows the run capacitor to radiate any heat that is built up inside it since it is connected to remain in the run winding circuit at all times.

From the diagram in Figure 11-20a you can see that the run capacitor is essentially connected to the capacitor start, induction run motor. This type of motor is used almost exclusively for hermetic compressor motors for air-conditioning systems. Since the run capacitor is connected in series with the run winding, it will be in the circuit during starting and remain in the circuit while the motor is running.

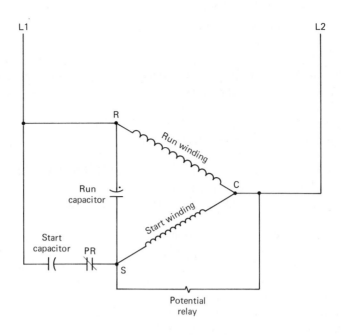

FIGURE 11-20 Capacitor start, capacitor run motor with potential relay connected to disconnect the start winding after the motor has started.

(a) Diagram of potential relay connected to a capacitor start, capacitor run motor

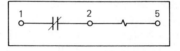

(b) Diagram of potential relay

During the time the motor is starting, the additional capacitor provides a little more phase shift than the start capacitor alone, which gives the motor more starting torque.

After the motor is running, the start capacitor is disconnected from the ciruit with the start winding, but the run capacitor remains in the circuit. If the load increases slightly because the compressor is trying to pump more refrigerant, the run capacitor will provide a small phase shift to give the rotor more torque and regain the loss of rpm quickly. The larger the load increase is, the more speed the rotor will lose, which will cause additional current to be drawn. The increase in current will pass through the run capacitor and cause it to provide a larger phase shift, which in turn provides the rotor with more torque. The run capacitor allows the speed of the motor to remain rather constant when the load is constantly varying.

Using a Potential Relay to Start a CSCR Motor

The capacitor start, capacitor run motor is used primarily in starting large single-phase hermetic compressors. Since the hermetic motor cannot use an end switch, a starting relay similar to the current relay must be used. The current relay requires that its coil be connected in series with the run winding of the compressor motor. In the CSCR motor, the run current can be exceedingly large when the motor is starting (LRA). This current is too large for the current relay coil and would tend to burn it out. For this reason a different type of starting relay, which senses the amount of counter EMF for actuation, is used to start the CSCR motor. Since this relay uses the difference of potential between the applied voltage and the counter EMF produced by the rotor, it is called a *potential relay*.

Figure 11-20b shows a diagram of a potential relay. Notice that its terminals are identified as 1, 2, and 5. It has a set of normally closed set of contacts between terminals 1 and 2 and a high-resistance coil between terminals 2 and 5. The second diagram in this figure shows the potential relay connected to a motor with the a start and run capacitor.

You should notice that the potential relay's contacts are connected in series with the start capacitor and the motor's start winding. The run capacitor is connected between the S and R terminals of the motor, where it will remain even when the motor is running. You should also notice the dot on the curved side of the capacitor symbol. It represents the side of the run capacitor that should be connected to the line side of the power supply. This dot corresponds to a red mark on one of the run capacitor terminals. This terminal is marked because it is the terminal that is closest to the outside of the metal can that the capacitor is mounted in. If the capacitor is shorting out, it will short to this side of the circuit and cause a fuse to blow and the motor would not be damaged. If this capacitor is connected in the circuit so that the identified terminal is connected to the start winding, short-circuit current would be drawn through the motor windings and the motor would be damaged.

The operation of the potential relay is controlled by the amount of counter EMF that the rotor produces. When voltage is applied to the motor through L1 and L2, the run winding is energized directly and the start winding is powered through the start capacitor and normally closed set of contacts of the potential relay. The applied voltage is not strong enough to activate the potential relay's coil.

When the motor starts and the rotor begins to spin, a counter EMF is produced between motor terminals S and C. The counter EMF will become large enough to energize the potential relay coil when the motor reaches approximately 75 percent rpm. When the coil is energized, it pulls its normally closed contact open, which deenergizes the start capacitor and start winding. This affectively removes the start winding from the source of applied voltage.

The run capacitor is connected between the run winding and the start winding and will allow a very small amount of current to flow through the start winding. When the load increases, such as when the compressor must pump more refrigerant, the rotor will begin to slow down and the counter EMF will be reduced slightly. This allows the run capacitor to provide a small phase shift, as it adds more current to the start winding. This increased current with the phase shift provides additional torque and the rotor's speed will again be increased to full rpm. This allows the CSCR motor to operate at a rather constant speed. The CSCR motor is not used in open motor applications, but the run capacitor concept is used frequently in a motor called the permanent split-capacitor motor.

PERMANENT SPLIT-CAPACITOR MOTORS

The permanent split-capacitor (PSC) motor uses only a run capacitor to provide the phase shift required to start the motor. Figure 11-21 shows a diagram of the PSC motor. The run capacitor is connected between the run and start winding and no disconnecting switch or relay is required to de-energize the start winding from the applied voltage when the motor has started.

When voltage is applied to the motor, current will flow through the run winding to the common terminal. At this same time current will flow through the run capacitor to the start winding. When the current flows through the run capacitor, it will provide a phase shift that is large enough to start the motor. As the rotor's speed increases, a counter EMF will be produced in the start winding that will limit current through it to less than 1 A when the motor reaches full speed. The small amount of current in the start winding when the motor is operating at full speed is small enough so that it will not cause the start winding to overheat.

When the motor shaft sees an increase in its load, it will slow down slightly. The decrease in the rotor's rpm causes a decrease in the counter EMF, which makes a larger potential difference between it and the applied voltage. The larger potential difference causes an increase in the current in the start winding, which will cause an increase in rotor torque that increases the rotor's rpm.

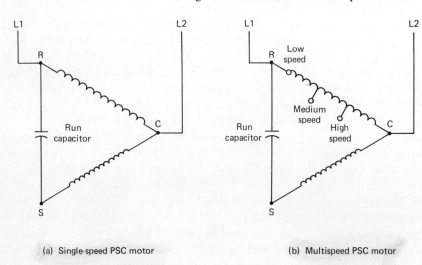

FIGURE 11-21 Diagram of the single-speed and multiple-speed PSC motor.

(a) Single-speed PSC motor

(b) Multispeed PSC motor

This characteristic allows the PSC motor to operate with a constant speed under varying load conditions without using any mechanical devices. The PSC motor is generally used for applications such as small hermetic compressors, blade fan loads, and other loads that require constant speed.

Connecting a PSC Motor for a Change of Speeds

The PSC motor is available as a variable-speed motor. Since the PSC motor does not require a centrifugal switch, it does not have an end plate or access plate. Instead, all the leads that the motor has provided are brought out of the motor together near the end plate. These leads are generally color coded to identify their speed. Figure 11-20b shows a diagram for a multispeed PSC motor. The color leads are connected at various points along the run winding coil. If all the poles of the run winding are used, the motor will operate at its lowest speed. If only two of the poles are used, the motor will operate at its highest speed. You should remember that the motor will lose torque as its speed is increased.

Changing Voltages or Direction of Rotation

Since the PSC is rather specialized and inexpensive, the motor cannot have its direction of rotation changed or be reconnected to operate on two different voltage. Instead, it is common practice to stock a clockwise and counterclockwise motor for each voltage application that you have in the facility. Since these motors are generally used for blade fan applications, they are commonly used for small air-conditioning units used to cool small offices erected on the factory floor or the main offices of the facility.

One problem that may arise with the PSC motor when it is used to drive a condensor fan on the outdoor air-conditioning unit is caused when it is exceedingly windy. The wind will blow across the fan blade when the motor is deenergized and cause the motor rotor to spin in the opposite direction that it would normally operate. When voltage is applied to the motor, the rotor will continue to rotate in the direction of rotation when power is applied. If the rotor is stationary when voltage is applied, the rotor will spin in the proper direction. If the wind is blowing the fan blade in the wrong direction, it will continue to spin in that direction when voltage is applied, which will cause the air conditioner to overheat and cause high pressure in its refrigerant coils.

A ratchet mechanism is mounted on the motor shaft of condensor fans when this is a problem. The ratchet allows the motor to spin in the correct direction and prevents the motor shaft from spinning in the wrong direction when the wind is blowing.

SHADED-POLE MOTORS

Shaded-pole motors are commonly found in applications that require light-duty fans such as small window air-conditioners and exhaust fans used in rest rooms. If you are a maintenance electrician or technician, you may be requested to service all of the electrical equipment in the factory, including the office areas. If this is the case, you will run into shaded-pole motors.

Figure 11-22 shows a diagram of the shaded-pole motor. From this diagram you can see that the motor has only one winding. It does not have a start winding and a run winding like other single-phase motors. Instead, it has a shading pole that provides the magnetic field phase shift that is required to start the motor. A diagram of the shading pole is also shown in this figure. You can see that it is a copper bar that is inserted around the front of the run winding. The bar is connected at the ends to make a complete circuit called a *pole*.

When voltage is applied to the motor to start it, current will flow through the run winding and build up a magnetic field. A current will be induced in the single winding of the shading pole, and it will cause a phase shift to occur that is larger enough to make the rotor start to spin. Once the rotor starts to spin, it will begin to build its own magnetic field and come up to full rpm.

The shading pole also helps the motor when its load changes at full rpm. If the motor shaft begins to

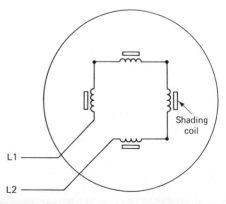

FIGURE 11-22 Four-pole shaded-pole motor. Notice that a shading coil is mounted on each pole.

slow down, the phase shift in the shading coil becomes stronger and provides enough torque to bring the rotor back up to full speed. Another unique feature of the shaded-pole motor is that it can withstand LRA for an extended period. Since the motor does not have a start winding, the run winding is large enough to carry locked rotor current if the rotor becomes stuck. This is important since it provides burnout protection without any additional devices or equipment being added to the motor.

Since the shaded-pole motor has these characteristics, it is commonly used for small fan applications. If the fan becomes immovable for any reason, such as lack of lubrication or dirt, the motor will become warm, but it will not overheat and destroy itself like the split-phase or capacitor start motors.

REPULSION START MOTORS

The repulsion start motor was the most common single-phase motor in use prior to the squirrel-cage motor. After the 1960s very few repulsion start motors were installed because they require brushes and a commutator to operate, which required too much maintenance. The rotor for this type of motor is slightly different from the squirrel-cage motor. It is shown in the diagram as a circle which represents the rotor and a wire running from the top to the bottom of the rotor, which the shorting mechanism used in conjunction with the commutator. Since this motor was designed before squirrel-cage motor theory and technology became prevalent, the rotor was patterned after the wound rotor that is used in dc motors. The rotor was made of laminated sections with coils of wire pressed into place and their terminal ends brought out to commutator segments.

When the motor was being started, current was directed to the rotor coils through the brushes. After the rotor was spinning fast enough, the brushes were disconnected from the applied voltage and shorted so that the rotor would act like an inductive rotor. In some motors, the brushes remained connected to the applied voltage, but they were lifted slightly so that they would not make contact with the commutator. At the same time a shorting mechanism would short the commutator segments to complete the circuit on each coil so that it could conduct the induced current like a squirrel-cage rotor in an induction motor.

In both of these types of motors the rotor would start the motor as a repulsion start motor, and after the rotor came up to speed the motor would operate like an induction motor. This would give the motor the maximum amount of starting toque.

Since the rotor required brushes and some kind of lifting or shorting mechanism, it would require a lot of periodic maintenance. This made the motor too expensive to maintain and they were soon replace with squirrel-cage motors.

Changing Voltages or Direction Rotation

When the motor is running in the clockwise direction, T1 is connected to L1, T8 is connected to L2, and T4 is connected to T5. When the motor is connected to operate in the counterclockwise direction, the start winding is reversed so that T5 is connected to L2 and T8 is connected to T4.

If the motor is capable of being connected for two voltages, its leads will be identified as T1 and T2 for the first coil and T3 and T4 for the second coil. When the motor is connected for the lower nameplate voltage, the two coils are connected in parallel so that terminals T1 and T3 are connected to L1 and T2 and T4 are connected to L2. If the motor is connected for the higher voltage, the two coils are connected in series so that T1 is connected to L1, T4 is connected to L2, and T2 is connected to T3.

TROUBLESHOOTING THREE-PHASE AND SINGLE-PHASE MOTORS

Three-phase and single-phase motors are similar in their operation and should be troubleshooted in a similar manner. Since each of the motors has some parts that are different, there are a few differences in the tests that should be made for each motor. The troubleshooting procedure should be broken into several sections. These sections are based on the types of symptoms the motor presents when it is not operating correctly. The procedures listed in this section are presented in sequence of the most likely to occur to the least likely to occur. You should adopt the same type of philosophy when you begin to troubleshoot the motor. You should always begin by looking for the faults that are most likely to occur, and move on to faults that are least likely to occur. You should also preform simple tests first and complex tests later. This type of procedure will allow you to find the majority of problems or faults with the simple tests.

Motor Will Not Turn When Power Is Applied

The most common problem that you will encounter with a motor is when voltage has been applied and the shaft does not turn. This can occur with the three-phase or single-phase motor. The following procedure will help you determine the location of the fault with a minimum of tests and provide you with information regarding the results of each test. It will also direct you to the next step that you should use. Since this procedure is written to cover every possible problem that you could encounter, you obviously will not need to complete every step; rather, you will make only the tests required, based on the previous tests.

The procedure is broken into two basic parts based on the results of the first test. A set of diagrams of a typical motor starter connected to a split-phase and three-phase motor is provided, as is the control circuit to the motor starter. All tests are referenced to the points identified on the electrical diagram.

Procedure I. The first test of this procedure will direct you to the correct process. Be sure to record the results carefully during each test because you will be directed to the next test based on the results of each test. Remember to be safety conscious!

1. Does the motor hum or draw any current when voltage is applied? (Current can be measured by placing a clamp on ammeter around L1 or L2.)
   a. If the answer to this test is *yes,* go to Procedure II and start with step 1.
   b. If the answer to this test is *no,* continue with this procedure and go to step 2.
2. Is voltage present at terminals L1, L2, and L3 of the motor starter?
   a. If the answer to this test is *yes,* go to step 4.

   Explanation: If voltage is present at these terminals but the motor does not try to start, there is an open in the wires connecting the motor to the starter, or in the motor winding. Remember that two of the three-phase wires or motor circuits must be open since the motor is not drawing any current. If the motor is a single-phase motor, an open in either one of the lines would cause this problem.
   b. If the answer to this test is *no,* go to step 3.

   Explanation: If voltage is not present, the contactor may not be energized, or voltage may be lost at the fuses or disconnects.

3. Test terminal L1, L2, or L3 for voltage at the top of the motor starter contacts.
   a. If voltage is present at these terminals, go to step 4.

   Explanation: If voltage is present at these terminals but not at the bottom contacts, the contacts are faulty or they are not closed. You may need to check the starter's status indicator or test for coil voltage to see if the contacts are closed. The overloads could also have tripped the coil circuit.
   b. If voltage is not present at these terminals, go to step 4.

   Explanation: If voltage is not present at the top terminals of the starter, it indicates that voltage has been interrupted at the source or at a disconnect or fuses between this motor starter and the power source for the factory. (Be sure to look around to see if this is the only machine affected by the loss of voltage, or is the entire section of the factory affected?)
4. Test for voltage at the line-side terminal of the fuse disconnect for this motor.
   a. If voltage is present at these terminals, the problem in your circuit is one of the following. Make the necessary tests, repair the faulty parts, and start the motor.
      (1) One of the single-phase fuses is blown, or two of the three-phase fuses are blown.
      (2) One of the single-phase wires or two of the three-phase wires are open between the disconnect and the motor starter.
   b. If voltage is not present at these terminals, the voltage has been interrupted between this disconnect and the power source for the building. Be sure to check the bus duct and any other disconnects. Remember: you are looking for two lines of the three-phase or one of the single-phase lines to be open.
5. Test the motor leads for voltage or an open.
   a. If the motor has voltage at all of the leads connected to the power source, test the motor for an open. Check the motor for overtemperature. Allow

time for any internal thermal overloads to reset.
   b. If any of the motor leads do not have voltage, you have lost voltage in the wire between the motor starter and the motor terminals.

      Explanation: If the contactor has voltage and the motor leads do not, one or more of the wires connecting the motor to the contactor are open and must be replaced. If applied voltage is present at the motor terminals but the motor does not draw any current, there is an open in the motor at a thermal overload or one or more of the motor coils is open. Use an ohmmeter to test for an open coil. If the motor has an open coil, the motor must be replaced.

6. Test the operation of the contactor and the control circuit for the coil.
   a. If the coil closes the contacts but voltage is not present at the load-side terminals, the contacts are faulty. Remove the contacts and inspect them visually. Change the contacts and try again to operate the motor.
   b. If the coil does not close the contacts, check the control circuit and voltage across the coil. If no voltage is present at the coil, check the control ciruit to locate the open. Use the voltage-drop method presented in Chapter 5 to locate the open in the control circuit. Be sure to check the overloads and the start/stop station.

      If a safety pilot device is preventing the coil from being energized, be sure the safety condition has been rectified before you try to energize the motor. If an overload is open, be sure to check the cause of the overcurrent.

      If voltage is present at the coil terminal but the coil does not close the contacts, the coil is defective and must be changed.

When you have found the problem in this circuit and made the appropriate repairs, be sure to check the motor circuit thoroughly when it is restarted. If the motor does not start after you have made the repairs, you may have a second problem, so start the procedure over again from the first step.

Procedure II. This procedure should be used if the motor will not start but is drawing current or humming loudly. Since the motor is drawing current, you can assume that the motor starter coil is energized and can rule out problems with the control circuit of the motor starter.

1. Determine if the motor is a single-phase or a three-phase motor. (Look at the motor's data plate to determine this.)
   a. If it is a three-phase motor, go to step 2.
   b. If it is a single-phase motor, go to step 5.

      Explanation: If the motor is a three-phase motor and is drawing current, it indicates that only one of the coils or one phase of the supply voltage is causing the fault. If the motor is a single-phase motor, it indicates that the motor is receiving supply voltage and you can assume that the coil has energized the contacts of the motor starter and both fuses are good.

2. Since the motor is a three-phase motor, test all three of the motor leads for voltage.
   a. If all of the motor leads have voltage, go to step 3.
   b. If voltage is missing at one of the leads, go to step 4.

      Explanation: If voltage is present at all three of the motor leads, the problem must be an open in one of the three motor coils. If voltage is not present at one of the lines, a contact, fuse, or wire is faulty between the power supply and the motor.

3. Test all three leads of the motor to see which one is not drawing current. (Use a clamp ammeter for this test.)
   a. Test the lead that is not drawing current for an open circuit. Be sure to account for an open thermal overload that may be mounted inside the motor. If the motor is warm, allow 2 to 5 hours for the motor to cool.

      If the motor coil is open, the motor will need to be replaced. If the motor has an open thermal overload that closes, be sure to inspect the motor and its load to see why the motor is overheating. If the motor is operational but continually overheats, you should suspect a bad bearing or faulty coil and the motor should be replaced. Make sure that the load is not binding or out of alignment, which would cause the overcurrent.

4. Since one of the three leads does not have voltage, the motor is trying to start on single-phase voltage. Make the following checks for voltage in the the phase that does not have voltage to locate the point where the voltage is lost: at the load side of the contacts, at the line side of the contacts, at the fuse disconnect. Continue making checks until you locate the missing voltage, and read the correct response that is relative to the point where you found the lost voltage. (Use a voltmeter and leave one of the probes on a terminal where you have voltage.)
   a. If voltage is found at the load side of the contacts, you have an open in the wire between the load-side contacts and the motor. Change the wire and restart the motor.
   b. If voltage is found at the line-side terminals but it was not found at the load-side terminal, the contact in the motor starter is faulty and should be replaced.
   c. If voltage is found at the load side of the fuse disconnect, the wire between the disconnect and the motor starter is open. Change this wire and restart the motor.
   d. If voltage is found at the line-side terminals of the fused disconnect, the fuse is bad and should be replaced; then restart the motor.

5. Since the motor is a single-phase motor, you must assume that voltage is getting to the motor leads. The hum you hear is current in either the start winding or the run winding. The motor will not start if either of these is bad. This does mean that the control circuit has pulled in the contacts of the starter and the two fuses are good. The problem is going to be with either the start winding or the run winding. Disconnect the motor from power and test the start and run windings. (Use an ohmmeter and test leads 1 to 4 and 5 to 8. Remember: the start capacitor should be shorted during the Ohm's test if the motor is a capacitor start motor.)
   a. If the start winding shows an open, one of the following components in the motor is open. Check the end switch or each coil of the start winding. (Check the capacitor also.) The end switch can be replaced, but if the coil is open, the motor will need to be replaced.
   b. If the run winding is open, be sure to test the thermal overload and the run winding coil. The overload can be replaced, but if the open is in the coil, the motor will need to be replaced.
   c. If both the run winding and start winding are good, the motor shaft or load is binding and not allowing the motor to start. If the motor shaft does not turn, test the bearings, make replacements as required, and put the motor back into service. If the motor spins freely and starts when it is removed from the load, replace the bearing in the load and put the motor back into service.

This concludes the procedures for troubleshooting a motor that will not start. The test should be repeated as many times as required until all problems are located. It is possible that the motor has more than one problem.

Motor Continually Overheats or Does Not Run Properly

If the motor is running but continually overheats, the following lists will provide you with some likely problems to test for. Again, the list is presented from the most likely to occur to the least likely. Also, some differences are noted between single-phase and three-phase motors. You may also need to review the theory of operation for some of the motors if you do not understand why the condition could exist or why the test is required.

Three-Phase Motor

1. Loss of one phase of voltage is due to blown fuse, open wire, or an open motor coil. (The motor probably will not restart if it is turned off.)
2. The load or the motor bearings are in need of lubrication.
3. The motor is experiencing low voltage. (Check the entire facility to see if the

brownout is in your system or caused by the utility; a dual-voltage motor could also be connected for the wrong voltage.)

4. The motor is too small, or the load is too large.
5. The motor has a loose or shorting coil.
6. The field wiring connections are not correct. (This could be a possibility only on new installations or where changes have been made in the terminal connections.)

Single-Phase Motors

1. The start winding is not being disconnected at the proper time by the centrifugal switch.

2. The load or motor bearings need to be lubricated.
3. The motor is experiencing low voltage. (This could include a dual-voltage motor that is connected for the wrong voltage.)
4. The motor is too small or the load is too large.
5. The motor has a loose or shorting coil.

When you find the problem you may be able to make repairs while the motor remains installed, or you may need to remove the motor and replace it with a new one.

QUESTIONS

11-1. Draw a diagram that shows the three phases of a three-phase electrical current. Label the degrees on the diagram to show the amount of separation between each of the phases.

11-2. Draw a diagram that shows the basic parts of an ac motor. Be sure to identify the major parts.

11-3. What is the function of the stator?

11-4. What is the function of the rotor?

11-5. Explain the operation of a simple three-phase ac motor.

11-6. What does the term *induction* mean in regards to the operation of the three-phase ac motor?

11-7. What did the experiment that showed ball bearings rotating inside the stator of a three-phase motor prove?

11-8. What determined the speed of the ball bearing as it rotated in the stator?

11-9. Explain what determines the speed of an ac motor.

11-10. Explain how you would reverse the rotation of a three-phase ac motor.

11-11. Draw an electrical diagram of a delta-connected three-phase ac motor.

11-12. Draw an electrical diagram of a wye-connected three-phase ac motor.

11-13. What advantage does a wye-connected motor provide?

11-14. What advantage does a delta-connected motor provide?

11-15. Sketch a typical data plate from an ac motor and explain what each of the categories of data listed on the data plate mean. (Example: What is SF? What is the frame size? What is temperature rise?)

11-16. Explain the operation of a synchronous motor and how it differs from an induction motor.

11-17. Explain the operation of a single-phase induction motor. Be sure to include the function of all the major parts, such as start winding, run winding, and end switch.

11-18. Draw a sketch that shows the position of the windings in a single-phase ac motor.

11-19. Draw an electrical diagram of a capacitor start, induction run motor and explain its operation.

11-20. Draw an electrical diagram of a capacitor start, capacitor run motor and explain its operation.

11-21. Draw an electrical diagram of a permanent split capacitor motor and explain its operation.

11-22. Explain why a centrifugal switch is required in a split-phase and capacitor start single-phase motors.

11-23. Explain why a single-phase motor needs a start winding and a three-phase motor does not.

11-24. Explain the operation of a thermal overload and where it would be found on a single-phase ac motor.

11-25. Explain what must be done to split-phase or capacitor start motors to allow them to be connected to a 120-V ac power source. Draw the diagram for this connection.

11-26. What must be changed to allow the motor in Question 11-25 to be connected to 230 V ac? Show the diagram for this connection.

11-27. Why must a current relay be used to start a split-phase or capacitor start motor used in a refrigeration compressor?

11-28. Explain why a PSC (permanent split capacitor) motor does not need a centrifugal switch to start.

11-29. Explain how you would troubleshoot a single-phase ac motor.

11-30. Explain why small ac motors are generally not repaired.

11-31. Explain how you would troubleshoot a three-phase motor.

11-32. Explain why a three-phase motor can continue to run if one of the three fuses protecting it fails, but the motor cannot be restarted with only two good fuses.

11-33. Explain the operation of a shaded-pole motor.

11-34. Give an application that uses shaded-pole motors. Why?

11-35. Explain the difference between an induction motor and a repulsion start motor. Why isn't the repulsion start motor used in modern industry?

Motor control circuits   12

INTRODUCTION

In this chapter we explain basic motor control circuits commonly in use on the factory floor. We also explain their theory of operation and the flow of current through each circuit. We also provide laboratory exercises that allow you to wire components commonly found on the factory floor so that you can see the operation of these circuits. Later in the chapter you will be able to have faults entered into the circuits as they would occur on the machinery, which will provide you with realistic circuits to troubleshoot and repair.

The circuits in this chapter are identified so that you will be able to come back and study them if you need to review. This will be extremely useful when you find machinery that incorporates four or five of these basic circuits with additional complex circuits. In these circuits you will be able to gain knowledge that you can transfer to more difficult circuits for troubleshooting and repair. Another important feature of this chapter is that it will be the first time that all the components of the circuit will be used. These include pilot devices, motor starters, as well as different types of motors. You should be able to deal with the operation of any of these devices as they are used in these circuits. If you do not understand their function or operation, you will need to review the chapters where they were introduced.

TWO-WIRE CONTROL CIRCUITS

The two wire control circuit is commonly used in applications where the operation of a system is automatic. This may include such applications as sump pumps, tank pumps, electric heating, and air compressors. In these systems you normally close a disconnect switch or circuit breaker to energize the circuit, and the actual energizing of the motor in the system is controlled by the operation of the pilot device.

The circuit is called a two-wire control because only two wires are needed to energize the motor starter coil. The circuit controls prior to the coil will provide the operational and safety features, while the circuit on after the coil will contain the motor starter overloads.

Figures 12-1 and 12-2 show a typical circuit of the pumping station and a circuit that controls an air compressor. In the pumping station diagram (Figure 12-1) you can see that the circuit is energized by a fused disconnect. After the disconnect is closed, the float switch in the diagram is in complete control of the motor starter. Notice that the float control is identifying a high-level control. When the level of the sump rises past the set point on the control, its contacts will energize and the motor starter coil will become energized. The coil of the motor starter will energize and pull in the contacts, which will energize the motor. As the motor operates, it will pump

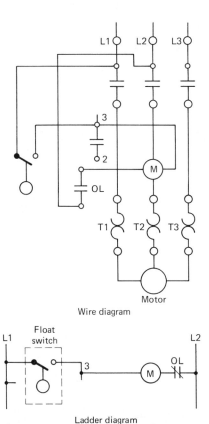

FIGURE 12-1 Elementary (ladder) diagram of a two wire control for a pumping station, the float control will energize the motor starter when the level in the sump raises to the predetermined setpoint. (Courtesy of Allen-Bradley)

down the level of the sump until it is below the low set point of the float control. When the float control reaches this level, it will open and deenergize the motor starter coil, and its contacts will open and deenergize the pump motor.

When the level of the sump rises again, the float control will close again and turn the motor on. This on/off sequence will continue automatically until the disconnect is switched off or unless the motor starter overloads are tripped. If the motor draws too much current when it is pumping, the heaters will trip the overloads and open the circuit. Since the overloads must be reset manually, the circuit will not become energized automatically when the overload is cleared. In this case, someone must physically come to the motor starter and press the reset button. At this time, the system should be thoroughly inspected for the cause of the overload. It would also be important to use a clamp-on ammeter to determine the full load amperage (FLA) of the motor and cross reference this to the size of the heaters in the motor starter to see that they match. The FLA should be checked against the motor's data plate. You should also take into account the service factor listed on the data plate when you are trying to determine if the pump motor is operating correctly.

Figure 12-2 shows a control circuit for an air compressor. This two-wire control circuit is used to turn the air compressor on automatically when the pressure drops below 30 psi and to turn off the compressor when the pressure reaches 90 psi. Pressure switch A in this circuit controls the operation of the control at these pressures. Its high and low pressures are adjustable so that the system could be energized and deenergized at other pressures if the need arises. A hand switch is provided in this circuit that allows the system to be pumped to a predetermined pressure. The hand switch is intended to be used when the auto switch is not functioning properly or if you need to test the system.

Pressure switch B in this diagram acts as a safety for this circuit. This pressure switch is set at 120 psi and is not adjustable. It is in the circuit to prevent the tank pressure from rising too high. This could occur if the operational pressure switch became faulty and would not open when the pressure reached 90 psi. It could also be used to protect the system against overpressure when the switch is in the hand position. Generally, the safety switch is meant to protect the system against component failure or control failure.

If a pressure control would fail, the compressor would continue pumping air into the tank, which would allow its pressure to raise to an unsafe level. Since the operational switch is still closed, the air pressure in the tank could be increased to a level where the pump would cause the motor to stall if the safety switch were not in the circuit. This could build up pressures to several hundred pounds, which could cause lines and fittings in the air system to explode.

The nonadjustable pressure switch would act as a backup to the operational switch and trip off any time the pressure reaches 120 psi. This switch is also different in that it is interlocked when it trips, so that it must be reset manually by having someone pressing the reset button. If you find the reset button activated on the safety switch, you must test the system thoroughly to determine why the operational switch did not control the circuit.

In normal operation, the motor starter would cycle the air compressor on and off to keep the pressure in the tank between the high and low set point that is set on the operational control. The motor starter overloads could also trip this circuit and require manual reset. This could occur if the motor was incurring overcurrent problems. The overcurrent could occur due to bad bearings or the

Two-Wire Control Circuits

Starter Operated by Pressure Switch or Thermostat with Manual Control Provided by a Selector Switch. High Pressure Cut-Out Switch Can be Added

The selector switch feature can be obtained as part of the starter.

The selector switch makes it possible to operate the starter manually for testing or in case of failure of the automatic pilot control. When a standard Bulletin 509 across-the-line starter without push buttons is used, connection "Y" is removed and the wiring follows the solid lines of the diagram.

If a "high pressure" cut-out switch is added, it should be inserted in the line leading from L1 to the "HAND" terminal of the selector switch.

"A" represents the thermostat or "low pressure" switch and "B" represents the "high pressure" cut-out or "safety" switch.

For the "HAND-OFF-AUTO" selector switch, either a Catalog Number 800T-R3SX (standard duty), a Catalog Number 800H-R3HA (heavy duty), or a Catalog Number 800T-R3TA (oiltight) can be used.

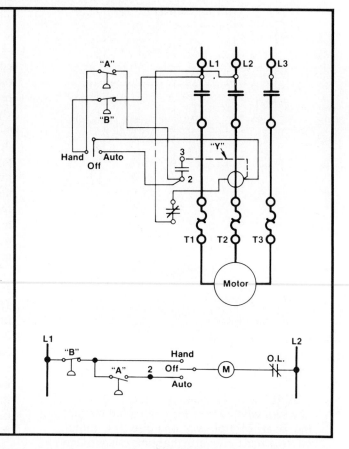

FIGURE 12-2 Elementary (ladder) diagram of a two wire control for a pumping station, the float control will energize the motor starter when the level in the sump raises to the predetermined setpoint. (Courtesy of Allen-Bradley)

pressure being set too high on the systems. It is very important to identify the cause of an overload condition and repair it before the circuit is put back into operation.

THREE-WIRE CONTROL CIRCUITS

Three-wire controls are so named because they have three wires connected to the motor starter coil. The extra wire in this case is the wire that connects the motor starter auxiliary contacts in parallel across the start pushbutton. A diagram of this circuit is shown in Figure 12-3. The circuit is shown as a wiring diagram and a ladder diagram. The operation of the three-wire circuit is easier to understand with the ladder diagram. From this diagram you can see that the start pushbutton is a normally open momentary pushbutton switch. This means that its contacts will stay closed only as long as someone physically depresses the switch. When this pushbutton is released, the normally open contacts return to their normally open state.

The normally open auxiliary contacts from the motor starter will close when the motor starter coil is energized. When they close they will provide an alternate path around the pushbutton contacts for current to get to the coil. These contacts are called *seal* or *seal-in contacts* when they are used in this manner. Sometimes the seal contacts are said to have memory, since they will maintain the last state of the pushbuttons in the circuit.

The three-wire control circuit is the most widely used motor control circuit. You should fully understand this circuit and learn to recognize it when it is shown as a ladder diagram and as a wiring diagram. In this way you will be able to understand the operation of the circuit wherever it appears.

One variation of the three-wire circuit is shown in Figure 12-4. In this figure, multiple start

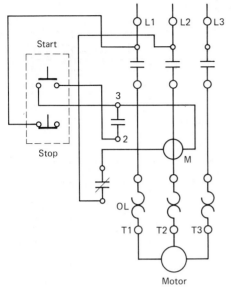

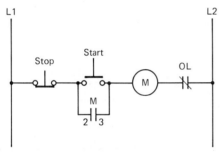

FIGURE 12-3 Three wire control circuit with a start/stop station controlling a motor starter. (Courtesy of Allen-Bradley)

and stop buttons are used. These multiple start and stop buttons allow additional control stations to be installed for the system. This type of application is usable where the operational system is installed over a large area, such as in a long conveyor system. If the conveyor is several hundred feet long, it would be inconvenient and unsafe to have one start and stop button at one end of the system. Pushbutton stations can be installed every 50 ft using the diagram shown in Figure 12-4. Each additional start button should be connected in parallel with the first start button, and all additional stop buttons should be connected in series with the original stop button.

Step-Down Transformer in the Control Circuit

In some systems it is important that the control circuit be powered by 120 or 24 V ac, so that all the pilot switches will be connected to low voltage.

This can be a problem in circuits where large motors are powered by 440, 550, or larger voltages. It would be impractical and unsafe to connect the control circuit directly to the high voltage.

In these applications, a control transformer is used to provide a lower voltage to for the control circuit. The primary winding of the control transformer is specified to operate on line voltage, and provide to the 120 or 24 V at its secondary. Figure 12-5 shows an electrical wiring diagram and a ladder diagram using a control transformer. The primary winding of the control transformers (H1 and H2) are connected to ac line voltage at terminals L1 and L2. Terminals H3 and H4 allow the control transformer to be connected to the high or low voltage specified on its data plate. Refer to Chapter 4 if you need help with this conversion.

Operation of the control circuit provided by the step-down transformer is identical to the circuit shown in Figure 12-4. This allows the circuit to be troubleshooted with the same procedures developed for the normal three-wire control circuit.

Three-Wire Circuit with an Indicating Lamp

An indicator lamp can be added to a three-wire circuit to show when the circuit is energized or deenergized. The indicator lamp can be green or red to show when the system is energized and red to show when the circuit is deenergized. The lamp is usually mounted where personnel can easily see it at a distance.

Some times the indicators are used where one operator must watch four or five large machines. After the machine has been set up, the operator will move on to the next machine. Since the installation is very large and spread out over a distance, the operator can watch for the indicators to see if the machine is still in operation.

The indicator lamps can also be used by maintenance personnel. The indicator lamps are wired into the circuit in such a way as to tell the maintenance person that the motor starter is either energized or deenergized. This will help them to begin testing for faults in the correct part of the circuit when the system has stopped or is not operating correctly. This is especially useful if the motor starter is mounted in a NEMA enclosure where its status is not easily verified.

Figure 12-6 shows a ladder diagram and a wiring diagram of a circuit with an indicator lamp connected in the control circuit. The lamps in this circuit are connected in parallel with the seal in contacts on the motor starter. When the motor starter closes, the lamp will be energized to indicate

Variations with START-STOP Stations

More Than One START-STOP Station Used to Control a Single Starter

This is a useful arrangement when a motor must be started and stopped from any of several widely separated locations.

Notice that it would also be possible to use only one "START-STOP" station and have several "STOP" buttons at different locations to serve as emergency stops.

Standard duty "START-STOP" stations are provided with the connections "A" shown in the adjacent diagram. This connection must be removed from all but one of the "START-STOP" stations used.

Heavy duty and oiltight push button stations can also be used but they do not have the wiring connection "A", so it must be added to one of the stations.

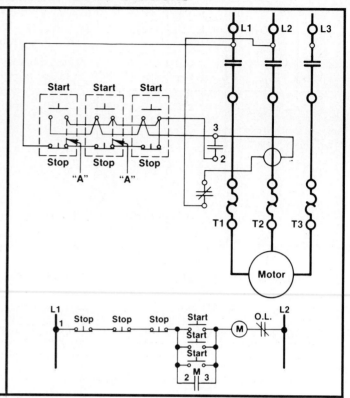

FIGURE 12-4 Multiple start and stop stations connected with a three wire control circuit. (Courtesy of Allen-Bradley)

that the motor starter contacts are closed. Any time the lamp is deenergized, the operator and maintenance personnel know that the motor starter is not energized. A press-to-test lamp could also be used in this circuit to allow the operator and maintenance personnel to press the lamp at any time to see if it is operational. If the indicator is energized most of the time, such as in a continuous operation, the press-to-test lamp may not be necessary since the indicator will be energized most of the time.

Indicator lamps are available for 120, 240, 480, and 600 V, which provides them for any control circuit voltage. A wide variety of colored lenses are also available to indicate other conditions with the machine. These indicators can be connected across different individual motor starters in the machine to provide other information, such as hydraulic pump running, heaters energized, conveyor in operation, and other conditions that are vital to the machine.

Three-Wire Circuit with Indicator Lamp Showing That Circuit Is Deenergized

The circuits shown in Figure 12-7 has a red indicator lamp connected in series with a normally closed set of contacts. Since the lamp is connected in series with the normally closed set of auxiliary contacts, it will be energized any time the coil is not energized. When the coil becomes energized and pulls the contact carrier to the closed position, the normally closed set of auxiliary contacts will be opened and the lamp will become deenergized.

This circuit is useful in applications where it is important to know when the circuit is off, such as in a pumping station. The circuit shown in Figure 12-6 could be used, but a problem could arise if the indicator lamp burned out. In this case it may be useful to use a lamp to show when the circuit is energized and deenergized. If both lamps are out, it indicates that one of the lamps is burned out. Generally, this circuit is used when it is the system spends more time in the deenergized condition than in the energized condition.

The indicator lamp in this circuit can be mounted remotely or on the cover of the enclosure, where it will be close to the motor starter. This allows the lamp to be used in a variety of applications. In fact, a second or third indicator could be connected in the circuit in parallel with the existing lamp, and these could be mounted at a remote terminal or remote panel board to provide other

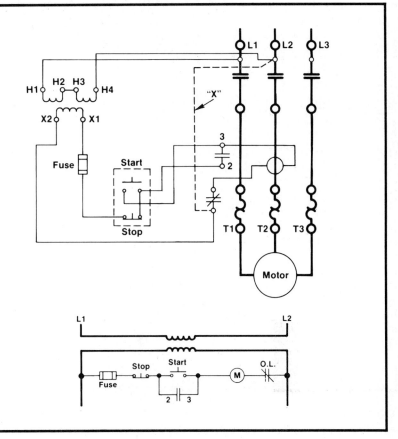

Step-Down Transformer in Control Circuit

Step-Down Transformer Provides Low Voltage for Control Circuit Wired for Three-Wire Control

The starter coil is to be operated on a voltage lower than line voltage. (Usually done for safety reasons.) This requires the use of a stepdown transformer in the pilot circuit. The starter is operated from a "START-STOP" push button station, either Bulletin 800S, 800H or 800T.

When a control circuit stepdown transformer is used with a standard Bulletin 509 starter, the wiring connection "X" must be removed. Note that a fuse is added to the transformer secondary.

FIGURE 12-5 A wiring diagram and a ladder diagram showing a step down control transformer used to provide low voltage for the control circuit. The ladder diagram shows the isolation that the transformer provides to the control circuit. (Courtesy of Allen-Bradley)

personnel with information regarding the status of the circuit.

The circuits shown in these figures show the components as a ladder diagram and as a wiring diagram. The ladder diagram shows the operation of this circuit, while the wiring diagram shows the location of the terminals where these connections should be made. The wiring diagram will also be useful when the circuit is being troubleshooted, as it will show the location of terminal connections that are necessary for making voltage tests.

REVERSING MOTOR STARTERS

From previous chapters you have found that dc and ac motors can be reversed. These chapters provided the terminal connections for each type of dc, ac, single-phase, and ac three-phase motors. Other circuits in Chapter 8 showed the control circuit required to operate the forward and reverse motor starter coils.

The circuit shown in Figure 12-8 will combine the information of the previous chapters to help you understand the operation of the complete circuit. When you must install or troubleshoot this circuit, you must treat it as an entity rather than as separated control and load circuits.

From this diagram you can see that the control circuit has a forward and a reverse pushbutton. The forward and reverse pushbuttons each have a normally open and a normally closed set of contacts. In the ladder diagram each button switch shows a dashed line which indicates that both sets of contacts are activated by the same button. The forward and reverse pushbuttons are better defined in the wiring diagram, which shows that each switch has an open set and a closed set of contacts. The stop button is wired in series with the open contacts of both of these switches, so that the motor can be stopped when it is operating in either the forward or the reverse direction.

The operation of the pushbuttons is best understood through the use of a ladder diagram. The

Reversing Motor Starters 321

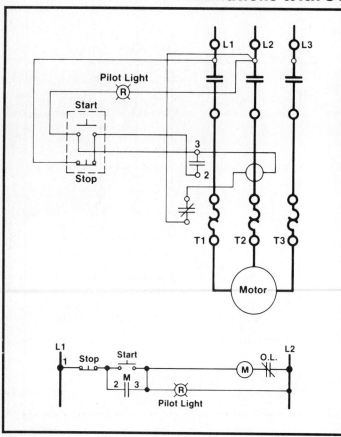

FIGURE 12-6 A three wire control circuit with an indicator lamp connected to the deal in contacts to show when the circuit is energized. (Courtesy of Allen-Bradley)

ladder diagram is used to show the sequence of the control circuit, while the wiring diagram shows the operation of the load circuit, which includes the motor and heaters for the overloads. The overload contacts are connected in the control circuit, where they will deenergize both the forward and reverse circuit if the motor is pulling too much current.

From the wiring diagram you can also see that two separate motor starters are used in the circuit. The forward motor starter is shown on the right side of the reverse motor starter. Each starter has its own coil and auxiliary contacts that are used as interlocks. The location of the auxiliary contacts is shown in the wiring diagram, but their operation is difficult to determine there. When you see the auxiliary contacts in the ladder diagram, their function can be more clearly understood.

This comparison of the ladder diagram and the wiring diagram should help you understand that you need both diagrams to work on the equipment. The ladder diagram will be useful in determining what should be tested, and the wiring diagram is useful in showing where the contacts you want to test are located.

You should also notice that the control circuit is powered from a control transformer that is connected across L1 and L2. The secondary side of the transformer is fused to protect the transformer from a short circuit that may occur in either coil.

Other Methods of Reversing Motors

Several other motor control circuits allow you to reverse the direction in which a motor is turning.

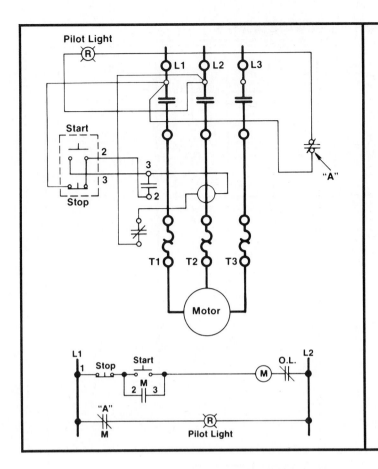

Station with Pilot Light to Indicate When Starter is Deenergized.

If it is necessary for a pilot light to show when the starter is de-energized, this requirement is most easily fulfilled by attaching a normally closed **Bulletin 595 auxiliary contact** to the starter and connecting it between L1 and L2 in series with the pilot light. "A" represents the Bulletin 595 auxiliary contact which can be added to any Allen-Bradley Bulletin 500 Line starter, sizes 0 through 4.

If the pilot light is to be included in the same enclosure with the start and stop buttons, any of the push button stations listed with drawing No. 20 can be used. The Bulletin 595 auxiliary contact has many other uses besides the ones shown here. It can also be used to operate other control circuit devices, interlock starters, etc.

FIGURE 12-7 A lamp used to indicate that a motor starter is deenergized. (Courtesy of Allen-Bradley)

You have studied the operation of the drum switch in Chapter 5 and studied methods of reversing three-phase and single-phase ac motors and dc motors in Chapters 10 and 11. As you know, the switch contacts are open and closed manually by moving the drum switch from the off to the forward or reverse position.

After you understand the operation of the drum switch in its three positions, and methods of reversing each type of motor, these concepts can be combined to develop manual reversing circuits for any motor in the factory as long as its full-load and locked-rotor amperage (FLA and LRA) do not exceed the rating of the drum switch.

These diagrams are especially useful for installation and troubleshooting of these circuits. The drum switch can be tested by itself or as part of the reversing circuit. The motors can also be disconnected from the drum switch and operated in the forward and reverse directions for testing or troubleshooting if you suspect the switch or motor of malfunctioning.

Adding Pilot Devices to the Control Circuit of a Motor Starter

Pilot devices can be added to the control circuit of any motor starter. They are generally added in series with the coil of the starter they are trying to control. When the contacts of the pilot device are opened, the coil is deenergized and the motor will also be deenergized. Since the pilot devices are located in the control circuit and have complete control over the motor, they are sometimes called *control devices*.

Figure 12-9 shows limit switches connected to the reversing motor starter control circuit. From this diagram you can see that a limit switch is placed in series with the forward and reverse motor starter coils. The operation of this circuit is difficult to determine from the wiring diagram for this circuit since it shows the limit switches and pushbuttons where they are located.

The ladder diagram for this circuit shows that limit switches will open and deenergize the motor

Reversing Motor Starters **323**

Push Button Station Variations

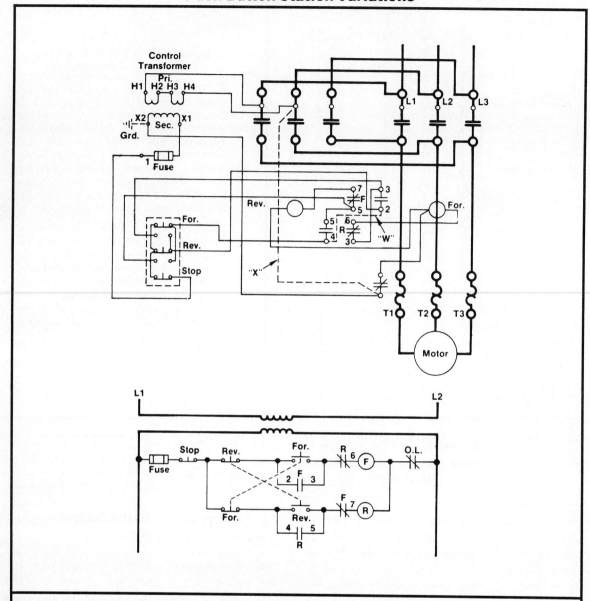

Push Button Wired so Starter Can Be Switched from One Direction to the Other Without Pushing STOP Button

This scheme allows immediate reversal of the motor when it is running in either direction. It is not necessary to depress the "STOP" button when changing direction. A standard Bulletin 505 reversing switch can be used if wire "W" is removed.

The diagram shows the control circuit set up for reduced voltage control, although this may not be necessary in many cases. Notice that wire "X" must be removed when reduced voltage control is used. The push button station can be a Catalog Number 800S-3SA (standard duty), 800H-3HA (heavy duty) or a Catalog Number 800T-3TA (oiltight).

FIGURE 12-8 Wiring diagram and ladder diagram of a reversing motor starter with its control circuit powered from a step down transformer. (Courtesy of Allen-Bradley)

3-Phase Starters

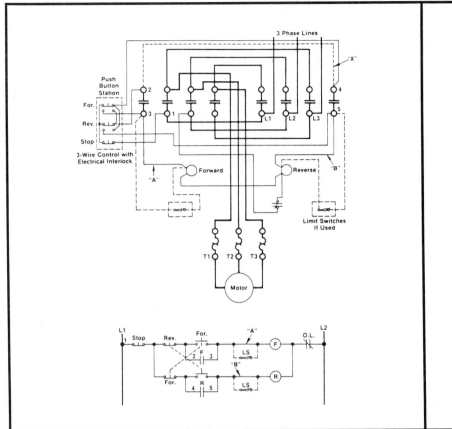

FIGURE 12-9 Limit switches added to the reversing motor starter to provide exact locating capability. (Courtesy of Allen-Bradley)

when load travels too far in either the forward or reverse direction. This circuit is very useful for applications where ball screw mechanisms are used for critical placement of a load, such as a robot arm. When the arm is moving in the forward direction, the motor will continue to move it until it strikes the forward limit switch. When the limit switch is opened, the forward motor starter is deenergized and the motor is turned off, which stops the load at the precise location. As long as the arm stays in this location, it will continue to keep the limit switch open, which disables the forward coil. When the reverse pushbutton is depressed, the reverse coil can be enabled since its limit switch is closed when the arm is against the forward limit switch.

As soon as the reverse coil is energized, the reverse starter will close and the motor will begin to move the ball screw in the reverse direction. This action moves the arm in the reverse direction until it moves against the reverse limit switch and opens it. When the reverse limit switch is opened, the reverse coil circuit is deenergized, which turns off the motor. This action stops the motor in the exact location that is required.

If the arm is slightly out of position when the motor is deenergized and the ball screw stops, the limit switch can be adjusted to stop the motor in the correct position. The position of the limit switch can be used to aid an operator who may not be able to see the exact arm location from the operator's station where the pushbuttons are located.

Other types of pilot devices could also be used in this type of application. A proximity switch or photoelectric switch could also be used to open when the mold is in the proper location. The operation, installation, and adjustment of these types of pilot devices are in Chapter 13.

Using Indicator Lamps to Indicate Direction of Operation of the Reversing Starter

You should see that concepts from several diagrams presented in this section can be combined to provide unique control applications. If you understand

Push Button Station Variations

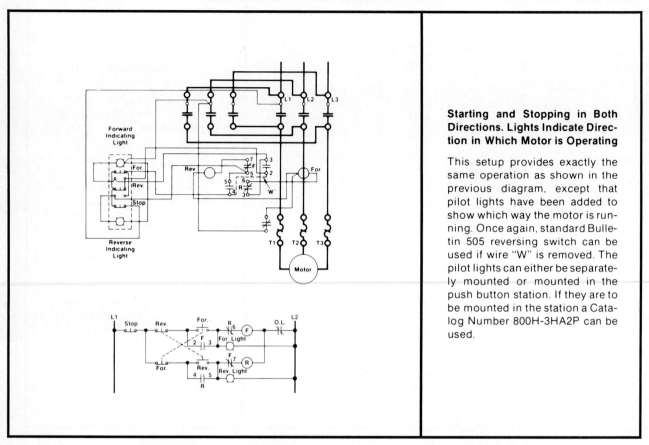

Starting and Stopping in Both Directions. Lights Indicate Direction in Which Motor is Operating

This setup provides exactly the same operation as shown in the previous diagram, except that pilot lights have been added to show which way the motor is running. Once again, standard Bulletin 505 reversing switch can be used if wire "W" is removed. The pilot lights can either be separately mounted or mounted in the push button station. If they are to be mounted in the station a Catalog Number 800H-3HA2P can be used.

FIGURE 12-10 Indicator lamp added to the reversing starter circuit. (Courtesy of Allen-Bradley)

the operation and function of each circuit as it operates on its own, it is easy to see how you could combine several of the controls to design a new circuit. The main point to remember is whether the components should be connected in series or parallel with the coil in the circuit, and if the normally open or normally closed contacts should be utilized.

Figure 12-10 shows indicator lamps added to the reversing starter circuit. Two different-color lamps are used to indicate when the motor is energized for the forward or reverse direction. You can see that the normally open forward contacts will energize the forward lamp, and the reverse normally open contacts will energize the reverse lamp.

This circuit combines the principles that were presented in Figure 12-6 and 12-7 on pilot lamps with the circuit shown in Figure 12-8. You should start to get an idea of how other features can be combined to design the circuit required for your applications. When you must troubleshoot a circuit that combines concepts from several separate circuits, you should consider each part of the circuit's operation individually, and then as they come together to function as one circuit. You should try to isolate each individual section to make troubleshooting easier, but you may have to look at the entire circuit at times.

JOGGING MOTORS

In some applications, such as motion control, machine tooling, and material handling you must be able to turn the motor on for a few seconds to move the load slightly in the forward or reverse direction. This type of motor control is called *jogging*. The jogging circuit utilizes a reversing motor starter to

Jogging

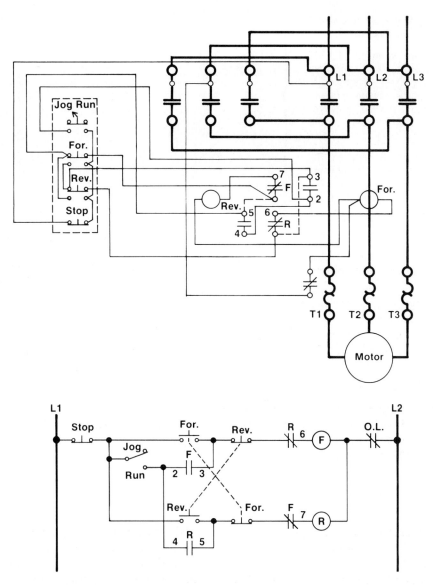

FIGURE 12-11 Electrical wiring diagram and ladder diagram of a jogging circuit added to a forward/reverse motor starter. (Courtesy of Allen-Bradley)

allow the motor to be moved slightly when the forward or reverse push button is depressed.

Another requirement of the jogging circuit is that the motor starters do not seal in when the pushbuttons are depressed to energize the motor when it is in the jog mode, yet operate as a normal motor starter when the motor controls are switched to the run mode. A diagram of a jogging circuit is provided in Figure 12-11. This diagram is shown as an electrical wiring diagram which shows the location of each component, and as a ladder diagram.

The wiring diagram gives you a very good idea of the way the jog/run switch operates. This switch is shown to the left of the motor starter in the diagram. You can see that it is part of the start/stop station. The jog/run button is a selector switch that

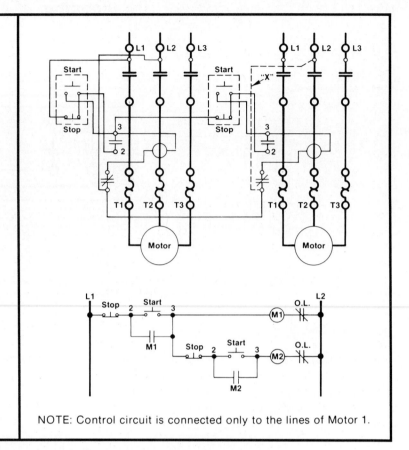

Starters Arranged for Sequence Control of a Conveyor System

The two starters are wired so that M2 cannot be started until M1 is running. This is necessary if M1 is driving a conveyor fed by another conveyor driven by M2. Material from the M2 conveyor would pile up if the M1 conveyor could not move and carry it away.

If a series of conveyors is involved, the control circuits of the additional starters can be interlocked in the same way. That is, M3 would be connected to M2 in the same "step" arrangement that M2 is now connected to M1, and so on.

The M1 stop button or an overload on M1 will stop both conveyors. The M2 stop button or an overload on M2 will stop only M2.

If standard Bul. 509 starters are used, wire "X" must be removed from M2.

NOTE: Control circuit is connected only to the lines of Motor 1.

FIGURE 12-12 Electrical wiring diagram and ladder diagram showing sequence control for 2 conveyors. Conveyor 1 must be operating before conveyor two is allowed to operate. (Courtesy of Allen-Bradley)

is mounted above the forward/reverse/stop buttons.

When the switch is in the jog mode, the selector switch is in the open position. From the ladder diagram, you can see that the jog switch is in series with both of the seal-in circuits, which prevents them from sealing in the forward or reverse pushbuttons when they are depressed. This means that the motor will operate in the forward direction for as long as the forward pushbutton is depressed. As soon as the pushbutton is released, the motor starter will become deenergized. This jog switch also allows the motor to be jogged from one direction directly to the other direction without having to use the stop button.

The motor is protected by the overloads that are connected in series with the forward and reverse motor starter coils. If the overload trips, the overload contacts in the control circuit will open and neither coil can be energized until it is reset.

These two diagrams will allow you to understand the operation of the jog circuit. You can make a forward and reversing motor starter circuit into a jogging circuit by adding the jog switch, but you must be sure that the motor and the motor starters are rated for jogging duty. Remember: some motor starters and motors cannot take the heat that will be built up when the motor is started and stopped continually during the jogging operation. The motor and the motor starters will be rated for jogging or plugging if they can withstand the extra current and heat.

SEQUENCE CONTROLS FOR MOTOR STARTERS

Sequence control allows a motor starter to be utilized as part of a complex motor control circuit that use one set of conditions to determine the operation of another circuit. Figure 12-12 shows an example of this type of circuit. The circuits in this figure are presented in wiring diagram and ladder diagram form. You will really begin to see the importance of the ladder diagram as it shows the sequence of operation, which would be very difficult to determine from the wiring diagram. The wiring diagram is still very important since it shows the field wiring connections and the locations of all terminals that will need to be used during troubleshooting tests.

The operation of the circuit in Figure 12-12 shows two conveyors that are controlled by two separate motor starters. Conveyor 1 must be operating prior to conveyor 2 being started. This is required because conveyor 2 feeds material onto conveyor 1 and material would back up on conveyor 2 if conveyor 1 was not operating and carrying it away. The ladder diagram shows a typical start/stop circuit with an auxiliary contact being used as a seal-in around the start button. When the first start button is depressed, M1 will be energized, which will start the first conveyor in operation. The M1 auxiliary contacts will seal the start button and provide circuit power to the second start/stop circuit.

Since the second circuit has power at all times after M1 is energized, its start and stop buttons can be operated at any time to turn the second conveyor on and off as often as required without bothering the first conveyor motor starter. Remember: this circuit requires conveyor 1 to be operating prior to conveyor 2, since conveyor two feeds material onto conveyor 1.

The circuit also protects the sequence if conveyor 1 is stopped for any reason. When it is stopped, the M1 motor starter becomes deenergized, and the M1 auxiliary contacts return to their open condition, which also deenergized power to the second conveyor's start/stop circuit.

You should also notice that the power for this control circuit comes from the L1 and L2 of the first motor starter. This means that if supply voltage for the first conveyor motor is lost for any reason, such as a blown fuse or opened disconnect, the power to the control circuit is also lost and both motor starters will be deenergized, which will stop both conveyors. If the first motor draws too much current and trips its overloads, it will cause an open in the motor starter's coil circuit, which will cause the auxiliary contacts of the first motor starter to open and deenergize both motor starters.

If you need additional confirmation that the belt on the first conveyor is actually moving, a motion switch can be installed on the conveyor, and its contacts would be connected in series between the first start button and M1 coil. This would cause the first motor starter coil to become deenergized any time the conveyor belt was broken or slipping too much.

Another application of sequence control is shown in Figure 12-13. This circuit shows several motors that require a pump or fan to be in operation any time one of them is running. One example of this application would be an exhaust fan that was required to remove fumes from an area any time manufacturing equipment was in operation. Another example would be multiple air-conditioning compressors that utilized the same water-cooled condenser. In this application, the water pump on the condenser must be operating prior to allowing any of the compressors to run.

This application would also be useful where water is required for a food process. In this system the water under pressure must be provided to each processing machine. The water is pumped into a tank under pressure. The operation of the water pump is controlled by motor starter M1 and the pressure switch. A conveyor is also used in this system and it must be operating any time either of the process machines is operating. The conveyor is operated by motor starter M4, and the process machinery is controlled by motor starters M2 and M3.

The ladder diagram in this figure shows the sequence of this operation. You can see that the master start circuit must be energized to allow any of the motors to operated. A control relay is used to create the master start circuit. When it is energized, its seal in contacts keep it energized as long as the stop button is not depressed. A second set of normally open CR contacts are also used to apply power to the remainder of the control circuit.

After the control relay is energized, the water pump will be energized any time the pressure in the tank drops below the set point. Motor starters M2 and M3 control two separate process machines and they can be started or stopped by their individual start/stop controls. When either M2 or M3 is energized, their auxiliary contacts will energize motor starter M4 and cause the conveyor to operate. If the master stop button is depressed, the control relay will deenergize and all the other circuits will become disabled.

Sequence Control with Time Delay. Another variation of sequence control uses a time-delay relay to prevent two motor starters from energizing their motors at the same time. If the two motors are powered from the same distribution system, they may draw too much LRA if they both try to start at the same time. This application uses one start/stop station to start both motors, and a small time delay is used to allow the first motor to start and come up to speed before the second motor is started.

Sequence Control

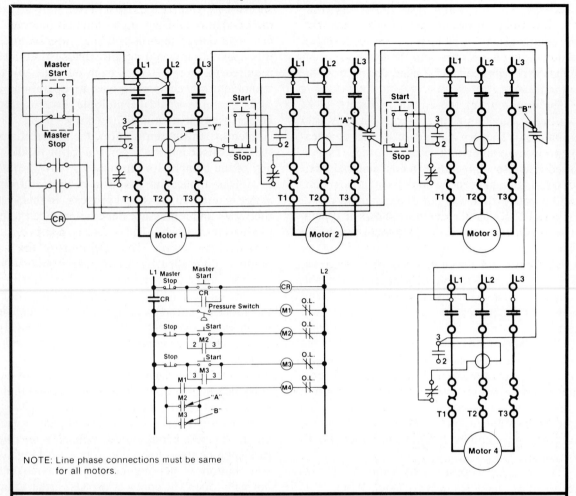

NOTE: Line phase connections must be same for all motors.

Operation of Any One of Several Starters Causes a Pump or Fan Motor to Start

Several motors are to be run independent of each other, with some of the starters actuated by two-wire and some by three-wire pilot devices. Whenever any one of these motors is running, a pump or fan motor must also run.

A master start and stop push button station with a control relay is used to shut down the entire system in an emergency. Control relay (CR) provides "three-wire control" for M1 which is controlled by a two wire control device such as a pressure switch. Motors M2 and M3 are controlled by START-STOP push button stations.

Auxiliary contacts on M1, M2, and M3 control M4. These auxiliary contacts are all wired in parallel so that any one of them may start M4. Bulletin 595 auxiliary contacts have been added to M2 and M3 for this purpose. The standard "hold-in" contact on M1 may be used as an auxiliary if wire "Y" is removed. "Hold-in" contacts are not required when a two-wire control device is used.

The Bulletin 595 auxiliary contacts are designed as "A" and "B" on the wiring diagram. These contacts are easily added to any Allen-Bradley Bulletin 500 starter sizes 0 through 4.

When this system is used, the phase connections on all of the starters must be the same. That is, L1 of each starter must be connected to the same incoming phase line, L2 and L3 of each starter must be similarly phased out.

FIGURE 12-13 Electrical wiring diagram and ladder diagram of a motor starter and control relays used to control the sequence of operation for a food processing system. (Courtesy of Allen-Bradley)

Figure 12-14 shows an electrical wiring diagram and a ladder diagram of this type of sequence circuit. From the ladder diagram you can see that both motor starters are controlled by one set of start/stop switches. A time motor (TR) is connected in parallel with the coil of the first motor starter. The coil of the second motor starter is connected in series with the contacts of the timer.

When the start button is depressed, the first motor starter coil (M1) is energized, which starts the first motor and closes the auxiliary contacts that act as a hold-in for the start button. Since the timer motor is connected in parallel the coil M1, it will become energized and begin to run its cycle. The time delay of this cycle will be adjusted so that it provides enough time for the first motor to reach full RPM.

After the time delay has expired, the timer contacts will close and energize the second motor starter. This allows the second motor to draw locked-rotor current after the first motor is fully up to speed. This causes less current demand on the power distribution system than if both motors were to draw locked-rotor current at the same time.

CONTROLLING CIRCUITS WITH A PROGRAMMABLE CONTROLLER

The motor control circuits shown in this chapter can also be controlled by a programmable controller rather than being hard-wired. Since the late 1960s programmable controllers have been used to provide control function that had previously been provided by control relays. This means that if you intend to install, troubleshoot, or repair motor

Variations With START-STOP Stations

Two Starters are to Be Operated from a Single Station, But a Short Delay Must Prevent Them from Being Energized Together

In some cases, the power distribution system does not have sufficient capacity to start several motors simultaneously. If several motors are to be started from the same push button station under these conditions a time delay can be provided between the operation of the motor starters.

When the START button is pushed, the first starter is energized along with a timing relay. When the timing relay times out, it operates a contact which closes the control circuit of the second starter. (The timer can be a Bulletin 849.)

If more than two starters are to be used, additional timers will have to be added in the same way as the one shown connected to M1 here.

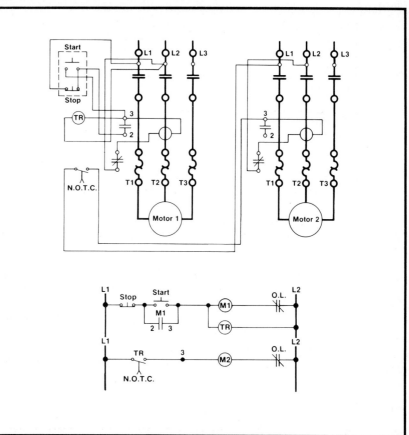

FIGURE 12-14 Two motors started from same control switch using time delay relay to prevent large locked rotor current. (Courtesy of Allen-Bradley)

controls in industry today, you must be able to recognize the basic control circuits as they appear in programmable controllers. In the remainder of this chapter we present the basic motor control circuits that appeared in Figures 12-1 through 12-14 as you would see them controlled by a programmable controller. A diagram of the input and output controls connected to their respective modules is presented with the control diagram. The control diagram and the I/O modules use addresses that represent Allen-Bradley and Texas Instruments address formats. If you are using another brand of controller, you may need to convert the addresses to the format of your control to enter these programs into your system and test them.

It is important that you understand that these circuits will provide the exact operation from the programmable controller that would occur if the circuit were hard-wired. Since the programmable controller provides this sequential function by means of a microprocessor, its control is called *logic*. If this control is programmed in the form of a ladder diagram, it may also be called *relay ladder logic* (RLL).

This section of the chapter is intended to show that a programmable controller can provide the same functions as those of a hard-wired control circuit. Some extra information will be provided that shows ways that the programmable controller circuit can be altered without additional hardware changes, which means that these changes could be added as conditions changed after the system was been installed.

P/C-Controlled Two-Wire Circuit

The operation of a two-wire circuit can be duplicated by a programmable controller very easily. A diagram of the control switches and motor starters connected to the input/output (I/O) modules is provided with the ladder logic program in Figure 12-15. This application is the same one as provided in Figure 12-1, which uses a float switch to control the level of water in a sump.

The float switch in the programmable controller circuit is connected directly to the input module at address 111/01. Notice that this switch is powered by 110 V ac. The motor starter coil M1 is connected to the output module at address 010/01, which is also powered by 110 V ac. The overloads for this motor starter are connected in series with the coil at the output module. The addresses used in this example assume that an input module and output module is installed at the addresses specified. If this circuit does not function properly, be sure to check the addresses.

The ladder logic diagram is shown below the hardware diagram. It will appear on the screen of the programming panel as you see it in this diagram. When it is entered into the P/C it may also be called a program or "the program." Notice that the float switch is shown in the program as a set of normally open contacts instead of the symbol of the switch. The contacts will be identified as 111/01, which is the address where the switch is connected to the input module. It may also be identified by the name float switch in some P/Cs. The addresses for this diagram use the Allen-Bradley format.

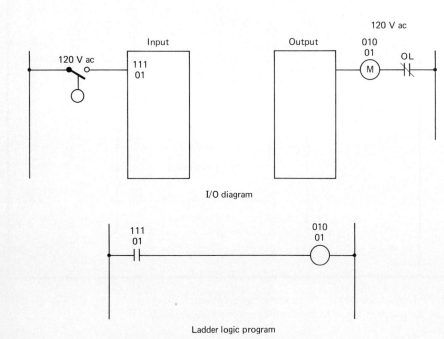

FIGURE 12-15 Two wire control provided by a programmable controller. This circuit shows a hardware diagram of the float switch connected to the input module and a motor starter connected to an output module, and a logic diagram as it appears in the programmable controller program.

The motor starter coil is identified in the program as an output coil. The address 010/01 is placed on this output in the program to indicate that this coil is connected to the output module at address 010/01. The motor that is connected to the motor starter is not shown in these ladder logic diagrams. It is assumed that you will locate the wiring diagram of the load circuit to determine the motor connections.

The operation of this circuit will be identical to the operation of the same components if they were connected together in a hard-wired circuit. When the P/C is set to the run mode, the program will execute. This means that any time the float switch is closed by the level of the water in the sump, the output in the program will be energized. When the programmed output is energized, the output module will energize the motor starter coil connected to address 010/01. When the motor starter coil is energized, the motor will begin to operate and pump the sump level down.

When the level of the water in the sump is lowered, the float switch will open, which will deenergize the input at address 111/01. When this occurs, the program at output 010/01 will also become deenergized and the motor starter will open, which turns off the pump motor. The operation of the pump that is controlled by the P/C will appear to operate exactly like the circuit where the float switch is connected directly to the motor starter coil.

P/C-Controlled Two-Wire Circuit with Two Input Switches

Figure 12-2 showed a two-wire control circuit that used two pressure switches to provide operational control and safety control for an air pressure system. The P/C hardware diagram and logic program that is equivalent to the circuit shown in Figure 12-2 is presented in Figure 12-16.

In the hardware diagram you should notice that pressure switch A, pressure switch B, and the hand/auto switch are all connected to the input module. These switches are connected to four consecutive input addresses from 111/01 through 111/04. The motor starter coil is connected to output address 010/01. The Allen-Bradley format is used for this diagram also.

Each switch in the ladder logic program is identified by a set of contacts. Notice that the input module address where each switch is connected is used in the program to identify the contacts. The operation of this circuit is identical to the operation of the hard-wired circuit shown in Figure 12-2. Pressure switch A is a pressure safety switch. When the hand/auto switch is in the hand position, the output at 010/01 will be energized as long as the pressure in the tank is below the maximum pressure set by the safety switch. When the hand switch is returned to the off position, the output will become deenergized and the motor starter will turn the compressor motor off.

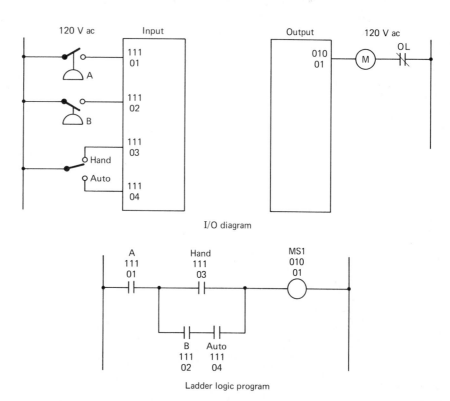

FIGURE 12-16 A programmable controller controlled two wire circuit that utilizes two pressure switches for operation.

Controlling Circuits with a Programmable Controller

When the hand/auto switch is moved to the auto position, the operational pressure switch (B) will cycle the output on and off as the pressure in the tank drops below and then above the operational pressure set point. If the operational pressure control would fail for any reason, the safety pressure switch would open and deenergize the output, which would turn the compressor motor off.

P/C-Controlled Three-Wire Circuit

The programmable controller begins to show its advantages over a hard-wired control system when it is used for three-wire control. Figure 12-17 shows a hardware diagram of the input and output modules along with the ladder logic program for a three-wire control circuit. This program duplicates the operation of the three-wire control circuit shown in Figure 12-3. The addresses in this figure use the Texas Instruments format, where inputs are identified by an X and outputs are identified with a Y.

From the diagram, you can see that the stop switch is connected to address X1, the start switch is connected to address X2, and the motor starter coil is connected to address Y9. The start and stop switch are represented by their electrical symbol in the I/O diagram, and they are represented by a set of contacts in the ladder logic program. It is important to understand the operational theory of the P/C and its input modules so that you will know why the stop switch is hard-wired as a normally closed pushbutton switch and the start switch is wired as a normally open pushbutton at the input module, but both switches are represented by a set of normally open contacts in the program.

The input module will send an input signal to the microprocessor when the input address terminal on the module receives voltage. This would occur when the start button is depressed since it is a normally open pushbutton, and the stop button would send voltage to its address when it was not depressed since it is a normally closed pushbutton.

When the P/C receives the input signal from its module, it will cause the contacts with that address in the ladder logic program to change state. This means that contacts that are shown in the ladder logic program as normally open contacts will be closed, and normally closed contacts will be opened.

In the case of the start/stop circuit in the three-wire control, the start switch is wired as a normally open push button at the input module and is shown in the program as a set of normally open contacts. The stop button is wired as a normally closed pushbutton and is shown in the program as a set of normally open contacts.

In operation, the normally closed stop button will pass power to its input module address at X1 as long as no one depresses it. In the program its normally open contacts at X2 will be energized closed by the microprocessor since a signal has been received at that input address. When the start button is depressed, it will also send an voltage signal to its X2 address on the input module. The microprocessor will recognize this signal and energize the normally open X2 contacts to the closed position in the logic program, which allows power

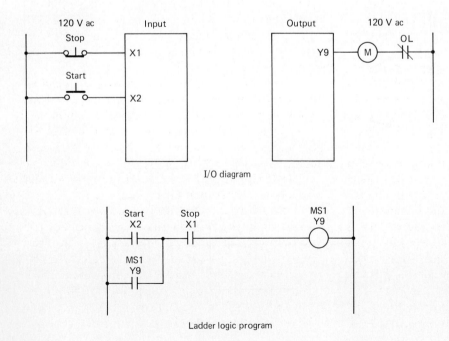

FIGURE 12-17 Wiring diagram and ladder logic program of a programmable controller controlled three wire control circuit.

to flow to the output in the program at Y9. When all the switches in a line (rung) of ladder logic are closed, all the outputs at the end of that rung will become energized. This condition is called *power flow* since all of the closed contacts allow power to flow to the outputs in that line.

The next important concept of the P/C program for three-wire control involves the way the ladder logic program holds in or seals in the start contacts. In the hard-wired circuit shown in Figure 12-3, a set of auxiliary contacts from the motor starter are used to hold in or seal in the start button when it is released. In the ladder logic program, a set of contacts marked Y9 will be programmed in parallel to the start contacts X2. Since the programmed contacts are identified as Y9, it means that they will become energized (change state) when output Y9 is energized. This means that the normally open Y9 contacts will become energized closed and provide a parallel hold in circuit around the normally open X2 start contacts, just as in the hard-wired circuit. Operation of the P/C-controlled circuit will be exactly the same as for the hard-wired circuit.

Another difference between the ladder logic program and the hard-wired control circuit is in the location of the start and stop contacts. In the hard-wired circuit diagram, the stop button is always shown first next to the power rail, and the start button is shown next to the coil. In the ladder logic program this may be reversed to show the start button next to the power rail and the stop button next to the output.

The reason for this goes back to the physical characteristics of the hard-wired components and the amount of memory in the programmable controller. In the hard-wired circuit, the auxiliary contacts are wired to one side of the motor starter coil with a jumper wire when the starter is manufactured. Since these auxiliary contacts are used as a hold-in circuit in parallel with the start button, one terminal of the start button can be connected to this terminal on the motor starter during field installation and save time and extra wiring. In the case of the start contacts coming before the stop contacts in the P/C program, this practice was first used to save an extra byte of memory that was needed to provide two vertical connections for the hold in contacts if the start button was programmed in the middle of the rung. When the start contacts were programmed against the power rail at the far left of the ladder logic rung, the power rail supplied one side of the two verticals, which saved memory. As P/Cs have become larger and have more memory, this is no longer critical, but the practice has continued since it also saves several keystrokes when the program is entered into the P/C memory.

The only difference is that the seal-in circuit around the start button is provided by memory contacts inside the P/C, which means they cannot arc or wear out from operation. These contacts are called *memory contacts* because they are actually one segment of memory called a *byte,* which will be energized or deenergized based on the signal it receives from the microprocessor. The status of all the contact and coils in the ladder logic program is displayed on the cathode ray tube (CRT) display as it is updated by the microprocessor. When a contact in the program is passing power, CRT shows that contact symbol highlighted. This allows the troubleshooting technician to check the CRT and determine the status of the control circuit. Other advantages that the P/C provides will be explained as the circuits become more complex.

P/C-Controlled Three-Wire Circuit with Multiple Start and Stop Switches

The P/C can also control circuits with multiple start and stop buttons. Figure 12-18 shows the wiring diagram and ladder logic program for this circuit. It performs the same operation as that of the hard-wired circuit shown in Figure 12-4.

In this figure you can see that three start buttons and three stop buttons are shown connected to the input module at addresses X1 through X6. The motor starter coil is connected to output address Y9. You should notice that the pushbutton start switches are wired normally open and the stop switches are wired normally closed. In the program, both the start and stop switches are shown as normally open contacts. The operation of the stop button is the same as discussed for the three-wire circuit.

The operation of this circuit is very similar to the circuit in Figure 12-17, where the P/C will energize the output at Y9 whenever the start button is depressed and the stop button is not. The Y9 contacts in the ladder logic are shown programmed in parallel with all the start buttons. They will act as a hold-in circuit again since all of the start buttons are momentary pushbuttons.

If additional start buttons are required in this circuit, they should be wired to the input module like the other pushbuttons and entered in the program as normally open contact in parallel with the other start contacts. If additional stop buttons are required, they should be entered in the program as normally open contacts and placed in series with the original stop contacts.

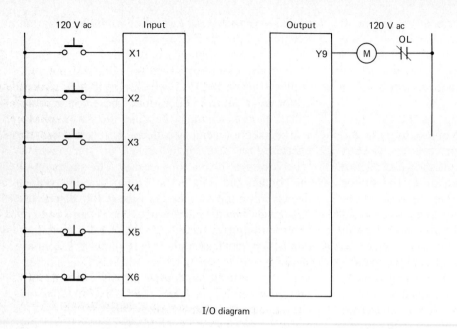

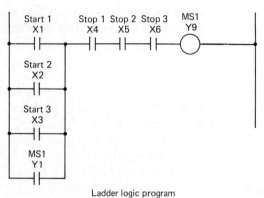

FIGURE 12-18 Programmable controller controlled three wire circuit with multiple start and stop switches.

P/C-Controlled Circuits Where the Switches Are Powered by a Lower Voltage Than the Motor

Another distinct advantage that the P/C provides is that the input switches and the output coils can be any voltage between 12 volts and 120 V dc or 12 and 220 V ac. Typical voltage for inputs are 24 and 120 V ac and dc. Typical output voltages are generally 120 V ac since the majority of existing motor starter coils are that voltage. Since modules are available in such a wide variety of voltages, the acceptable practice has been to match the voltages for controls used in a P/C circuit to those found elsewhere in the factory. This means that you will not need to stock a wide variety of special switches or motor starters just for the P/C circuits; instead, this will allow you to consolidate your stock.

An example of this functionality can be shown by comparing the hard-wired circuit shown in Figure 12-5 to the circuit shown in Figure 12-18. In the hard-wired circuit, a step-down transformer was required to produce the control voltage for the start/stop circuit since supply voltage for the motor was over 400 V. In P/C diagram you can see that the voltage for the modules is rated for 110 V. This means that a control transformer was provided for all the P/C inputs and outputs also. This means that a P/C could be installed to control the same circuit, with similar components.

A programmable controller would not be installed to control just one three-wire control circuit; rather, it would be used to consolidate all of the control circuit for a particular manufacturing process, assembly process or one large machine, such as a punch press or injection molding machine. Since it is possible to find inputs that require traditional 110-V ac and output that requires dc voltages, the programmable controller can easily make this interface of different input and output voltages. In fact, it is possible for a P/C to have over 10 different types of input modules to receive signals from a variety of voltages and be able to energize as many output modules, requiring four or five different ac and dc voltages.

P/C-Controlled Three-Wire Circuits with Pilot Lamp Indicators

The P/C also provides several methods of adding pilot lamps to circuits to show their condition. Figure 12-19 shows an example of this type of circuit. It will operate exactly as the hard-wired circuits shown in Figures 12-6 and 12-7. In the P/C circuit you can see that the start and stop pushbuttons are connected to the input module at addresses 111/01 and 111/02, and the motor starter coil and two pilot lamps are connected to the output module.

The ladder logic program shows the start/stop circuit with its hold-in contacts controlling output 010/01. Any time the start button is depressed, the motor starter coil will be energized and the memory contacts of output 010/01 will seal in the start contacts. The output will energize the motor starter, which will start the motor. Output 010/02 that energizes lamp 1 is connected in parallel with output 010/01 that energizes the motor starter coil. This means that lamp 1 will be on any time the coil is energized, indicating that the circuit is operating.

A second indicator lamp can be added to this circuit to make it operate exactly like the hard-wired circuit in Figure 12-6. The extra lamp in the ladder logic program is controlled by output 010/10. It is programmed on the second rung of the diagram in series with a normally closed set of memory contacts 010/01 that are controlled by output 010/01, which also energizes the motor starter coil. Since the normally closed contacts exist only in the P/C memory, they cannot wear out, as they provide the same function as auxiliary contacts found on the motor starter.

Another variation of this circuit could incorporate a separate test circuit which would allow the operator or technician to test all of the lamps in the system by pressing a single pushbutton. The pushbutton would need to be connected to the input module at address 111/03. You should notice that each output that controls an indicator lamp is programmed on a separate rung. The normally open 010/01 contacts are programmed in series with lamp 1, and normally closed 010/01 contacts are programmed in series with lamp 2 on the third rung. When output 010/01 is energized by the P/C, the motor starter coil will receive power. The normally open 010/01 contacts will have power flow and energize output 010/02 for lamp 1, and deenergize output 010/03 and turn off lamp 2. When the stop button is depressed and the motor starter output is deenergized, the normally open contacts in the program will be deenergized and return to their

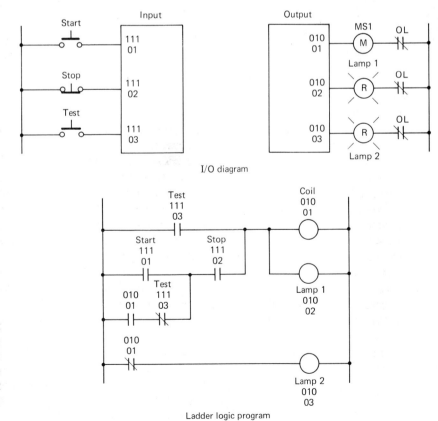

FIGURE 12-19 Diagrams of P/C controlled three wire circuits with pilot lamps. This diagram shows a press to test a circuit that can be added to allow a technician to test all lamps in the circuit by pressing one button.

Controlling Circuits with a Programmable Controller

open state, which will interrupt power flow to output 010/12, which will turn off lamp 1. At the same time, the normally closed contacts 010/01 will be allowed to return to their normally closed state and output 010/03 will receive power flow to energize lamp 2. This means that lamp 1 will be illuminated whenever the motor starter is energized, and lamp 2 will be illuminated whenever the motor starter is deenergized.

The contacts for the test pushbutton will be identified as 111/03. A set of these normally open contacts will be programmed in parallel with the other contacts in both of the lamp control circuits. When the pushbutton is depressed, its contacts will provide power flow to each of the lamps regardless of the condition of the motor starter. If either of the lamps does not illuminate, the technician will know that a problem exists in that lamp and the bulb should be replaced. If a panel has several dozen indicator lamps, they can all be tested in the same manner by entering a program similar to this one. *Be sure that each lamp is on a rung by itself to use a test lamp circuit. If the test lamp contacts were added in parallel to the lamp in Figure 12-19, the motor starter coil would also be energized and sealed in when the test button was depressed.*

P/C-Control Reversing Circuits

The programmable controller is especially well suited for more complex circuits such as reversing circuits that required multipole pushbuttons. Figure 12-20 shows three diagrams of P/Cs controlling reversing circuits. Figure 12-20a shows a P/C circuit that will operate exactly like the hard-wired circuit shown in Figure 12-8.

In the P/C diagram shown in Figure 12-20a you can see several differences in the pushbutton switches to control the reversing circuit. The forward and reverse pushbuttons in the hard-wired circuit must have one set of normally open contacts and one set of normally closed contacts. The pushbuttons used in the P/C circuit need only to have one set of normally open contacts. Another change in the I/O structure is the location of the overload contacts. In the hard-wired circuit the overloads were connected in series with both motor starter

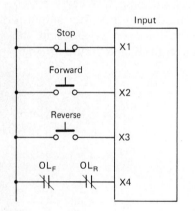
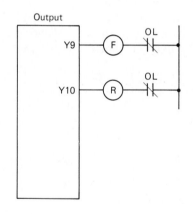
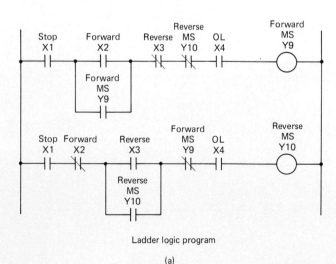

FIGURE 12-20 a-c Three diagrams showing P/C's being used to control reversing circuits. Figure 12-20a shows a circuit where the stopbutton must be used prior to changing directions. Figure 12-20b shows a reversing circuit where limit switches have been added for deenergizing the motor starter at the end of its travel. Figure 12-20c shows a reversing circuit used for jogging a motor.

coils. In the P/C circuit the overloads of both motor starters are connected in series with each other and connected as an input at address X4. This allows them to be used in the ladder logic circuit to protect the motor in either direction.

The ladder logic program shows the location of each set of contacts to provide the same logic function as that of the hard-wired circuit. Notice that the stop contacts (X1), forward contacts (X2), reverse contacts (NC X3), overload contacts (X4), and normally closed contacts from the reverse coil (Y10) are all in series with the forward coil (Y9). A set of forward contacts (Y9) are placed in parallel with the forward pushbutton contacts (X2).

Any time the forward pushbutton is depressed, the forward output will receive power and energize the forward motor starter coil. If the stop button is depressed or the overloads trip, the coil will become deenergized.

The second rung of the program shows the stop contacts (X1), normally closed forward contacts (X2), reverse contacts (X3), overload contacts (X4), normally closed contacts from the forward coil (Y9) are all connected in series with the reverse motor starter coil (Y10). A set of normally open contacts marked (Y10) are connected in parallel with the reverse pushbutton contacts (X3). Any time the reverse button is depressed, the reverse motor starter coil will be energized by output Y10.

The normally closed contacts in each rung act as electrical or logic interlocks. This means that they will prevent both the forward and reverse motor starters from being energized at the same time. The normally closed contacts from each pushbutton will also cause the stop button to be depressed to prevent the circuit from energizing the reverse coil while the motor is operating in the forward direction, and vice versa. In this circuit the

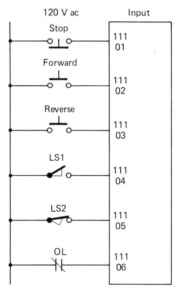
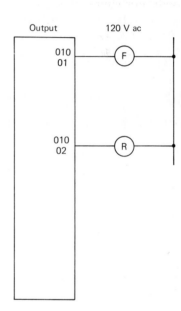

I/O diagram

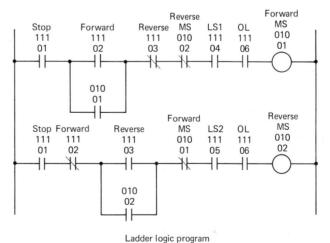

Ladder logic program

(b)

FIGURE 12-20b P/C being used to control reversing circuit. (Continued)

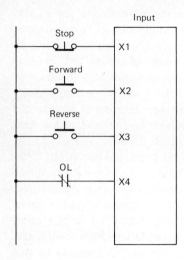
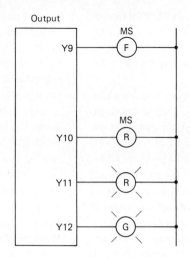
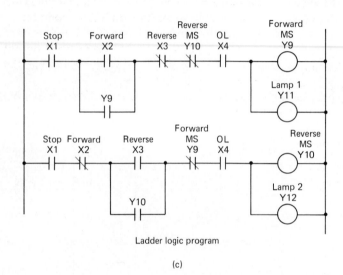

FIGURE 12-20c P/C being used to control reversing circuit. (Continued)

stop button must be used before the starters can be reversed.

The second circuit shows two limit switches added to the input module to be used to deenergize the motor starter when the motor has moved its load full travel. This circuit is very useful in motion control applications that use a lead screw mechanism. The switches in the I/O diagram for Figure 12-20b are the same switches as those in the previous circuit with two limit switches added. The diagrams in Figure 12-20c are shown using the Texas Instruments addressing format, and the diagrams in this diagram use the Allen-Bradley addressing format.

In the ladder logic program the forward limit switch is added in series to the forward control circuit and the reverse limit switch is added in series to the reverse control circuit. When the motor is activated in the forward direction, the forward motor starter is energized and the motor will operate in the forward direction. This will cause the load to move in the forward direction until it moves against the forward limit switch and causes it to open. When the forward limit switch is opened, the forward control circuit is deenergized.

When the reverse pushbutton is depressed, the reverse output is energized and sealed in. This energizes the reverse motor starter and the motor starts to move the load in the reverse direction. When the load has moved full distance in the reverse direction, it will cause the reverse limit switch to open, which will deenergize the reverse output and turn off the motor, and sets the control circuit to be operated in the forward direction when the operator presses the buttons again.

A modification can be made to this circuit to make the motor change direction automatically when it reaches its maximum travel in either direction. This change would involve adding a stop/start memory coil to the circuit. The contacts from this memory coil will take the place of the stop contacts in each circuit. When the start button is depressed,

340   Chap. 12 / *Motor Control Circuits*

it will energize the memory coil, which will be sealed in by a set of parallel contacts. The contacts from this coil will also enable both the forward and reverse circuits so that they will be able to operate automatically.

Another addition to this circuit would include placing a normally open set of forward limit switch contacts (111/04) in the reverse circuit and placing a set of normally open reverse limit switch contacts (111/05) in the forward circuit.

When the motor is traveling in the forward direction, the load would move until it reaches the forward limit switch. When the forward limit switch is opened, it will deenergize the forward motor starter output and energize the reverse motor starter circuit. This moves the load in the reverse direction until it presses against the reverse limit switch and opens it. When the reverse limit switch is activated, it will open the reverse motor starter and energize the forward circuit. This forward and reverse switching will continue automatically until the stop button is depressed. Since the stop button is now a sealed circuit, any time it is depressed, it will remain in the off state until the start switch is depressed again.

P/C-Controlled Reversing Circuit with Indicator Lamps

Another variation of the reversing motor starter can easily be controlled by the P/C. The diagram and program in Figure 12-20c duplicate the operation of the hard-wired circuit shown in Figure 12-10. This circuit adds lamps to the circuit to indicate which direction the motor will be traveling.

The I/O diagram in this figure uses the address format for Texas Instruments P/Cs. You can see that the inputs are identical to the circuit in Figure 12-20a, while two lamps have been added to the output module at addresses Y11 and Y12. One lamp has been added to the forward and reverse rung of the ladder logic program. The red lamp (Y11) is connected in parallel with the forward output, and the green lamp (Y12) is connected in parallel with the reverse output.

When the forward motor starter is energized,

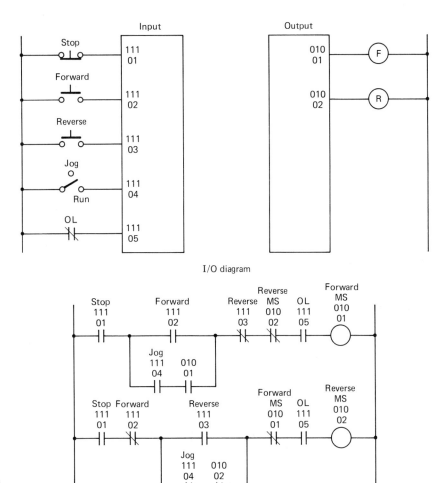

FIGURE 12-21 Programmable controller used for control of a jogging circuit.

the input module has two start and two stop switches connected to addresses X1 through X4. Two motor starter coils are connected to addresses Y9 and Y10.

The ladder diagram is programmed to require motor starter 1 to be started before motor starter 2 is allowed to start. Remember that this circuit is designed to make one conveyor be started prior to starting the second conveyor since the second motor starter controls a conveyor that feeds material onto the first conveyor. This ensures that the lead conveyor is running so that the feed conveyor does not jam up.

From the ladder logic program you can see

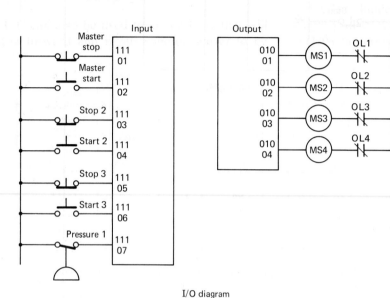

FIGURE 12-24a & 24b Several motors that may operate independently of each other after a master circuit has been energized. Figure 12-24b shows a timer used to provide a delay between two motor starters that are energized by the same start switch.

344 Chap. 12 / *Motor Control Circuits*

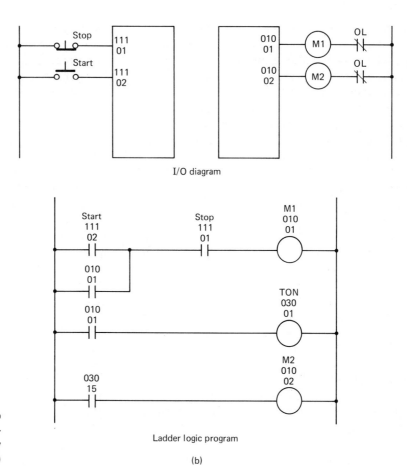

FIGURE 12-24b A timer used to provide a delay between two motor starters that are energized by the same start switch. (Continued)

that the output that controls motor starter 2 cannot be energized until the output at Y9 is energized. After Y9 is energized, its memory contacts enable the second circuit. This allows the second motor starter to be energized at any time after the first one has been started.

The circuit shown in Figure 12-24a shows another sequential application that can be programmed into the programmable controller. The ladder logic program will provide operation that is exactly the same as that of the hard-wired circuit shown in Figure 12-14. You should recall that this application required a master control relay to be energized prior to allowing any motor starters to be operated. This application also required that a water pump maintain a specified pressure in a tank before either of the two process machines was allowed to operate. A conveyor is controlled by motor starter 4 and it must run any time either of the process machines are in operation.

The I/O diagram shows start and stop pushbuttons and the pressure switch connected to addresses 111/01 through 111/07. The four motor starter coils are connected to output addresses 010/01 through 010/04. In the ladder diagram output 012/01 is used as a memory coil since no output modules are located at this address. Its contacts will be used like a control relay, even though they exist only in the processor's memory.

The operation of the circuit requires the master start switch to be energized to provide power flow to output 012/01. This memory coil has three sets of normally open contacts. As you know, it could have as many sets of normally open or normally closed as are required to make the circuit function properly. The only limiting factor in the number of contacts that could be programmed is the size of the P/C memory.

In this application, each of the rungs that control the first three motor starters has a set of the 020/01 normally open contacts connected in series with the other controls. Motor starter 4 is energized by the contacts of 010/02 or 010/03, since they are both in parallel with each other in rung 5. The pressure switch at 111/07 allows the water pump at 010/01 to become energized any time the water pressure drops below the set point. Any time the stop button is depressed, the control relay is deenergized and all the motor starters will be turned off.

Another variation of the sequence control is shown in Figure 12-24b. The ladder logic program provides the same functionality as the hard-wired circuit shown in Figure 12-14. This circuit is de-

Controlling Circuits with a Programmable Controller 345

signed to prevent two motor starters from starting at exactly the same time. As you know, if two motors are started at the same time, their locked-rotor current will be too large for the power distribution system.

The hard-wired circuit used a motor-driven timer to provide the time delay, which allows the first motor to come fully up to speed before the second starter is allowed to be energized, even though they are both controlled by one start and stop switch. The circuit in the programmable controller uses a programmed timer that is in the P/C memory to provide the time delay between the two motor starters.

The contacts that the timer operate are identified by the same address that the timer uses (030). The contact number 17 indicates that these contacts are controlled by the "done" bit (17) of timer 030. The timer will energize bit 17 after it has timed out. This set of contacts will operate exactly like the motor-driven timer contacts.

Whenever the TON instruction is energized, the timer will begin to operate. A preset value is entered into the program at the time the program is entered into the P/C. If the contacts close too soon, time delay can be added to the timer preset value.

Troubleshooting P/C-Controlled Systems

The programmable controller is designed to aid the troubleshooting procedure as well as to provide control for the sequence of operation. This means that you will be able to locate the status indicators on the input and output modules and see if they are energized. If the input status indicator is illuminated, an input signal is present at that terminal. If an output status indicator is illuminated, you can begin testing the load, as the proper amount of voltage should be present at the output address. In some systems, the output status indicator is on the logic side of the circuit, and it may glow to indicate that a logic signal is present, but the voltage may not be passing completely through the module. This is easy to test for by testing for voltage with the voltmeter and comparing the results to the status indicator. The CRT of the P/C can be used in conjunction with a printed copy of the program that is in the system. These will help you determine which components should be on or off to provide the proper logic to energize an output. The CRT provides several methods of indicating if a device is on or off, which include a highlight function on the screen. If a device is energized, and if a particular line of logic is passing power, those components and lines will be highlighted so that the troubleshooter will know which devices in the circuit are not functioning properly. If a device is not highlighted, is should be suspected of being faulty, and you should test it with the conventional tests used for pilot devices and loads. In Chapter 16 we provide more details about using the P/C to aid in the troubleshooting procedure.

You can now see that you will encounter typical motor control circuits that operate as stand-alone control circuits or as part of a programmable controller circuit. You will soon realize that the devices and components will operate the same and they can be installed and troubleshooted using the same information and principles.

QUESTIONS

12-1. Draw an electrical diagram of a two-wire control system and explain its operation.

12-2. Draw an electrical diagram of a three-wire control system and explain its operation.

12-3. Draw a three-wire control circuit that uses multiple start and stop buttons.

12-4. Draw an electrical diagram of a three-wire control circuit that uses an indicator lamp to show when a motor is energized.

12-5. Draw an electrical diagram of a forward and a reverse motor starter.

12-6. Draw the electrical diagram for a jog button used in a conjunction with the regular start and stop controls on a motor starter.

12-7. Draw an electrical diagram of a control circuit that allows conveyor 1 to start before conveyor two is running.

12-8. Draw an electrical ladder diagram of the relay logic for a two-wire control circuit that is controlled by a programmable controller.

12-9. Draw an electrical ladder diagram of the relay logic for a three-wire control circuit that is controlled by a programmable controller.

12-10. Draw an electrical ladder diagram of the relay logic for a three-wire control circuit that includes a pilot lamp to indicate that the load is energized. This circuit is also controlled by a programmable controller.

12-11. Explain how the programmable controller controls the circuits that you have drawn in the preceding questions. Be sure to include the functions of the status indicators in the input and output modules.

12-12. Explain how you would troubleshoot the control circuits controlled by the programmable controller.

Photoelectric and proximity controls

13

INTRODUCTION

Photoelectric switches are used extensively as motor control devices. They provide on/off (discrete) control that can be amplified to control larger ac and dc voltages at pilot-level currents. Some photoelectric devices control relay and motor starter coils so that even larger load currents such as motors can be controlled through its contacts. Photoelectric switches are very versatile in that several different models are available to operate in a variety of industrial environments that have differing amounts of background lighting. These switches are made to be used as stand-alone controls, or they can be connected directly to robots or programmable controllers. Some of the terms used in this chapter will be new to you, so a small list of terms with their definitions is provided.

There are two basic classes of photoelectric controls used today: the incandescent type and the modulated light source (MLS). *Incandescent controls* have been in use for many years, while the MLS type has been developed around solid-state electronic technology. The incandescent type uses a beam of incandescent light focused on a receiver that is light sensitive, such as a photovoltaic cell or photoresistive cell. The earliest types of these controls used photoemissive vacuum tubes.

When solid-state technology provided devices such as light-emitting diodes and phototransistors, new controls using the modulated light source technology were designed. *MLS controls* are designed around three basic principles of transmitting and receiving light. The three basic designs are the retroflective scan, through-beam scan, and the diffuse scan.

Figure 13-1 shows the basic MLS types of photoelectric types of controls. The *retroflective scan* device sends out a beam of light that is reflected off a target back to the receiver. When something breaks the beam of light, the contacts in the photoelectric switch changes position. The contacts of the switch may be connected to devices such as a counter to count parts being manufactured, or a robot to indicate that a part is in place, or a programmable controller to detect the presence of a part.

The *through-beam scan* sends a beam of light directly to a receiver unit instead of using a reflected beam. When the beam of light reaches the receiver it will trigger the contacts there, which can be used to provide control to pilot devices. As with retroflective control, when something interrupts the beam of light, the contacts will change position.

The *diffuse scan* switch works on the principle of reflecting light in a diffuse pattern off a specific type of material used as a background. Since different material will reflect light in a variety of ways, the receiver can be adjusted to the proper angle to

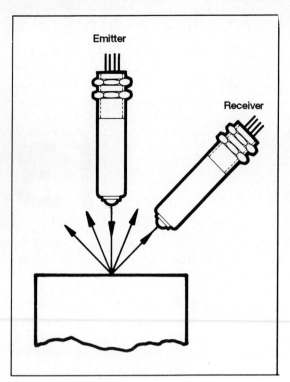
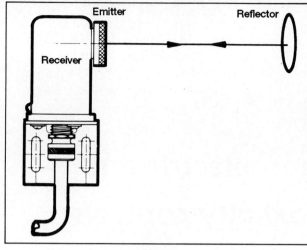
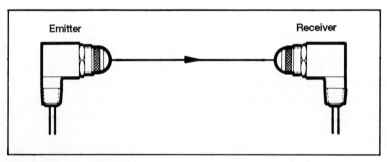

FIGURE 13-1 Modulated light source types of photoelectric controls. (Courtesy of Micro Switch a Honeywell Division)

collect light that is reflecting off the material at a predetermined angle.

Each of these types of switches can be adjusted to energize the controlling contacts to the on position in either the presence or absence of the light beam. A detailed discussion of the incandescent and MLS types of photoelectric controls is presented in this chapter. We include the theory of operation, wiring diagrams for installation and interfacing, adjustment and calibration methods, and troubleshooting procedures for all types of photoelectric controls.

INCANDESCENT PHOTOELECTRIC CONTROLS

Early photoelectric switches used a variety of methods of sensing the presence or absence of light. One of these devices was the photoemissive tube. The tube works on the principle of photoemission. When light strikes the cathode element of the tube, it emits electrons. The stronger the light source, the more electrons are emitted from the cathode. Another element is placed very near the cathode to collect the emitted electrons. This element is the anode. The anode will begin conducting current as it absorbs more electrons. This means that the anode will conduct current in proportion to the strength of the light source until it is saturated.

In most applications of this type of switch a beam of light is focused on the cathode, which causes current to flow in the anode circuit. The anode current was used as the control current to power a relay coil. If something interrupted the beam of light, the relay contacts would change from closed to open to perform the actual control function.

This type of control was very popular before

the advent of solid-state devices, but it presented some problems when background light caused false triggering. Early solid-state devices such as cadmium sulfide cells, photovoltaic cells, and transistors were easily adapted for use as photoelectric devices and began to replace photoemissive tubes because their sensitivity could be adjusted to account for background lighting.

Photovoltaic Cells

The photovoltaic cell is used as a light detector because of its ability to produce a small voltage when light is shined on it. The photovoltaic cell is also called the *solar cell*. The small voltage it produces is in the microvolt range, so like a transistor, it must be connected to an amplifier to provide control of pilot-level voltages. Since this type of switch includes a transistor for control, some amount of adjustment can be made to compensate for background light or other problems, such as suspended dirt and dust in the air. The adjustment involves applying a small amount of bias voltage to the transistor base. The bias could be positive or negative to change the sensitivity. If the bias voltage was positive, less voltage would be required from the photovoltaic cell, thus making the switch more sensitive. If the bias voltage were negative, the photovoltaic cell would need to produce more voltage to trigger the transistor into conduction.

Cadmium Sulfide Cells

A similar type of control is the photoresistive cell. This device is made from cadmium sulfide (cad) cells. When light strikes a cad cell, it changes its internal electrical resistance. In total darkness the cad cell has very high resistance, in the range of several hundred thousand ohms. In direct light the resistance in the cad cell drops below 100 Ω. In most applications the cad cell is connected to the base of a transistor. When the cad cell is sensing direct light, the resistance in the transistor base circuit drops and base current flow causes the transistor emitter–collector circuit to go into conduction. The collector current is large enough to power most pilot devices. This type of sensor can also be adjusted for sensitivity by applying a small bias voltage to the base.

The cad cell has some inherent problems in that it has difficulty distinguishing control light and ambient light or background light. This is a problem in most industrial areas, since background light is ever-present from a variety of sources: overhead lights used to light work areas, reflection of light off shiny surfaces, and light from processes such as pouring molten steel, arc welding, and opening and closing outside doors.

A new generation of photoelectric switches are required that are more sensitive and more accurate than previous types. They are used in controlling robotic work cells and totally automated processes and must have response times fast enough to interface to programmable controllers and other microprocessor controls.

ANALYSIS OF LIGHT

If you are to fully understand the design differences of photoelectric switches, you must first understand some basic concepts about light. Light travels around us in the form of waves, and it is identified by its wavelength. Figure 13-2 shows a table indicating the various types of light and their wavelengths. The wavelength spectrum shows that ultraviolet light is at the low end and infared light is at the high end of the spectrum. You can determine

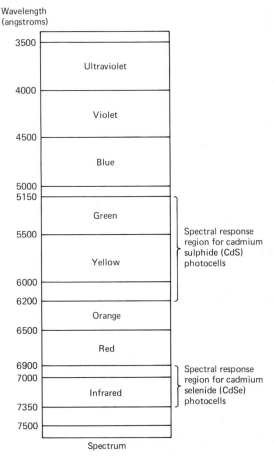

FIGURE 13-2 Spectrum that shows the wavelength of various types of light sources used in photoelectric sensors. (Courtesy of Micro Switch a Honeywell Division)

the response range for sensors from this graph. The cadmium sulfide (cad) cell is sensitive to green and yellow wavelengths. This means that it can distinguish light with this wavelength from all other light. It is very important for the cad cell to sense only the source of light (green or yellow) that is emitted from its source and not the light that is in the background, such as overhead lighting. Another type of cad cell, the cadmium selenide (CdSe) cell, is only sensitive to light in the infared spectrum range. This allows the selenide cad cell to be used where the cadmium sulfide cell cannot be used.

MODULATED LIGHT SOURCE DEVICES

The majority of photoelectric devices used today operate on a principle known as modulated light source. This involves the use of a light-emitting diode (LED) as a light source, a phototransistor as a receiver, and a control circuit driven by the phototransistor. The LED is modulated at a specified frequency by a solid-state circuit called an oscillator. The phototransistor utilizes a circuit that tunes in the oscillated frequency that the LED is sending and filters out all other frequencies of light waves received. Since each type of light such as ultraviolet, infared, and regular colors, from the color spectrum oscillates at a different frequency range, it is rather simple to filter out unwanted light signals. This is known as *light rejection*. The phototransistor and the LED have such a good frequency match that MLS sensors can be used in the presence of strong ambient light sources in an industrial environment.

Pulsing the LED also allows the light source to be more powerful, since less power is consumed during pulsing than if the source was powered continuously. Heat from the devices is also dissipated more easily when the LED is pulsed. Actual on time for the LED may only be several milliseconds, while the off time may be over 100 ms. This provides an off-to-on duty cycle of better than 10:1. This also means that the signal pulse may have the strength of several hundred milliamperes, yet consume an average current of only 30 mA.

THROUGH-BEAM SCAN PHOTOELECTRIC SWITCHES

The through-beam scan photoelectric switch may also be called direct scan or separate control. This type of photoelectric switch is made of two separate parts. One part is the source for a beam of light, which is developed from an LED modulated light source. This part is called the emitter or source. The source focuses the light beam on the second part, called the receiver or detector (see Figure 13-3). When the beam is focused on the receiver, it triggers the phototransistor into conduction. The phototransistor may control a solid-state or relay output. The output will be able to operate solenoids, motors, or other types of loads. When the beam of light is interrupted, the phototransistor drops out of conduction and the output that it is driving changes state.

Figure 13-3 shows the electrical block diagram for an Allen-Bradley Bulletin 880L through-beam control. Other companies use similar circuits in their through-beam switches. Both the modulated light source and the receiver require a power supply. The power supply for the source provides voltage to the modulator circuit and to power the pulse of LED. The modulator circuit is solid state, so it needs a regulated dc voltage source. Since industrial voltage is ac, the power supply will use a rectifier to produce the dc voltage. This allows the source to be connected directly to a 110- or 220-V ac circuit in the factory. The output power from the

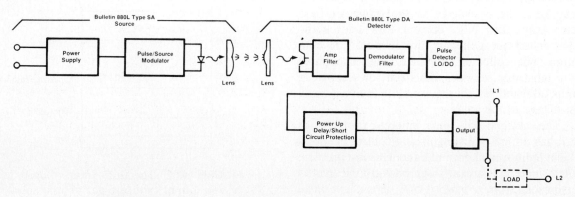

FIGURE 13-3 Electrical block diagram of Allen-Bradley bulletin thru beam type control. (Courtesy of Allen-Bradley)

rectifier to pulse the LED needs to be strong enough to pulse the LED for several microseconds during its on cycle.

The receiver is comprised of five basic circuits. Each of these circuits use solid-state components, so dc voltage is supplied from a rectified ac voltage as a power supply. The modulated light beam from the LED is focused on a phototransistor. This causes the phototransistor's circuit to modulate. The modulated signal is very weak, so it is sent to an amplifier and filter circuit. The filter allows only signals within a narrow frequency to pass to the amplifier. This means that any other light source striking the phototransistor will cause it to oscillate at a frequency outside the narrow band, and the receiver will filter it out.

The amplified signal is sent to a demodulator and additional filter to be conditioned further before it is sent to the pulse detector state. This provides the detector with a sensitivity range so that the switch can be adjusted for a variety of environmental conditions. The pulse detector stage also has an external switch that can be set so that the output stage is light or dark activated. The fourth stage of the receiver is a protection circuit. This circuit protects the switch against a short circuit on its output and against false pulses on power-up or due to signal transients.

The last stage of the receiver is the output stage. The output stage will act like a switch in that it will conduct and allow current to flow to the load, or it will open and stop current flow. This part of the control is a solid-state device triggered by the amplified demodulated pulse from the phototransistor. The output portion of the control is wired in series with the load. This means that the output acts like a set of contacts that will close to allow current to flow to the load when the photoelectric control is triggered. The output would operate in the opposite way (would be turned off when light is sensed) if the device were set for dark operation rather than light operation.

Most companies offer several choices of output voltages and interfaces. These include both ac and dc pilot-level voltages. PNP and NPN transistors and SCR are used to control the dc outputs; triacs and bridge rectifiers are used to control ac outputs. Figure 13-4 shows several types of output control circuits. Remember: these circuits are part of the photoelectric device and do not have to be constructed during installation. All the electrician must do during installation is read the diagram and make the proper connections between the photoelectric switch, the load device, and the correct supply voltage.

From Figure 13-4 you can see that the first four sections of the photoelectric switch from Figure 13-3 are now represented by a box marked "photoelectric switch circuit." The output part of each switch is now shown by discrete components. The signal from the switch circuit is sent to the base of the transistor or to the gate of the SCR. This signal's polarity will be adjusted so that the base of the NPN transistor and gate of the SCR will receive a positive voltage and the base of the PNP transistor would receive the negative voltage. These circuits are typical of all photoelectric switches. Some companies use more complex circuits, but the theory of operation is similar. The solid-state circuits inside these devices are not meant to be repaired. If you understand their theory of operation, you will be able to use the troubleshooting techniques provided at the end of this chapter to determine if a device is operational. It is more important to identify the faulty motor control device and change it as quickly as possible rather than to spend time trying to repair a faulty device.

Through-beam switches come in several different configurations for the multitude of industrial applications. This type of switch needs a source and a receiver. If they are mounted any distance apart, they will require two separate sources of voltage, which may become a problem for some applications, so the detector part of the receiver and the LED for the source may be made very compact so that they can be mounted remotely. The remote sensor and source are wired to the amplifier, which may be located in the control cabinet some distance away. Figure 13-5 shows this type of photoelectric switch.

From the picture of these devices you can see that you will still get the advantage of the through-beam device, in that scanning distances can be up to 70 feet with high gains, yet the problem of running supply voltages to two separate locations is solved. Another type of through-beam device is shown in Figure 13-6. This type of device has the source and receiver mounted on opposite sides of a C-shaped housing. The operating circuitry for the system is contained right in the same housing. This allows the device to be more compact, for applications that require the sensors to be mounted very close to the material to be sensed. Having the source and receiver very close to each other also allows the switch to have very good sensitivity. This type of switch is often used to locate color marks on clear or opaque material.

Figure 13-7 shows an application where a photoelectric switch is used to detect color marks in transparent film. The film is used for wrapping finished products, and the color mark is used to align trade marks and labels that are printed on

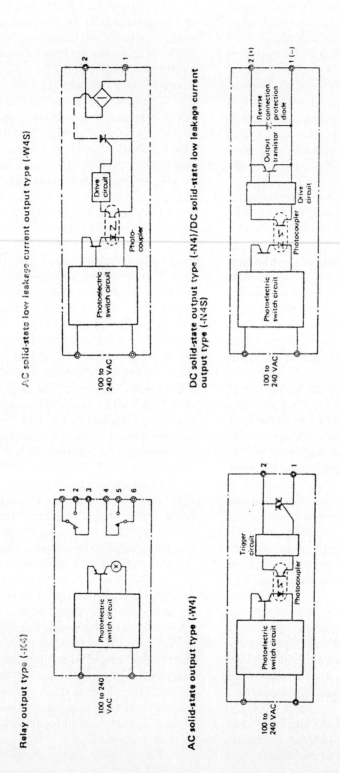

FIGURE 13-4 Types of output circuits for photoelectric switches. (Courtesy of Omron)

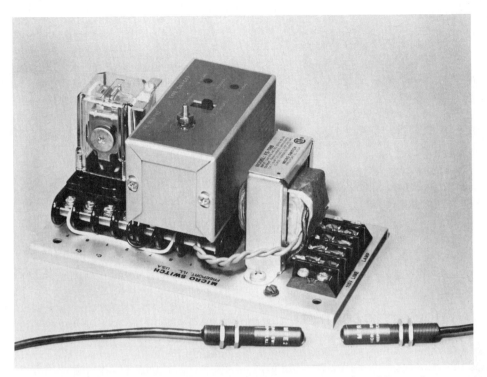

FIGURE 13-5 Separately mounted receiver and source with amplifier base. Courtesy of Micro Switch a Honeywell Division)

the film and must appear on the front of the product.

The photoelectric switch in this application senses a difference in the frequency of the light passing through the clear film and the frequency of light passing through the color mark. Since the source and receiver are very close to each other, the sensitivity can be set very high and the filter can be adjusted to sense only a very narrow frequency, indicating light passing through clear material. When the light passes through the color mark, it will cause the receiver to sense light having a slightly different frequency. This would act in the same way as if a solid object has passed between the source and the receiver.

This type of application is very susceptible to dust or other particulates in the air. If small amounts of dirt accumulate on the receiver or

FIGURE 13-6 "C" photoelectric controls with diagrams. (Courtesy of Omron)

Through-Beam Scan Photoelectric Switches   **353**

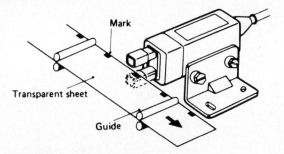

FIGURE 13-7 "C" type thru-beam scan photoelectric switch used for sensing color marks. (Courtesy of Omron)

source lens, the gain of the device will be altered and affect the detecting resolution and repeatability.

FIBER OPTIC PHOTOELECTRIC CONTROLS

Another solution to the problem that requires both the source and the receiver to have a power supply is the use of fiber optic technology. The fiber optic through-beam scan photoelectric switch operates very similarly to the regular type with a source and a receiver. The difference can be seen in Figure 13-8. Instead of having two separate housings for the source and receiver, the circuitry for both are contained in one housing. The light from the source is generated in the module housing and is transmitted to a lens through a fiber optic cable that will focus it. The fiber optic cable is very durable, yet is pliable enough to be routed through conduit. It also keeps the signal from receiving electrical interference or noise.

The fiber optic switch has circuits that are similar to those of the regular through-beam switch. The main difference between fiber optic photoelectric switches and regular modulated-light-source photoelectric switches is in the fiber optic cables. The cables come in precut lengths that must be specified when ordering. Since the cable is made of a glass fiber, you must be careful not to step on the cable or bend it sharply during installation, as the glass will shatter.

RETROREFLECTIVE PHOTOELECTRIC SWITCHES

The retroreflective photoelectric switch has the modulated light source and the receiver in the same housing. A LED is used to develop the modulated light source, which is focused on a target that has light-reflecting properties. The reflector, called a target, will reflect and refocus the light back to a lense, where it is directed to the detecting circuit to be filtered and demodulated. Figure 13-9 shows a photograph of this type of photoelectric switch.

Since the source and receiver are in the same housing, installation and wiring is easier. The target, or reflector, for this type of switch is rather inexpensive and is easily mounted to most surfaces. Figure 13-9 also shows material that can be used as reflectors or targets. The main property of the target is that it will reflect and refocus the beam of light that has been sent from the source.

Once a retroreflective switch is installed, it can easily be adjusted to focus the reflected light beam back into the receiver. If the reflector is large enough, the alignment may be up to 15 degrees out of perpendicular to the receiver. Since the alignment is not crucial for this type of switch, it makes it a very good choice for applications where vibration of the sensor is a concern. After the control is mounted you should check the exact location where the object will break the beam and trigger the output, and make minor adjustments if needed.

Since this sensor operates from reflected light it will also work very well for applications where clear objects such as glass bottles or clear plastic containers must be sensed. Light from a through-beam scan source may be strong enough to penetrate the clear material and trigger the switch. By comparison, a reflected light will be scattered and its signal strength will be diminished when it passes through clear objects, which will allow them to be detected. Figure 13-10 shows several of these applications.

Another application for the retroreflective switch is to have a reflective surface or reflective patch on the product being sensed. If alignment of the product can be assured, the source beam can be reflected from the product as it passes the switch.

Figure 13-11 shows a block diagram of a retroreflective photoelectric switch. This diagram is of an Allen-Bradley type RL switch. Other companies make retroreflective photoelectric switches that use very similar circuits and operating principles.

In this type of switch only one power source is required since the source circuit and receiver circuit are in the same housing. The source (emitter) circuit receives power from the regulated power supply. The main function of this circuit is to establish a frequency to modulate the LED. You should remember that the pulsing will allow the LED to be pulsed at full power to provide the strongest beam of light, yet not overheat the device. This also

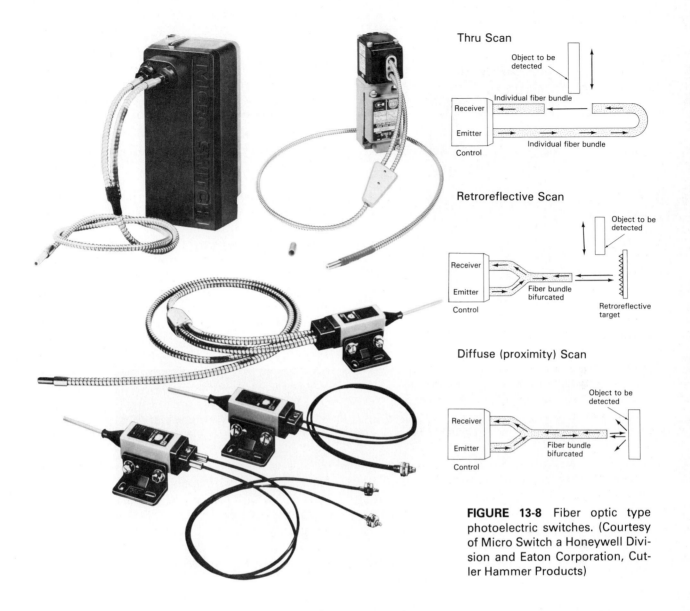

FIGURE 13-8 Fiber optic type photoelectric switches. (Courtesy of Micro Switch a Honeywell Division and Eaton Corporation, Cutler Hammer Products)

allows power consumption of the source circuit to be as small as possible.

The modulating beam of light is directed through the lens toward the reflector. From Figure 13-11 you can see that the lens for this type of device actually has two separate sections. One part of the lens is ground to make it a perfect emitter of light. The other part is ground to make it a perfect receiver of light. Once the source beam is adjusted and focused, the reflector will direct the reflected light back into the receiver part of the lens. This reflection process does not disturb the frequency on which the beam is modulated.

Once the beam is focused back into the receiver, it is amplified and filtered. The filter does the same job that it did in the through-beam device, in that it eliminates any signals from light that is outside the frequency bandwidth. From this section the signal is passed through a second filter. This filter is digital and improves the selection process, so that the device will have greater sensitivity. If the light causing this signal has originated from the modulating LED source, it will be allowed to pass through the filter and trigger the output. If the light is modulating at a different frequency, it will be rejected by one of the two filters.

After the signal is passed through the digital filter it is demodulated and sent to the output amplifier. The output section is very similar to the output section in the through-beam device and requires that the load be connected in series with it. Light operation or dark operation is selectable through an externally mounted switch as in the through-beam type. The retroreflective switch also has built-in protection against short circuits in the output stage, reverse polarity of dc power supply

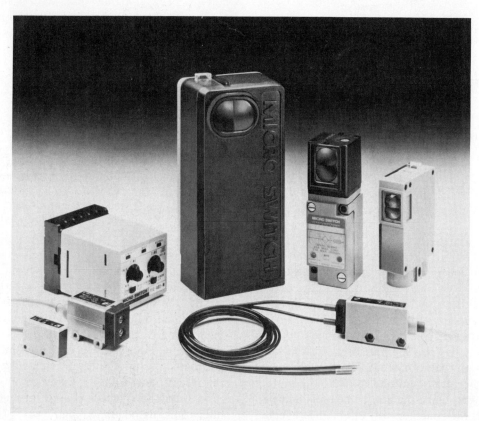

FIGURE 13-9 Retroreflectrive type photoelectric switches. (Courtesy of Micro Switch a Honeywell Division)

356   Chap. 13 / *Photoelectric and Proximity Controls*

╔══════════════════════ **Application Examples** ══════════════════════╗

• When the sensor is susceptible to the reflection from background object surface

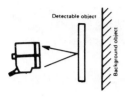

Typical examples
(1) Sensing of thin objects on the conveyor line.
(2) Sensing of objects in the presence of a background object with high reflection factor such as rollers, metallic plates, etc.
(3) Sensing of the residual quantity in a hopper or a parts feeder.

• Sensing of level or height

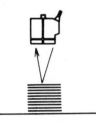

Typical examples
(1) Sensing the height of stacked plywood, tiles, etc. from above.
(2) Monitoring and control of the liquid level from above.
(3) Determination of the heights of objects on a conveyor line.
(4) Sensing of slack in sheets from above.

• Sensing of objects traveling in contiguous succession

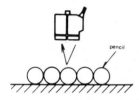

Typical examples
(1) One-by-one sensing of pencils or metallic bars traveling successively or in contiguous succession.
(2) Similarly, one-by-one lateral sensing of bottles or cans traveling in contiguous succession.

• Sensing of small, slender or fine objects

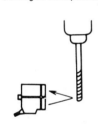

Typical examples
(1) Sensing of broken drill bits.
(2) Sensing of small parts such as electronic components.
(3) Sensing of the presence or absence of bottlecaps.
(4) Sensing of fine mesh.

• Sensing of small holes, narrow openings, or unevenness

Typical examples
(1) Sensing of holes in flat board.
(2) Sensing of protrusions.

• Sensing of objects utilizing their difference in luster

Typical examples
(1) Identifying the face or back of tiles.
(2) Identifying the face or back of lids.

• Sensing of transparent objects

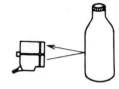

Typical examples
(1) Sensing of transparent or translucent objects.
(2) Sensing of transparent glasses, film or plastic plates.
(3) Sensing of the liquid level.

• Sensing of objects through a transparent cover

Typical examples
(1) Sensing of the contents in a transparent case.
(2) Sensing of the position of meter pointer.

• Sensing of the edge of object

Typical examples
(1) Positioning control of plywood
(2) Positioning control of various other products.

FIGURE 13-10 Applications for retroreflective type photoelectric switches. (Courtesy of Omron)

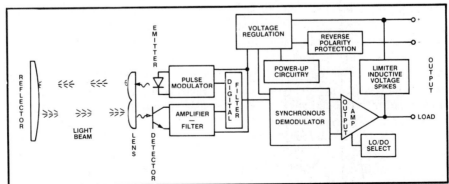

FIGURE 13-11 Diagram of an Allen-Bradley retroreflective type photoelectric switch. (Courtesy of Allen-Bradley Company)

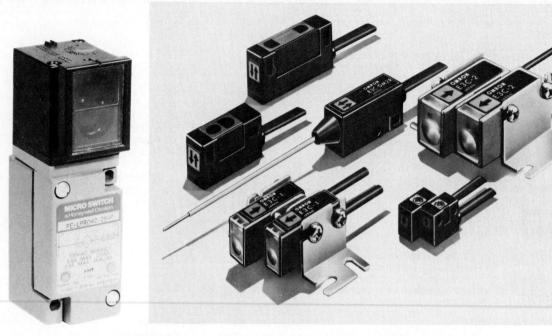

FIGURE 13-12 Self-contained type photoelectric switch. (Courtesy of Micro Switch Honeywell Division)

voltage, and a circuit that provides the voltage-limiting characteristic of inductive spikes in the control circuit.

Fiber Optics and Remote Control Bases. Some companies make a the retroreflective photoelectric switch that uses fiber optic cable to allow the control base to be mounted in a electrical enclosure located for easy access during troubleshooting. The part of the sensor that is mounted near the process is very compact, which allows the sensor to be placed exactly where it can perform the best job of detecting, and the bulky control base can be mounted remotely. As with other fiber optic cables, care must be taken during installation to protect the cable from becoming crimped or shattered.

Another way to solve the problem of mounting the sensor as close as possible to the object or material being sensed is to use compact units that have a very small amplifier built in. These devices can control output directly or be connected to a remote base control that can power a relay. Figure 13-12 shows an example of these compact sensors with self-contained amplifiers. This type of configuration also works well for applications where explosion-proof controls are required. The source/receiver unit, which must be rated for explosion-proof applications, is installed in the explosive environment and the control base is located in an approved cabinet outside the restricted area.

DIFFUSE SCAN PHOTOELECTRIC SWITCHES

The diffuse scan control and specular scan control are very similar in operation. Some switch manufacturers consider them the same device for applications involving different types of reflective surfaces and mounting positions. These types of photoelectric switches operate on the principle of emitting a beam of light on an object rather than on a reflector. The receiver is mounted at a specified angle to the surface to collect the reflecting light.

The main differences between the specular scan and diffuse scan are shown in Figure 13-13. The emitter and receiver for the specular scan application are both mounted at a specified angle. The angle must be identical so that the receiver will receive only light that is reflected from the surface of the conveyor. When the object being detected moves into position, the emitted beam of light will strike the top of the object and reflect at a different angle and miss the receiver. Since this type of control operates from light reflected from the surface of a part, it is also called *proximity* photoelectric control.

Operation of the diffuse scan switch is very similar to that of the specular scan in that the source light beam is focused where it will be reflected by either the background or the material being sensed. The receiver portion of the switch is mounted at an angle where it will receive the reflected light. For diffused scan the source part of the switch is

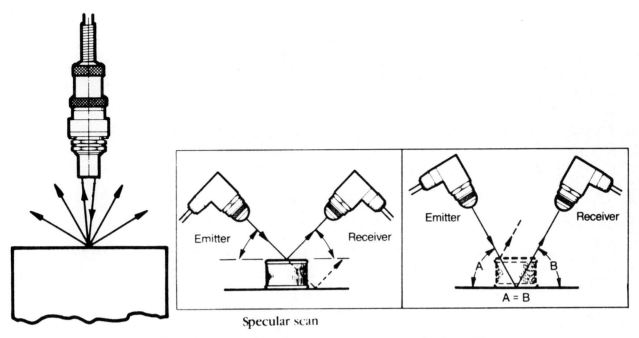

FIGURE 13-13 Diffuse and specular scan type applications. (Courtesy of Micro Switch Honeywell Division)

mounted so that the beam of light strikes the reflective surface at exactly 90 degrees. This will provide the best reflective action for surfaces that do not reflect well.

The diffuse scan works well in applications where the object being sensed is made of paper, mat, cardboard, or similar materials. Since these materials do not reflect light as well as does shiny material such as foil, the sensitivity on the switch will need to be adjusted for proper switch triggering.

The biggest problem with the diffuse scan and specular scan photoelectric switches is that the beam of light must travel twice as far as with through-beam devices. Also, the reflected light may lose some of its intensity when it is reflected from material that may absorb part of the light. Generally, the switches' sensitivity can be adjusted close enough to provide precision control action. A lens can be included with this type control to increase the sensing distance.

The circuitry required for operation of the diffuse scan and specular scan devices is very similar and in some cases identical to the circuitry for retroreflective devices. These controls come in several styles for various applications, just like the other types of controls. These styles include self-contained types and discrete types using a remote base for control.

OUTPUT STAGE FOR PHOTOELECTRIC SWITCHES

The output stage of the photoelectric switch is the most important part of the device when it comes to interfacing the switch to relays, motor starters, programmable controllers, or robots. The reason the output is so important is that it is the active part of the control that must switch voltage on or off to meet specific logic conditions. It is also the part of the device that must match the voltage, polarity, and signal type of the rest of the control circuit. The voltage may be ac or dc and its level may vary from TTL levels of dc through 220 V ac. The device may also need to interface with any number of logic level signals for robot or programmable controller applications. These may range from current sourcing or current sinking to a CMOS interface. In this part of the chapter we provide examples of all the common types of outputs provided and some example of common circuits.

AC Voltage

One type of output is made for controlling 100 to 240 V ac at currents up to 200 mA. On/off control of ac current can be obtained several different ways, which are shown in Figure 13-14. The first way (Figure 13-14a) uses a bridge rectifier, SCR, zener

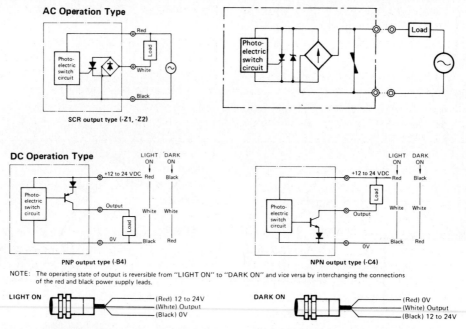

FIGURE 13-14 AC and DC output circuits for photoelectric switches. (Courtesy of Omron)

diode, and varistor. The load is connected in series with the ac part of the bridge rectifier and the ac supply voltage (100 to 250 V), while the SCR is connected in series with the dc side of the bridge rectifier. The SCR has control over the bridge rectifier. If current is flowing through the SCR, ac current can flow through the ac side of the rectifier and through the load device. The load device will be some type of ac relay, motor starter, or solenoid. The gate to the SCR will receive a signal any time the photo portion of the switch is triggered. The signal to the gate comes directly from the amplified circuits in the photoelectric switch. The zener diode will provide protection against overvoltage on the dc side of the bridge, and the varistor will protect against overvoltage on the ac side.

The second type of ac control is similar to the first except that the output from the photoelectric switch's amplifier goes directly to an opto coupler (Figure 13-14b). That will isolate the amplifier and switch from the ac circuit. The output of the optocoupler sends a drive signal to the gate of the SCR. The SCR has control over the dc side of a bridge rectifier, just as in the previous circuit. The ac part of the bridge rectifier will be wired in series with the ac supply voltage (100 to 250 V) and ac load device, which again could be an ac relay, motor starter, or solenoid.

Another type of output circuit for controlling ac loads, uses a triac controlled by a optocoupler. The ac load and power is connected in series with the triac main terminals. When the triac's gate receives a signal from the optocoupler it causes the triac to go into conduction and pass current to the ac load. The optocoupler receives the signal from the amplifier of the photoelectric device. The optocoupler provides isolation between the ac load circuit and the amplifier and optical portion of the photoelectric switch. It also allows the amplifier signal, which is dc, to interface with the triac, which is an ac device.

DC Voltage

Another type of output control is for dc voltages. One variation of the dc-controlled circuit uses a PNP transistor for current-sourcing applications, while another uses an NPN transistor for current-sinking applications. The current-sourcing circuit has a dc load connected in series with a PNP transistor and diode. The amplifier stage of the photoelectric switch provides a small dc voltage signal to the base of the transistor any time the switch has been triggered. The signal on the base of the transistor causes current to flow in the emitter–collector circuit. The diode provides reverse polarity protection.

Circuits for interfacing the photoelectric output to TTL or CMOS circuits are also available. The voltage for this circuit must be supplied outside the photoelectric switch to isolate the switch from the TTL or CMOS devices. This type of circuit is very useful to interface the rugged photoelectric switch with small microprocessor-controlled circuits.

Since programmable controllers and industrial robots are specifically designed to interface with traditional motor controls, they will have input/output modules available for direct interface to all ranges of ac and dc voltages, and this type of circuit would be unnecessary.

Another type of output uses a built-in relay for control. This type of photoelectric switch is different in that the relay's coil is wired directly to the transistor in the output stage and the relay is actually mounted and housed in the switch. The contacts of the relay are isolated from the coil and are available for control of load currents up to 10 A. This type of control allows the photoelectric switch to be used without an additional relay. Remember, if an extra relay is required, as in the ac and dc solid-state controlled outputs, it will require cabinet space for mounting.

Some companies, such as Micro Switch, use plug-in relays for output control. The types of relays available for plug-in mounting are the reed relay, solid-state relay, and electromechanical relay. Operation of these relays was covered in previous chapters. These relays provide control for a variety of different voltages and applications with easy mounting for installation.

The plug-in relay requires a control base for proper mounting. The control base will also house a power supply, which is a step-down transformer that will convert 110 or 230 V ac to low-voltage ac or regulated 12 V dc. This is the same control base that is required for photoelectric switches that do not have self-contained power supply and control circuitry.

The control base can also support a logic card that provides a wide variety of logic functions. This means that the photoelectric switch can execute complex control functions as a stand-alone control or in conjunction with robots and programmable controllers. These functions include time delay-on, time delay-off, latched output, nonrepeat electronic latch one-shot pulsed output, zero speed detector, and division counter (similar to a four-input AND gate or OR gate, where all four inputs are considered to get an output). Each of the types of cards except the zero speed detector also include a switch to select light or dark operation and a sensitivity adjustment that must be set at the full clockwise position if they are used with modulated LED controls. They use an edge connector for easy push installation.

Two-Wire and Three-Wire Control

Another way to classify outputs for photoelectric controls is by the number of wires that are required to make them operate correctly (Figure 13-15). The two-wire control (Figure 14-15a) is self-contained control. The retroreflective and diffuse scan controls are available as small stand-alone two-wire controls. This control is connected in series with the load and will receive operating power from this voltage. When the proper amount of light is received, the switch portion of the control will become energized and the current needed to power the load will be allowed to flow.

A diagram of three-wire control is shown in Figure 13-15b. The control receives power to operate from a separate wire than that from which the load receives its power. When the photoelectric switch energizes, the load current will flow through the control to power the load. Each of these types of controls provides a smaller control that is usable in stand-alone applications. In each control the amount of load current is limited to 300 to 500 mA. This type of device is ideal for interface to programmable controllers and robots.

SELECTING A PHOTOELECTRIC SWITCH FOR YOUR APPLICATION

The installation of any photoelectric switch starts with selecting the correct type of switch for your application. The selection of a switch type involves

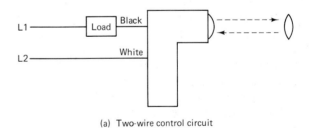

(a) Two-wire control circuit

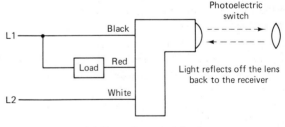

(b) Three-wire control circuit

FIGURE 13-15 Diagram of 2 wire and 3 wire photoelectric controls.

APPLICATION EXAMPLES

BOTTLING MACHINE

Label detection

Three sensors are used to detect the front and back labels of a bottle approaching on the conveyor. When no label is discovered, a defect signal is output for a predetermined period from the S3D sensor controller

- Sensors used: Photoelectric switches
  - IN1: } Reflection type (LIGHT-ON mode: Model with suffix E1 in
  - IN2: } type number)
  - IN3: Separate type (DARK-ON mode: Model with suffix E2 in type number)

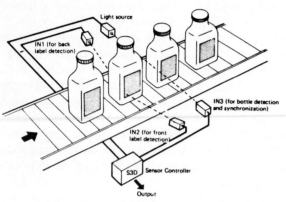

NOTE: Sensors IN1 and IN2 must be installed oblique to the bottle surfaces to avoid reflections.

FOOD PROCESSING MACHINE

Detection of liquid level in the tank

A glass bypass pipe is provided on the tank to detect liquid level. The actual level of liquid with foam (such as beer) can also be detected.

- Sensors used: Proximity switches
  - IN1: Capacitive type
    (Type E2K-C25ME2: NC contact type)
  - IN2: Capacitive type
    (Type E2K-C25ME1: NO contact type)

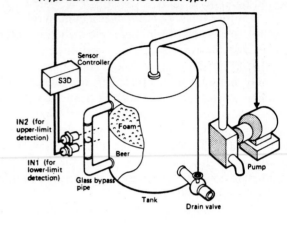

TEXTILE MACHINE

Detection of feeding speed of tape, paper, or cloth

Uneven, insufficient, or excessive feeding speed of paper tape, rolled paper, or cloth can be detected and stabilized using sensors. A change in the feeding speed of materials is detectable from the tension roller position in feeding which varies according to the feeding speed.

- Sensors used: Photoelectric switches
  - IN1: } Separate type
  - IN2: } (DARK-ON mode: Model with suffix E2 in type number)
  - IN3: }

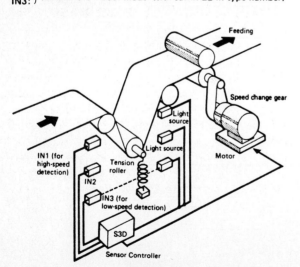

FEED CONVEYOR

Determination of dimensional accuracy

Limit switches are used to determine whether the dimensions of a work are within the prescribed limits. If they do not meet the specifications, the pusher removes the work as a defect.

- Sensors used: Limit switches
  - IN1: }
  - IN2: } Limit switches (with SPST-NO contact output)
  - IN3: }
  - IN4: }

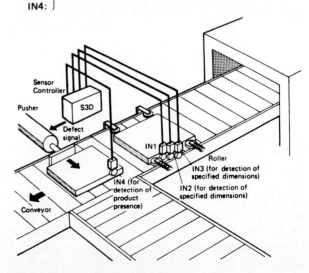

FIGURE 13-16 Applications of photoelectric switches. (Courtesy of Omron)

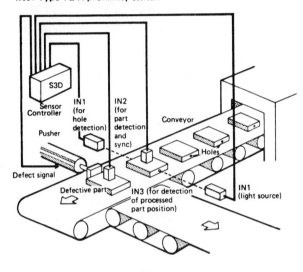

MACHINE TOOL

Detection of defective pressed parts
Three sensors are used to detect the presence or absence of a hole in each pressed metal part on the conveyor and to remove the defective part (without hole) from the process line. A photoelectric switch is used to detect the presence of a hole and a pusher is used to remove the defective part.
- Sensors used: Proximity and photoelectric switches
  IN1: Separate type photoelectric switch (DARK-ON mode: Model with suffix E2 in type number)
  IN2: Type TL-X proximity switch
  IN3: Type TL-X proximity switch

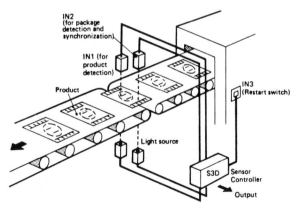

PACKAGING MACHINE

Detection of a product in the package
The product in a package is checked after processing by the automatic packaging machine. When a package without product is detected, the packaging machine is stopped. Restarting the machine is performed manually with the Restart switch after removing the defective package.
- Sensors used: Photoelectric switches
  IN1: } Separate type photoelectric switches
  IN2: } (DARK-ON mode: Model with suffix E2 in type number)
  IN3: Pushbutton switch for restart

FIGURE 13-16 Applications of photoelectric switches. (Continued)

determining the object to be sensed, the distance the source must be mounted from the receiver or reflector, the response time, the type of ambient conditions (dust, dirt, or other particulate in the air), and the voltage type and level of the load devices. Each of these conditions can be addressed by carefully checking the specification sheets and data sheets listing switch characteristics for each of the applications being considered. The specification and data sheets are provided in manufacturers' catalogs and product literature.

Figure 13-16 is presented to give you a better idea of types of applications where photoelectric switches can be used to sense parts being present or in the correct location. These are just a few examples of the many applications where photoelectric switch can be used.

Ambient Conditions. Part of the selection process includes determining the ambient conditions where the switch must operate. Tables are available to show the amount of contaminate in a particular environment and an example of where this type of environment may be found. The suggested amount of gain required for each application is also provided. This type of table is used to determine the estimated amount of particulate and what type of photoelectric switch will operate best in these conditions.

You may find that several different types of photoelectric switches will work well for your application. If this is the case, other conditions such as pricing, stocking, and parts replacement, should enter into your decision for selection.

INSTALLING THE PHOTOELECTRIC SWITCH

After you have determined which type of switch to use, the mounting part of the installation can begin. For this example a three-piece photoelectric switch is being used (Figure 13-17). The three parts are the receptable, sensor body, and the photoelectric sensor head. The receptacle is the part that is actually mounted solidly to the bracket or support. All wiring to the switch is connected to screw terminals in the receptacle. The terminals are numbered or identified to match the wiring diagram of the switch.

The sensor body is the part of the switch that contains the electrical controls, power supply, and output devices. It is connected to the receptacle part of the switch once the control wiring has been connected to the receptacle's terminals. The sensor body has electrical pins that mate with pins in the receptacle to complete the control circuitry. The sensor body is connected to the receptacle in this

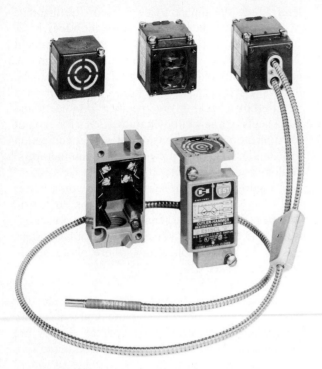

FIGURE 13-17 Photoelectric sensor ready for installation. (Courtesy of Eaton Corporation, Cutler Hammer Products)

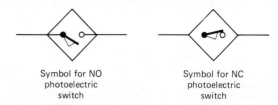

Notice that limit switch symbol is placed inside the symbol for static control device

FIGURE 13-18 Photoelectric switch symbols.

manner so that the electrical part of the switch can be changed quickly, without rewiring any part of the switch.

The photoelectric head is the part of the switch that contains the photoelectric sensors. This part of the switch is also connected in a way that it can easily be changed if problems arise. This also allows the sensor body and receptacle to be used as a base for other types of switches, such as inductive or capacitive proximity switches. This allows the user to stock fewer parts and still be able to provide a quick change out of any inoperative components.

Wiring the Switch. Once the switch is mounted the control wiring must be completed according to the control wiring diagram. The wiring diagram uses symbols to show the location of terminals where field wiring is connected and shows whether the switch contacts are normally open or normally closed. The symbol for the photoelectric switch can be shown in several ways. One of the symbols in Figure 13-18 shows a box setting on its corner with switch contacts inside. This is the symbol for all types of proximity switchs. Since photoelectric switches are used to detect the location of objects, they are considered proximity switches.

Another type of diagram used by the electrician is the ladder diagram of the photoelectric switch (Figure 13-19). Remember: the purpose of the wiring diagram is to show the location of the switch and its terminals, while the purpose of the ladder diagram is to show the sequence of operation of the circuit. This means that the contacts of the photoelectric switch are located on the line of the load (relay coil, solenoid, or motor starter) that they are controlling.

Troubleshooting Photoelectric Switches. The photoelectric switch is easy to troubleshoot. You must consider problems concerning both the electrical portion of the switch as well as the optical part of the switch. Figure 13-20 shows a typical troubleshooting chart that you will use on the job.

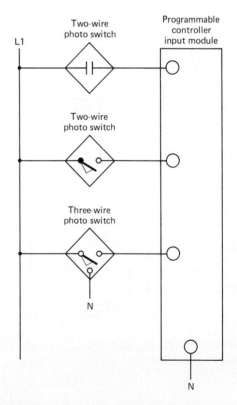

FIGURE 13-19 Photoelectric switch interfaced to a programmable controller.

364 Chap. 13 / *Photoelectric and Proximity Controls*

Symptom	Output Configuration				Possible Cause	Solution
	Current Sink and Dark Operate	Current Sink and Light Operate	Current Source and Dark Operate	Current Source and Light Operate		
	Output LED Indication					
Load Will Not Energize	On A, B, C Off D, F, H	On A, B, C Off D, E, G, I	On E, G, I Off A, B, C, D	On F, H Off A, B, C, D	A. Power supply voltage incorrect. B. Wires to the load or switch loose. C. Load device faulty or incorrect. D. Power supply off or polarity incorrect. E. Switch and reflector not aligned. F. No target in front of the reflector, or target too small. G. Target or other object in detectable position. H. Target or background too reflective. I. Reflector missing, wrong reflector, or reflector out of range.	A. Apply correct voltage. B. Reconnect and tighten wiring. C. Replace with correct load. D. Apply power or correct polarity. E. Align switch and reflector. Refer to **INSTALLATION-Initial/Final Alignment**. F. Move target over reflector; increase target size, or install smaller reflector at closer range. Refer to **SENSING CHAR.-Target Size**. G. Remove target or object. H. Change target or cover with a non-reflective material; cover background. I. Install correct reflector at proper range. Refer to **SENSING CHAR.-Operating Range**.
Load Will Not De-Energize	On D, F, G Off A, B	On C, E Off A, B	On A, B Off C, E	On A, B Off D, F, G	A. Load miswired or bypassed directly to power supply. B. Load device faulty or incorrect. C. No target in front of the reflector, or target too small. D. Switch and reflector not aligned. E. Target or background too reflective. F. Target or other object in detectable position. G. Reflector missing, wrong reflector, or reflector out of range.	A. Correct wiring. B. Replace with correct load. C. Move target over reflector; increase target size, or install smaller reflector at closer range. Refer to **SENSING CHAR.-Target Size**. D. Align switch and reflector. Refer to **INSTALLATION-Initial/Final Alignment**. E. Change target or cover with a non-reflective material; cover background. F. Remove target or object. G. Install correct reflector at proper range. Refer to **SENSING CHAR.-Operating Range**.
Load Energizes and De-Energizes Intermittently	On & Off Intermittently A, B, C	On & Off Intermittently A, B, C	On & Off Intermittently A, B, C	On & Off Intermittently A, B, C	A. Loose wire/terminal connections; broken wire. B. Switch or reflector loose and vibrating out of alignment. C. Target or other objects fluctuating in and out of the detection area.	A. Tighten connection, re-wire. B. Refer to **INSTALLATION-Initial/Final Alignment**. Re-align switch and reflector and tighten to each respective mounting. C. Secure target to minimize movement; remove other objects from detection area.

FIGURE 13-20 Troubleshooting guide for photoelectric switches from Allen-Bradley. (Courtesy of Allen-Bradley Company)

MAINTENANCE — Recommended maintenance to the switch consists of periodic cleaning of the lens with a soft **dry** cloth. DO NOT use solvents or cleaning agents on the lens. If the lens is damaged, it can easily be replaced. Remove the screws with a No. 1 cross-recessed screwdriver. The replacement lens (Part No. 77101-802-51) can now be inserted. Tighten the screws until they are snug.

PROXIMITY SWITCHES

Proximity switches are used in a variety of applications for sensing the presence of parts and locating fixtures. These applications include sensing distance or location, and sensing the presence or absence of parts at high speeds. One application is the sensing of missing parts. The proximity switch in this application is set up to sense missing aluminum caps on a bottling line. This particular switch senses the presence of the caps. If a cap is missing, the sensor trips the output and the bottle is moved off the main line to a recapping line. Proximity switches are particularly well suited for these applications because they can sense a variety of ferrous and nonferrous metal products as well as a variety of nonmetallic materials and liquids. The proximity switch provides functions that are similar to a limit switch in that it can detect the presence or absence of an object without making contact with the part. For this reason it is called a noncontact limit switch. These switches are based on several different theories of operation used for a variety of sensing techniques and include the eddy current killed oscillator (ECKO), high-frequency capacitive oscillator, Hall effect transducer, and reed relay. Figure 13-21 shows several examples of proximity switches and a typical wiring diagram.

Proximity switches are readily adaptable to today's control systems because they are compact and totally solid-state devices. This allows them to be mounted very close to the material they are sensing, in highly automated locations. They come in a variety of self-contained sensor bodies as well as separate sensors with remote bases for providing power supplies, amplifiers, and outputs.

Proximity Switch Operation

Proximity switches are actually a broad family of switches that are used as sensors in industrial control systems. They include the photoelectric switches discussed earlier in this chapter and a variety of circuits that can sense the presence of various materials. The different classes of material require different sensing circuits.

One class of material that is commonly sensed by proximity switches is ferrous and nonferrous metals, including iron and steel products, aluminum, brass, and copper. The main type of circuit used to sense these materials is the *eddy current killed oscillator* (ECKO). An ECKO proximity switch is made of three basic parts: the sensor, which is part of a tuned *LC* tank and oscillator circuit, a solid-state amplifier, and a switching device.

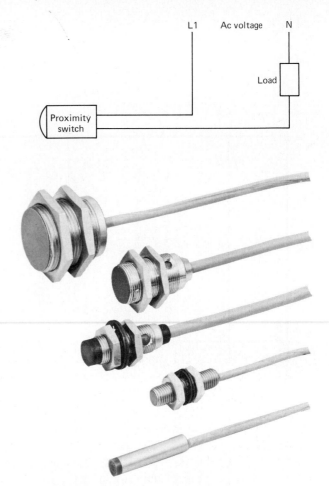

FIGURE 13-21 Typical proximity sensing circuit with typical proximity switches. (Courtesy of Eaton Corporation, Cutler Hammer Products)

Figure 13-22 shows a diagram of an ECKO proximity switch. Honeywell Micro Switch describes the operation of their switch this way. Dc voltage enters the switch through the regulator circuit. The regulator will maintain a constant supply voltage. The oscillator generates a radio-frequency (RF) field out the end of the sensor. This is accomplished by mounting the coil part of the tank circuit near the end or head inside the proximity sensor. The tuned tank circuit oscillates at a predetermined frequency to produce the RF field. When metal is brought near the RF field (at the sensor end of the switch), eddy currents are formed in the metal that kill the oscillations in the tank circuit. The integrator part of the circuit converts the sine-wave signal generated by the oscillator into a dc signal. The dc signal, which varies in amplitude with the amplitude of the oscillation, is sensed by the Schmitt trigger and converted to a digital (off/on) signal. The digital signal from the Schmitt trigger is used to power the output transistor.

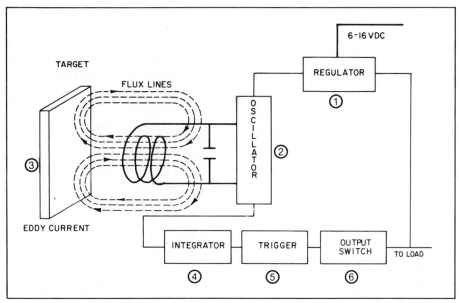

Eddy Current Killed Oscillator

FIGURE 13-22 Description of operation for Eddy Current Killed Oscillator. (Courtesy of Micro Swtich a Honeywell Division)

A similar type of proximity switch that is powered by ac voltage is shown in Figure 13-23. This type of switch is connected in series with the ac load it is controlling. The hot line of the ac voltage is wired to the load. From the load the voltage is sent to a noise suppression circuit in the proximity switch. The noise suppression part of the switch consists of an *RC* circuit for filtering and a metal-oxide varistor that is used to shunt unwanted spikes on the ac line.

The voltage passes from the suppression circuit to a full-wave bridge rectifier. In this case the rectifier is acting as an ac switch that is being controlled by the dc side of the circuit. A SCR is used to switch the current in the dc circuit on or off. If dc voltage is allowed to flow through the bridge rectifier, ac current will also flow through the ac side of the bridge. If current is interrupted on the dc side of the bridge by the SCR, it will also interrupt current on the ac side of the bridge and thus switch the current off and on to the ac load.

The oscillator and sensing coil at the end of the switch operate like a dc ECKO proximity switch. Whenever a metal object enters the RF field produced at the end of the sensor, the oscillations are killed and the Schmitt trigger activates the

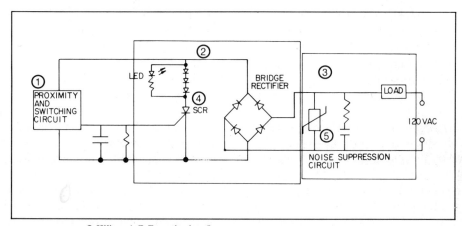

2-Wire AC Proximity Sensor

FIGURE 13-23 Diagram of AC type, Two wire ECKO Proximity Switch (Courtesy of Micro Switch a Honeywell Division)

Proximity Switches 367

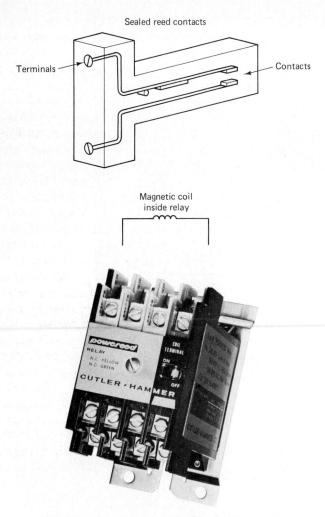

FIGURE 13-24 Ferromagnetic actuated reed type relays. (Courtesy of Eaton Corporation, Cutler Hammer Products)

amplifier circuit. The amplifier in the ac switch is used to trigger or gate the SCR. An LED is connected with the SCR circuit to indicate when the switch is on. The LED in this case will cause a small voltage drop to the load, but it is very useful as an external indicator of the switch's condition and can be seen at some distance by the electrician or technician during troubleshooting. You may also need to take into account that a small amount of leakage current will pass to the load when the switch is in the off state. This leakage current is passing through the suppression circuit. If it is important to have absolutely no leakage current at the load, you can use the coil of a relay as the load to the proximity switch and through the relay's contacts, control the load that cannot tolerate leakage current.

Another type of switch is the *capacitive proximity switch*. This switch works on a principle similar to that of the inductive switch. In the capacitive proximity switch, the tank circuit, which includes the sensor, is tuned to the dielectric field immediately surrounding the outside of the sensor head. Any changes in the dielectric field (capacitance of the area around the sensor) will cause the switch to trigger. During installation the sensitivity of this switch can be set to sense the presence or absence of the material as it moves through the sensor's field. As material (the target) moves through the field, the field's capacitance is changed.

The sensitivity of the switch is adjusted so that the switch will not trigger in ambient air. This means that when no part or material is in the sensor field, the switch will not trigger. As soon as material enters the field and changes the dielectric characteristics, the switch can be adjusted to trigger on that material.

This type of proximity sensor is well suited to measuring liquids since the sensor can be fully immersed in the liquid. Since liquids have different dielectric characterisitcs than those of air, the switch can be sensed to trigger on the liquid when it covers the sensor. If the level lowers, and the sensor is now exposed only to the air, the dielectric field will change and the switch will turn off.

The sensor does not have to be immersed to work well with liquids. A glass tube can be used to make the measurement. When the liquid level in the tank rises, it will also rise in the glass tube. When the liquid level in the tube rises to a point where it moves past the sensor, the capacitance in the field in front of the sensor changes and triggers the switch. The sensitivity of the switch must be set so that it will not trigger when the liquid level in the glass tube is below the sensor.

The sensing distance of the capacitive proximity switch may be affected by elements in the air, such as humidity and dust concentration. If the material being sensed is moving very rapidly, the pulse will be of short duration and may not be long enough to trigger the output circuit. By adjusting the sensitivity, the capacitive proximity switch can be set to operate correctly under these conditions. If the pulse is still too short, a latching output can be used to overcome the problem.

Since the capacitive proximity switch can be adjusted to trigger on any change in the dielectric field around the sensor, it is capable of detecting ferrous and nonferrous metals as well as nonmetallic materials. This makes the capacitive proximity switch a universal type of sensor.

The main problem with capacitive proximity switches is that they may be very sensitive to background or ambient conditions. This means that changes in the air due to contamination or electrical fields or even strong RF signals may cause the switch to trigger falsely. This problem also prohibits locating sensors too close to each other. The main correction for this condition is to use sensors that have built-in shielding. This shielding helps to focus the capacitive field where the switch is sensing. In this way the sensor can be focused directly on the area where the target (parts or material that is being sensed) will pass through.

Applications for Proximity Switches

Proximity switches presently use as industrial controls for detecting motion on machinery and robotic work cells are also being used to detect parts in place for automated applications. Figure 13-24 shows applications using proximity switches as level indicators.

Motion detectors are also available in bistable styles. The bistable proximity switch is capable of sensing motion in two directions. This is useful in automated applications where cylinders must move out and back or where robots, mills, and lathes use X,Y axis motion. Other applications include label detection, parts counting, missing parts detection, broken tool detection, and machine on/off control. A full list of applications would involve almost every parameter of automation control.

Proximity switches are available in a variety of models and are usable under nearly all conditions in industry. Among these are watertight, oiltight, and totally submersible models. Other switches are available for high-pressure applications (0 to 5075 psi), high-temperature applications (32 to 392°F), and for use where extremely large magnetic fields are present, such as in welding applications (30,000 A). Proximity switches have also been very successful in applications in hazardous locations such as explosion-proof environments.

Hall effect sensors can be used with other mechanical and electronic components to provide very accurate sensing capabilities. Figure 13-25 shows several of these applications. The first application shows a circular vein with windows. This looks like a cap with holes cut in it. The cap or vein rotates on a shaft. the Hall effect sensor detects the presence of the tooth or window. In this manner it is able to count the teeth as they go by. This count is sent to a logic circuit that is able to produce an rpm value for the shaft. This type of sensor is used quite often to measure shaft speeds in motors and also as flow-rate indicators by measuring the speed of a flow-rate turbine.

Types of Proximity Switches

Proximity switches come in several types or models for a variety of industrial applications, including both self-contained and remote sensors (Figure 13-26). The *self-contained sensor* has the sensor, oscillator, amplifier, and output all in one body. The *remote sensor* has the sensor mounted with the oscillator part of the circuit mounted near the parts that are to be detected. The base, which includes the power supply, voltage regulator, amplifier, and output, is mounted in the electrical cabinet, where it can be troubleshooted. Both the self-contained and remote styles of proximity switch are available in several designs. The designs include the tube type, which is threaded; the rectangular type, which is mounted similarly to a photoelectric switch; and the pancake style, which is surfaced mounted. The tube switch can be mounted in vertically or horizontally for various detecting applications. Tube sensors are available in long-barrel or short-barrel applications. Long-barrel and short-barrel sensors are especially useful in setting proper sensor position. In most

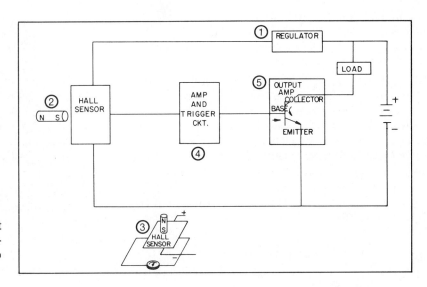

FIGURE 13-25 The Hall Effect principal and examples of Hall Effect switches. (Courtesy of Micro Switch a Honeywell Division)

FIGURE 13-26 Typical styles of inductive and capacitive proximity switches. (Courtesy of Micro Switch a Honeywell Division and Eaton Corporation, Cutler Hammer Products)

cases these sensors can be used through any thickness of material, such as mounting plates, tables, machine beds, or similar applications, with little adjustment required.

Another switch is the *slot* or *grooved head* type. This type of proximity switch has the sensor on one side of the grooved head and the target on the other side of the head. Any material passing through the groove will trigger the proximity switch. This type of switch is generally used in applications where positioning is critical. For example sheet material such as aluminum or steel and material such as conveyors can be aligned through the slotted head, and any variation in the travel between the grooved head will cause the switch to deflect. Also, any time material would be absent, such as when a belt breaks or the foil breaks in a roll of aluminum, the switch would trigger.

Operating Characteristics

To fully understand all of the operating characteristics for the proximity switches, it is important that you understand the terminology. In the first part of this section we define several terms common to proximity switch characteristics.

Detecting distance The distance at which a proximity switch operates. This distance is measured from the tip of the sensor to the target (the target is the part being detected: for example, bottles on a bottling line). The detecting distance is calculated in the specifications by using a standardized-sized target.

Standard target A piece of material whose size, shape, and type are controlled for test measurements. This standard is identified for measurement conditions only. Differential travel is the distance measured from the point where the proximity switch first activates on picking up the target, to the point where the target is located when the sensing element deactivates the switch. The differential travel will very slightly depending on how close the target is to the sensor when it is passing the detecting location.

Response time The time it takes a proximity switch to locate the target and activates its output.

Response frequency The time it takes the switch to cycle. This is the time a switch is activated and then deactivated and ready to be activated again. This particular term is critical in applications where proximity switches are sensing vein- or tooth-type revolving targets.

Horizontal detecting distance The distance measured horizontally out from the sensor to the target.

Accuracy The ability of a sensor to determine the actual value being measured.

HALL EFFECT SENSORS

The Hall effect sensor works on a rather simple electromagnetic principle. The Hall effect is made from a thin sheet of conductive material cut into a square and it has two main terminals connected to opposite sides of the square, which allows current to flow through it. Each of the remaining two sides has a small terminal connected to it that will be used to remove the Hall effect potential (voltage). These terminals are marked + and −.

A Hall effect voltage is produced at these two terminals when current is passed through the material via the main terminals and a magnetic field is passed across the sheet of material at right angles (perpendicular). The amount of voltage produced at the voltage terminals will be proportional to the amount of current flowing through the material and the strength of the magnetic field passing perpendicular to the material.

This principle was first observed by Edwin Hall in 1879 when he was completing doctoral research at Johns Hopkins University in Baltimore, Maryland. He made this observation while studying an earlier theory on electron flow proposed by

Kelvin. Through his experiments using gold foil as the current-carrying material, Hall found that the effect could produce only small amounts of voltage, in the range 20 to 30 µV. At that time, this small amount of voltage had no useful purpose. In the 1950s the experiments found their first applications when semiconductor material was substituted for the gold foil. These applications were limited by the lack of a well-developed solid-state amplifier. In 1965, Everett Vorthmann and Joe Maupin, who were senior engineers with Micro Switch, produced the first Hall effect application, incorporated on a single silicon integrated circuit. This provided Micro Switch with a functioning circuit that has been incorporated into solid-state keyboards and many proximity switch applications.

The Hall effect is used in both digital (on/off) and analog sensors. In the analog applications, the sensor is generally set up with a constant current passing through the semiconductor material, called the Hall element. The voltage produced is considered to be proportional to the strength of the magnetic field passing perpendicular to and across the Hall element. The maximum amount of Hall voltage, approximately 20 µV, will be produced when the magnetic field has saturated the Hall element.

In this chapter we have provided all the information required to understand the operation and troubleshooting of photoelectric and proximity controls. If you have trouble understanding the operation of the device or the troubleshooting procedures, you should review the information provided in the chapter. If you need aditional help in troubleshooting these control circuits, you should review the testing procedures provided in Chapter 8.

QUESTIONS

13-1. Explain the operation of a modulated light source photoelectric control.

13-2. List the three types of photoelectric controls.

13-3. Explain how a cadmium sulfide cell works.

13-4. Explain the advantage of modulating a light source used in a photoelectric control.

13-5. Sketch a diagram of the components used in a through-beam photoelectric control.

13-6. Explain the advantage of using a C-shaped photoelectric sensor.

13-7. Explain how fiber optics are used in photoelectric controls.

13-8. Sketch a diagram of the component parts of a retroreflective photoelectric control.

13-9. Explain the advantage of using a remote control base.

13-10. Explain the advantage of using a self-contained photoelectric control.

13-11. Explain the operation of a diffused scan photoelectric control. Where would you find this type of control used?

13-12. Draw an electrical diagram of the output circuit for a photoelectric control for an ac interface and a dc interface.

13-13. Why would you use a relay as the output stage of a photoelectric control?

13-14. Draw an electrical diagram of a two-wire and a three-wire control used in photoelectric switches.

13-15. List four applications of photoelectric switches and state why the photoelectric switch works well in these applications.

13-16. Draw an electrical diagram that shows a photoelectric switch interface with a programmable controller as an input.

13-17. Explain the steps in installing, adjusting, and testing a photoelectric control.

13-18. Why must the sensitivity of a photoelectric switch be adjusted?

13-19. Explain how you would troubleshoot a photoelectric control.

13-20. Explain the operation of a proximity control.

13-21. Draw a sketch of the internal parts of a proximity switch.

13-22. Draw the electrical symbol for a photoelectric and proximity controls.

13-23. What does "ECKO" stand for?

13-24. Explain the difference between an inductive proximity switch and a capacitive proximity switch.

13-25. Explain how you can quickly test a proximity switch to determine if it is operating correctly.

13-26. Explain the operating characteristics that are provided with the data sheets for a proximity switch. Be sure to include why each characteristic is important to the selection of a control.

13-27. Explain how you can adjust a proximity switch to ensure that it is sensing correctly.

13-28. Draw an electrical diagram of a two-wire and three-wire proximity control connected to a motor starter as a load.

Advanced motor control circuits 14

INTRODUCTION

At times you will be asked to install or troubleshoot more advanced motor control circuits that are used to control the way a motor is started or stopped. These circuits are called acceleration and deceleration circuits. *Acceleration circuits* are used to start a motor without allowing the locked-rotor current to become too large. The locked-rotor current for a motor during starting can be up to six times the normal operational current. This can cause problems if the distribution system is loaded to near capacity, because the excessive current draw can cause interruption to the whole system. The excessive starting current can also cause the demand factor on the electric meter to become too large, which doubles or triples the electric bill.

Another problem created by the large locked-rotor currents is the wear and tear on switchgear. When motors are allowed to draw maximum current, they cause arcing and heat buildup that stress contacts and switchgear. This stress causes the equipment in bus ducts, disconnects, and motor starters to wear out prematurely. If locked rotor current is limited during starting, the life of the switchgear can be extended more than enough to pay for the more complex circuits required to control the motors.

A problem may also arise when loads are started with full torque. The starting torque of a squirrel-cage motor can be as large as 140 percent of the normal operating torque. This may become a problem with loads such as a conveyor. When the large torque is applied during starting, the material on the conveyor belt may be spilled.

It is also important to control the time it takes a motor to stop. In some applications it is important that the load stop at exactly the time and location when the motor is deenergized. In normal motor operation, when a motor is deenergized, the load is allowed to coast to a stop, which means that the larger the load is, the longer the coasting time. This causes the load to be located at random, which may be unacceptable in motion control applications. In other applications it is unsafe to allow the load to coast to a stop. This is true where large cutting blades are turned at high speeds in machine tool or wood-cutting applications and are allowed to continue to rotate after power to the motor has been deenergized.

In all these cases, motor control circuits or hardware must be provided to bring the motor to a stop quickly when power is deenergized. These circuits, called *deceleration circuits,* may involve shorting terminals of the motor to cause regeneration of voltage back into the rotor, or may involve hardware such as some type of mechanical or electrical brakes.

In this chapter we provide information with diagrams to explain operation of these types of circuits. Other information regarding wiring of basic components will aid in installation and troubleshooting of each circuit.

ACCELERATION CIRCUITS

Acceleration circuits allow motors to come up to speed without drawing excessive locked-rotor current (LRA). These circuits are designed for dc and ac three-phase motors. They can utilize several theories of operation. One way to reduce the LRA is to reduce the amount of voltage applied to the motor while the motor is starting. This can be accomplished by using resistors in series with the supply voltage to the motor, or reducing the amount of voltage supplied by the transformer. Another way to reduce the amount of LRA is to change the winding configuration of the motor during the starting. When three-phase windings are connected in a wye configuration, they will draw less current than when they are connected in delta. After the motor is up to speed, the motor windings can be converted to delta to provide better running torque and operation.

Another method of reducing current involves using only part of the motor winding when the motor is started. After the motor is running, the remainder of the winding is reenergized so that the motor is operating at full horsepower. This type of acceleration requires the motor load to be reduced during starting, and it can be increased after the motor is up to speed.

REDUCED-VOLTAGE STARTERS

Five types of reduced-voltage starting circuits are presented and explained in this chapter. These circuits are compared to each other and to an across-the-line starter with regards to the amount of voltage applied to the motor at starting, the line current, the starting torque, the type of transition, the cost, advantages and disadvantages, and applications.

Figure 14-1 shows a table with these comparisons. The solid-state reduced-voltage starter is not included in this table, but we discuss it at the end of this section, following our discussion of the primary resistor, autotransformer starter, wye–delta starter, part winding starter, and secondary resistor. Two graphs are also presented in this figure to compare line current versus speed characteristics and torque versus speed characteristics. These ratings are based on a typical NEMA design B motor. The information in these graphs is presented in this section to provide you with a comparison of these types of starting circuits. Operation of these circuits is also explained.

Open and Closed Transition Starters. The classification of open and closed transition starters are used to explain the operation of the switching contacts in these types of starters. Each of these types of starters provides a means of starting the motor on the reduced-voltage circuit, and then switching to the normal across-the-line configuration to allow the motor to go into its run mode. If the operation of the contacts is such that the reduced-voltage circuit is completely disconnected before the run circuit is connected, the starter is called an *open transition starter*. During the time when the first set of contacts have been opened and the run set have not yet closed, the motor is completely disconnected from power—and hence the name "open transition."

If the operation of the contacts allow the run set of contacts to close before the first set have been fully disconnected, the starter is called a *closed transition starter*. During the time when the run set of contacts are closed, the starting circuit is still connected. This means that the motor is never fully disconnected from the power source except when the control circuit is completely disconnected.

PRIMARY RESISTOR STARTERS

The primary resistor reduced-voltage starter allows a large resistor to be connected in series with the supply voltage for a specified amount of time until the motor is running approximately 65 percent of full rpm. It provides approximately 42 percent of normal starting torque and limits starting current to 65 percent. This type of starting configuration provides smooth acceleration since voltage will increase with the motor's speed. It also provides a high power factor during starting and is less expensive than other types of reduced-voltage starters. This type of starting can utilize up to five separate resistors to provide stepped acceleration. This type of starting circuit is useful for belt and gear drives, conveyors, and textile machine application.

Primary Resistor Starter Operation

Figure 14-2 presents a wiring diagram and a ladder diagram that shows the basic components and the sequence of operation for this type of circuit. From this wiring diagram you can see that one resistor is used in each phase of the supply voltage. The resistors are called primary resistors because they

TYPE OF STARTER	STARTING CHARACTERISTICS IN PERCENT OF FULL VOLTAGE VALUES			STANDARD MOTOR	TRANSITION	EXTRA ACCELER. STEPS AVAILABLE	COST OF INSTALLATION	ADVANTAGES	DISADVANTAGES	REMARKS	APPLICATIONS
	VOLTAGE AT MOTOR	LINE CURRENT	STARTING TORQUE								
ACROSS-THE-LINE A10	100%	100%	100%	Yes	None	None	Lowest	• Inexpensive • Readily available • Simple to maintain • Maximum starting torque	• High inrush • High starting torque		Many and various
AUTO-TRANS-FORMER A400	80% 65% 50%	64% 42% 25%	64% 42% 25%	Yes	Closed	No	High	• Provides highest torque per ampere of line current • 3 different starting torques available through auto-transformer taps • Suitable for relatively long starting periods • Motor current is greater than line current during starting	• In lower hp ratings is most expensive design • Low power factor • Large physical Size	• Most flexible • Very efficient	Blowers Pumps Compressors Conveyors
PRIMARY RESISTOR A430	65%	65%	42%	Yes	Closed	Yes	High	• Smooth acceleration — motor voltage increases with speed • High power factor during start • Less expensive than autotransformer starter in lower HP's • Available with as many as 5 accelerating points	• Low torque efficiency • Resistors give off heat • Starting time in excess of 5 seconds requires expensive resistors • Difficult to change starting torques under varying conditions	• Can be designed so starting characteristics closely match requirements of load	Belt and gear drives Conveyors Textile machines
PART WINDING A460	100%	65%	48%	❶	Closed	Yes (but very uncommon)	Low	• Least expensive reduced voltage starter • Most dual voltage motors can be started part winding on lower voltage • Small physical size	• Unsuited for high inertia, long starting loads • Requires special motor design for voltage higher than 230 • Motor will not start if the torque demanded by the load exceeds that developed by the motor when the first half of the motor is energized • First step of acceleration must not exceed 5 seconds or else motor will overheat	• Not really a reduced voltage starter. Is considered an increment starter because it achieves objective by reconnecting motor winding.	Reciprocating compressors Pumps Blowers Fans
WYE DELTA A490	100%	33%	33%	No	Open ❷	No	Medium	• Suitable for high inertia, long acceleration, loads • High torque efficiency • Ideal for especially stringent inrush restrictions. • Ideal for frequent starts.	• Requires special motor • Low starting torque • During open transition there is a high momentary inrush when the delta contactor is closed	• Same as part winding (above) • Very efficient	Centrifugal compressors Centrifuges

❶ Standard dual voltage 230/460 volt motor can be used on 230 volt systems.
❷ Closed transition available for about 30% more in price.

FIGURE 14-1 Table showing the comparison of reduced voltage starting circuits. (Courtesy of Eaton Corporation, Cutler Hammer Products)

are in the supply side of the motor circuit. The resistor in the first phase is identified by the numbers R1 and R2, while the second resistor is numbered R11 and R12. The last resistor in the third phase is numbered R21 and R22. It is difficult to see the sequence of operation in this diagram since it is only supposed to show the location of each component.

The ladder diagram in this figure shows the load circuit and the control circuit. The load circuit shows the connection of the resistors in reference to the motor. From this part of the diagram you can see that one resistor is in series with each phase of the voltage that is supplied to the motor. A set of shorting contacts are connected in parallel with each resistor. These contacts are identified by the letter A, which indicates that they are controlled by contactor A in the control circuit.

The control circuit has a normal start/stop circuit with a set of hold-in contacts around the start button. The contacts for the hold-in circuit are the instantaneous contacts (1B) of the timer motor.

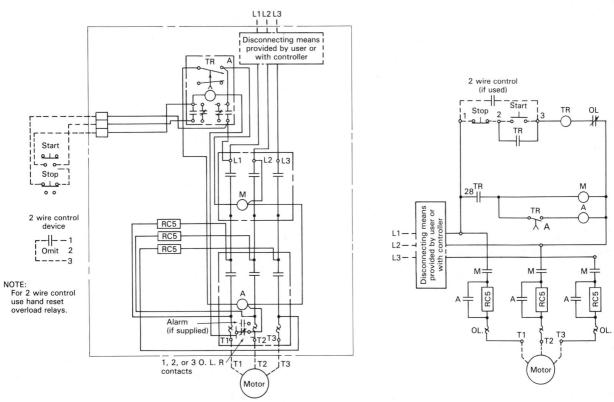

Sizes 3 and 4, 3 Phase Primary Resistor Starter

FIGURE 14-2 Electrical wiring diagram of a primary resistor reduced voltage starter circuit. (Courtesy of Square D Company)

Primary Resistor Starters

These contacts will close immediately when the timer is energized by momentary depression of the start button.

A second set of instantaneous timer contacts (2B) are used to energize the main motor starter. When the coil of this motor starter is energized. it closes the M contacts and provides voltage to the motor through the primary resistors. Since this voltage flows through the fixed resistors, a voltage drop of about 20 percent is produced.

The amount of locked-rotor current is determined by the amount of voltage applied to the fixed resistance of the motor stator. Since the voltage has been reduced by the fixed resistors, the amount of starting current will be reduced proportionately. As the amount of starting current determines the amount of starting torque the motor will be able to produce, the starting torque of this type of starting circuit will be reduced up to 56 percent.

After the reduced voltage has been applied to the motor, the rotor will begin to accelerate. As the rotor accelerates, it will produce an increased amount of counter EMF, which will cause the amount of current the motor is drawing to be reduced. The motor will accelerate to approximately 75 to 80 percent full rpm while the reduced voltage is applied. When the motor reaches this speed, the time-delay contacts from the timer are closed and the A contacts close to short out the primary resistors.

When the A contacts close, they provide a alternative path for current to reach the motor windings without going through the resistors. This connects the motor windings directly to the supply voltage, which allows them to receive full voltage. The increased amount of voltage provides an increased amount of current and torque while the motor is at full speed. The amount of time delay can be adjusted for more or less time to ensure that the motor is at the proper speed before the A contacts are closed and the motor is allowed to run on full voltage.

When LRA and the reduced voltage starter are compared, you can see that the amount of current the motor uses to get up to full speed is much less with the reduced-voltage starter circuit than when the motor is started with full voltage.

Sizes of Primary Resistor Starters

The primary resistor reduced-voltage starter is available in a variety of voltages, including 200, 230, 380, and 575 V. The starters are rated for 5- to 600-hp loads. These starters provide control through motor starters and contactors rated in sizes ranging from a NEMA 1 through a NEMA 7. This provides a complete selection of controls to fit motors for a large variety of applications.

A photograph of a typical primary resistor type starter is included in Figure 14-2. You can see that the resistors are mounted right in the enclosure with the contactors. Since they are mounted in an enclosure, a disconnect and fusing is also available as part of the control.

AUTOTRANSFORMER STARTERS

One problem that occurs with a primary resistor reduced-voltage starter is that all the voltage that is dropped through the resistors is turned into heat. The amount of heat may become very large and cause problems when it cannot be removed quickly enough. A autotransformer reduced-voltage starter is able to provide the same type of voltage reduction without building up large quantities of heat. This type of starting provides the largest amount of torque per ampere of line current. These controls have multiple taps that allow modification of the current or torque characteristics. They also have the ability to provide relatively long startup times. Another advantage of this type of starter is that the autotransformer can produce currents on its secondary side to the motor that are larger than the current in the supply voltage side. Since the current in the supply side is measured by the electric demand meter, this means that the autotransformer starter can produce a larger current without loading up the demand meter.

Figure 14-3 show a wiring diagram and an electrical ladder diagram of the autotransformer starter. A multitapped autotransformer is provided for each phase of the supply voltage. One transformer is placed in series with each phase of the supply voltage. The taps on the transformer are set at 50, 67, and 84 percent of the supply voltage. The tap that is connected will drop the voltage to that percentage of supply voltage. The taps allow the voltage to be raised or lowered in reaction to the load and locked-rotor current. If the load needs more torque, a higher tap would be used to provide the motor with more voltage and more current. If the motor was drawing too much LRA but was still starting satisfactorily, the next-lower tap could be tried.

The control circuit for this application is very similar to the control circuit of the primary resistor since it utilizes a motor timer to provide the time delay for the circuit. The timer controls the amount of time delay for the circuit. When the timer delay

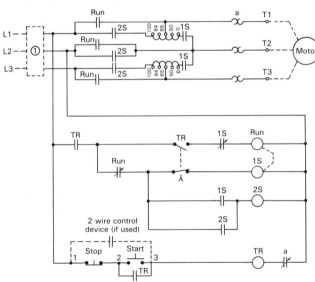

FIGURE 14-3 Wiring diagram for a autotransformer type reduced voltage starter. (Courtesy of Square D Company)

has elapsed, the timer contacts will close and energize the 2S contacts that are in parallel with the autotransformer. When these contacts close, they will short out the transformer circuit and apply full voltage to the motor. Remember: this is a shunt circuit, so this short circuit around the transformer is not dangerous or harmful to the autotransformer or the motor.

Autotransformer Starter Operation

You can follow the sequence of operation of this circuit by referring to the ladder diagram (elementary diagram). From this circuit you can see that when the start button is depressed, the timer motor is energized and the timer's instantaneous contacts at TR 1B and TR 2B are both closed immediately. The timer's delay contacts will close after the preset time delay has elapsed. The TR 2B contacts act as a hold-in circuit for the start button, while the TR 1B enables the rest of the control circuit and supplies power to the delay contacts. The time-delay contacts of the timer have direct control over the main motor starter and two auxiliary contactors. Their coils are identified as Run, 1S, and 2S.

The Run motor starter controls the contacts that supply full voltage to the motor after it has been started by the autotransformer. The 1S and 2S contacts are located at each end of the autotransformer, so that when they are open, the transformer is completely disconnected from the circuit and isolated.

The contacts in this type of control operate in close transition. This means that the second set of contacts will close while those in the first set are still energized. This does not cause problems in this type of starter since the run contacts merely apply line voltage in parallel with the autotransformer.

After the timer TR 1B instantaneous contacts have closed, the normally closed Run contacts allow the 1S coil to be energized. The 1S coil closes all three sets of contacts that are connected to one end of each autotransformer. An auxiliary set of contacts are used in the control circuit to energize the 2S coil. An mechanical interlock is also provided between the Run motor starter and the 1S contactor so that they both cannot be energized at the same time.

When the 2S coil is energized, it closes its three sets of load contacts, which are connected to the opposite end of the autotransformer. These contacts complete the circuit and connect the autotransformer in series with the three main motor windings. A set of 2S auxiliary contacts are used to hold in the 2S coil in the control circuit.

After the motor has started running and reaches the percentage of predetermined speed, the time-delay contacts close and energize the Run motor starter coil. This coil closes its load contacts, which connects the motor directly across the line, which provides the windings with full voltage. The auxiliary Run contacts deenergize the 1S and 2S coils in the control circuit at the same time that the load contacts are connecting the motor to full voltage. This deenergizes the autotransformer and completely isolates it from the circuit. After the motor is connected to full voltage, it can operate for as long as necessary, just like a normal across the line starter.

When the motor is connected to the autotransformer it will draw all its starting current through the transformer windings, which tends to heat up the autotransformer. If the motor took too long to come up to speed or the time delay did not deactivate the starting circuit, the transformer could overheat and become damaged. To prevent this from occurring, a thermal overload device may be installed in the transformer winding when it is manufactured. This overload device is connected in series with the timer motor, which would deactivate the complete starting circuit and require someone to come to the control and depress the start button to restart the circuit after the transformer cooled down.

Sizes and Applications for Autotransformer Starters

The autotransformer starter is available in a variety of sizes for different applications. Typical supply voltages for the control include 200, 208, 440, 550, and 575 V. Typical horsepower ratings include sizes from 15 through 400 hp. The contacts are available in NEMA sizes from 1 through 6, which can carry up to 540 A. The coils for these controls are available in sizes from 208/240 through 600 V. These controls are available in a variety of enclosure types.

This type of reduced-voltage starter is used for applications where compressors, blowers, pumps, and conveyors are started infrequently, but allowed to run for long periods. One reason it is generally used for these types of applications is that the control has a low power factor rating, which would cause problems to the power distribution system's correction factor if the motor required frequent starting.

WYE–DELTA STARTERS

Another type of motor starting circuit allows a three-phase ac motor to start without drawing excessive locked-rotor current by starting the motor when it is connected in wye configuration and then switching to delta configuration during run. You should recall that a three-phase motor will use less LRA when it is started as a wye-connected motor. It will have less starting torque in this configuration, but it is normally enough torque to move the load. This type of starting arrangement requires one motor starter and two contactors. This type of control is also called a *star–delta starter*.

An advantage of using this type of starting is that it is capable of supplying frequent starts to the motor. It is also well suited for high-inertia loads and long acceleration times. The ability to keep inrush or LRA to a minimum makes it ideal for many applications where the power distribution system is nearly overloaded.

Some disadvantages may require close comparison to other types of starters, since this type of control requires that the motor be six-terminal delta-connected so that the changes can be made. Another problem involves the reduction of starting torque when the motor is connected in the wye configuration. You should recall from Chapter 11 that the reduction in torque is caused by a change in the type of connection. When the motor is connected in a wye configuration, one end of each of the three motor windings is connected with the others at the wye point. When a supply voltage of 208 V is connected to any two of the three leads, the two windings that are powered make a voltage divider, so that each of the two coils splits the full 208 V. Instead of splitting the voltage directly in half, it is split at a rate of $\sqrt{3}$, since it is supplied by three-phase. This means that each winding receives 58 percent of 208 V, which is equal to 120 V, and the windings will automatically draw less current since less voltage is applied. Since less current is flowing, the windings will have a reduction in the amount of torque that can be supplied. You can figure the torque will be 33 percent of what it normally would be.

If the motor is connected in delta, voltage is supplied at each of the three corners of the delta so that each winding of the delta receives the full 208 V. Since each winding of the motor is now receiving 208 V instead of 120 V, it will be draw much more current and be able to supply much more torque. Remember, the load will determine the exact amount of torque and current the motor will draw. If the load is large, the torque will also be large, and if the load is reduced, the motor will automatically draw less current and provide less torque.

Figure 14-4 shows a wiring diagram and a ladder diagram for this circuit, and a photograph of the hardware that is used. From the wiring diagram you can see that the motor starter's contacts connect T1, T2, and T3 directly to L1, L2, and L3 of the power supply. It is difficult to determine the connection of the motor winding with this diagram, so two separate diagrams are provided to show only the connections that are used to connect the motor in wye and then in delta.

—Open Transition Type
WYE-DELTA Motors

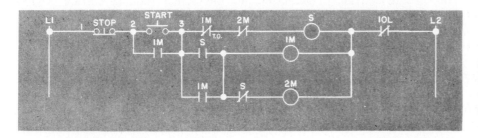

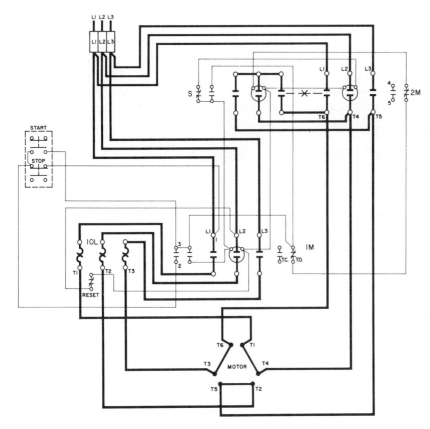

FIGURE 14-4 Wiring diagram diagram of a wye-delta starter. (Courtesy of Eaton Corporation, Cutler Hammer Products)

From the simplified wye diagram you can see that terminals T4, T5, and T6 all need be connected together, and terminal T1 should be connected to L1, T2 to L2, and T3 to L3 for the motor to operate as a wye-connected motor. You should notice that this would involve the 1M motor starter and the S contactor being closed. The 1M motor starter takes care of the connection between T1, T2, and T3 with the supply voltage of L1, L2, and L3, and contactor S provides a set of jumpers to connect T4, T5, and T6 together.

The ladder diagram also has been modified to show only the part of the circuit that is needed to make the starter and contactor energize at the same time during starting. You can see that this involves only the top two rungs of the circuit. When the start button is depressed, the S coil is energized immediately. The S auxiliary contacts provide voltage to

Wye–Delta Starters   **379**

energize the 1M coil, which will also become energized as soon as the S contacts have closed.

The 1M contacts provide a hold-in circuit around the start button. This means that the 1M coil will remain energized even after the start button is released. Since the S coil is not part of this hold-in circuit, the S coil will stay energized only as long as the start button is depressed. When the start button is released, the S coil becomes deenergized, and the normally closed S contacts will return to their closed condition and energize the 2S coil.

The operation of the circuit requires that the operator remain at the start/stop station while the motor is brought fully up to speed since the motor will stay on the wye starting circuit only as long as the start button is depressed. As soon as the start button is released, the S contactor will deenergize and the 2S contactor will energize, causing the motor to switch to delta connection for running.

Simplified Delta Connections

You can also see from the simplified delta connections that terminals T1 and T6 should be connected together with L1, and T2 and T4 should be connected together with L2, while T3 and T5 should be connected together with L3.

The 2S contactor provides a set of contacts that connect T6 back to the terminal block of the supply voltage at terminal L1. Since 1M connects terminal T1 to the same supply voltage terminal, this effectively connects T1 to T6 and to the supply voltage at L1. The same is true of the contact connection that 2S provides between T4 and L2 and between T5 and L3. Since 1M connects T2 directly to L2, and T3 to L3, all the proper delta connections are made when the 1M motor starter is energized at the same time as the 2S contactor.

You should remember that this will occur after the start button has been released once the motor is brought up to speed with the wye winding. When the start button is released, the hold-in circuit provided by the 1M contacts also provides voltage to the 2S coil through the normally closed S contacts.

Since the operator presses the start button to start the motor as a wye-connected motor and releases it to allow the motor to run in delta connection, the start/stop station does not have to be attended to allow the motor to continue to operate as a delta-connected motor.

The reduction in the amount of locked-rotor current during starting is significant enough to provide a savings in the maximum load a power distribution system will have to supply. This also prevents a large amount of heat from building up in the switchgear and contacts, which allows these components to last much longer. This reduction also allows the fuses and circuit breaker to be set closer to the actual running current of the motor, which will provide adequate overcurrent protection.

Sizes of Wye–Delta Starters

The wye–delta starter is available for use on motors with voltages from 200 to 600 V ac. This provides control for motors that are rated from 10 through 1500 hp. The starters and contactors are rated from NEMA size 1 through NEMA size 8 to handle the currents for this control. The control is available with coil voltages rated at 120, 208, 240, 480, and 600 V. This allows the control circuit to operate on the voltage that is available at the installation location. If low voltage is required, control transformers can be used to step down the higher line voltages.

The start/stop controls for this type of circuit can be mounted directly in the door of the cabinet or as a remote station. It is important to remember that since there is such a large difference between the amount of current the motor will draw during starting and during run, this type of motor-starting control requires specially sized heaters for overload protection.

PART-WINDING STARTERS

The part winding reduced-voltage starter utilizes the fact that a motor can be connected so that the coils within each winding can be split into two groups and be connected in parallel with each other. Since these coils are in parallel, the motor can be started using one group from each winding, bringing in the remaining coil after the motor is up to speed. This can be accomplished when the motor is wired for wye or for delta.

When only one group of motor windings are used, less current is drawn because the resistance of a single winding is larger than when the two equal groups are connected in parallel as one winding. (This is the same theory that causes total resistance to drop as resistors are connected in parallel with each other.) Since the amount of LRA is caused by the Ohm's law relationship of the supply voltage and the amount of resistance in the winding when the rotor is not turning, we know that the lower the resistance of the winding, the higher the LRA current will be.

This type of control provides the least expensive method of starting a motor with reduced voltage, since no hardware is required. This means that

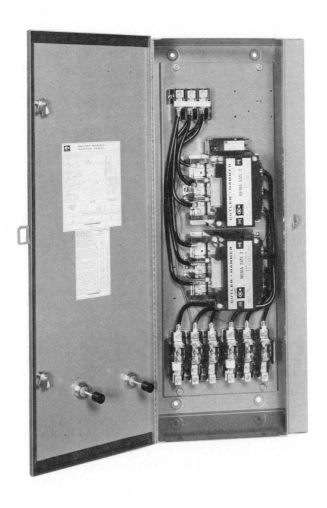

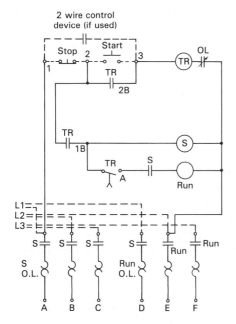

MOTOR LEAD CONNECTIONS TABLE						
PART WINDING SCHEMES	LETTERED TERMINALS IN PANEL					
	A	B	C	D	E	F
1/2 Wye or Delta 6 Leads	T1	T2	T3	T7	T8	T9
1/2 Wye 9 Leads (1)	T1	T2	T3	T7	T8	T9
1/2 Delta 9 Leads (2)	T1	T8	T3	T6	T2	T9
2/3 Wye or Delta 6 Leads	T1	T2	T9	T7	T8	T3
2/3 Wye 9 Leads (1)	T1	T2	T9	T7	T8	T3
2/3 Delta 9 Leads (2)	T1	T4	T9	T6	T2	T3

(1) Connect terminals T4, T5 & T6 together at terminal box
(2) Connect terminals T4 & T8, T5 & T9, T6 & T7 together in three separate pairs at terminal box.

FIGURE 14-5 Picture and diagram with terminal connection tables for a part winding start motor. (Courtesy of Eaton Corporation, Cutler Hammer Products)

Part-Winding Starters 381

it can be used with most dual-voltage motors as long as the lower voltage is used. Another advantage of not using any extra hardware is that this control will be smaller than other types of controls, which is important where panel space is restricted.

This type of starting does present several disadvantages, including not being suited for high-inertia or long-starting loads. It cannot start loads where the starting torque is very large; instead, the load must be energized with a clutch or by a valve if it is too large. Another problem associated with this type of control is that the motor will overheat if the motor starting time for acceleration exceeds 5 seconds.

The most typical applications for this type of starting include reciprocating compressors that can be unloaded by valves during startup. After the motor is up to speed and under power from the full motor winding, the valves can be closed to bring the compressor under load. Other applications includes pumps, blowers, and fans that can be started in an unloaded condition. It is possible to accomplish this with clutches or by arranging dampers and valves to unload the motor during start.

Figure 14-5 presents a photograph and a set of wiring diagrams, ladder diagrams, and terminal-connection diagrams for use with part winding start circuits. The elementary (ladder) diagram shows the control circuit and the load terminals for this circuit. To prevent the diagram from becoming too complex, not all terminals are shown for each motor. Instead of showing the motors, a table is presented to show the motor terminal connections that you would make to connect the motor for half-wye or half-delta operations. The connections in this table list the motor terminals that should be connected together, and the connections that should be made to the terminals of the motor starter and contactors. The terminals for the S contactor are listed as A, B, and C; the terminals for the run motor starter are listed as D, E, and F. For field wiring connections, you should look at the table and make the motor connections to the contact terminals as shown for terminals A through F.

The operation of energizing the coil of the S contactor and the run motor starter is controlled by the top part of the ladder diagram, called the *control circuit*. The control circuit utilizes a motor timer to provide a time delay to allow the motor to start on the part winding circuit and then shift to the full winding after the motor is running approximately 50 to 60 percent of full rpm. The timer motor will be energized when the start button is depressed. The timer's instantaneous contacts, TR 1B, will provide a hold-in circuit for the start button. This means that the start button only needs to be depressed momentarily and then released for the motor to begin its starting operation on the part winding.

The second set of instantaneous timer contacts, at TR 2B, provide voltage to the S coil as soon as the timer motor is energized. After the motor has started running and the timer delay has expired, the timer's delay contacts will close and energize the Run motor starter coil. When the Run motor starter contacts are closed, the remainder of the motor windings are connected back into the circuit and the motor will operate as a normal wye or delta motor.

This type of starting configuration allows the motor to start with a minimum of locked-rotor current. That this also means that the motor will be able to provide only minimal torque during starting. For this reason this application is used only where the motor load can be switched on and off with a transmission, clutch, or by similar means. If the load is some type of pump or large compressor, solenoids can be used to bypass all the pump's flow or unload compressor cylinders until the motor is up to full speed.

Sizes of Part Winding Starters

Part winding start motor starters are available in voltages of 200, 230, 460, and 575 V. These starters can control motors from 10 up to 1400 hp. They can contain NEMA-rated starters from size 1 through size 8, controlled by coils powered with voltages ranging from 120 through 600 V. This allows this type of control to be used in a large variety of applications. The control is available with a fusible or circuit breaker disconnect.

SECONDARY RESISTOR STARTERS

Secondary resistor starters are used exclusively for starting wound rotor motors. Although there are not a large number of wound rotor motors in use, this information will help you install, troubleshoot, and repair these controls when you come across them.

A diagram of a secondary resistor motor starter is provided in Figure 14-6. The stationary winding of the motor is connected directly across the line with S contacts and it will receive full voltage when the S contacts are closed. The rotor is a three-phase wye-connected rotor that has two sets of resistors connected to it. When the rotor is shorted out, it will operate like a squirrel-cage rotor, but when resistors are connected between the ends of the windings, it lowers the amount of current that is

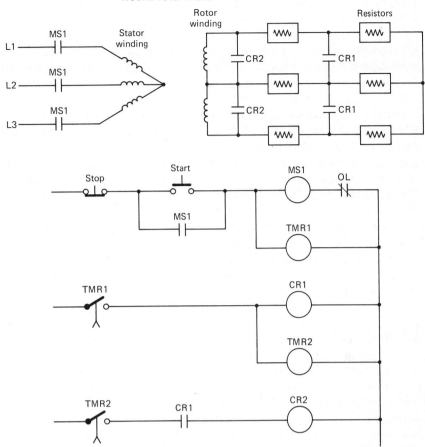

FIGURE 14-6 Diagram of solid state reduced voltage starter. (Courtesy of Eaton Corporation, Cutler Hammer Products)

allowed to flow in the circuit and the motor's speed is reduced. Since all current in the rotor is induced, the resistors will directly control the amount of induced current that is allowed to flow. The resistors in this type of motor are called *secondary resistors* because they are located in the rotor circuit rather than in the line circuit.

When the motor is started, all the resistance is connected in the circuit to keep the current flow to its lowest value. Since the rotor's current is very low, the amount of torque the motor can produce will be very low, and the amount of current the motor will draw will be very small. This is controlled by closing the S motor starter, which applies line voltage to the main motor winding, keeping the R1 and R2 contactors deenergized, which allows all the resistance to be in series with the rotor winding.

The timer motor 1 is energized at the same time as the S motor starter coil. This timer provides approximately 5 to 10 seconds of time delay before the first set of resistors is shorted out, which bypasses the amount of resistance they provide to the rotor. When the TMR1 times out, its contacts will energize contactor R1, whose contacts will close to short out the set of resistors farthest from the rotor. Since the rotor now has less resistance in its circuit, more current is allowed to flow and the motor will pick up speed.

When TMR1 contacts close, TMR2 is energized. This causes TMR2 to run for approximately 10 more seconds before it closes its contacts, which shorts out the remainder of the resistors. When this occurs, the three rotor windings are connected directly to each other, which allows maximum current to flow in the rotor. This allows the motor to reach maximum rated speed.

A selector switch could be used to allow the motor to be started manually rather than with the timers. The selector switch would also allow the motor to be operated continually at any of the speeds provided by the secondary resistors. This provides a multispeed control for this type of motor.

Since the wound rotor uses slip rings and technology has improved the amount of current that can be induced in a squirrel-cage rotor, the round rotor has largely been replaced with other types of three-phase motors. Replacement has been hastened by the introduction of solid-state controls.

Secondary Resistor Starters **383**

TROUBLESHOOTING REDUCED-VOLTAGE STARTERS

All reduced-voltage starters can be troubleshooted similarly since they all operate on similar principles. The first step in the troubleshooting process is to determined if the motor is having problems when it is running or when it is starting. It is also possible that the motor is having trouble during starting and when it is running. It is important to remember that each type of starters has two separate circuits that are used to make the starter operate the way it was designed. The control circuit will provide current to the motor starter and contactor coils at the appropriate time, while the load circuit will contain the multiple sets of contacts to direct current to the correct starting component or motor winding.

When you begin the troubleshooting procedure, you should break down the type of problem into one of the following categories, which will help you eliminate some of the possible problems. The first determination that you should make is whether the motor is trying to start when the start pushbutton is depressed. If the motor does not try to start, you should use the diagram to determine which type of reduced-voltage starting circuit the system is using. Next you should make several tests of the control circuit to see if the first coil is pulled in. You may need to refer the sequence of operation for the type of control that is being used. This test would include the loss of three-phase voltage to the system.

If the control circuit has voltage and the coils will not pull in, you should use the test method to determine where the voltage to the controls is being lost. Other problems to test for include a bad motor starter coil and open overloads.

If you have voltage at the coil of the motor starter but no voltage through its contacts, you may have a faulty contact or starting device, such as primary resistors or the autotransformer. Also test for the presence of voltage at the motor terminals. Be sure to measure the exact amount, since this is a reduced-voltage circuit. If you have any voltage, you can assume that the starting control is operating and that the problem may be caused by an open winding in the motor.

If the motor will start at the reduced voltage but will not come up to full speed, or trips out prior to reaching run speed, you should check the control circuit to determine what controls are used to place the motor across the line. After you have determined what contactors must be energized to switch the motor to the run mode, try to restart the motor and watch these devices closely to see if they are being energized at the correct time. Use a voltmeter and ammeter to watch the voltage and current to determine if the timers are set for the proper amount of time delay. Be sure that you are not trying too many starts too quickly, as the motor and control are rated for the maximum number of starts that they tried per hour.

It is also possible that the load or switching contacts are not allowing the motor to come up to full speed. One way to test these possibilities is to connect the motor temporarily for across-the-line operation. Be sure to check to make sure that other loads are off the power distribution system and that safeguards have been provided where high starting torque can cause damage to the load. *It is important that you do not attempt to run the motor in this configuration for more than one or two starting tests since the contacts in the starter are rated for the reduced-voltage application and cannot accept the high inrush current on a continual basis.*

Once you have made these precautions, start the motor with full applied voltage and watch its response. Be sure to monitor voltage and current on each phase. If the motor operates correctly during this test, you have narrowed the problem down to the switching circuit that changes the motor connections between the starting phase of the control and the running phase.

If the motor does not come up to speed properly during the across-the-line test, you have narrowed the problem down to the motor being faulty or its load has become too large. Remember that the torque for these applications is severely limited during starting and it is possible to have bad bearings or a load that has become too large from overloading. If the application uses valves, solenoids, or clutches to apply full load to the motor once it is started, be sure to test them for proper operation. If the load is a belt-driven load, you may loosen the belts temporarily to reduce torque during the troubleshooting to pinpoint the problem. If the load is a conveyor or machine that should start with no load, be sure that the load has been removed during these tests. After you have determined the problem, make corrections and test the circuit several times for proper operation before you sign off the job.

SOLID-STATE REDUCED-VOLTAGE STARTERS

Reduced-voltage starting can also be provided by solid-state controls. These circuits uses three pairs of SCRs (silicon-controlled rectifiers) to control the amount of voltage that is provided to start the motor. The reduced voltage from the SCR control will have the same effect on the motor as primary resistor control. Solid-state control is better suited for this application since it can control the voltage through an entire range of voltages rather than two or three stepped voltages like the primary resistor control.

The solid-state starter can provide voltage to the motor in small increments to allow the motor to accelerate smoothly. These controls can provide ramping through a potentiometer which increases the speed of the motor at a set rate. Other circuits allow the control to provide constant or varying current throughout the starting cycle.

Figure 14-7 shows a picture and diagram of a solid-state reduced-voltage starter. From this diagram you can see that this type of control uses two SCRs connected inversely and parallel to each other. This allows them to control ac current.

You should remember that the SCR is used to control voltage in one direction, such as a dc signal. The amount of time that the SCR is in conduction will determine the amount of voltage that is allowed to pass. Since one SCR of the pair controls the positive half-cycle of the ac and other one controls the negative half-cycle, the maximum amount of voltage will be conducted when the SCR is in conduction for the full 180 degrees. If the SCR is turned on at the 90-degree point, it will conduct half the voltage. The turn-on point is referred to as the *conduction angle*.

Since two SCRs are paired off to control each ac phase, their control is matched so that each SCR will fire at the same conduction angle. This ensures that each of the paired SCRs is conducting the same amount of voltage. The control circuit can adjust the conduction angle to control the amount of voltage the reduced-voltage starter will provide to the motor.

Control of the SCRs is accomplished through the control module box. This box contains several solid-state circuits that use feedback from current transformers and preset values from potentiometers on the control panels. This control also provides several sets of contacts that can be connected into the hard-wired control circuit. These contacts will energize when current is too high or when the motor hits the end of limit. The end-of-limit circuitry senses the current the motor is drawing and indicates when the current is returning to FLA after the motor is up to full speed at the end of the starting cycle. This is useful if a series of multiple motors must be started in succession. As soon as the first motor is up to speed and its current is near normal FLA, these contacts are closed, which can energize the control of the next solid-state starter.

A shunt trip is also provided that will allow the motor to be disconnected when an SCR has

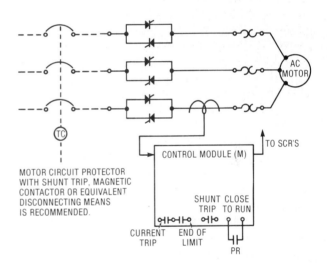

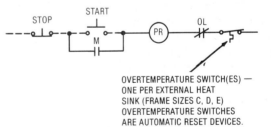

FIGURE 14-7 Diagram of solid state reduced voltage starter. (Courtesy of Eaton Corporation, Cutler Hammer Products)

Solid-State Reduced-Voltage Starters   385

shorted. If the SCR is shorted, the amount of current it is conducting cannot be controlled, and it acts as a shunt or parallel conductor (wire). When an SCR shorts, the phase of voltage it is controlling will act like an across-the-line contact, and it will allow more current to flow.

The amount of voltage that is supplied to the motor can be preset into the controller in one of several ways. One way is to determine the rate of change of current flow, which is called *ramping*, and another way is to use a fixed turn-on time. Another way to control motor speed during startup is to set the maximum or minimum current that is allowed to flow. The theory of operation of each of these controls is explained below so that you will understand the total operation of the solid-state reduced-voltage starter.

Solid-State Starter Operation

Figure 14-7 shows a typical control diagram used to start the solid-state control. You can see that the control module provides a set of terminals marked "close to run" which have a set of contacts connected across them. Since these terminal do not need an input voltage, the contacts will simply complete the control circuit inside the control module. The contacts are identified as PR, which means they are controlled by the PR relay in the control circuit. When the start button is depressed, the PR relay is energized and its auxiliary contacts will hold in or seal the start button. Any other pilot devices can be connected in series with this circuit for safety or operation. The overload contacts in this circuit can be controlled by the current trip from the control.

Another method of starting a motor with solid-state control requires a contactor to be connected in series. The contactor should always be connected between the solid-state control and the power supply. In this case the "close to run" terminals should be jumpered so that control will begin to operate as soon as the contactor is closed. The contactor's coil can be connected to a two-wire or three-wire control circuit. Since the solid-state reduced-voltage starter provides motor protection against overloads, the only other protection that should be supplied to the circuit is fuses for short-circuit protection for the supply voltage wires. If the motor has internal thermal protection, its contacts can be connected in the control circuit for the coil.

When the solid-state starter is energized, it will begin to allow only the preset amount of current to flow to the motor. This current will cause the motor to come up to full speed, in accordance with the preset conditions. The current transformers act as current sensors for the control, which will react automatically to the amount of current the motor is drawing at each step of the acceleration process. If the load presents any changes to the motor during the acceleration process, the motor will react with a change of current, which will be detected by the current transformers. The control can respond to these changes by adjusting the amount of current that is allowed to flow through the SCRs.

Specialized Control Circuits

One of the specialized control circuits provided by the control module is the fixed turn-on ramp. Figure 14-8a is a graph showing the amount of motor current that is allowed to flow during the fixed ramp. This ramp is not adjustable, but it is designed to provide an increasing amount of current at a fixed rate that will increase torque slowly to remove slack or backlash from the load. Other control functions are adjustable and can override the fixed ramp.

The starting current can be adjusted to provide enough current to allow the load to break away. This is called *breakaway torque*. Since this torque will vary from load to load, breakaway torque is adjustable through a potentiometer that allows starting current to be adjusted 100 to 400 percent.

If currents in excess of the fixed turn-on ramp are required, another circuit, called the *adjustable ramp*, may be used. The potentiometer for this circuit allows the current to be steadily increased over a longer period. Figure 14-8b and c present two graphs which show the point where the fixed ramp ends and the variable ramp picks up. The adjustable ramp time can be varied from 2 to 30 seconds to allow larger loads more time to come up to speed. The amount of time it takes the motor to come up to speed will determine the slope of the ramp. This adjustment should allow the motor to start the load to remove slack or backlash with a minimum of disruption.

The current limit adjustment controls the maximum amount of current that the control will allow the motor draw. The maximum current can be set at any amount from 100 to 400 percent. The limit affects only the amount of current flow during starting. This is a useful feature where the maximum amount of current must be limited because of an overloaded distribution system, or because several motors must be started at the same time.

Pulse start allows an extra pulse of full current to be sent to the motor for 2 seconds when current is first applied. The full current is rated at 400 percent, which allows the motor to provide full torque to the load to help it break away. At the end

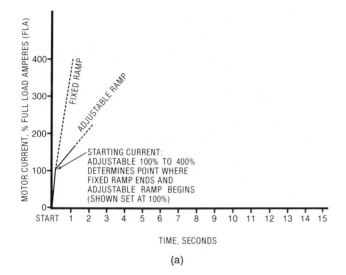
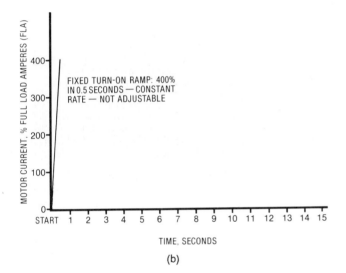
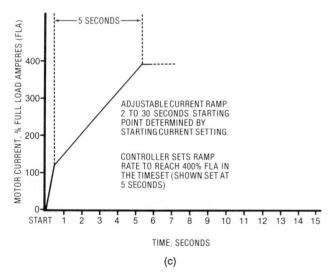

FIGURE 14-8 a-c Graphs of various motor control conditions provided by adjusting potentiometers on the solid state control. (Courtesy of Eaton/Cutler-Hammer)

of 2 seconds, the current is returned to the ramping level and the motor is brought up to speed under the ramp control.

Other features provided in the solid-state control include a current trip circuit, transient protection, and a power saver circuit. The current trip circuit monitors the amount of current the control is sending to the motor. During the starting operation, the current trip circuit is set to 50 percent above the maximum allowable current that the control will allow to flow. This makes the circuit inoperative during startup. Once the motor is running on full-load current, the circuit is again connected into the circuit. It is adjustable between 50 and 400 percent of the full-load current. This allows the control to sense either overcurrent or undercurrent conditions. An overcurrent condition can occur when the load is excessive or through a malfunction that causes the load to jam. An undercurrent may occur when a belt breaks or a clutch drive malfunctions. In either case, the current trip contacts can be used to disconnect power to the motor. The transient protection circuit uses metal-oxide varistors (MOVs) to provide protection against line voltage surges. This provides protection to the SCRs and to sensitive solid-state components in the control unit.

The power saver circuit reduces the amount of voltage to the motor after it reaches full speed. This option may be used only with motors rated for increased temperatures, since the reduced voltage will cause the motor to operate at a slightly higher temperature. The voltage reduction will not be large enough to affect the operation of the load. If the motor is lightly loaded, the power saver circuit will provide a small power factor correction.

Troubleshooting the Solid-State Starter

After the starter is installed and operating, it may have problems. The source of the problem must be identified so that corrective action can be taken. As a troubleshooting technician, you will naturally suspect the solid-state portion of the system because its operation cannot be observed without using test equipment. This is why it is important to use the troubleshooting procedure to identify the problem.

Figure 14-9 is a troubleshooting table for the solid-state reduced-voltage starter. This table will lead you through tests from the most likely to the least likely cause. It will also provide you with a list of suggested actions to take to make repairs. You should always have a strategy in mind when you are troubleshooting so that each test you make will verify the operation of the portion of the circuit that

Symptom	Most Likely Causes	Recommended Action
HMCP trips (electrically operated disconnecting means opens) as it is closed.	Motor is not connected	Connect motor to starter.
	Three phase power not connected or terminal loose	Switch two incoming power lines, corresponding motor leads must also be switched to maintain the same motor rotation.
	Shorted SCR	Perform shorted SCR check.
	SHUNT TRIP contact between terminals closes due to shorted SCR	Perform shorted SCR check.
	Shunt trip latching relay not reset (if used)	Reset STR relay by pushing reset button on relay.
HMCP trips while motor is running HMCP trips when STOP is pressed	Microprocessor detected non-standard operation. Voltage regeneration of motor causing zero voltage drop across SCR (looks like shorted SCR).	Install time delay relay — consult factory.
Starter does not pick up and maintain	Line voltage is not applied	Check incoming lines for proper voltage.
	Overload is not reset	Reset overload.
	120 V control voltage is not present	Check control transformer fuse. Check control circuits.
	Heatsink overtemperature switches are open	Check for continuity through overtemperature switches. Check that all fans are rotating freely. Check for excessive motor current draw.
Starter maintains but motor does not accelerate or does not attain full speed (stalls)	Current Limit is too low	Increase Current Limit setting clockwise. Set Ramp Time to minimum (counterclockwise).
Insufficient Torque	Improper current feedback	Check Current Feedback Resistor for proper calibration. See page 6.
	High breakaway torque required	See Pulse Start Section on page 12. Load is not suitable for reduced voltage starting.
Motor accelerates too slowly	Current Limit is too low	Increase Current Limit setting clockwise.
	Ramp time is too long	Decrease Ramp Time setting counterclockwise.
	Broken current feedback resistor	Check by recalibrating CFR. See page 6.
Motor accelerates too quickly	Current Limit is too high	Decrease Current Limit setting counterclockwise.
	Improper current feedback	Check current calibrator for proper resistance value. See page 6.
	Pulse start setting is too long	Decrease Pulse Start time counterclockwise.
	Broken current feedback transformer wire	Check for a broken Current Feedback Transformer wire. Frame Size A and B the Current Feedback Transformer is contained within the Logic Module.
Current Limit adjustment has no effect during acceleration	Improper current feedback	Check current calibrator for proper resistance value. See page 6.
Starter or motor is noisy or vibrates when starting	Single phasing due to open phase	Check wiring and overload heater coils.
	Single phasing or unbalanced current due to non-firing SCR	Check gate lead wiring to SCR's.
	Defective motor	Check motor for shorts, opens, and grounds.
Mechanical shock to machine	Current increases too quickly	Decrease Starting Current counterclockwise. If necessary increase Ramp Time clockwise.
End of Limit contact does not close	Starter is in current limit	Perform SCR Full Voltage Test. See page 15.
Current Trip contact does not close when current goes above the trip setting	Improper current feedback	Check current calibrator for proper resistance value. See page 6.
Current Trip contact closes when current is below trip setting	Improper current feedback	Check current calibrator for proper resistance value. See page 6.
Motor current, voltage and speed oscillate	Power Saver is misadjusted	Turn Power Saver adjustment CCW until oscillations cease.
Overload relay trips when starting	Incorrect heater coils (melting alloy)	Check heater coil rating.
	Loose heater coil	Tighten heater coil.
	Long starting time (high inertia applications may require slow trip overload and oversize starter)	Motor and starter thermal capabilities must be evaluated before extending overload trip times.
	Mechanical problems	Check machinery for binding or excessive loading.
	Single phasing	See "starter or motor is noisy" symptom in troubleshooting chart.
	Excessive starting time (current limit may be set too low)	Increase Current Limit setting clockwise.
Overload relay trips when running	Incorrect heater coils (melting alloy)	Check heater coil rating.
	Mechanical problems	Check machinery for binding or excessive loading.
	Single phasing	See "starter or motor is noisy" symptom in troubleshooting chart.
Heatsink overtemperature switch opens	Excessive current	Check motor current draw.
	Defective heatsink fan	Check that all fans are rotating freely.
Erratic operation	Loose connections	Check all connections.

FIGURE 14-9 Troubleshooting chart for a solid state reduced voltage starter. (Courtesy of Eaton/Cutler-Hammer)

you are testing. Most of these tests lead you through the control circuit in the same way that current is passed. This will help you verify if each section of the control is operating correctly.

ACCELERATION CONTROL FOR DC MOTORS

Dc motors can be accelerated slowly in much the same way as ac motors by changing the amount of voltage and current that they use during starting. You should recall that the speed of a series motor is controlled by the amount of load, but the speed of shunt and compound motors can be controlled in this manner. The current can be controlled by using primary resistors, rheostats, or SCR solid-state controllers.

Since these types of controls can also be used to control the speed of the motor, the circuits are generally used as double purpose for starting and controlling the speed of the dc motor. In recent years, SCR control has been used extensively, since it serves two distinct purposes. First, SCRs are used to rectify the ac current to dc current for use in the motor circuit. Prior to the use of the SCR or diode, dc current was produced by a motor–generator set. The motor was powered by ac current and the generator produced dc current.

Current in these systems was controlled by adjusting the field current of the generator, which determined the amount of voltage the generator produced. The voltage generated was sent directly to the dc motor. If the motor speed was controlled, a field rheostat was placed in the dc generator field to adjust the amount of generated current. The dc motor in these systems was accelerated with primary resistors. Usually, several banks of resistors were used to provide smooth acceleration.

When SCRs were introduced in the late 1950s and early 1960s they were not rated large enough to control voltage and current to a large dc motor individually. In these early applications, several SCRs would be paralleled to carry the larger currents. Today, SCRs can be produced that will easily carry several hundred amperes of current.

The second duty of the SCR is to adjust the amount of dc voltage and current that is sent to the motor. This type of control is very similar to the solid-state reduced-voltage starter except that SCRs are connected as a bridge rectifier, which produces a full-wave pulsing dc current. The firing angle of the SCRs is controlled to adjust the amount of voltage they will allow to conduct. As the firing angle is increased, more voltage is conducted to the motor.

The ramp and slope of the motor acceleration can be fixed or controlled by external potentiometers in the control. This and similar features, such as current limit, provide the SCR controller with the flexibility to control the motor's acceleration and speed under a variety of different loads.

DECELERATION AND BRAKING METHODS

It is just as important to bring a motor to a safe stop as it is to start with acceleration control. Some large motor loads develop high inertial forces when they are operating at full speed. If voltage is simply disconnected from the motor, the load may coast several minutes before the shaft comes to a full stop. This is true in applications such as those involving large saw blades and grinding wheels. If someone becomes injured or stuck in the machinery, the motor load must be stopped immediately. It is also important to bring these loads to a quick, smooth stop when the operator does not have time to stay with the equipment as it comes to a stop.

In other load applications, such as elevators and cranes, the location where the load stops is as important as moving the load. This means that the motor shaft must stop moving at the precise time to place the load in its proper location.

In some of these applications, the motor can be decelerated quickly by reconnecting the motor to operate in the reverse direction at the time when the stop button is depressed. This connection is only allowed for a second, while the motor shaft comes to a stop. The circuit is disconnected before the motor actually begins to run in the opposite direction. This type of circuit is called a *plugging circuit*.

In another type of circuit, dc current is applied to the stationary field of an ac motor when the stop button is depressed. Since the field is fixed and it replaces the rotating ac field, the rotor is quickly stopped by the alignment of the unlike magnetic fields between the rotating and stationary winding. The attraction between the unlike fields is so strong that the rotor is stopped quickly. This type of deceleration is called *dc braking*. A similar type of braking, called *regenerative braking,* is used in dc motors and some ac motors. The motor is reconnected as a generator when the stop button is depressed. When the stop button is depressed, it disconnects the motor leads from the power supply and reconnects them as a generator. The armature of the motor (generator) is connected directly to a small resistor to load the generator as heavily as possible. This heavy load brings the rotor to a smooth stop within several rotations.

The third type of deceleration circuit used in industry today is the electromechanical *clutch and brake assembly*. The clutch and break assembly uses combinations of disk brakes, drum brakes with shoes, and electromagnets to bring the motor to a quick stop. Since the mechanism can control the deceleration of the motor by allowing a specific amount of slip, the speed of the load can be adjusted through this type of clutch mechanism.

Plugging Circuits

One of the easiest way to bring an ac or dc motor to a quick, safe stop is to disconnect power from the motor and reconnect it so that the motor will rotate in the opposite direction. This can be accomplished in applications where the motor is operating in only one direction or in motors where they are operated in either direction. This type of circuit can also be used to apply current to the reconnected motor after the load has coasted for several seconds, so that the shock to the load is not so severe.

Figure 14-10 shows a diagram of a plugging circuit. The motor in this diagram is an ac motor, so only two of the three leads need to be reversed to cause the motor to rotate in the opposite direction. When the motor is started, the contacts of the motor starter connect the motor directly to the supply voltage so that the motor will run in the forward direction. The forward motor starter controls the set of contacts on the right side of the diagram.

When the stop button is depressed, the forward motor starter is deenergized and the reverse motor starter is energized. The reverse motor starter switches L1 and L3, which causes the magnetic fields in the stator to rotate in the opposite direction. The reversed magnetic field quickly brings the rotor to a stop and tries to start it in the reverse direction. If this circuit did not have any other controls, the motor would begin to run in the reverse direction.

Since we are interested in stopping the shaft rotation, a centrifugal switch is added to the circuit. This switch, which senses the rpm of the shaft and opens its contacts when the rotor approaches low rpm, is called a *plugging switch* or *speed switch*. Its symbol incorporates a double-pole switch with an arrow to indicate that the switch is rotary in nature. The centrifugal switch is adjustable so that its contacts deenergize the reverse motor starter coil at just the right time to make the shaft stop rather than coast or begin to turn in the reverse direction. Since the size of the load will also determine how fast the shaft can be brought to a stop, the plugging switch can be adjusted to fit the application.

The circuit shown in this diagram allows the motor to be stopped anytime the stop button is depressed. For this reason it is also called a *continuous plugging circuit*. The circuit could be modified so that the plugging circuit would be energized only when the emergency stop button was depressed, and would coast to a stop if the normal stop button was depressed.

Other modifications can be made to the circuit so that the motor can be operated in the reverse direction without the plugging circuit stopping the motor shaft. In this type of circuit, a lockout solenoid can disable the plugging switch so that when the reverse pushbutton is depressed, the motor is allowed to change direction and accelerate in the reverse direction. It is important to remember that split phase and capacitor start motors cannot be used with plugging circuits because they need to use their own centrifugal or (end switch) for starting, and it will not close until the motor is nearly stopped.

Figure 14-11 shows another plugging circuit. In this circuit, the motor can be plugged while it is operating in either direction. The first plugging circuit would operate correctly only when the motor was rotating in the forward direction. In some applications, the motor must be operated and stopped while it is running in either direction.

In this type of circuit, the plugging switch is set to sense the motor rpm in either direction. The motor can be started in either direction by pressing the forward or the reverse pushbutton. When the stopped button is depressed, the interlocks set up a condition where the opposite motor starter is energized to bring the motor to a rapid stop. When the rotor decelerates to near zero, the plugging switch disconnects all power from the motor, and the control circuit is ready to start the motor in either direction.

Using a Time-Delay Relay for Plugging. The circuits shown in Figures 14-10 and 14-11 could use a timer delay relay instead of the plugging switch. The time delay contacts would be located where the plugging switch contacts are in these diagrams. The timer motor would be connected to the reverse motor starter coil where the motor was designed to operate in only one direction, and it would be connected where it would be energized by either motor starter when the stopped button was depressed in circuits where the motor is allowed to operate in either direction. As in the plugging switch, the amount of time delay can be adjusted for different-size loads.

Plugging

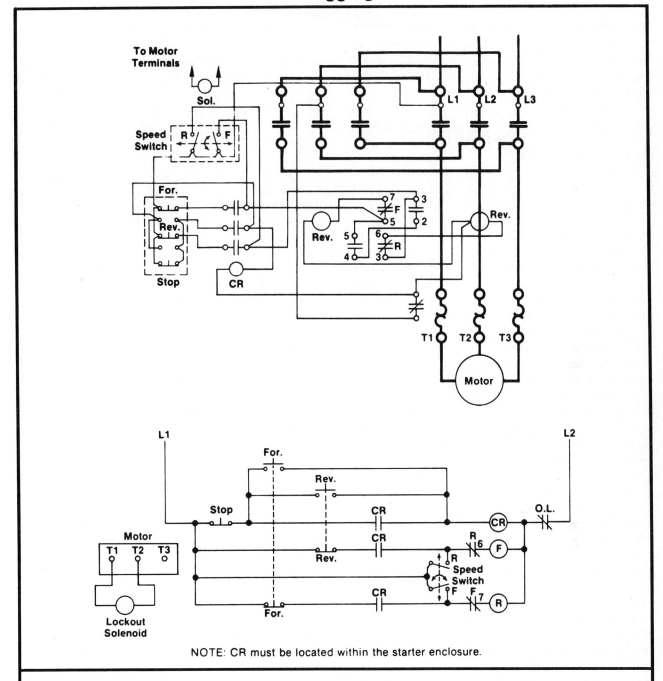

FIGURE 14-10 Plugging circuit for motor that can run in forward or reverse direction. (Courtesy of Allen-Bradley)

Plugging

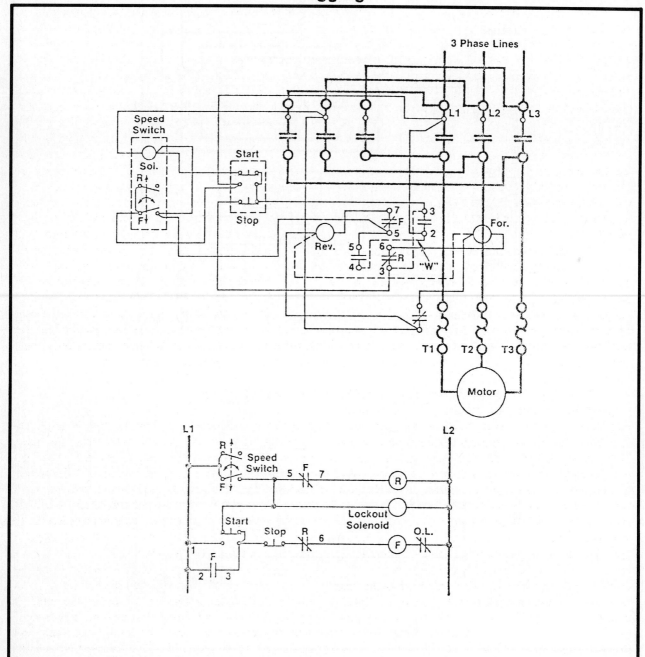

Plugging a Motor to a Stop from One Direction— Lockout Solenoid Provided

This system is for a motor that is to run in one direction only and must come to an immediate stop when the stop button is pressed. The reverse contactor of the Bulletin 505 reversing switch is used only for plug-stopping and not for running in reverse. When a standard Bulletin 505 is used, wire "W" and all wires represented by dotted lines should be removed.

The lockout solenoid is built into the Bulletin 808 Speed Switch and its function is to guard against an accidental turn of the motor shaft closing the speed switch contacts and starting the motor. This protective feature is optional and the speed switch can be furnished without lockout solenoid if desired.

FIGURE 14-11 Electrical diagram that shows a method of plugging a motor in either the forward or reverse direction. (Courtesy of Allen-Bradley)

Antiplugging Circuits

Another method of bringing a motor to a smooth stop is called *antiplugging*. This type of circuit allows the motor to coast for a short period before reverse voltage is applied. The short time the load is allowed to coast prior to applying the reverse current provides a small amount of cushion to bring the load to a smooth stop.

The centrifugal switch in this circuit prevents the reversing circuit from energizing until the motor's speed has slowed to the predetermined rpm. Since the centrifugal switch is used to prevent the reversing circuit from energizing until the motor has reduced its rpm, the switch is called an *antiplugging switch*.

The motor is allowed to coast before the reversing circuit is energized. This is accomplished by the set of contacts that the switch opens, as long as the rpm is too fast. When the motor is deenergized, it is allowed to coast while its rpm slows. The centrifugal switch will close the open contacts when the speed is reduced to the preset level. At this time, the reversing circuit is energized and the motor is brought to an immediate halt.

Figure 14-12 shows a diagram of an antiplugging circuit. From this circuit you can see the operation of the reversing switch. When the motor is turning at full rpm, the centrifugal switch dissables the reversing circuit. When the motor is deenergized, the load will cause the rotor to coast as the rpm is reduced significantly enough to allow the switch contacts to close. At this point the reversing circuit is energized and the motor is stopped. The centrifugal switch is adjustable to energize the reversing circuit at any rpm below full speed. You can also see that a time-delay relay could also be used in this circuit to delay the reversing circuit until the motor has slowed significantly.

Electric Braking

Another popular method of decelerating a motor or providing instantaneous stopping is called *electric braking*. The principle of electric braking is based on the theory of disconnecting the applied voltage that produces the rotating magnetic field and replacing it with a fixed magnetic field produced by dc current. When the dc current is applied to the motor's stationary field, the fixed north and south polarity of the magnetic field will attract the opposite field of the rotor and quickly bring it to a stop.

Figure 14-13 shows an electrical diagram of a dc electric braking circuit. The dc voltage for the braking circuit is produced through a set of diodes or SCRs which are connected as a single-phase bridge rectifier. If the circuit uses regular diodes, the amount of dc voltage provided for the braking circuit will be fixed. If the dc voltage is provided by SCRs, the gate firing circuit can be regulated to adjust the amount of braking action.

From the diagrams you can see that the dc current is applied to the stationary winding of a single-phase and a three-phase motor. This circuit requires a contactor or motor starter with one or two sets of normally closed auxiliary contacts. When the motor starter is deenergized, the applied voltage is removed from the stator as the main contacts are opened, and the dc current is connected through the auxiliary contacts.

If the full dc voltage is applied, the rotor will be brought quickly to a stop. If the motor needs some time to coast so that the load is stopped more smoothly, the amount of dc voltage may be reduced by resistors or by adjusting the gate firing circuit of the SCR. A time-delay circuit could also be used to allow the motor to coast for a time before the braking circuit is applied, which will make it operate similar to the antiplugging circuit.

Dynamic Braking

Dynamic braking uses a principle that is similar to that of dc electric brakes. This circuit is used primarily on dc motors, but it could be used on wound-rotor and some synchronous motors. In this type of braking circuit, the applied voltage is removed from the motor when the motor starter is deenergized. Since the armature will continue to spin as it coasts to a stop, it will produce voltage in a manner similar to a generator.

The generator action can be used to bring the rotor to a quick stop by loading it with a small resistor. The small resistor will cause the rotor to generate very high levels of current, which produces reversed magnetic forces on the shaft and causes it to stop quickly. The inertia that is built into the rotating shaft is quickly dissipated by the strong generated current since the rotor is no longer under power. If the shaft needs to be stopped more quickly, a jumper can be used to short the output of the rotor, which will cause a much larger current and a stronger magnetic field to be produced. A diagram of this circuit is shown in Figure 14-14. You can see that the resistor is connected in the circuit by the normally closed set of auxiliary contacts in parallel with the armature. These contacts will disconnect the resistor when power is applied to the motor, and reconnect at the instant when the motor starter coil is deenergized. This circuit is called a *dynamic braking circuit* since it utilizes the energy

"ANTI-PLUGGING"

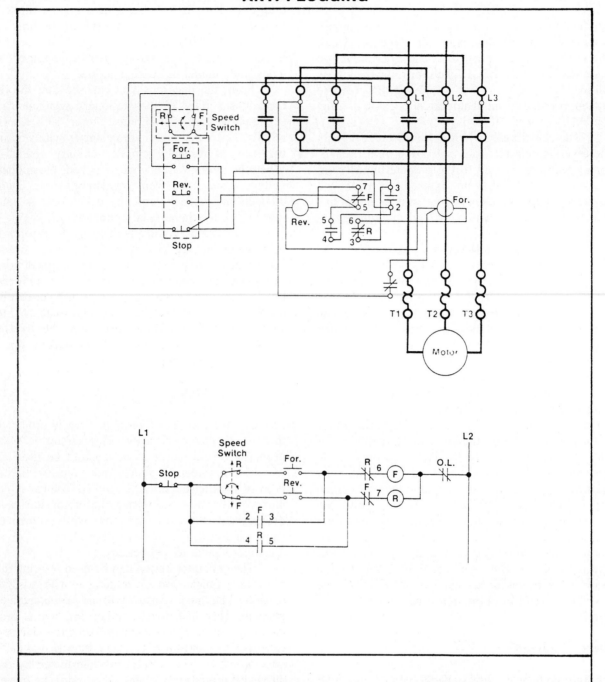

"ANTI-PLUGGING" — Motor Is to Be Reversed, but It Must Not Be Plugged

A Bulletin 808 Speed Switch with normally closed contacts is used for anti-plugging. The schematic diagram shows that with the motor operating in one direction, a contact on the Speed Switch opens the control circuit of the starter used for the opposite direction. The open contact will not close until the motor has slowed down, and thus, the reversing switch cannot be energized to change the direction of the motor until the motor is moving slowly.

A standard Bulletin 505 reversing switch can be used with this application. The push button station can be a Catalog Number 800S-3SA, or a Catalog Number 800H-3HA, or a Catalog Number 800T-3TA.

FIGURE 14-12 Electrical diagram of an anti-plugging circuit. (Courtesy of Allen-Bradley)

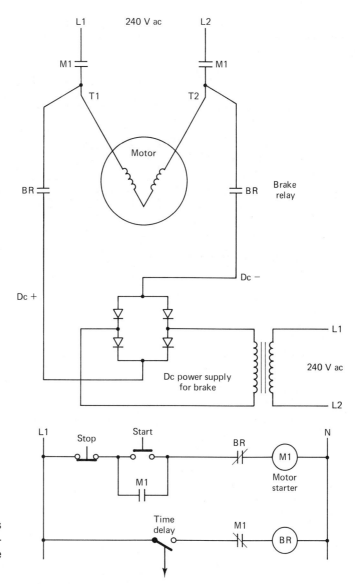

FIGURE 14-13 Circuit diagrams of a DC electric braking circuit connected to a three phase and single phase motor.

that is stored in the rotor to cause the braking action.

BRAKES AND CLUTCHES

Brakes and clutches allow the speed and torque of a load to be controlled while the motor is operating at full rpm. In these types of systems, the speed and torque of the motor is not controlled; instead, the motor is allowed to run at full speed while adjustments are made to the clutch or brakes to accelerate, control operating speed, or decelerate and stop the load. This means that mechanical and electromechanical systems are used between the shaft of the motor and the shaft of the load to provide this control.

One of the most simple methods of controlling the deceleration and stopping of a load is with an electrically operated mechanical friction brake. Figure 14-15 shows a photograph and electrical diagram of this type of friction brake. You can see that the brake mechanism utilizes brake shoes that are very similar to the type found on the rear brakes of an automobile. This system is slightly different in that these shoes apply force inward against a drum, whereas automotive brakes apply force outward against a drum.

In this system, the brake shoes are fitted around a drum that is fitted onto the shaft of the load or motor. A large spring is provided for adjustment to set the amount of pressure that the shoes will apply against the drum. A linkage is designed to connect an actuator arm to powerful solenoid coil that will open the brake shoes slightly when it is energized. When the coil is deenergized, the spring will again cause the shoes to apply force to the drum.

Brakes and Clutches **395**

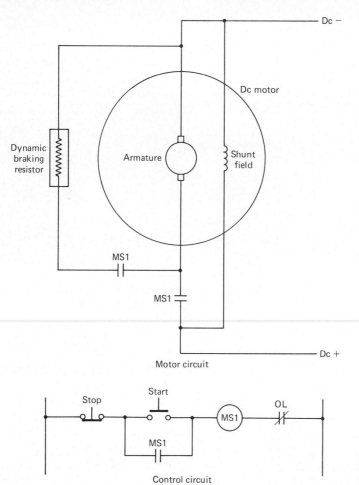

FIGURE 14-14 Electrical diagram of a dynamic braking circuit.

Friction Brake Operation

The operation of the friction brakes can best be understood by an electrical diagram, provided with Figure 14-15. The solenoid coil for the brakes is connected across the supply voltage lines for the motor, which means that the coil will be energized any time the motor is energized. When the solenoid is energized, it pulls the linkage to cause the shoes to open slightly and remove all friction to the drum. This allows the motor to operate freely and to accelerate to normal speed and run for as long as the application requires.

When the motor starter is deenergized, the voltage to the motor is disconnected, which also deenergizes the solenoid to the brakes. When the solenoid is deenergized, the spring is allowed to apply full force to the shoes so that they create maximum friction with the drum. This friction causes the drum to slow down and stop quickly. Since the drum is rotating with the motor and load shaft, they are also stopped. A screw is provided to adjust the amount of pressure the spring can apply to the brake shoes. If the pressure is maximum, the drum will stop immediately when power is removed from the motor and coil, and if the pressure is reduced, the motor can be allowed to coast slightly as it comes to a smooth stop. This configuration of control also provides fail-safe operation, so that if power is lost for any reason, the brakes will be applied automatically by the spring tension.

The adjustment of brake shoes is very important since they determine the amount of friction that is applied to the drum when the brakes are applied. If they are set too loose, they will not apply enough friction to stop the load correctly, and if they are set too tight, they will become overheated and wear prematurely.

Applications for Drum and Shoe Brakes

The drum and shoe mechanical braking system is generally used on cranes and elevators, where positioning is critical and slippage when power is disconnected cannot be tolerated. Many other applications also require that the motor hold a load in position after it has been deenergized. Some larger robots and transport automation may use dc electric

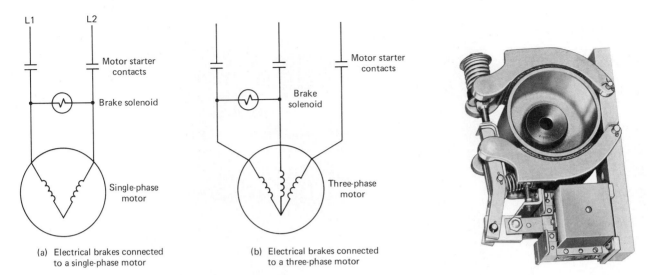

FIGURE 14-15 Friction type brakes. Notice the adjustment screw to set the amount of friction and the linkage to the solenoid that opens the shoes during the run mode. (Courtesy of Eaton Corporation, Cutler Hammer Products)

braking to stop the motor and brakes to hold it in location when the emergency stop button has be depressed or when the motor has been deenergized.

Other Types of Friction Brakes

Other types of friction brakes utilize a slightly different theory of operation. Figure 14-16 shows several of these types of brakes and you can see that their design utilizes a rotating disk and pads to cause friction. Figure 15-16a shows a sketch of a simple friction brake. This brake utilizes a rotating disk and friction pads. The pads apply friction to the disk with electromagnetic force. The magnetic force can be varied with current. In some styles of brakes a permanent magnet is used to created the magnetic field, and electric current is applied to nullify the strength of the permanent magnet. This means that the stronger the current, the weaker the pressure is to provide friction to the brakes.

In other styles, an electromagnet is used to cause the pressure against the pads. In this styles of brake, the stronger the current, the more friction the brakes will apply. The brake shown in Figure 15-16b may be the easiest to understand since most of its parts are open and easy to view. This system is very similar to the front disk brake system on an automobile. The rotating disk is connected directly to the motor or loads shaft by a keyway. This means that it will rotate at the same speed as these shafts. The pads are connected to the stationary part of the brake shown at the top of this system.

In this system, the pad applies pressure to the rotating disk when the electromagnetic coil is energized. The more current that is applied to the electromagnet, the more pressure the pads will apply to the rotating disk, which will bring it to a stop faster. If a small amount of current is applied to the magnet, the pads will apply a small amount of pressure and the friction between the pad and disk will be small. This will allow the load to coast slightly as it comes to a stop. The electromagnet assembly is available with 1 to 16 magnets to apply the amount of force required to stop the load. In smaller applications the disk will have one friction face, and for larger loads it will have two.

The friction face of the disk in this type of brake is replaceable. The face is bolted to the disk with four small bolts. This allows them to be replaced easily when they begin to show excessive wear. The brake pads that rub against the friction face on the disk are also easily replacable while the brake is installed on the factory floor.

The brakes shown in Figures 14-16c and d are called *fail-safe brakes* because they operate similarly to shoe brakes. This type of brake operates slightly differently, in that a very powerful ceramic magnet is used to apply the braking force. This magnet is a permanet magnet and current is applied to nullify its field when the brakes are dissengaged to allow the load to turn freely. When power is removed from the brake, the permanent-magnet field is allowed to activate the brake disks so that they rub against each other and cause sufficient friction to stop the load.

Other styles of brakes are also available for a wide variety of applications. Figure 14-17 shows four different designs for braking applications. Fig-

FIGURE 14-16 Diagram of simple friction brakes and disk brakes. (Courtesy of Warner Electric Co.)

ure 14-17a shows a motor with an electric brake mounted directly to the motor shaft. One end of the shaft for the brake slips over the motor shaft when it is mounted, and the other end provides a mounting for the load to be mounted. This type of brake is compact brake and air cooled and will operate in a fail-safe mode, providing braking action as well as holding power when the brake is deenergized.

Figure 14-17b shows the same type of brake connected to the end of the motor that is opposite the drive end. This allows the drive end of the motor to have pulleys connected to operate a belt-driven load, and have a brake mounted to the opposite end. This makes maintenance of the belts and the brakes much simpler since they are not in each other's way. This type of application also prevents the two systems from adding heat to the other system.

Figure 14-17c shows the brake module mounted on the motor shaft prior to it entering a right-angle speed reducer. This provides braking action and holding action to loads that are driven by

398   Chap. 14 / *Advanced Motor Control Circuits*

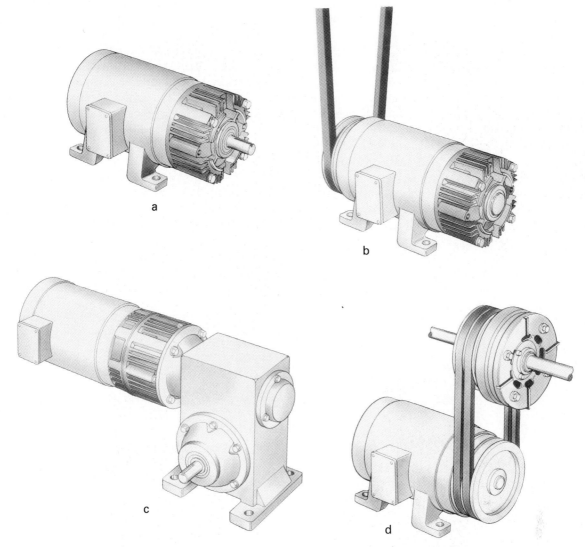

FIGURE 14-17 Diagram and pictures of brake applications. (Courtesy of Warner Electric Co.)
a. Electro module brake on C face motor.
b. Fan cooled electro module motor brake.
c. Electro module brake between motor and reducer.
d. Electro brake on through shaft.

the motor through a speed reducer. Since the brake is self-adjusting, it does not require much maintenance, and it can easily be replaced by unbolting it from the motor and speed reducer.

In Figure 14-17d the brake module is mounted on the through shaft. This allows it to hold the load at the shaft end where there may be more room. In some applications it is also more desirable to have the brake on the load shaft rather than on the motor shaft since a belt may break or slip and allow the load to run free. Since the brake in this application is mounted on the load shaft, the load will always be held, regardless of the condition of the belts.

These types of brake can handle rapid cycling applications. The friction pads are designed to self-adjust as they wear. This provides a system that does not require constant maintenance for adjustments. Since the brake incorporates fail-safe operation, it will be able to stop and hold loads when the power is removed from the brake. This makes them useful for most applications.

Electromechanical Clutch Used as a Motor Drive

Another method of controlling the speed of a load during acceleration or deceleration is through the use of a clutch mechanism. The clutch operates on the theory that the motor's speed is not going to be controlled, so it will be assumed to operate at full speed. The clutch will allow the load to slip from 100 percent to 0 percent, which will provide speed control. If the application is looked at from the load's point of view, the operation of the speed control will be the same as if the motor speed were controlled.

Figure 14-18 shows several photographs of clutch drives and mechanisms. Figure 14-18a shows a simple photograph of the clutch. The clutch consists of rotating members that are connected to the motor shaft and to the load shaft. The motor shaft drives the clutch, and as electric current is applied, the clutch plates are engaged and the load beings to turn. As control current is increased to the electromagnet in the clutch, more pressure is applied to the friction plates. The stronger the pressure is at this point, the smaller the amount of slip will be. If the clutch does not have any slip, the load will rotate at the same speed as the motor shaft. If the clutch provides 50% slip, the load will turn at half the rpm of the motor. The main function of clutches in industrial applications today is to allow the motor to start and accelerate in the unloaded condition. When the motor is operating at full speed, the clutch can be engaged and the load can smoothly be brought on line. This allows the motor to start the load with a minimum amount of locked-rotor current. This arrangement also allows acceleration systems to be used that do not provide a lot of starting torque.

This figure also shows other clutch applications. In Figure 14-18b, a clutch is shown with a two-speed belt and pulley drive. A seperate clutch is mounted on each of the shafts, which allows both high speed and low speed to be engaged separately. Figure 14-18c shows two clutches used with a reversing belt drive. In this application, the clutches are mounted to energize the load in both the clockwise or counterclockwise directions. The last figure shows a foot-mounted clutch that is chain

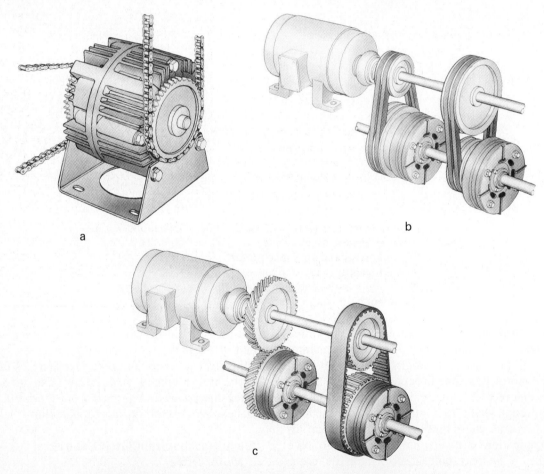

FIGURE 14-18 This figure shows a diagram of the clutches operation and pictures of several clutch applications. (Courtesy of Warner Electric Co.)
  a. Foot mounted electro module clutch.
  b. Two speed drive with electro clutch.
  c. Reversing with electro clutches.

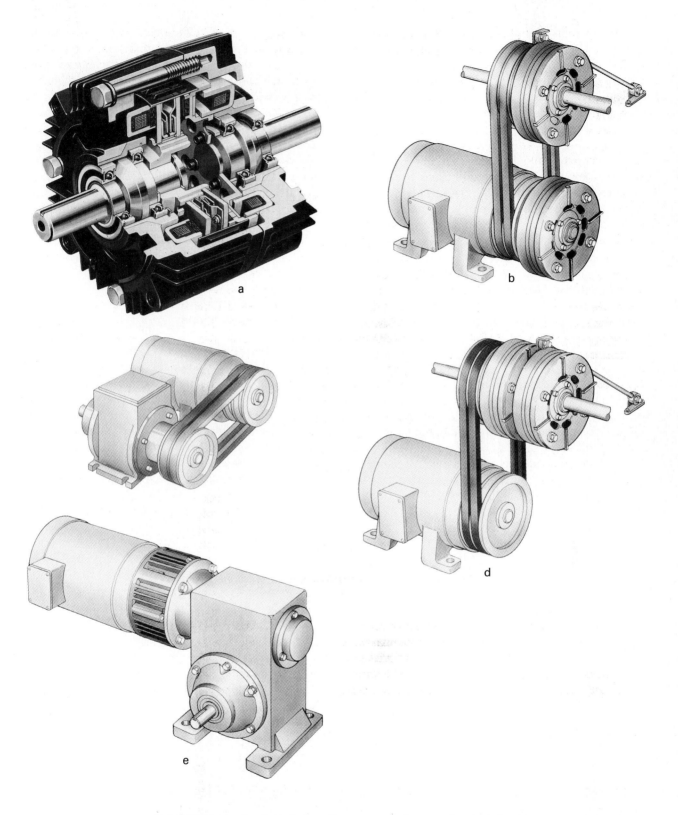

FIGURE 14-19 Exploded view diagram and pictures of typical clutch and brake modules. (Courtesy of Warner Electric Co.)
a. Exploded view of clutch brake.
b. Electro clutch and brake on a through shaft.
c. Electro pack right angle drive clutch and brake.
d. Clutch on motor and brake on the shaft.
e. Clutch and brake between motor and reducer.

Brakes and Clutches

driven. This provides a way to limit the torque as the load is engaged smoothly. Notice that this clutch provides vanes on the exterior of the housing to help dissipate heat.

Using Clutch and Brake Functions in the Same Device

Since the operation of the clutch and brake are very similar, they can be combined into one device and mounted on most motors or loads. This allows a compact device to be used to provide two functions rather than using two separate devices. Figure 14-19 shows a cutaway view of a clutch and brake assembly and several photographs of these devices in various applications. From the cutaway view (Figure 14-19a) you can see that the pads of the brake assembly are mounted very similar to the stand-alone brake. The clutch mechanism is mounted on the shaft of the device. When the clutch is energized, the two shaft sections begin to operate as one, and when the motor is stopped, the brake pads apply friction to stop and hold the load. Figure 14-19b shows the clutch and brake module mounted on a through-shaft application. In this application the clutch can engage the load once the motor has reached full speed, and the brake can stop and hold the load when the motor is deenergized. If the belts become loose or broken, the brake part of the module will be able to hold the load in the fail-safe mode.

Figure 14-19c and d show the clutch and brake module used in a right-angle drive and a speed reducer. In these types of applications, the drive can be belt driven through a right-angle drive with the clutch and brake mounted in between them, or with a speed reducer where the clutch and brake are mounted between the motor and load. In both of these applications, the load can be engaged smoothly once the motor is brought up to speed, and can also be used to stop and hold the motor when power is removed. Figure 14-19e shows a clutch attached to the motor and the brake attached to the load. This design provides a fail-safe function for the load if the belts become loose or if they are broken. The clutch is mounted on the motor, where more room may exist for installation and maintenance.

Sizing and Selecting the Proper Clutch and Brake Assembley

Several charts and graphs to aid in the selection of a clutch and brake based on the measured rpm of the shaft at the clutch or brake and the size of the load (hp) are available. The formulas allow you to calculate the amount of clutch torque and the amount of brake torque required. These selection charts show typical models and their overall ratings and dimensions. If you were to select a different brand name of clutch and brake, the size of the unit that is required will be the same, but each manufacturer will provide their own product selection charts which list their products by size.

QUESTIONS

14-1. Explain why acceleration circuits are required for larger loads.

14-2. List the five types of reduced-voltage starting circuits.

14-3. Explain the difference between open and closed transition starters.

14-4. Draw an electrical diagram of a primary resistor reduced-voltage starter and explain its operation.

14-5. Draw an electrical diagram of an autotransformer reduced-voltage starter and explain its operation.

14-6. Draw an electrical diagram of a wye–delta reduced-voltage starter and explain its operation.

14-7. What advantage is provided by the wye circuit, and what advantage is provided by the delta circuit?

14-8. Draw an electrical diagram of a part winding reduced-voltage starter and explain its operation.

14-9. **Draw** an electrical diagram of a secondary resistor starter and explain its operation.

14-10. Draw a diagram of a solid-state reduced-voltage starter and explain its operation.

14-11. What advantages do the solid-state reduced-voltage starter provide?

14-12. Explain how dc motors are accelerated.

14-13. Explain why motors may need deceleration or braking controls.

14-14. Draw a diagram of a plugging circuit and explain its operation.

14-15. Explain why you need a time-delay relay or a centrifugal switch in a plugging circuit.

14-16. Explain what an antiplugging circuit is and how it operates.

14-17. Draw a sketch of mechanical brakes (drum and shoe type) and explain how they operate.

14-18. Draw an electrical diagram of a motor using electrical brakes. When are the brakes energized? When the brakes are energized, do they allow the motor shaft to rotate, or do they prevent the motor shaft from rotating?

Dc and ac electronic motor drive systems

15

INTRODUCTION

One of the most rapidly changing areas of motor control systems is that of dc and ac electronic motor drive systems. These drive systems include dc variable-voltage and variable-current drives, eddy current drives, ac variable-voltage and variable-frequency drives, stepper motor systems, and servomotor systems. As a technician, you will be expected to install, troubleshoot, and repair these drive systems as well as the switches and motor starters that control them. You will see that as an integral part of the motor control system, it does not matter whether the system is inoperative because of a broken limit switch, motor starter, or electronic motor drive; instead, the system must be repaired as quickly as possible. Production managers do not distinguish between those parts of a system that are operative or inoperative; rather, they see that a system is either operating or faulty. When you are called to troubleshoot and repair the system, you will have to make a judgment about which part of the system is inoperative and make the tests as quickly as possible, so that the system can be returned to operation.

In this chapter we describe different types of drives and drive systems and explain the basic parts and the theory of operation behind them. Typical applications for each type of drive system are given, as are their strengths and weaknesses.

As you will see in this chapter, motors can be controlled by drives using stand-alone speed and torque control, or they can be part of a motion control system, such as a stepper motor system or a servo system.

In each of these systems specific information is provided to help you understand the similarities and differences between drives and control systems. This information will help you select the type of system that can be used for your applications, or will help you install, troubleshoot, and maintain the system that you have on the equipment with which you must work.

DC MOTORS AND DRIVE SYSTEMS

The dc shunt motor and permanent-magnet (PM) motors are the most widely used motors in dc control systems. The speed, torque, voltage, and current of these motors are generally controlled for specific load applications. A diagram of a shunt motor and permanent-magnet motor is provided in Figure 15-1. The permanent-magnet motor uses a permanent ceramic magnet to produce the magnetic field for the motor, while the shunt motor uses a field winding to produce the magnetic field. The difference in these two motors should be discussed when you intend to utilize a dc electronic motor drive to control their speed and torque.

Shunt-Wound

Shunt-wound motors have the armature connected in parallel across the field winding. With constant armature voltage and field excitation, the shunt-wound motor offers relatively flat speed-torque characteristics. The shunt-wound motor offers simplified control for reversing, especially for regenerative drives

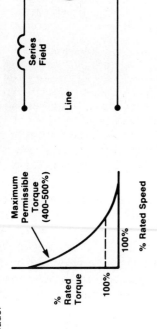

Shunt-Wound DC Motor Characteristics

Compound-Wound

The compound-wound DC motor utilizes a field winding in series with the armature in addition to the shunt field, to obtain a compromise in performance between a series and a shunt wound type motor. The compound-wound motor offers a combination of good starting torque and speed stability.

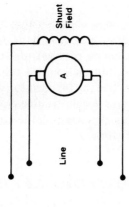
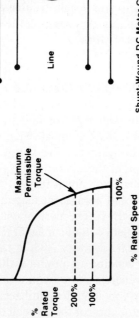

Series-Wound

The series-wound motor has the armature connected in series with the field. Although the series-wound motor offers high starting torque, it has poor speed regulation. Series-wound motors are generally used on low speed, very heavy loads.

Permanent-Magnet

The permanent magnet motor has a conventional wound armature with commutator and brushes, permanent-magnets replace the field windings. This type of motor has excellent starting torque, with speed regulation slightly less than that of the compound motor. Peak starting torque is commonly limited to 150% of rated torque to avoid demagnetizing the field poles.

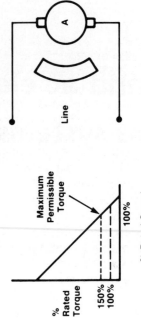
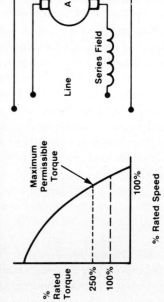

FIGURE 15-1 Graphs used to compare the speed and torque characteristics for permanent magnet and shunt wound DC motors. (Courtesy of Eaton Corporation, Electric Drive Division)

A graph that compares the torque and speed for the PM motor and the shunt motor is provided with a graph for two other types of dc motors, the series wound and the compound wound motors. From these diagrams you will see that the shunt wound and the PM motors are best suited for drive control. The graph of the PM motor shows a straight-line function which indicates that the motor is not affected by the armature reaction that causes the flux field to distort. As you know from Chapter 10, when the flux field becomes distorted, it is weakened and the shunt motor will stall prematurely under high-torque demands. In comparison, a PM motor's torque is constant at all speeds and is proportional to the amount of current at all speeds. In fact, you can get 4 to 11 times the normal torque at peak times, which may allow you to use a slightly smaller size motor with the drive control than you would be allowed to with a shunt wound motor. Since the torque is not proportional in a shunt wound motor, it will affect the amount of current limiting that should be used on SCR drives. The speed characteristics of the PM motor will allow speed ranges of up to 60:1, which may allow the motor drive to eliminate the need for transmissions or speed reducers.

The torque of a shunt wound motor can be calculated for determining the amount of current the drive will have to provide to control the torque at speeds. A formula is provided in Figure 15-2. From this figure you can see that torque is found by the formula $T = K \phi I_a$, where K is the motor constant, ϕ is the flux density, and I_a is the armature current. Since the flux density is a product of the number of turns of wire in the field coil and the amount of current flowing through it, the strength of the field is related directly to the torque the motor can provide. It also becomes

---

Torque = T (lb-ft)   Note: 1'hp = 3 lb-ft
Motor constant = K Flux density ϕ
 Armature current I_a
T = $K \phi I_a$
Torque = $\dfrac{hp \times 5252}{rpm}$
Rpm = $\dfrac{hp \times 5252}{T}$
Hp = $\dfrac{T \times rpm}{5252}$

FIGURE 15-2 Calculations used to determine the torque, speed or HP of a shunt type motor. A table is also provided to show the amount of current required to produce a given amount of torque. (Courtesy of Warner Electric, SECO Electronics Division)

---

apparent that since K remains constant and the flux density is usually at its rated value, the formula can be reduced to $T = I_a$. Another formula is also provided in this figure to calculate the torque when the hp and rpm are known and which can be transposed to calculate the rpm for a given motor by dividing the (hp × 5252) by the torque.

SCR MOTOR SPEED CONTROL

The SCR speed control is the most common type of speed and torque (load) control that is used in industry today. It has become an economical method of controlling dc motors since it can be connected directly to a three-phase or single-phase power distribution system in the factory. In years past, a source of dc voltage had to be produced before any type of dc motor control could be considered. Diagrams of typical SCR controls are provided in Figure 15-3.

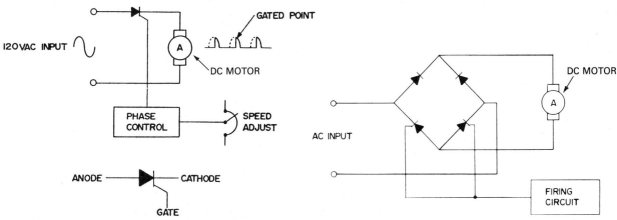

FIGURE 15-3 Diagram of a single phase, half wave SCR DC motor control, and a full wave bridege rectifier control. (Courtesy of Warner Electric, SECO Electronic Division)

From these diagrams you can see that the SCR will provide two essential functions for the drive. The first function is the ability to rectify ac current to pulsing dc current. The SCR is a silicon-controlled rectifier that can change ac to dc like a regular diode. The second essential function that the SCR provides is the ability to be controlled through its gate firing angle from 90 to 180 degrees, which will control the amount of current it is allowed to conduct.

Figure 15-3a shows a single SCR providing half-wave rectification used to control armature current in a small dc motor. The phase control module consists of an RC network where the capacitance is constant and the resistance is changed by an external potentiometer. When the resistance is increased, the time constant is also increased, which increases the firing angle and allows more current to flow. As you know, the output of the capacitor is not connected directly to the SCR gate; rather, it is shaped into a sharp pulse with a unijunction transistor or a programmable unijunction transistor (PUT). In some circuits, the unijunction signal is boosted by small SCRs that are used to control gate currents for larger SCRs, or a pulse transformer is used to increase the signal directly to the gate of the large SCR. You should also remember that since the pulsing dc current returns to zero at the end of each 180-degree cycle, the SCR will automatically be commutated (turned off) without the use of any additional electronic components.

Figure 15-3b shows a full-wave bridge rectifier that uses two rectifier diodes and the two SCRs. Since the SCR and diode are in series, only one SCR is required for each cycle of the ac current. When the firing angle of the SCR is set, it will also regulate the current through the diode. A second set of diodes are used to provide full-wave fixed dc current to the field of the motor. As you know, by keeping the current to the field constant and varying the voltage to the armature, the motor speed can be adjusted while the torque remains constant.

Speed Control Process

Speed control can be achieved by adding several other small circuits to the original circuit. In Figure 15-4, an additional diagram is provided to add speed control to the circuit. From the diagram in this figure you can see that a set of resistors are placed across the SCR output to produce a voltage divider that will be used as a reference voltage. A compa-

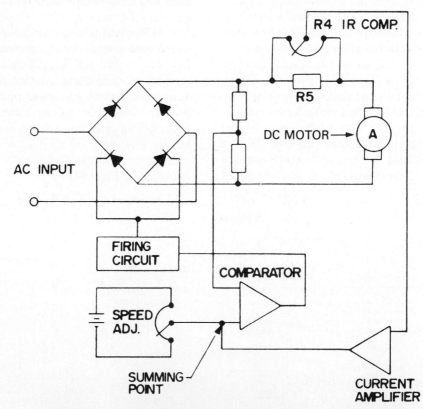

FIGURE 15-4 Diagram of advanced SCR motor speed controls that include comparison signals for voltage and current feedback that is used to provide voltage and current regulation. (Courtesy of Warner Electric, SECO Electronic Division)

rator circuit is also provided by an op-amp that receives a voltage from the speed control potentiometer and is referenced across the SCR output. The output of the comparator is sent directly to the firing circuit to provide changes in the firing angle. In this way, the speed can be set manually by adjusting the R3 resistor, and any time a change occurs at the output of the SCR, such as a drop in the control voltage, the motor would tend to slow down. The op-amp is continually comparing the output across the R1, R2 resistors to the set-point voltage provided by the reference supply and the speed pot. When the change occurs, the op-amp output at the comparator is changed to counteract the change. If the motor speed is dropping, the firing angle will be increased until it returns to the predetermined set point. If the motor is tending to overspeed, due to a change of load, the firing angle will be reduced slightly to slow the motor.

The SCR control can exercise a degree of control over the speed that is called the *percent of regulation*. The actual value of this control is determined by the formula

$$\frac{(\text{no-load speed} - \text{full-load speed}) \times 100}{\text{motor rated speed}}$$

The second variation of this circuit uses a second op-amp as a current amplifier. This op-amp is connected to a potentiometer that is wired across an armature resistor. This resistor is placed in series with the armature so that a reference voltage will be produced across the resistor that is proportional to the amount of current the armature is using. As you know, the amount of armature current is related to the amount of torque the motor is providing. From the graphs provided in Figure 15-1 you can see that the relationship between torque and current is nearly proportional.

A tachometer can be used to provide feedback for a velocity loop. The signal from the tachometer is produced by a tach generator that is mounted directly to the motor shaft. The output of the generator is proportional to the speed of the motor. The diagram of this circuit is shown in Figure 15-5, where the tach generator is shown connected to the shaft of the dc motor.

The output signal from the generator is shown as a single conductor since the other side of the circuit uses ground as a return. The voltage produced by the generator is sensed by the comparator, which uses the voltage from the speed adjustment circuit and the current feedback loop as a reference. The tach generator signal replaces the voltage divider, which provides closer control of the motor's speed since the output of the comparator op-amp is sent directly to the firing circuit of the drive SCRs. This circuit can provide up to ±0.5 percent regulation on the velocity of the motor at high and low speeds.

You should also understand the importance of the current-limiting portion of these circuits since they can be preset to protect the SCRs of the drive control and the armature of the dc motor from drawing too much current. The current limit set point potentiometer is generally set to match the maximum current value of the armature or the drive, whichever is smallest. When the current limit

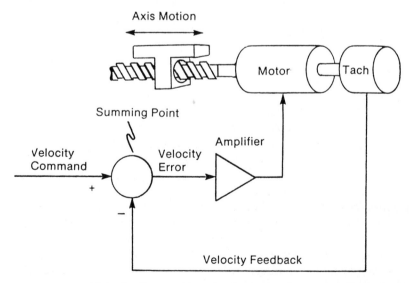

Velocity Error = (Velocity Command) - (Velocity Feedback)

FIGURE 15-5 Circuits that provide tachometer feedback signals that are usable for closed loop velocity control. (Courtesy of Allen-Bradley)

is reached, the signal at the current op-amp begins to lower the firing angle of the SCRs, which will lower the amount of voltage that is sent to the drive. Since the amount of current can be controlled by the amount of voltage, the drive will be able to bring the current bank into limits.

Block Diagram of the Feedback Loop

Figure 15-6 shows a block diagram of a typical closed-loop system with speed feedback signal. From this diagram you can see that the position controller or programmable controller can put the reference signal into the system. This is the desired speed at which the system motor should turn the load. The signal is sent to the motor speed drive, which will use the signal to help determine the gate firing angle for the SCRs. The drive will also use a feedback signal from the tach generator to compare the desired speed to the actual speed the motor shaft is turning.

The point where these two signals are compared is called the summing junction. The signal that comes from the operator panel will be positive and the feedback signal will be negative. The resulting signal is the mathematical sum of these two signals, and it is called the *error*. If the error is positive, the firing angle is increased to cause the motor to turn faster, and if the feedback is larger because the motor is turning too fast, the error will be negative and the firing angle will be decreased to slow down the motor.

The error signal is sent directly to the gate firing circuit, which controls the main power SCRs. This part of the controller is called "fixed power source." As you know, this is also where the three-phase voltage is connected and where the SCRs converts it to adjustable dc voltage.

The diagram shows a single line from the power source to the motor. This indicates that the power which is being controlled is sent directly to the motor at this point. You will also notice that a tach generator is connected to the motor at this point to sample the shaft speed. The tach generator signal is called the *speed feedback signal,* and it is an option for the system. If the feedback signal is not used, the loop becomes an open loop, where the operator puts a speed value in the controller and manually makes changes if the speed is not at the desired level.

In some applications a gear reducer may be connected to the drive. The gear reducer will change the speed of the motor before it gets to the machinery. Since the speed is being reduced, the changes in the signal will not be as pronounced as a direct-drive system. The feedback signal can be connected to the machinery shaft if the speed there must be critically controlled and there is the possibility of any slip between the drive and the machinery. Regardless of where the tach is connected, it will help the drive control to sense when the actual speed is higher or lower than the desired speed and make adjustments while the system is operating.

External Signals from Programmable Controllers

In all the diagrams presented thus far, the speed of the motor has been set by an external potentiometer. This means that someone must physically change the potentiometer for each different load application. It is possible to provide a voltage or current signal from an external source, such as an output module from a programmable controller (P/C). The analog output module can provide a 1- to 5-V or 4- to 20-mA signal to set the speed for the dc drive. The strength of this signal is determined by P/C program logic, which can sample a variety of input sensors to determine the number of boxes on a conveyor, or the amount of water that should be pumped to keep a tank full, and convert this information to the appropriate 4- to 20-mA signal. If a tach generator is also used to provide a feedback signal, a closed loop can be developed for the speed of the motor at any set-point velocity the P/C indicates.

The analog signal from the P/C output module uses two wires that are connected as the reference voltage or as a replacement for the speed-adjustment potentiometers. When the P/C analog signal is used as a reference voltage, the firing circuit will increase the firing angle of the SCRs until the motor speed is sufficient to cause the tach generator to produce a voltage that will offset it. This means that the higher the analog voltage or current from the P/C, the higher the firing angle of the drive and the faster the motor will rotate. The voltage and current feedback circuits that were introduced earlier will remain in the circuit to provide a regulated speed under varying load conditions.

The P/C permits the dc motor speed drive to be fully automated, which allows it to respond to any conditions that control circuit senses. Since the value for the speed control is stored in the P/C program, it can also be changed from external means. You will learn about local area networks (LANs) in Chapter 16 and will see that the value for the motor speed can also be sent from another

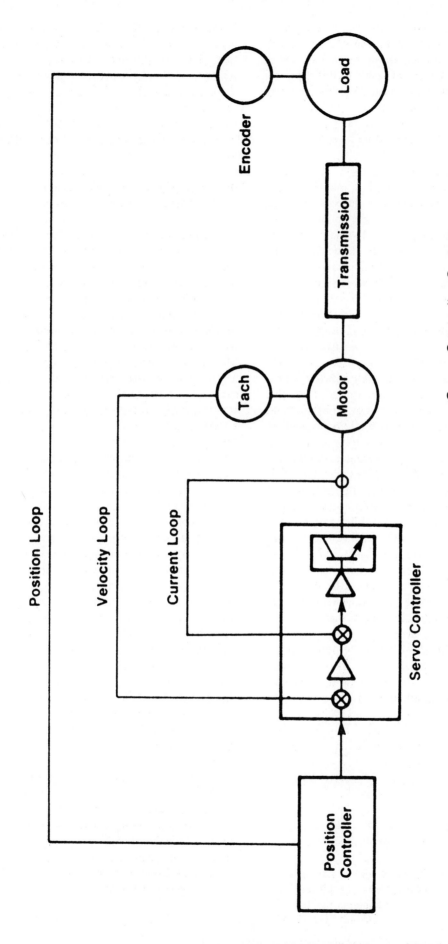

FIGURE 15-6 Block diagram of a closed loop speed controller. (Courtesy of Allen-Bradley)

computer, such as a mainframe that is the master of the network, or it can come from another P/C that is controlling another facet of the automated systems, such as a robot or an assembly cell. In this way the P/Cs can regulate the speed of conveyors or other motors in the system in reference to the speed of production and other machine cycles.

Output Signals From the SCR Speed Control

At times you will be required to determine if the SCR is controlling the voltage for the dc motor. The easiest way to test the SCR drive for proper operation is to use an oscilloscope to check the waveform of the output signal. Figure 15-7 shows several waveforms that you would see if the drive is operating properly. *You must use an isolated case oscilloscope or one with a differential amplifier when reading the signal from the dc SCR drive so that you do not create a short circuit that will severely damage the SCRs.*

Since the voltage from the drive is provided by a single-phase or three-phase ac voltage, both the supply and the return of the dc circuit from the SCR output are at a potential that is well above ground reference. When a normal grounded case oscilloscope is used to make the measurement anywhere in the circuit, the terminal of the scope that is at ground potential will cause a violent short circuit that will damage the SCRs. A plastic case oscilloscope or one with an isolated amplifier can be used to make these measurements because their leads are not at ground potential.

The first waveform in this figure shows the voltage measured at the armature of the dc motor when the drive is producing full voltage (full speed). Notice that the complete half-wave of the pulsing dc is allowed to flow to the armature. The other wave forms in this diagram show the SCR output at 50 and 25 percent. You should notice that the SCR turns on later as the amount of voltage is reduced. You will need to adjust the speed control potentiometer to make the output change to 50 and 25 percent. You could also use a voltmeter to make these readings. You will need to adjust the speed pot to 100 percent and record the voltage, and then make comparison readings for the 50 and 25 percent speeds. You can compare these voltages to the specifications provided for the drive. It is also possible to use a dc ammeter to measure the amount of current that the drive is providing at each of these levels. The amount of current for the drive is also specified in the drive documentation.

Some drives will list the waveforms that should be observed when the oscilloscope leads are placed directly across each SCR in the bridge. The waveforms in this case will be the complement of the waveform shown in the diagram. This means that the waveform measured across an SCR that is at 75 percent conduction will look like the waveform that is measured across the load where the SCR is at 25 percent conduction.

When you are testing the drive you can determine that something is wrong by observing the operation of the motor when the speed control is adjusted. If the motor does not operate at all, or if it does not come up to full speed, you should test the ac supply voltage to determine that both lines of a single-phase power source and all three lines of the three-phase source are operational. If a fuse is blown in any of the lines, the motor drive will become partially or completely inoperative.

If the motor does not change speeds as the control indicates, you should test for the proper voltage, current, or waveform at the load side of the circuit. You can also test for the proper input gate firing signal. If the proper input signal is present but the output waveform is not, you can assume that one or more of the SCRs are faulty. If the input signal is not providing the proper phase shift, the firing circuit must be replaced before further tests

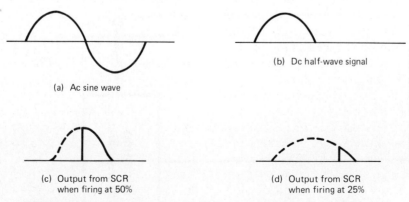

FIGURE 15-7 Waveforms for various output levels of the SCR type DC motor speed control.

are made. If a P/C is responsible for providing the milliampere signal for this system, you will need to test for the proper amount of current from the output module prior to making any other tests.

EDDY CURRENT DRIVES

The eddy current drive operates on the principle of electromagnetic field that is developed by placing a soft-iron bar into a coil of wire. Figure 15-8 shows several diagrams. As you know, the strength of the magnetic field can be varied by changing the current in the wire or the number of turns in the coil of wire. Another way to increase the strength of the field is to increase the number of bars in the core. An example of adding extra bars is shown in the second diagram in this figure. The shaft that is inserted through the middle of the core, allows the entire assembly to rotate.

The current in the coil of wire is controlled by varying the amount of dc voltage that is applied to the circuit. When current is passed through the coil of wire, each of the iron bars will become magnetized like a bar magnet. Since the bars are shaped like a U, the lines of force will flow from the south pole to the north pole of the bar.

A soft-iron ring that looks like a piece of metal pipe 1.5 to 3 inches long is placed so that it can rotate around the magnetic field. A small air gap is provided between the rotating ring and the magnetic poles. The gap is small enough that the full force of the magnetic field is passed to the rotating ring. Since the ring can rotate separately from the magnetic core, it will make up two of the three parts of the magnetic clutch.

A more detailed diagram of an actual eddy current drive is shown in Figure 15-9. The drive has three basic parts: the driving member, called the drum; the driven member, called the rotor; and the magnetic member, called the field coil assembly. In this diagram the shaft of the ac motor is shown connected to the drum portion of the eddy current drive. This means that the drum portion of the clutch will rotate at the same speed as the shaft of the motor. The motor in this case will be a constant-speed ac squirrel-cage induction motor.

The load is connected to the rotor of the eddy current drive. When the motor is started, it will come up to speed and rotate at a constant speed. This causes the drum of the drive to rotate at the same speed as the motor shaft. If the magnetic field is at zero, the load will not turn at all, and when the magnetic field is increased it will place a force on the drum which causes the load to begin to rotate.

If the magnetic field is weak, the load will rotate at a much lower speed than the motor. The difference between the motor's constant speed and the speed of the load is called *slip*. The slip in this case is caused by the weak strength of the magnetic field, and it does not cause any heat to build up since none of the parts are touching or rubbing together.

When the field is increased to full strength, the force between the drum and the rotor will increase to a point where the load is rotating at the same speed as the motor shaft. Since the strength of the magnetic field is controlled by the dc current flowing in the coil, the amount of slip can be adjusted by adjusting the amount of dc voltage that the clutch receives. The relationship between the speed of the motor and the speed of the load is called the *linkage effect*. When the current is controlled, the linkage effect is also controlled and the torque that is transmitted to the load is also controlled.

Since the control of the eddy current drive is accomplished by varying the amount of dc voltage that is sent to the magnetic coil, a SCR amplifier can be used to produce the dc voltage from single-phase or three-phase ac voltage. This allows the eddy current drive to be used with either ac or dc constant-speed motors that exist throughout the plant. This means that the eddy current drive can be added to any of these constant-speed motors and provide a variable-speed load control.

The speed of the load can be adjusted in manner similar to controlling the speed of the dc SCR motor drive. The speed can be set with an external potentiometer that is adjusted by the machine operator, or a tach generator can be added to make the system a closed-loop control. The firing angle of the SCR amplifier can also be set by an analog output signal that is sent from a P/C module. This allows the P/C to provide speed control of the load while it is connected to a constant-speed motor. Since the ac motor is generally a constant-speed motor, this application is generally used to connect ac squirrel-cage motors to loads that need to have a varying speed.

AC VARIABLE-SPEED DRIVES

The ac variable-speed drive provides the same type of motor speed control as the dc drive, but the means of controlling the ac motor's speed is different. In the dc drive, the motor's speed was controlled by raising or lowering the voltage sent to the motor. In an ac drive, the frequency of the power must be changed to make the motor speed change. As you know, the speed of an ac motor will change when the number of poles in the windings are

NEMA B Squirrel Cage Induction Motor

AC motor

The eddy-current drive converts ac line power directly to rotational power through the use of standard squirrel cage induction motors. Using a 60-Hertz induction motor as reference, the basic output speeds can be 3600 rpm synchronous for a two-pole design, 1800 rpm for a four-pole design, 1200 rpm for a six-pole design, etc.

These base speeds are relatively fixed. The standard squirrel cage induction motor has a full load speed two to four percent lower than the synchronous speed. In the case of the four-pole design, which will be used as reference, the synchronous speed would be 1800 rpm, and the full load speed would be approximately 1750 rpm (50 rpm slip).

To obtain a variable output speed from the relatively fixed ac motor speed, an eddy-current clutch is added to the power transmission train.

Motor and Clutch Assembly

Drum

The input member of the eddy-current clutch is essentially a steel drum driven directly by the ac motor.

Rotor

The output member of the eddy-current clutch, a steel rotor, is free to rotate within the clutch drum.

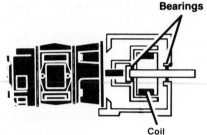

Bearings / **Coil**

The third element is a coil of wire wound on the rotor of the clutch. There is no friction contact between the input member (drum) and the output member (rotor). The torque is transmitted by an electrical (flux) linkage across the air gap. The fixed air gap is maintained by shaft-mounted bearings.

Torque Development

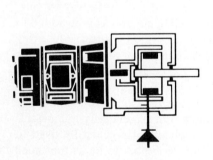

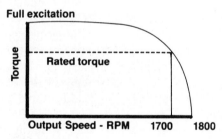

(Typical Curves)

To establish this electrical linkage, excitation level as low as two percent of the total power to be transmitted from the ac line to the output shaft is rectified to dc from the ac source. An isolation transformer is normally provided.

By applying this excitation level power to the rotating coil, a flux pattern is established through the drum and rotor.

With relative motion between the drum and the rotor, eddy-currents are generated in the drum. These currents produce the resultant electrical (flux) linkage between the drum and the rotor.

FIGURE 15-8 Basic principals of operation of an Eddy Current clutch. (Courtesy of Eaton Corporation, Electric Drive Division)

With a fixed amount of excitation applied to the coil, the output speed of the clutch will vary as the load is increased or decreased (points 1 and 2 on the Torque-speed curve). While this feature is desirable in some applications, the criterion of an adjustable speed drive is its ability to maintain a set speed with a varying load.

By varying the level of excitation to the coil, the amount of linkage effect, and consequently, the amount of torque transmitted from the AC motor to the output shaft, can be varied. (See points 1 and 2 on the Torque-speed curve.) Varying the output shaft torque enables the adjustable speed drive to maintain a set speed with a varying load.

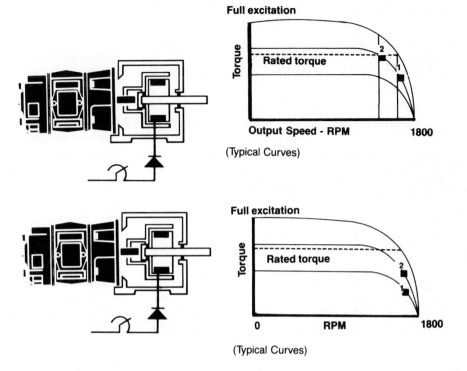

Speed Control

A tachometer generator of the permanent magnet brushless type, mounted integrally with the output shaft, provides a signal proportional to the output speed.

By comparing this signal to a setable speed reference signal, and amplifying the difference or error, the level of excitation to the clutch coil can be adjusted or corrected to realign the actual speed to the set speed when the load is varied.

The degree of accuracy of realigning the output speed to the set speed is a measurement of the performance of a drive regulator.

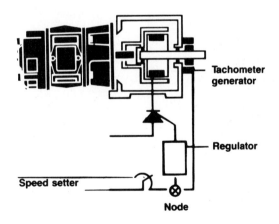

FIGURE 15-9 Cut away diagram of a motor connected to a load that is controlled by an Eddy Current drives. (Courtesy of Eaton Corporation, Electric Drive Division)

AC Variable-Speed Drives **413**

changed or when the frequency of the power is changed. Since the number of poles that a motor is using is limited and must be changed with relays, that type of speed control usually consists of several step speeds, such as high, medium, and low.

When a variable-frequency drive is used, the speed of the motor can be controlled from zero through full speed in smooth increments. Since the drive will vary the frequency, any three-phase squirrel-cage induction motor can be used as long as its horsepower is large enough to move the load, but does not exceed the size limitation of the drive.

The benefits of using an ac variable-speed drive include the ability to provide a better operation efficiency for the motor, which will save money on the electric bill. The drive also provides a means of starting the motor with just enough torque to cause the load to start moving. This is called *breakaway torque,* and the process of starting the motor with minimal torque is called *soft start.* The soft start prevents serious damage from occurring when the motor applies full torque to a load that is setting at rest. As you learned in Chapters 10 and 11, if a motor applies full torque to a load that is at rest, the shaft and its bearings are quickly worn out.

Another benefit the drive provides is the ability to set the maximum amount of torque the load will receive. If the load requires more torque, the drive will assume that the load is not functioning properly and protect the system by reducing the torque or deenergizing. The drive also provides a method of synchronizing the speed of multiple motors in an application where several motors must operate at speeds determined by another motor's speed. An example of this application is where multiple conveyors must provide the same speed even though they are loaded differently, because they must feed into a main conveyor or must cycle in unison with other machinery.

The ac drives in use today use one of three methods to electronically control the frequency of the power sent to the ac motor. These methods include the variable-voltage input (VVI), the pulse-width modulation (PWM), and the current source input (CSI). The theory of operation that each of these drive utilizes to control the speed of a motor is explained in detail in the following material.

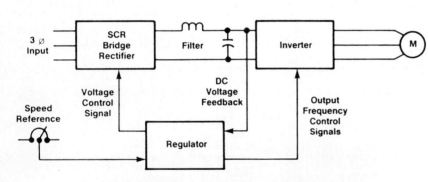

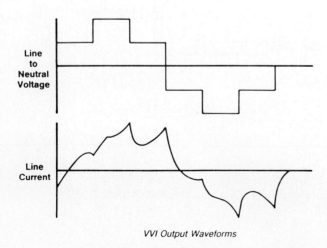

FIGURE 15-10a Diagram of VVI type control with voltage and current waveforms. (Courtesy of Eaton Corporation, Electric Drive Division)

Variable-Voltage Input Drives

Figure 15-10a shows a block diagram of a typical VVI drive. From the diagram you can see that this type converts three-phase voltage to dc and then shapes the waveform of the ac signal that is sent out to the ac motor. The dc converter is similar to the one used in the dc variable-speed motor drive, where SCRs in a bridge circuit or a diode bridge with a chopper circuit are used to rectify the three-phase voltage. A voltage regulator is used to control the level of the dc voltage that is provided at the output of the bridge circuits. Since this voltage is going to be used as the input to the inverter, its level will have a direct bearing on the level of the ac voltage sent to the motor.

The inverter uses a set of solid-state devices such as SCRs or triacs to switch on and off at appropriate times to reshape the dc voltage into waveform that is similar to the sinusoidal waveform of ac voltage produced from an alternator. Some of these devices are used to produce the positive part of the ac wave and an identical set is used to produce the negative part of the waveform. A microprocessor is used to control the exact timing of these components to produce the desired waveform.

An example of the voltage and current waveforms is also shown. You can see that the voltage waveform is shaped by six-step transitions. For this reason it is commonly called a six-step waveform. You can also see that both the voltage and current waveforms look similar to a normal ac sine wave in the respect that part of the signal is positive and part of the signal is negative and its frequency can be controlled from zero to 60 Hz. As long as the signal has the sinusoidal characteristic, it will be sufficient to cause the magnetic field in the ac motor to rotate. You should also remember that since the motor is three-phase, three of these waveforms will be produced at 120 degrees apart. If the motor is a single-phase motor, only one waveform is required.

As you know, the higher the frequency of the waveform, the faster the motor will rotate. The torque of the motor can also be controlled by varying the amplitude of the waveform. This can be adjusted by the switching circuitry and by controlling the amount of ac input voltage that is converted to dc by adjusting the firing angle of the SCRs or changing the input control to the chopper circuit. Figure 15-10b shows a photograph of a typical variable frequency drive.

Pulse-Width-Modulation Drives

The PMW drive utilizes a theory of operation that is similar to the VVI drive. Figure 15-11 shows a block diagram of the components in the drive and the voltage and current waveforms that are provided from the output to the motor. From the block diagram you can see that the drive uses three-phase ac line voltage. This means that the input line voltage should be 60 Hz. The ac line voltage is connected directly to a diode bridge rectifier. This means that the dc voltage that comes from the bridge will be fixed. Since the voltage sent to the

FIGURE 15-10b Picture of variable speed drive. (Courtesy of Eaton Corporation, Electric Drive Division)

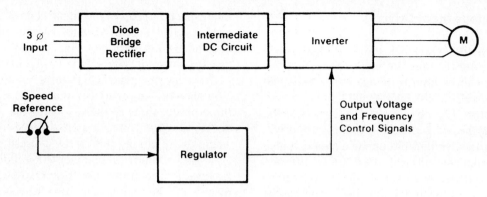

Power Conversion Unit (PWM)

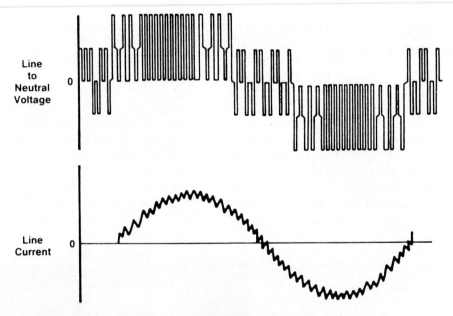

PWM Output Waveforms

FIGURE 15-11 A block diagram of the PWM drive with voltage and current waveforms that are provided to the AC motor. (Courtesy of Eaton Corporation, Electric Drive Division)

inverter is fixed, the amplitude of the output waveform will also be fixed. To change the speed of the motor, the pulse width of the output signal will have to be adjusted.

The dc voltage that leaves the converter is pulsing dc, and after it passes through the filter, it will be smoothed to pure dc. Once it enters the inverter, solid-state switching devices are utilized to convert it back to a quasi-sinusoidal waveform that will have varying pulse widths. The waveform is made wider by increasing the zero voltage interval. In this way, both the average output voltage and the frequency of the signal are controlled at the same time. As you know, the longer a sinusoidal waveform spends at the zero voltage level, the smaller the average voltage will become. Since the amplitude of the voltage remains constant, the average voltage is controlled directly by increasing or decreasing the zero voltage interval.

The shorter the zero voltage interval is, the larger the average output voltage. This also increases the frequency of the waveform, which will cause the motor to rotate at a higher rpm. You should remember that the speed of the motor is a direct result of the frequency of the voltage that is powering it. Since the number of poles in the motor are fixed in this application, the higher the frequency of the motor, the faster the motor will run. When the frequency is exactly 60 Hz, the motor will run at its nameplate rated speed. As the frequency is decreased, the motor will slow down. Since the inverter is controlling the zero voltage integral, the average voltage will also be changed, which will provide current control. As you know, the amount of current available to the motor will directly affect the amount of torque the motor can produce to turn the load. This is especially useful at very high load speeds, as the torque can be regulated to allow the motor to move the load without stalling.

A speed reference signal can be sent to the voltage and frequency control circuit of the drive. An external potentiometer is used to provide a reference current signal, which could also originate from the analog output module of a programmable controller. This allows the motor speed to be fixed by the operator or controlled by the P/C automatically. If the P/C is used, a tach generator could be used and the P/C logic could design a closed-loop algorithm that could vary the input signal as load conditions changed the speed of the motor shaft.

Current Source Input Drives

The CSI drive is the third type of variable-frequency drive commonly used in industrial motor speed controls. A block diagram of the CSI drive circuit is shown in Figure 15-12. The drive uses three-phase ac voltage at 60 Hz to power the drive circuit. The line voltage is connected to a converter circuit that changes the ac voltage to dc voltage in a SCR bridge circuit or a diode bridge with a chopper circuit. This circuit is the same type of circuit that is used in the dc variable-speed drive and the VVI drive. Both the SCR and the chopper allow the dc voltage from the circuit to be controlled from zero to maximum in a smooth transition rather than rough steps.

When the dc voltage leaves the converter, it passes through a filter to smooth the pulsing dc to pure dc, where it is placed on the dc bus. The dc bus provides voltage to the inverter circuit, where it will be changed back to a sinusoidal waveform. A current sensor is used to monitor constantly the amount of current on the bus. The amount of current at this point in the drive circuit is proportional to the ac current the motor is using. The current-sensing signal is sent to the current regulator, where it will compare the actual current flow to the maximum limit that is preset into it from an external potentiometer. If the ac motor is drawing too much current, the regulator circuit sends a signal to the SCR firing circuit or the chopper control circuit to change the amount of current to correct the overcurrent condition.

A speed control or voltage control circuit will be used to set the amount of voltage the drive will send as it varies the step current waveform of the drive signal. The speed control circuit senses both the output of the inverter as a feedback circuit, and the preset value from the speed adjustment pot to determine the value of the signal that is constantly being sent to the frequency control circuit. This circuit is also used to adjust the amount of current on the dc bus, which will provide torque limiting. Since an external potentiometer is provided to allow the operator to dial in the preset speed for the system, the signal could also come from a P/C analog output module which would generate a variable-current (4 to 20 mA) signal.

The frequency control circuit provides the control signal to the inverter that determines the size of the steps in the current waveform. The size

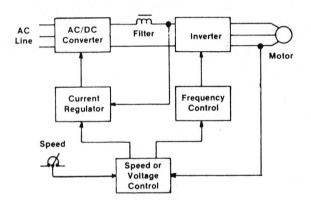

Block Diagram for a Typical CSI Drive

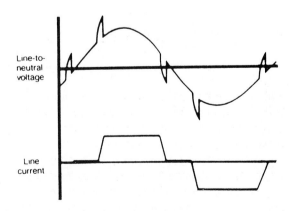

Typical CSI Voltage and Current Waveforms

FIGURE 15-12 Block diagram with current and voltage waveforms for a CSI drive. (Courtesy of Eaton Corporation, Electric Drive Division)

AC Variable-Speed Drives **417**

of the steps will determine the frequency and the amount of average current that the ac motor will see. The inverter uses solid-state switching devices that can shape the pure dc input signal into the stepped current waveform shown in Figure 15-12 with the block diagram. These devices may be transistors, SCRs, or gate turnoff thyristers (GTOs).

SPEED AND TORQUE CHARACTERISTICS OF VARIABLE-FREQUENCY DRIVES

Since the speed and the torque of a motor can be controlled with a variable-speed drive, the system is ideal for motor and load protection during startup. This means that the drive circuit can provide a soft start to the load that will save wear and tear on the motor and load shafts and bearings. Since the current and the frequency can be controlled over the entire range from zero to maximum, the motor can be smoothly accelerated with sufficient current to provide adequate torque to the load. This allows the motor to operate smoothly even at low speeds with a heavy load.

Figure 15-13a shows this torque and frequency relationship. You can see from this diagram that the amount of current is directly related to the torque at any given speed, over the frequency range of 0 through 60 cycles. You can see from this graph that adequate torque is available (100 percent) even at the lowest speeds.

Figure 15-13b shows the effects of increasing the frequency above 60 cycles. You can see that when the frequency is held below the 60-cycle limit, the drive can provide a constant voltage per hertz ratio, which means that it produces constant torque. When the frequency is raised above the 60-Hz level, the motor will loose some torque capability, but it will continue to have a constant-horsepower relationship. As you know, the motor speed is proportional to the frequency of the voltage it receives, so when the frequency of the drive voltage is increased above 60 Hz, the motor speed will be increased above the nameplate rating. This may be quite useful in some applications, but you must be ware of other conditions that may change in the motor, such as thermal buildup, which will require a special motor or extra cooling capacity.

Figure 15-13c shows a method to extend the motor's performance by increasing the frequency above the base frequency of 90 percent of 60 Hz. This graph allows you to size the motor used for the application slightly smaller. Since the motor's performance can be extended by increasing the frequency above the 60-Hz level, you can control a

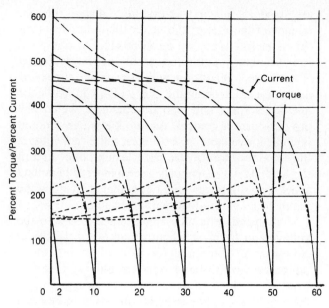
Figure 15-13a

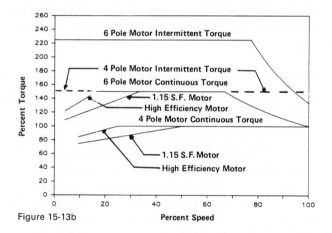
Figure 15-13b

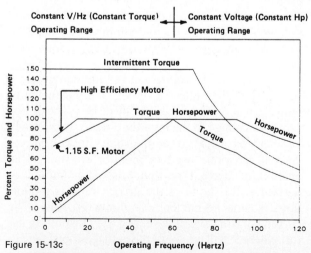
Figure 15-13c

FIGURE 15-13 Three graphs are provided to show the relation between motor performance and frequency. (Courtesy of Eaton Corporation, Electric Drive Division)

smaller motor to carry a load that is larger than its rating on an intermittent basis. This is similar to using a motor with a larger service factor to provide increased load capability on a limited basis instead of investing in a larger motor. You must understand the importance of this feature on larger motors. When the application uses motors less than 3 hp, this is not as critical, because the price difference between the motor that is selected and the next larger size motor is minimal. When large motors above this horsepower rating are used, the next-size motor may cost up to 60 percent more, which means that the investment in the variable-speed drive will be paid for from the increased capacity of a smaller motor.

Constant-Torque Load Applications

The applications for variable-speed ac motors is very diverse. Some of the applications require the motor to provide constant-torque characteristics, while others require variable torque. The same is true about the horsepower and voltage characteristics.

A constant-torque application is one where load demand is constant over the entire rpm range of the motor. These types of loads are generally friction loads which remain constant. Even though the torque remains constant over the entire range, the horsepower characteristic will vary. A graph of this condition is provided in Figure 15-14a. As you can see from this graph, the horsepower that the load requires is constant as the speed increases. As you know, horsepower is calculated by determining the weight of the load, the distance it must be moved, and the time that the movement takes place, whereas torque is calculated only from a force multiplied by a distance; time (speed) is not considered.

Constant-Horsepower Load Applications

Figure 15-14a also shows the torque response of a constant-horsepower load. This type of load application is found on machine tool spindles, extruders, and some types of mixers. In this type of application you can see that the horsepower requirement remains constant, but the torque will decrease as

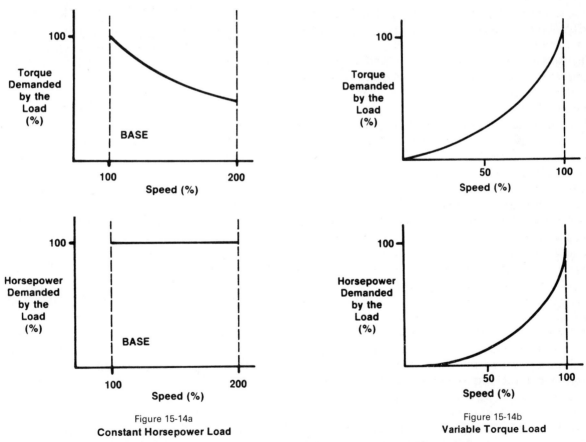

Figure 15-14a
Constant Horsepower Load

Figure 15-14b
Variable Torque Load

FIGURE 15-14 Graphs of constant torque, constant horsepower and variable torque load applications. A table is also provided to show the type of load characteristic and the application it is used for. (Courtesy of Eaton Corporation, Electric Drive Division)

Speed and Torque Characteristics of Variable-Frequency Drives

the speed of the load is increased. This is the result of the load requiring large starting torque to provide rapid acceleration. Normally, the torque in this application is inversely proportional to the speed. From the graph you can see that the torque is decreased as the speed is increased.

Variable-Torque Load Applications

The graphs in Figure 15-14b shows the relationship between speed and torque and speed and horsepower when the load requires a variable-torque characteristic. In this type of application you can see that the torque is directly proportional to the mathematical square of the speed, and the horsepower is proportional to the mathematical cube of the speed. This type of load characteristic is useful in large centrigual fan and pump loads where the load can be added to the motor as the speed is increased. Generally, loading vanes or valves are used to increased the load gradually as the motor reaches a higher speed. This type of load also exists on high-inertia loads such as flywheels for punch presses.

SERVO SYSTEMS

The servo system uses a motor and sensor to indicate position or velocity. This type of system is used in applications where motion control is required to provide critical positioning ability. The positioning system can utilize a linear actuator such as a ball screw mechanism or a rotary actuator such as a motor or gearbox. These motion control systems are used for machine tool, robotic, and automatic positioning applications.

In the previous information about variable motor speed drives, the only part of the system that is controlled is the velocity (speed) of the load. The position of the load is not considered and is generally not important in these applications. An example of the speed control application is a conveyor that must be operated at constant speed. In this closed-loop system, the speed is controlled but the position of the load on the belt is not important.

In the motion control application for a servo positioning system the location of the load is the most important part of the system. In a typical servo system the position that the load is moved to may be in a straight line from where it was, or it may be at some angle if it were rotated about a circular axis. Figure 15-15 shows an example of a servo system that is using a ball screw to move a load to any position along the length of a 3-foot ball screw. This application would be used by an automated arm (robot) that was used to pick parts out of a mold in a plastics press. The robot arm has two other axes that are moved by air cylinders. When the air solenoids that control the cylinders are given the command, one cylinder will move the gripper down into the mold, and the servo system will move the ball screw mechanism ahead so that the part can be gripped. When the part is gripped, the ball screw is turned in the reverse direction so that the part can be pulled from the mold. When the part is clear of the mold, the air cylinder is switched to the up position and the part is pulled upward to its original position above the press. A second air cylinder is then energized that moves the robot arm in a horizontal direction until it is over a conveyor. When the arm reaches the cylinder stop, the down solenoid is activated again, and the vertical cylinder lowers the arm to the conveyor, where the part is released. After the part is released, the vertical arm is returned to the up position, and the horizontal cylinder is retracted to bring the arm back to its starting position for the start of a new cycle.

The vertical and horizontal cylinders can only be moved to either end of their stroke. Their motion is controlled by two on/off solenoids, which means that they can only be extended or retracted. Since the length of their stroke lines up perfectly with the centerline of the mold and the centerline of the conveyor, no other positions are needed. The depth of the part may vary, so the ball screw travel is needed to be able to locate the part accurately even when different mold patterns are used.

Servo Positioning Loop Operation

In the previous example the servo positioning system will be able to locate the exact position to move forward to pick the part out of the mold and retract far enough to be able to clear the mold when the vertical cylinder pulls the arm up. Figure 15-15 shows a detailed diagram of the ball screw mechanism that the servo loop will control. The ball screw is belt driven by a dc servomotor.

If the ball screw is rotated in the clockwise direction, the arm will move forward, and if it is rotated in the clockwise direction, the arm will move in the reverse direction. You should also notice that a resolver is connected directly to the motor shaft to indicate the number of entire turns the load has made during any move. The resolver is also able to indicate the position (0 to 360 degrees) where the shaft has stopped within any one complete revolution.

The block diagram in Figure 15-16 shows the operation of the servo positioning loop. For the purpose of this example the arm can be moved from

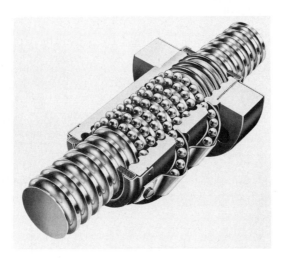

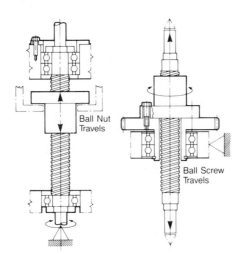

FIGURE 15-15 A close up view of the ball screw mechanism used for servo positioning. (Courtesy of Warner Electric Company)

Ball Groove or Ball Track
The helical groove of a ball nut and ball screw shaft which transmits the load reaction between the ball nut, ball screw and balls.

Ball Nut Body or Ball Nut
The nut, single or multiple ball circuits, assembled with balls, recirculation system and accessories.

one end of the ball screw to the other in 100 revolutions of the motor shaft, and the length of the ball screw is 20 inches. This means that the arm will travel 1 inch for every four revolutions of the motor, or the travel can be computed at $\frac{1}{4}$ inch for every revolution. The arm will need two positions (P1 and P2) that are located at 4 inches and 16 inches to move forward to pick up the part and then return to pull it from the mold.

From the block diagram and operational sketch of the ball screw you can see the operation of this system. When the system is first energized, the motor will go through a homing routine to locate the home limit switch. After the servo has located the home limit switch, it will be ready to accept commands to locate position P1 or P2. The servo is only required to locate the home position once when it is powered up, and then the system is ready to move accurately to any position along the ball screw.

The first command the servo system receives is to move to position P1, which places the robot arm directly above the open mold cavity. Since the position of P1 is at the 2-inch mark, and the home position is located at the 1-inch mark, the command signal will be an analog voltage of 1 V positive. From the block diagram you can see that if the motor has not moved yet, and the command voltage is 1 V, the summing junction will take the mathematical sum of the 0 V from the resolver, and the 1+ V from the command.

The summing junction sends the result of this calculation, which is called *error,* to the amplifier. The amplifier converts the error signal to a larger voltage that the power converter uses to send a

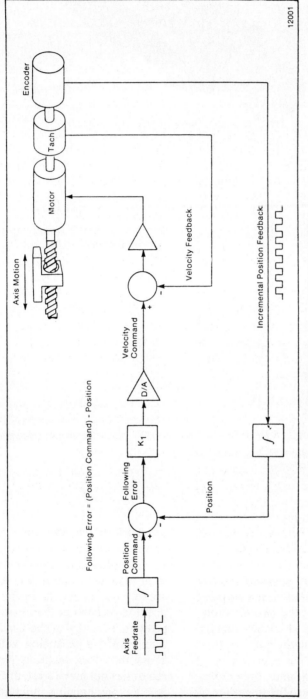

FIGURE 15-16 Servo system with a velocity loop control added to the positioning loop. (Courtesy of Allen-Bradley)

120-V signal that the motor. When the motor receives this signal, it begins to turn in the clockwise direction and the ball screw begins to move the arm at a rate of $\frac{1}{4}$ inch per revolution. The resolver begins to produce a negative voltage signal as soon as the motor begins to turn.

This signal is produced at the same ratio as the command signal, so when the motor has moved the load one revolution, the resolver is sending a negative 0.25-V signal to the summing junction. At this point the command voltage remains at +1 V, so the sum of the two signals is +0.75 V. Since the error signal is still positive, the motor will continue to rotate in the clockwise direction.

When the motor has rotated through its second revolution, the ball screw will have moved 0.5 inch. The resolver is producing an analog signal, so it is constantly changing. When the ball screw has moved the load exactly 0.5 inch, the resolver will send −0.5 V to the summing junction. This time when the comparison is made, the error signal is +0.5 V.

As you can see, as the motor continues to move the ball screw in a clockwise direction, and the load moves forward, the resolver sends a more negative signal. As the resolver's feedback signal becomes more negative, the error signal becomes smaller. When the motor has moved the load four complete revolutions, the load will be at position P1. The error signal at this point will be −1 V and the command signal remains at +1 V. The error at this time is now 0 V. When the amplifier receives the 0-V error signal, it will deenergize the power converter and the motor is stopped.

You can see that the summing junction is continually comparing the signal from the command voltage to the feedback signal of the resolver. As long as there is any error, the power converter will continue to send voltage to the motor to cause it to rotate. When the load reaches the set point, the error becomes zero and the power converter stops sending voltage to the motor.

When the arm reaches the P1 position, it sends a command signal to the controller that the arm is at its start position. When the press completes its cycle, it will open its mold and signal the motion control circuit that the part can safely be picked up. At this time the controller sends a new command signal to the summing junction to move to position P2. The new command signal causes the error to become positive again, and the amplifier and power converter again send a positive dc voltage to the motor. As the motor begins to move in the clockwise direction, the resolver signal becomes more negative, which begins to make the error smaller. When the arm reaches the P2 position, the error signal will be 0 and the motor will receive 0 volts, which stops it exactly on the mark.

Typical Resolution and Accuracy

The resolver in this application has an accuracy of 0.01 V. This means that every 3.6 degrees the shaft turns, the resolver will produce another 0.01 V. This provides the servo system with resolution of $\frac{1}{100}$ of $\frac{1}{4}$ inch. In decimal form this means that the accuracy of the ball screw will be approximately 0.0025 inch. This means that as long as the arm is not within 0.0025 inch of the P2 position, the summing junction will continue to calculate error, which will be amplified and passed to the motor. As the load moves closer to the position, the smaller the error becomes, the smaller the voltage to the motor becomes. As you know, when the motor receives less voltage, it will slow down. By keeping the field strength constant or by using a permanent-magnet servomotor, the motor will be able to move the load even when the error is relatively small. This allows the servo system to move the load rapidly when the error is greatest, and to slow the load down as it approaches it new location.

The actual resolution and accuracy of any servo system will depend on the quality of each component in the system. The accuracy of these components must be coordinated, or the entire system cannot be any more accurate than the least accurate part of the system. The repeatability of the system is the ability to reach the same position time after time under varying load conditions. Generally, the repeatability will be nearly as concise as the resolution.

Driving the Servo System in the Reverse Direction

After the load is positioned at point P2, it will need to return to location P1 before the vertical cylinder can be retracted. To accomplish this move, the controller will send a command signal to the summing junction to cause the motor to run in the counterclockwise direction. Since the resolver is already sending a signal equal to −16 V, the signal from controller will need to be +2 V. The error at this time will be −14 V, which will be amplified and converted to a negative motor voltage. Since the servomotor is a dc motor, it will begin to run in the counterclockwise direction. As the load approaches the P1 position, the resolver voltage will continue to move from −16 V to the −2 V that it will send when it reaches point P1.

When the load reaches point P1, the resolver voltage will be −2 V and the command voltage signal will be +2 V, so the error is again zero and the motor is turned off. The controller will continue to send the command signals to move the load back and forth from point P1 and P2. The commands are synchronized with the operation of the press, so the robot arm can move into and out of the mold to pick up the part without being damaged by the press.

Digital-to-Analog Servo System

The analog system described previously has been in use since prior to World War II. Several enhancements have been made to the servo system in the past 10 years that has made it more widely accepted. The first enhancement has been the advent of the digital-to-analog converter and the analog-to-digital converter. These devices have allowed an encoder to be used in place of the resolver to provide the feedback circuit for the servo system. Since the encoder provides the signal as a digital value, it can be read directly by a digital control device such as a programmable controller or microprocessor chip. The second enhancement has been the use of power transistors and other power thyristors that allow the use of large integral horsepower ac motors as the servo drive motor.

A servo system can also use an encoder as a feedback device. This type of control servo system uses a digital computer (microprocessor chip) that will also receive the command signal as a digital value. The command signal, which may also be called the set point, will always be positive in relation to the feedback signal.

The operation of the encoder is explained in detail later in this chapter with information about the resolver. At this time all you need to understand is that the encoder will send a value to the processor (at the summing junction) as a value between 0 and 2047. The value 2047 has been chosen since the processor is using 12 binary bits to transmit and store values. If 16 bits were used, the value would be 32768 (rounded to 32000), or if 12 BCD would be used the value would be 999. Each type of controller will use the value that they feel is most compatible with the resolution of their equipment. Regardless of the maximum value, it will range from minimum to maximum or in both the positive or negative value if a bipolar system is used.

The controller will compare the command signal to the feedback signal at the summing junction. The error will be the result of this summation, and it will be sent to a controller logic circuit. The output of the controller will be sent to a digital-to-analog converter, which will change the digital signal to an analog value. The analog value will be amplified and converted to a dc power voltage that is usable to the motor. At this point the circuit is very similar to the traditional analog servo system. If the command signal indicates that the motor must move forward, the signal will be a positive value that will cause the amplifier to send a positive dc voltage. As the load moves closer to the set point, the feedback value will become larger, which will make the error become smaller. When the load reaches the set point, the error will be zero and the amplifier will no longer send a voltage to the motor. This causes the motor to stop when the load reaches the set point.

Adding a Velocity Loop

Since the digital computer or microprocessor is used as the controller, a velocity loop control can be added to the original circuit shown in Figure 15-16. A tach generator can also be added to provide a velocity feedback signal. This feedback signal is sent directly to a second summing junction after the D/A converter. The positioning loop must operate in conjunction with the velocity loop. The velocity loop will determine the amount of voltage that is sent to the motor, which will directly control the dc motor's speed. When the positioning error is large, the velocity signal will be maximum, and when the load is approaching the set-point position, the velocity will be reduced to provide more accuracy in positioning. By allowing the speed to operate at higher values when the error is large, the load can be moved into position in the shortest period.

The velocity loop must also be used when multiple-axis servo systems are used. This is important when precise motion control is being attained. Another advantage of the velocity servo loop is that a gain factor can be included in the velocity calculation that the controller executes. This allows the response of the velocity loop to be predetermined by changing the gain factor.

In the circuit shown in Figure 15-16 an encoder is used to produce a digital feedback signal. The encoder was chosen for this application because the controller is a digital computer that uses digital values to produce the value for the output signal. The command signal is also digital for this circuit. Since the microprocessor chip has become so reliable and inexpensive, it has been widely accepted and used in most types of servo systems. It is also selected because it can provide other functions, such as memory storage, where a set of moves to provide a machine operation may be stored. Since the storage is in random access memory (RAM), it can easily be changed and saved on floppy disk or magnetic tape for future use. This also makes the servo system easy to integrate into a programmable controller system. Since the microprocessor operates at a high speed, it is normally placed in a servo positioning module that will be mounted in the programmable controller rack and update the processor within the normal scan time. This allows the servo module to provide reliable motion control and receive new move sets from the P/C only as the need arises. The P/C can also send signals to the servo module start cycle, hold cycle, or jog, which allows the servo to be integrated with other parts of machine operation that are under direct control of the P/C.

The resolver was not used in this example because its output signal is analog, which is not directly compatible with the digital computer. In the last 15 years, the analog-to-digital converter has been perfected and integrated into this type of application. This allows a device such as a resolver to be connected to the converter that changes the analog signal to a digital value that is compatible to the digital computer.

In the previous circuit, the A/D converter would be used between the resolver and the summing junction. This would provide instantaneous converting of the signal as change in the analog level take place. The A/D converter is built directly into the circuitry of some encoders, and in other cases, the encoder can be connected directly to an analog input module that resides in the P/C rack. The operation of the analog module is to provide analog-to-digital conversion that is compatible with the data bus of the microprocessor. This module provides the needed interface to allow the resolver to be used with the digital servo system controller.

Types of Servomotors

The servomotor is a specially designed permanent-magnet or shunt wound dc motor. Other types of motors, including ac motors, can also be used, but they are generally limited to larger loads. Of the two dc servomotors, the permanent-magnet (PM) motor is the more popular. It is able to provide a two-wire connection that simplifies installation connections, and it provides exceptionally high torques for frequent starts and stops that are common in servo positioning. The PM motor is very compatible with the SCR drive used in most servo systems. Not all motors are compatible with the SCR drive because of the tendency toward heat buildup. The PM motor does not need major modification to be used as a servomotor. Some detail is given to changes in the design since the motor normally can provide up to 10 times its duty rating under normal servo applications. The PM motor can also operate at slightly cooler temperatures since it does not have a wound rotor that produces additional heat. The fact that PM motors provide between 10 and 15 percent more efficiency than other types of motors make them the most usable motor for the servo system.

You should also understand that the servo principal can be applied to other electrical loads, such as hydraulic valves or pneumatic values. Since the valves are proportional in nature, they can be controlled as a closed-loop or open-loop servo system. In this case the load will be moved by a hydraulic cylinder or a hydraulic motor instead of an electric motor.

RESOLVERS AND ENCODERS

The operation of a resolver and the encoder is essential when you need to troubleshoot the servo system or other types of positioning systems. Once you understand the fundamental principles involved, you will be able to measure the input voltage to these devices and determine whether or not they are operating normally. The resolver uses the difference of angular displacement of two coils and a rotor to produce a signal voltage, while the encoder uses a photoelectric device to pass light through a transparent disk to produce a series of pulses. The pulses by themselves are unusable until they are sent to a decoding device that will turn the pulse signal into a digital value. You will see that these devices are able to convert the actual shaft position into an analog or digital value that is usable to the servo controller.

Resolvers

A diagram of a resolver is shown in Figure 15-17. A sketch of a resolver is also shown in this figure, where you will notice that it looks like a small motor connected to the shaft of the drive motor. In the diagram you will notice that the resolver has two stator windings and one rotor winding. The stator windings remain stationary while the shaft rotates the rotor winding. The speed of the rotor winding will be the same as the motor speed if they are coupled directly, and the speed will be some ratio of the motor speed if the resolver is driven from a set of gears.

The easiest way to understand the operation of the resolver is to restrict the first example of its operation to one full revolution. This type of application is useful where the motor is to turn a load such as the base axis of a robot. A photograph of the robot and its axis is also provided in this figure. You can see that the maximum amount of travel

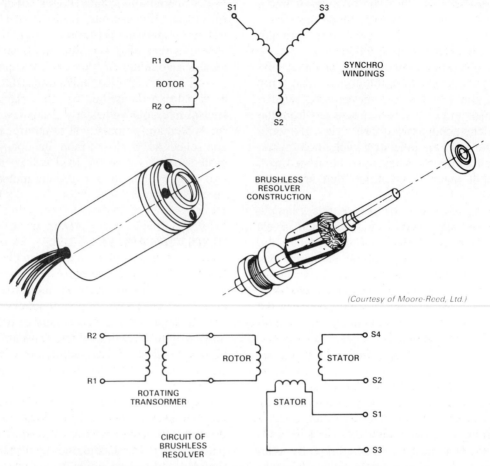

FIGURE 15-17 Sketch of a resolver and diagram of its internal windings. (Courtesy of Analog Devices)

that the base can rotate is 360 degrees. This means that the resolver that is tracking the robot's base travel will also be restricted to one revolution. In actual application, the resolver can be driven from the motor or from the load, whichever the design application calls for.

Since the resolver can rotate only 360 degrees, the amount of angular displacement of the signal produced by each stator in reference to the rotor can be calculated in degrees. The two stator windings are mounted at 90 degrees to each other. The rotor is connected directly to a source of ac current to provide an excitation field. Since this is used as a reference, it can be identified as $A \sin \omega t$. The voltage that is measured from the two stationary windings can be compared by the relationship of S1 to S3 as $V \sin \omega t \sin \theta$, and the relationship between S4 and S2 can be expressed as $V \sin \omega t \cos \theta$, where theta ($\theta$) represents the shaft angle of the resolver. This is a simple relationship for use in the explanation. Other resolvers may use additional rotor windings to provide more accuracy, but the relationship is still confined to the rotor's displacement to the stator field within the 360 degrees.

If you plotted the output waveform of the stator voltages you would see that they produce a sinusoidal voltage. Since the two stators are displaced by 90 degrees, their voltages will not be of the same magnitude at any given time. By comparing the magnitude of each of their voltages and determining if they are positive or negative, the actual location of the rotor can be determined by using them.

This output voltage is used as the feedback signal to the servo system to determine the following error. If you remember the first servo example in this chapter, you will remember that the command voltage was a positive amount of voltage. Since the shaft of the load did not move, the voltage from the resolver was zero, which made the following error positive. Once the load began to move, the resolver began to produce a voltage that increased in amplitude as the rotor moved within the resolver. When the resolver's voltage became equal to the command voltage, the following error became zero and the motor stopped moving the load.

Resolver Applications for Multiturn Systems. The previous example showed the resolver's rotation confined to one revolution. It is more common to find the resolver used in applications where the motor shaft will turn many times to move the load to its position. When the resolver is used in multiturn applications or when more detailed accuracy is required, it may become necessary to use a coarse and a fine resolver to determine more precise locations.

The two resolvers are driven by the same shaft, but a gear ratio of 36:1 is used to produce a difference in accuracy. For example, if the resolvers were measuring an angle of 148.3 degrees, the coarse resolver would show the angle between 140 and 150 degrees, and the fine resolver would show the angle at 8.3 degrees. The output signals of the two resolvers are usually sent to a digitizing circuit that determines the composite voltage and converts it to a digital value (word) that is usable by the digital servo controller. Since the gear ratio of the fine resolver is so high, a switching network is usually provided to bring the fine resolver value in as an update, only when the following error is within a specified small value, which will occur when the load has nearly reached its set-point position.

Troubleshooting the Resolver. Even though the resolver's operation is quite complex, it is rather simple to troubleshoot. Since it operates similarly to a normal generator, all you need to do is measure the reference voltage to determine that it is within the specified limits. If the reference voltage is missing, the resolver will not operate, and if the voltage is not within tolerance, the resolver will not be accurate.

After you have determined that the reference voltage is present, the next test is to determine that the shaft of the resolver is actually rotating. Since the resolver shaft must be coupled to the motor or load, it is possible that the shaft is slipping, or that the coupling has broken and the shaft is not turning at all.

After you have determined that the shaft is rotating, you can place an oscilloscope across the resolver's output and observe the waveform. A digital voltmeter could also be used to determine that the voltage is present at the output when the resolver is rotating. If no voltage is present, the resolver should be removed and replaced. If a voltage is present but the servo system is not accurate, you should plot the output to determine if it is functioning in synchronization with the motion of the load. You may also try to isolate other causes of the inaccuracy and change the easiest parts of the servo system until the fault is corrected. Some manufacturers of motion control equipment will provide output waveform diagrams that can be used as a comparison for troubleshooting. If this documentation is not available, the parts substitution method should be used.

Encoders

Two basic types of encoders (absolute and incremental) are used in positioning control. The incremental type encoder is used to indicate the distance that the shaft has moved from the previous position, and the absolute encoder indicates the exact location the shaft has moved to within its range.

Figure 15-18 shows an incremental encoder and a diagram of the location of the encoder with its optical device. As you can see, a light source is mounted on one side of the encoder disk and it is directed at the encoder surface. A light receiver is mounted on the opposite side of the encoder surface, where it will receive any light that passes through it. The incremental encoder disk is manufactured so that it has alternating transparent and opaque segments. When a transparent segment passes under the light source, light passes through that segment and energizes the light receiver. When an opaque segment passes under the light source, the receiver does not receive any light and it is deenergized. This causes the light receiver to send a pulsing signal similar to the one shown in the figure.

Since the signal transitions from off to on, it is easy to detect and count. A high-speed counter circuit is used to count the number of pulses in the light receiver's signal. The number of transparent segments in the encoder disk will determine the number of pulses per revolution and the resolution of the encoder. The typical optical encoder contains between 100 and 6000 transparent segments.

If the encoder has 100 transparent segments, it will produce 100 pulses per revolution. This means that each pulse will indicate that the encoder shaft has moved 3.6 degrees. This allows the linear position to be measured by calculating the distance the load will move during one revolution of the motor's shaft. For example, if the load will move 3.6 inches during one revolution, each pulse of the encoder means that the load has moved 0.01 inch. The total distance that the load has moved can then easily be calculated by multiplying the number of pulses by 0.01 inch. If the total number of pulses for a given move is 100, the load has moved 1 inch (100×0.01).

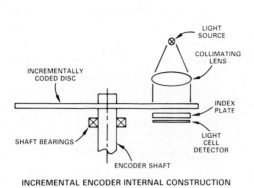

Incremental encoder

Pulse train for incremental encoder

FIGURE 15-18 Sketch and diagram of a incremental encoder with an example of the pulse train that is produced by a light source. (Courtesy of Analog Devices)

The main principle of the incremental encoder is that it can only determine the distance the load has moved from the previous position. This means that if the load is required to move to three separate locations 2 inches apart, the encoder will only know that the second location is 2 inches from the first location and that third location is 2 inches from the second. The incremental encoder will not be able to determine that the third position is 4 inches from the first position.

Even though the incremental encoder does not know its absolute coordinates, it is still useful as a feedback signal when a simple step program is required to allow the load to move to several difference locations that can be referenced from each other.

Absolute Encoder. The absolute encoder is designed to determine the number of pulses that the receiver detects when light passes through the encoder disk, and the direction in which the encoder is revolving. It is also designed to keep track of each full revolution. A sketch and diagram of two types of encoders are shown in Figure 15-19. The absolute encoders look different from the incremental encoder in that the absolute encoders have several sets of opaque and transparent segments. From the figure you can see that the simplest incremental encoder has three sets of segments. Two of the sets have an identical number of transparent and opaque segments, but they are offset from each other by 15 degrees. The outset set of segments provides a pulse train called the channel A signal, and the inner segments provide a pulse train called channel B. If channel A's signal leads channel B's signal, the encoder is rotating in the clockwise direction. If channel B's signal is leading channel A's signal, the encoder is rotating in the counterclockwise direction. An op-amp is used to detect the phase angle difference between channel A and channel B and determine the direction of shaft rotation.

A third segment is provided on the innermost ring of the encoder. Only one transparent segment is provided in this ring to produce the signal called the *command signal*. Since there is only one segment, it will produce a pulse once during each revolution. This allows the absolute encoder to determine the number of full revolutions and decimal parts.

Since the range of the pulse counts is known, after referencing, any point along the positional axis can be identified as an exact location. The example in this figure also shows the positions indicated on the axis. The positions can be indicated as a bipolar value that uses zero is the midpoint. All locations to the right of zero are positive locations and all points to the left of zero are negative. Another method of indicating the positions requires the initial position on the far left end of the axis to be identified as zero, and all positions to the right of this location will be positive. By counting the number of command

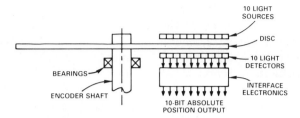

ABSOLUTE OPTICAL ENCODER INTERNAL CONSTRUCTION

Absolute encoder construction and disc

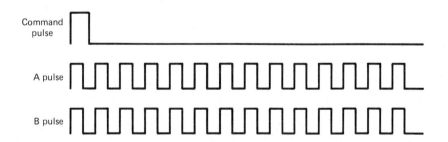

Pulse from absolute encoder

FIGURE 15-19 Two types of absolute type encoders. One type of encoder uses an A, B, and commad pulse to indicate position, while the other encoder uses a gray code or BCD code to indicate accurate position. (Courtesy of Analog Devices)

pulses and the number of pulses of the channel A signal, the exact location can be determined. The difference between the channel A and the channel B signals will determine if the shaft has turned in the clockwise or the counterclockwise direction.

The other type of encoder that is shown in this figure uses 10 concentric rings of segments on the encoder disk. The 10 segments provides data for a 10-bit data word. From the sketch you can see that the number of segments in each ring is twice the number of the next inner ring. This allows the encoder to provide 10-bit Gray code, BCD, or binary data. Other encoders are available with up to 16-bit resolution.

The output data word from this type of encoder will be used to count up or down in code. Since the segments will increment when the encoder rotates clockwise, and decrement the count when it rotates counterclockwise, an absolute position can be calculated from these data. It can also be interfaced directly with other digital controllers, which can be used to control open-loop or closed-loop servo systems.

Troubleshooting the Encoder. An encoder can be troubleshooted easily by determining if the optical source has the correct amount of power. If the optical source is powered, the receiver should produce a signal as the encoder shaft moves. The signal will be a series of high-speed pulses that cannot be detected by a voltmeter; instead, an oscilloscope or high-speed digital counter should be used to detect the presence of the pulses. The major malfunction that occurs with the encoder is a defect with the optical source, or the optical receiver. Addition defects include the loss of the power supply or signal due to a broken conductor. It is also possible that coupling between the encoder and the shaft of the motor is defective and the encoder shaft is not able to turn.

In some cases the main object of the troubleshooting exercise is to determine if the encoder is operational. If it is not functioning properly, it should be replaced so that the servo or stepper system can be made operational again. The inoperative encoder can be sent for repair or a replacement part may be ordered as a substitute.

STEPPER MOTORS

The stepper motor is a specialized motor that can be used in open-loop or closed-loop positioning systems. The stepper is designed to move a specific number of degrees when a pulse of dc current is supplied to its windings. The amount of movement is determined by the number of angle steps that are built into the motor.

The theory of operation of a simplified stepper motor is very easy to understand. An example motor is shown in Figure 15-20a. From this diagram you can see that the motor has four separate windings called phase A, B, C, and D. These phases are placed in the motor at 90-degree angles to each other during manufacturing. The rotor for this motor is a permanent magnet with one north and one south pole. The magnet in the rotor will be attracted and repelled by the motor poles as they change polarity.

When voltage is first applied, phase B is energized, so it is negative, and phase D is energized, so it is positive. This will cause the magnet to be attracted to these poles, which will make the rotor move 90 degrees in the clockwise direction.

When the rotor has moved 90 degrees, the polarity of each of the phases must change in sequence to make the rotor continue to rotate. This is accomplished by turning the voltage off to phase B and phase D after the rotor passes them. Phase A is then energized positive and phase C is energized negative to turn the rotor another 90 degrees. The polarity of the phases are controlled by sending a pulse train (Figure 15-20b). From the diagram you can see that the polarity of each pulse is alternated positive and negative. From this diagram you can also see that five positive and five negative pulses have been sent to the motor windings. Since there

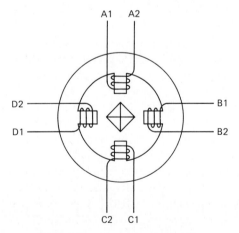

(a) Windings of a simple stepper motor

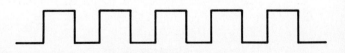

(b) Pulse train for stepper motor

FIGURE 15-20 Basic diagram to show the theory of operation of a stepper motor.

are four windings, they are connected to make two groups. Each of the groups will be energized with a similar pulse train.

Since each group of windings received five negative and five positive pulses, the rotor would rotate 2.5 revolutions (90 degrees × 10 pulses/360). This allows the position of the motor shaft to be predicted accurately for each group of pulses that are sent to the motor. If the motor has more windings, it will turn a smaller number of degrees when a pulse is received. For this reason each motor is identified by the number of degrees its shaft will turn when a pulse is received.

If the rotation of the motor needs to be reversed, the sequence of the pulses are reversed so that the polarity of the voltage is opposite to the sequence explained above. Since the D phase is powered with negative and the B phase is powered with positive, the rotor will be forced to rotate in the opposite direction. A stepper motor drive circuit is used to produce the stepper motor voltages. This drive circuit is designed to produce the correct sequence of voltages to make the motor rotate the desired direction. The speed of the pulses will determine the speed of rotation of the motor. The torque of the stepper motor can be controlled by varying the amount of current that is provided within each pulse.

Typical Stepper Motor Step Angles

Stepper motors are designed for 90, 45, 15, 7.5, 1.8, and 0.9 degrees of revolution per pulse. The resolution of these motors can be determined by dividing the stepper angle into 360 degrees. For example, the 0.9-degree motor will be able to provide 400 steps per revolution, and the 15-degree motor is able to provide 24 steps per revolution. The larger the number of steps the motor provides per revolution, the better the resolution it can provide. It should also be pointed out that the higher the resolution, the slower the speed of the shaft.

The number of steps per revolution can also be used to determine the resolution of the stepper positioning system. If a 15-degree stepper motor is used to drive a ball screw mechanism that has a pitch of 4 threads per inch, the load can be positioned with 0.01-inch accuracy. The amount of accuracy and the speed of the motor shaft will depend on the application of the load.

Types of Stepper Motors

There are three basic types of stepper motor commonly used in industrial positioning applications today: the variable-reluctance (VR) motor, the permanent-magnet (PM) motor, and the hybrid motor, which combines the characteristics of the VR and the PM motor.

Variable-Reluctance Motor. The VR motor utilizes a multipole rotor and a wound stator. This means that the rotor has the same number of teeth or spokes as the number of stator windings. From the diagram you can see that the motor has three sets of windings, called phase A, phase B, and phase C. Phase A has four separate stator teeth that are mounted at 90 degrees to each other. When phase A is energized, voltage is put into the winding at point A and returns from the series winding from point A1. This means that all the pole teeth will be the same, polarized magnetically at the same time.

When phase A is energized, the spokes will move (step) one pole in the clockwise direction. At the instant in time when phase A is energized, the other two phases are deenergized. After phase A has been energized, and the rotor has stepped once, phase B will be energized and phases A and C will be deenergized. Since phase B poles are mounted directly to the right of phase A poles (going clockwise), the rotor will continue to step in the clockwise direction.

After phase B has been energized, phase C will be energized and phases A and B will be energized. This will cause the rotors to continue to spin in the clockwise direction. By controlling the pulse train in the order A, B, C, A, B, C, the motor will continue to step in the clockwise direction and the rotor will step the same number of times as the number of total pulses that the phases receive.

This type of servo controller must use an electronic circuit to provide the series of pulses in the proper order. If the motor is required to rotate in the opposite direction, the pulse train will be C, B, A.

Permanent-Magnet Stepper Motors. The PM motor is different from the VR motor. The PM motor's rotor does not have teeth or spokes. Instead, the rotor is made of perment magnet and its magnetic poles will be attracted or repelled by the polarity of the poles of the phases. This motor has four phases, called phase A, phase, B, phase C, and phase D. In this case each phase has only one pole, so the motor will step 90 degrees when the phase sequence is implemented. If each phase in the motor has two poles, the motor would step 45 degrees for each pulse that was received. This type of motor generally provides 90- or 45-degree steps, which provide a higher speed of shaft rotation, with slightly less resolution. Since there are only four poles in the motor, they can also be slightly larger and can carry more current, which will provide more torque than the variable reluctance motor.

Hybrid Stepper Motor. The hybrid stepper motor uses some of the VR characteristics and some of the PM characteristics. The design of the motor is very similar to the VR motor in that the rotor has multiple spokes, but it is manufactured as a multitooth permanent-magnet rotor. Each of the stator spokes have multiple teeth. These teeth will allow the motor to provide exceptionally high step rates with only eight major windings. As you can see, each spoke of the stator has a coil from two different phases. Each coil (phase A, phase B, etc.) uses a continuous series circuit with the start of the winding at the point marked with only the letter, and the end of the coil marked with a letter and a number. This type of winding is called a bifilar winding, which allows a single power source to provide voltage for all the phases in their proper sequence.

The desirable characteristics of the hybrid motor is the strong dynamic torque, excellent holding torque, and a high detent torque. This allows the motor to move larger loads at high stepping speeds, which means that the shaft speed is relatively fast also. The holding torque is especially important for motion control circuits, where the load may be stepped to several locations where it must perform some additional operation such as machine tooling when it reaches each position. The detent torque and dynamic torque are essential since most loads are started and stopped frequently.

Open-Loop and Closed-Loop Stepper Motor Drives

Since the rotation of the stepper motor's shaft is predictable for a given number of pulses, it is quite reliable for open-loop systems. Figure 15-21 shows a diagram of this system. This type of system will generally use a home position that is determined with limit switches to orient the load to the work envelope. After the home position is determined, the controller can send the motor a specific number of pulses to cause it to move to a predetermined location. In the next move, additional pulses can be sent to the motor to send it farther in the same direction, or the sequence of the pulse train can be reversed to send the motor in the opposite direction. The stepper motor driver receives a number from a programmable controller. The number represents the number of pulses and the direction the motor should move (+ or −). This value could also be provided by any other type of computer or microprocessor-controlled system. The stepper motor driver has a communications board that changes the numerical value that is received from the programmable controller into a set of pulses. These pulses are sent to the amplifier part of the stepper drive, where the signal-level pulses are converted into power-level pulses at higher dc voltages. A TTL-level signal is sent to the directional control circuit. If this signal is a 1, the amplifier will send the pulses in a sequence that will cause the motor to rotate in the clockwise direction, and if the digital signal is a 0, the amplifier will send the pulse in a sequence that will make the motor rotate in the counterclockwise direction.

Since the motor is driven by an open-loop system, the complete positioning control is provided by the numerical values sent by the programmable controller. This system assumes that if a pulse is received, the motor will move that step increment, which will move the load the designated distance. The only time that this system will not be accurate is if the stepper motor develops a fault or if the load is obstructed in any way.

The stepper motor drive system has several safety circuits provided in it to make the system quite reliable as an open-loop drive. An overcurrent circuit is provided to protect the motor against overcurrent conditions that may develop if the load does not allow the motor shaft to step the correct

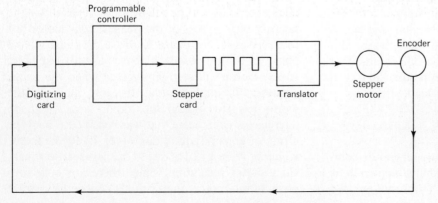

FIGURE 15-21 Diagram of a stepper motor drive system that receives a new position destination from a programmable controller. The stepper drive circuit translates the new positional data into a number of steps and a sequence that causes the motor to rotate in either the clockwise or the counterclockwise direction.

number of times. When the motor does not move the number of prescribed steps, it is said to "cog out." The term *cogging* refers to the condition when a motor does not step smoothly, or is not allowed to move, due to conditions at the load that may restrain the shaft rotation.

When the overcurrent condition is sensed, the move set is interrupted by the programmable controller, and the system must be reset. If the motion of the load is critical, such as applications where a robot arm must move very close to tooling fixtures or mold and die cavities, it may be necessary to "home" the motor at the beginning of each cycle. This can be accomplished as the last step of a move set or as the first step of the set. Since the motor is homed to a limit switch once during each cycle, the worst-case error would remain in the system for only one move set. At the end of the set, the error would be removed when the motor's home position was reset again.

Ramping and Speed Control of a Stepper Motor

The speed of a stepper motor can easily be controlled by controlling the rate at which the pulses are sent to the motor. The rate control can be set by the stepper motor drive circuit and is determined by values that the programmable controller or other computer sends with the number of pulses. The ramping rate for acceleration and deceleration can also be set by a digital value or by presetting the resistor and capacitor time constant on the circuit board. This is accomplished by adjusting the acceleration and deceleration ramp-rate potentiometers.

If the stepper is used in a closed-loop system, an encoder will be added to the circuit to provide feedback data on the actual position of the load. If the velocity is to be controlled as a closed loop, a tach generator will also be used to provide data concerning the speed of the shaft. These circuits will operate similar to the closed-loop servo systems in that they can provide continual error correction. Since the stepper motor's shaft rotation is dependent on the pulse train that it receives, the closed-loop type of control is used in less than 25 percent of stepper motor applications.

Troubleshooting the Stepper Motor

The stepper motor system is fairly easy to troubleshoot since you understand the function of each part of the system. The motor can be tested by energizing the drive system, including the drive amplifier. When the motor is at rest, you should not be able to turn the drive shaft because the amplifier should be sending a holding voltage to poles. The holding voltage causes the rotor spokes to align with the pole of the last step. Since the polarity of the rotor spoke and the field pole are causing a magnetic attraction, the next-nearest pole in either direction will be repelling the movement of the rotor spoke. This means that the shaft should not be able to turn in either direction.

If the shaft can be turned, the motor should be disconnected from the drive amplifier and you can test the fields for continuity. If the field winding is continuous, you should replace the drive amplifier, and if any of the field windings have an open circuit, the motor should be replaced.

If you are not able to move the motor shaft, you should begin to suspect the pulse generation circuit of the driver control or that the stepper driver is not receiving a proper signal. You can test for a low-voltage pulse signal with a simple LED (light-emitting diode) connected in series with a resistor (sized for the signal voltage level). Touch one side of the LED circuit to the signal and the other side to the circuit common, being sure to observe the polarity of the circuits. If a signal is being transferred correctly, the LED will glow continuously or flicker as the pulses are received. If the LED does not light, be sure to check the documentation for the system to ensure that all "dip switches" and jumpers on the electronic boards are in their proper locations.

At this point the best strategy is to begin substituting the driver circuit and amplifier for known good ones until the problem is found. When the suspect part is located, it should be removed from the system and sent for repair. It is also good practice to keep a spare amplifier and driver module on hand if the stepper system is used for production applications. This allows you to have parts for substitution tests, and it also allows the system to be returned to operation as soon as possible.

BALL SCREW MECHANISMS

The *ball screw* mechanism, also called a *lead screw*, is a most useful device to convert the rotary motion of a motor shaft to linear motion. The resolution of the motion that a ball screw provides can be as close as other types of positioning equipment. Several previous positioning applications in this chapter have used ball screws to provide linear motion.

Figure 15-22 shows a typical ball screw. From the diagram you can see that the shaft of the ball screw is threaded with a predetermined number of threads per inch. The number of threads per inch is known as the *pitch* of the ball screw. A bearing

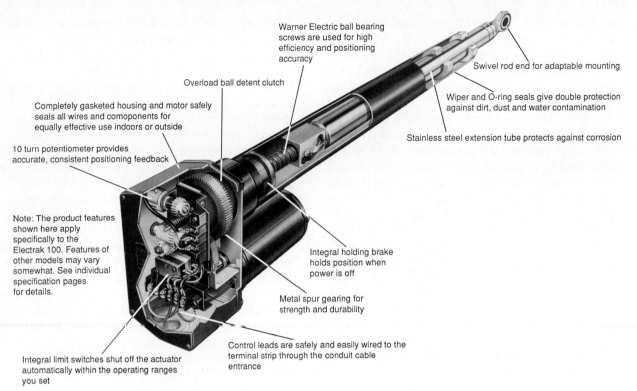

FIGURE 15-22 A ball screw mechanism used in a linear actuator. (Courtesy of Warner Electric Company)

assembly that contains multiple ball bearings is mounted so that the balls will fall into the grooves provided by the threads. The shaft is mounted in a bearing assembly on each end. These bearings will support the weight of the shaft and load and provide a bearing surface for the torque forces that are applied to the shaft.

The shaft can be driven directly by the motor through a coupling, or it can be driven by belts or gears in an indirect fashion. When gears or belts are used, the shaft can rotate at a different speed than the motor and it can be geared up for speed or geared down for torque. When the shaft rotates, the ball bearings in the ball screw move through the grooves of the threads in the shaft. The balls are allowed to move from one end of the carrier to the other. When a ball reaches the end of a carrier, it will move out through the cross over the tube, which delivers them to the entrance of the carrier again. This keeps the ball bearings in constant movement when the shaft is rotating. This constant movement provides less friction so that the load can move easily along the shaft.

The ball screw mechanism can be mounted on any length of shaft. This is especially useful for applications where linear motion of up to 6 feet is needed. The torque that is available at the load can be calculated from the pitch of the shaft, the length of the shaft, and the horsepower of the motor. Since the shaft can be sized to any length, the ball screw can be designed to custom fit the application.

The ball screw mechanism can also be mounted directly in a linear actuator. The motor is mounted directly below the actuator and the shaft is gear driven. When the shaft rotates, the ball bearing housing moves along the length of the shaft and extends the rod. The rod can be connected to any load, such as opening and closing a machining vise, control valve, or damper on an air movement system for HVAC. Since the motor is mounted as an integral part of the actuator, the ball screw mechanism is easy to mount for many applications. The ball screw also provides a servo positioning capability that allows the rod to be located anywhere within its range. The pitch of the ball screw and the number of teeth in the gears of the motor drive will determine the amount of resolution of the actuator's positioning capability.

QUESTIONS

15-1. Explain why a permanent-magnet motor is best suited for drive controls.

15-2. List several applications where motor drive controls may be used.

15-3. Explain the two functions the SCRs provide in the motor drive.

15-4. Why is the function of rectification so important when a dc drive is used in a factory?

15-5. Draw a block diagram of the feedback loop used in drives.

15-6. What purpose does the feedback loop serve?

15-7. What will happen to the drive if the wire providing the feedback signal is broken? Why?

15-8. Explain how a signal from a programmable controller can be used to control the motor drive system.

15-9. Draw a sketch of an eddy current drive and explain its operation.

15-10. List the three types of ac drives.

15-11. Provide a block diagram of the PMW drive and show a typical output waveform. Explain how the motor speed is controlled by this type of drive.

15-12. Provide a block diagram of the CSI drive and show a typical output waveform. Explain how the motor speed is controlled by this type of drive.

15-13. Explain the difference between a constant-horsepower application and a constant-torque application.

15-14. Show a block diagram of a typical servo system and explain its operation.

15-15. Explain why a servo system can use a dc or an ac motor.

15-16. Explain how a hydraulic cylinder can be connected to a servo control for position and speed control. (An example of this type of system is used in hydraulic robots and plastic presses.)

15-17. Draw a sketch of a ball screw mechanism and explain its operation.

15-18. What is meant by the pitch of a ball screw?

15-19. Explain the operation of a digital-to-analog servo system.

15-20. Draw a sketch of an absolute and an incremental encoder and explain their operation.

15-21. Draw a sketch of a resolver and explain how its output waveform is used to determine position.

15-22. Draw a block diagram of the components used in a stepper motor, and explain their operation.

15-23. List the typical stepper motor step angles.

15-24. Explain the difference between an open-loop and a closed-loop stepper control.

15-25. Explain how you would troubleshoot a stepper motor system.

Programmable controllers 16

INTRODUCTION

Programmable controllers have made the largest impact on motor controls since they were first introduced into industries in 1969. The programmable controller, which was called a PC long before the personal computer and professional computer, is now referred to as a P/C. The first units were nothing more than a replacement for relays, which could be reprogrammed. One specification of these early units which ensured their success was that their program was made to look exactly like an elementary electrical diagram, which we presently call a ladder diagram (ladder logic).

Since the programmable controller was a computer that used ladder diagrams, they were readily accepted on the factory floor, where they have become a troubleshooting aid as well as a machine control. The P/Cs of today are available in small, medium, and large sizes. The smallest units are dedicated since they usually have approximately 20 inputs and 12 outputs. They also have a limited set of instructions that they can execute. The medium-sized programmable controller allows the user to select the amount of memory the system can have and the number of inputs and outputs it can control. Typical medium-sized units can have memory up to 10 kilobytes and they can control up to 1000 inputs and outputs.

Large programmable controllers today have many of the functions of a mainframe computer, and their instruction set may include more powerful instructions in Fortran or C programming languages; they are also able to control thousands of inputs and outputs. The larger systems are capable of controlling entire factory automation, or complex cells. They are also used for complex process control systems that control refineries, food processing, and pharmaceutical processes.

The P/Cs that you will find in industry today are all very similar even though they are made by over 50 different manufacturers. It is easy for the first-time user to become overwhelmed with all the differences in programming formats and options that are available. The best advice you can receive at this time is that the P/C is generally doing nothing more than emulating motor control circuits that have been in existence for many years. When these circuits are programmed into the P/C, they may not look exactly as they do in the ladder diagram that is provided when the circuit is hard wired. These differences are generally slight and are due to the type of microprocessor chips the P/C is utilizing to execute its program.

In this chapter we introduce and explain the basic sections of a programmable controller. We then explain the basic programming functions that you will find in all brand names of P/Cs. In the remainder of the chapter we show you how to read the program that is shown on the screen and how to use the P/C to troubleshoot problems that occur on the factory floor.

Several basic programs will be provided through out this chapter, with all the keystrokes for

Allen-Bradley, Gould Modicon, and Texas Instruments controllers. These programs will allow you to enter the programs into your P/C and see exactly how they function. It is important to remember not to become discouraged by small problems that occur when programs are entered into the P/C. The P/C is a computer, so you must put the program into its memory in exactly the way its format expects to see it. If you make a minor mistake, it will display an error message to tell you why it cannot run the program. You may need to double-check these small format errors and write them in a notebook or logbook since you will be making program changes for the rest of your life if you work on motor controls. After several hours of practice, you will encounter fewer of these types of problems. I have found that you cannot begin to understand the P/C fully until you are forced to explain it to others. You should also remember that you may be hired into companies where you will be expected to teach apprentices and other technicians how the P/C is used to control and troubleshoot the system, so you should get all the practice you can at this time. One last very important point: when you are learning to program and use a P/C for the first time, you will probably use the "monkey see–monkey do" method. This is a good method for the first several hours, but you should begin to understand the concepts involved. For example, if you understand the concept of a programmed timer, you will be able to understand the small differences you will encounter when you try to program the timer into different models and different brand names. If, instead, you try to memorize keystrokes for each different brand name, you will quickly become confused and the P/C will seem very difficult to understand, when in fact it is rather easy. It will also help to understand that P/Cs use microprocessors that are similar to the ones used in the personal computer or professional computer that you use in other courses. If you have taken a course in microprocessors, you will recognize some of the functions of assembly language. Once you understand the function of all the basic parts of the system, you will see that all systems are very similar even though they are made by different manufacturers.

BASIC COMPONENTS OF THE P/C

The basic parts of the programmable controller are similar regardless of the brand name. Some manufacturers combine several basic parts into one section or module, but the function of each of these systems is still similar. Figure 16-1 shows two typical programmable controllers. A block diagram is also included to give you an idea of how these parts operate together as a system. These basic sections include the processor, which uses a microprocessor chip that executes all instructions in the program; the memory, which stores the user program; the input module, which conditions the signals to lower voltages so that they can be sent to the microprocessor; and the output module, which raises the microprocessor voltage to a level that can power solenoids and coils. The cathode ray tube (CRT), which is similar to a television screen, is used to display the program. Some smaller P/Cs provide LED displays instead of CRT display to provide a lower-cost system.

It may be easier to understand the operation of the programmable controller if you follow a signal as it is received at the input module and then a signal is sent out through the output module to energize a motor starter coil. The start/stop circuit is easy to use for this example. Figure 16-2a shows the hardware portion of the system. This includes the start and stop switches connected to the input module, and the motor starter coil connected to an output module.

Figure 16-2b shows the ladder diagram program that is stored in the P/C's memory. This diagram looks like a ladder diagram that is used in hard-wired systems, except that all switches are shown as a set of normally open or normally closed contacts. It should be pointed out at this time that regular JIC symbols are not displayed on CRT because they require too much memory.

You should notice that the JIC symbol for the stop button and the start button are used in the wiring diagram, which shows these switches connected to the input module. The terminals where these components are connected are numbered so that the computer will recognize where the signal is coming from. This number is called an *address*. The addresses are shown at the input and output modules and they match addresses in the programmable controller's image register (Figure 16-2c). The image register is part of the P/C's memory, where the status of each input and each output is stored and updated as switches open and close. The concept of the image register is important since all P/Cs must update or keep track of the status of input switches that are connected to their input modules. When these switches are closed, they send a voltage to the input module which converts the signal to computer-level voltages. The computer continually reads each input address regardless of whether a switch is connected to it or not. When a switch is passing power to an address, a "1" is registered into that address location in the image register.

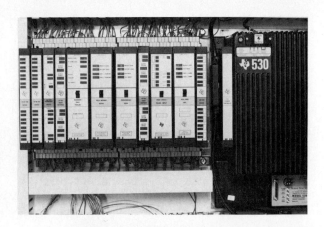

(a) Typical programmable controllers with input and output modules in their racks (bases)

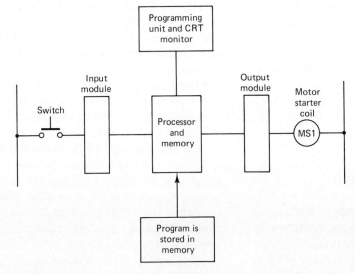

(b) Block diagram of typical programmable controls

FIGURE 16-1 Texas Instruments 560 and 530 programmable controllers with block diagram of a typical programmable controller. (Courtesy of Texas Instruments)

438   Chap. 16 / Programmable Controllers

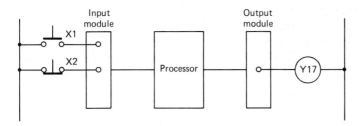

(a) Diagram of hard-wired switches and relay coil connected to programmable controller

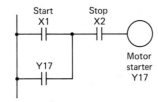

(b) Program in programmable controller

	X image register		Y image register
X1	1	Y17	1
X2	1	Y18	0
X3	0	Y19	0
X4	0	Y20	0
X5	0	Y21	0
X6	0	Y22	0
X7	0	Y23	0
X8	0	Y24	0

(c) Image register in a programmable controller

FIGURE 16-2 Typical start-stop circuit used in a programmable controller.

When the switch is not passing power to an address, a "0" is placed in the image register address. This means that each address in the image register, a "1" or a "0," will be displayed in every address.

The P/C also has an image register for each output address, which it also updates continually. If the computer places a "1" in an address for an output, that output address at the module will be energized. This will cause any coil or solenoid connected to the output address to be energized, too.

When P/C's processor checks its input signals to update its input image register, it is called a *read cycle*. When the P/C writes values to addresses in its output image register, it is called a *write cycle*. In this fashion the P/C reads its input image register and writes to its output image register.

The processor will write a "1" or a "0" to an address based on the results of the ladder logic. For instance, in our example of the start/stop circuit, when the start and stop buttons are both closed, a "1" will be written into each of their addresses in the input image register. When the processor checks the logic in its program, it will see that both switches need to be closed before the output should be energized. Since image register addresses of both switches contain a "1," the processor will energize the output address and write a "1" in its output image register address. Every address in the output image register that contains a "1" will send a signal to that respective address in the output module, which will provide a current at that output terminal. The motor starter coil that is connected to this module address is energized when this signal is present. If the stop button is depressed, it will interrupt the voltage to the input module. This will cause the module to stop sending a signal to the processor and a "0" is placed in the input image register's address. When the processor reads this "0," its logic will determine that the output in the program should be deenergized because the stop switch is open. The next time the processor updates the output image register, it will write a "0" into the output address and the output module will deenergize the signal at that address terminal and the motor starter coil will deenergize.

PROGRAM SCAN

Each time the processor reads its inputs, solves its logic, and writes to its outputs, it is called a *scan*. The time it takes a P/C to scan its program depends on how many I/Os the system has and how large the program in its memory is. Typical scan time for a P/C is between 10 and 30 microseconds (μs). Later in this chapter some problems with extended scan times will be explained. Since you are just learning about P/Cs, you can assume that the inputs and outputs are scanned constantly so that when a switch is closed, it will immediately be read into the P/C, and after the logic is solved, the appropriate output will be energized immediately. In fact, if you place a hard-wired start/stop circuit next to a P/C-controlled start/stop circuit, you would not notice any difference in their operation even though the P/C takes several microseconds longer to energize its motor starter coil.

ADDRESSES

When you begin to write simple programs to enter into the P/C, you must be aware of the address to assign to each input and output. The address that is used for the start and stop switches in the previous examples will depend on which input module the switches are physically connected to. A good way to understand how each manufacturer identifies their addresses is to present three of the most common and show the same switches connected to each of them. Figure 16-3 presents an example of a rack of modules for an Allen-Bradley system, a Texas Instruments system (TI), and a Modicon system. One difference between these systems is the name that is used when input and output modules are mounted. Allen-Bradley uses a *rack*, TI uses a *base*, and Modicon uses a *housing* to mount their modules. For simplicity, these terms will be used interchangeably so that you become familiar with each.

Figure 16-3a represents eight modules mounted in an Allen-Bradley rack controlled by a PLC 2/30 programmable controller. Addressing for their PLC-3 and PLC-5 family of processors will be slightly different, but they should easily be understood once you see the method that the PLC 2/30 uses. Each input and output module for the PLC 2/30 has eight addresses. Eight modules can be mounted into each rack, and the system can have up to seven racks (the first rack, called rack 0, is used for system functions and is not usable for inputs and outputs). The diagram in this figure shows eight modules (four input and four output) mounted in one rack. This provides a total of 32 input addresses (circuits) and 32 output addresses (circuits).

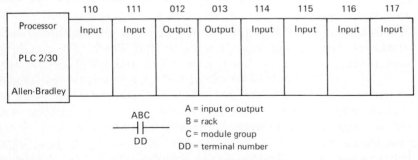

(a) Addressing example for Allen-Bradley

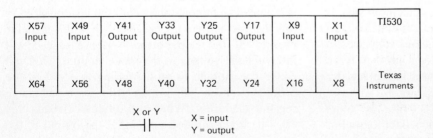

(b) Addressing example for Texas Instruments

FIGURE 16-3 Addressing examples for Allen-Bradley PLC-2/30 and Texas Instruments 530 programmable controllers.

From Figure 16-3a you can see that a pushbutton switch is connected to the first input module address, and a motor starter coil is connected to the first output module address. The numbering for the input address is shown as 110/00, and the number for the output is 013/00. A small diagram that shows the symbol of a contact and output is also provided to help you understand the addressing. The letters ABCD are shown on these symbols and an explanation of their use will help you see that the system is rather simple. From this diagram you can see that A can be a 1 or a 0, B can be 1 to 7, C can be 0 to 7, and D can be 00 to 17 (octal).

You will see that the addressing formats for the contact and the coil are similar. The letter A indicates whether the symbol is an input or an output, where 0 is used for outputs and 1 is used for an input. From the ladder diagram in this figure, you can see that the contacts for the start and the stop button both have a 1 as the first number of their address. This indicates that the contacts in the program will be controlled by a real switch. You should notice that the seal-in contacts connected in parallel with the start contacts have a 0 as their first number. The 0 indicates that these contacts are controlled by an output in the program. In other words, the contacts in the program that are identified by a 1 are controlled by real switches. When the real switch is closed, the programmed contact will be energized, and when the switch is open, the programmed contacts will be deenergized.

The contacts that are identified with the 0 as the first number will be energized only when the output in the program with the same number is energized. The 0 means that the programmed output controls the action of these contacts. This is very similar to identifying contacts from a pushbutton switch differently from contacts controlled by a relay coil.

The letter B represents the location in the address where a number 1 to 7 will be used to identify which rack the modules are mounted in. This number 1 in the address of the inputs and output of this example indicates that the modules they are connected to are mounted in rack 1. The PLC 2/30 system can have up to seven racks, which are numbered 1 to 7.

The letter C indicates the module address within the rack. The first module address is 0, and the last address in every rack is 7. This means that the Allen-Bradley system uses an octal numbering system to identify its addresses. The letter D in this format is used to identify the terminal address of each module. Again an octal numbering system is used, as addresses are numbered from 00 to 07 and 10 to 17. Remember that these numbers are meant to identify addresses, so they are not very useful in counting the total number of addresses, as the decimal numbering system is.

Texas Instruments Addressing System

The addressing system for the TI system is much simpler to understand since all inputs are identified by the letter X and all outputs are identified by the letter Y. This addressing system is similar to others in that each module has eight terminals or circuits. Each base can house up to eight modules (a larger base is available into which 16 modules can be mounted), and up to eight bases can be controlled by the TI 530 system. This means that the TI 530 system can control up to 512 inputs and 512 outputs. As you can see in Figure 16-3b, if the first module is an input module, its terminals will be numbered X1 through X8. Since the input is using the first address, the output cannot use it. This means that since an input is numbered X1, there cannot be an output Y1. An easier way of looking at this is to identify all addresses as a number from 1 to 1024. If an X is used for an address, a Y cannot be used for that same address. Since the modules use groups of eight addresses, the same would hold true for all eight addresses. This makes the TI system very simple to understand.

In the ladder diagram you can see that the start and stop buttons are numbered X1 and X2, while the output is connected to the fifth module, so it is connected to address Y40. Since the hold-in contacts are identified by address Y40, they are controlled by output coil Y40 in the program. Whenever Y40 is energized in the program, the Y40 contacts will be closed. It is important to remember that the output will energize the circuit in the output module where the motor starter coil is connected, and it will also control any sets of contacts in the program that are identified as Y40.

Modicon Addressing System

The Modicon 484 addressing system is very similar to the one used by Texas Instruments. In this system all inputs are identified by the first number in the address being a number 1, and output addresses start with a 0. This means that the start and stop switches are connected to the input module, so their addresses are 1001 and 1002. A four-digit address is used because the Modicon 484 can control up to 1024 I/O, as in the TI 530.

The output in this circuit is identified as 0017 since it is connected to the fifth module. From the Modicon diagram in this figure, you can see that the 484 system uses four terminals per module and this system refers to the place where the modules are mounted as a housing. Newer Modicon system such as the 584, 884 and 984 all use 8 terminals per module.

The important part of the addressing scheme for any P/C is that it is used by the processor to send and receive signals for inputs and outputs, and it is used by the technician to identify the proper terminal to connect an input or an output. Another important concept that you should understand is that a contact in a program can be controlled by a real switch or by some other device in the program, such as an output. When a set of contacts like the seal-in contacts for each of these examples are controlled by the output in the program, they are called *memory contacts,* since they exist only in the P/C's memory.

PROGRAMMED FUNCTIONS

The main feature of programmable controllers that make them so useful is their programmed functions. These functions include timers, counters, sequencers, math, and other motor control functions. All of these functions are programmable, reside in the P/C memory, and are executed by the processor. This means that they will not wear out like electromechanical devices, and the values they control can be adjusted as the system is operating. For instance, a timer in the programmable controller can have its preset value (set point) changed by conditions that are occurring in the system, whereas an electromechanical timer must be adjusted manually if a change is required.

Each brand of programmable controller uses a slightly different method of programming and displaying these functions. You should understand the concept of these functions; then the format that is used will be easier to understand. In some cases a programmable controller company use a different programming format for each different model they manufacture; therefore you should not become overly concerned with these slight format differences.

When you are learning about the various programming functions you should try to equate their operation with the hard-wired control function with which you are familiar. You should also try to understand the operation of each of these functions in regard to what input conditions must be met before the P/C will execute the function. Usually, the conditions are programmed on the left side of the function and they will act as permissives. When all the input conditions have been met, the programmed function will be executed and change data or activate an output. It is also important to understand how each of these functions manipulates data within the system, and then you will be able to understand the operation of the complete control circuit.

Memory Coils (Outputs) and Contacts

The contacts explained so far in this chapter have been controlled by real switches that are connected to the input module, such as the pushbutton start and stop switches. The contacts in the program that have the same address as the start switch will become energized in the program any time the real switch is pressed closed in the real world.

Another type of contact that is commonly used in a P/C program is the *memory contact*. These contacts will be identified by addresses that belong to an output coil in the program. All the memory contacts with the same address as the coil will energize in the program any time the memory coil address is energized. An example of a set of memory contacts is the set of contacts used to seal in the start button in the program. These contacts will have the same address as the output that the start/stop button controls. It is important to remember that this output does energize a signal to an output module which will turn on a real motor starter, but it will also energize the set of contacts in the program at the same time. Since these contacts exist only in the P/C memory, they are not considered to be real and are called memory contacts.

Memory contacts can also be controlled by outputs that do not send a signal to an output module. When an output is not used to send a signal out to an output module, it is called a *memory coil* or a *programmed control relay*. The program control relay acts like a hard-wired control relay, in that it is used to set up logic sequence and control other lines of program logic. Most P/Cs identify their control relays so that you will readily identify them. Texas Instruments uses the simplest method of identifying the control relays. The TI 530 identifies all its control relays with the letter C. If you see an output addressed as C20, you will know that output C20 is not controlling a signal to an output module but will have control over every set of normally open and normally closed contacts marked C20. Allen-Bradley and Modicon use a slightly different method of identifying their control relays (memory coils). Any address of an output that is higher than the last output module address is considered to be a memory output. This means that if the highest output module address in an Allen-Bradley system is 013/17, any output with a number that is higher than that is considered to be a memory output.

442   Chap. 16 / Programmable Controllers

Modicon uses the same format. If the last Modicon output module is mounted in address 0400, every output address above that address is considered to be a memory output address, which can be used as a control relay.

If you need extra memory addresses in the Allen-Bradley and Modicon systems, you should remember that the address or slot in a rack can only be used to house an input module or an output module. This means that if you have an input module in the first slot, the addresses 110/00 through 110/17 will be connected to real input switches, so the system can no longer use the first slot for an output module and the first 16 output addresses (010/00 through 010/17) can be used as memory addresses. This can be confusing when you are looking at the program and do not remember what types of modules are in each slot.

The memory coil and contact are one of the strongest features of the P/C. Once you have a memory coil in your program, you can use it to control as many sets of normally open or normally closed contacts as you would like to place in the program. You could imagine trying to purchase a real control relay that has 10 sets of normally closed contacts and eight sets of normally open contacts and then try to keep track of all of them when you would have to connect the hard wire to them. In the P/C, all wiring to the memory coils is taken care of when you enter the coils and contacts into the program. When you look at the CRT and see all the contacts and coil connected, this symbolizes that the wiring connections have been made in the P/C memory by the processor. Another nice feature about memory contacts is that they do not wear out. If the coil is shown on the CRT as energized, all its contacts will be activated. The only exception to this rule is if the P/C is not in the "run mode." When a P/C is not in the run mode, it will not execute the instructions of its program, so outputs and their contacts will not be energized. When the P/C is not in the run mode, it is similar to having an automobile in neutral, so that the engine is running but the wheels are not turning.

You should remember that the output in a program will be considered as a memory coil if it does not control a signal to an output module address. If it does control a signal to an output module address, it will be considered a real coil. Any contacts in the P/C program that are controlled by a real coil will still be considered as memory contacts since they do not exist anywhere except in the P/C's memory. Do not let this confuse you, since it is important only during troubleshooting. When a system is being troubleshooted, you will only need to check to see if real inputs have failed, as memory contacts will follow the condition of the coil that is controlling them.

Programmable Timers

One of the simplest functions that a P/C can execute is a programmable timer. These timers have a programmable preset value, and the microprocessor uses one of its clock frequencies as the time base. The timer can be programmed to operate like the on-delay timer, the off-delay timer, the reset timer, or an accumulating timer. You should be familiar with the operation of the electromechanical type timer that provides the same function, and it will be easy to equate this to the function of programmable timers.

Timer Operation. Texas Instruments and Modicon use a similar format, which presents the timer as a box instruction. An example of the Texas Instruments timer is shown in Figure 16-4a. The timer instruction consists of several parts. The box instruction is identified with the letters TMR to indicate that this is a timer instruction. The "TMR" is a mnemonic for the instruction. This means that the letters that are used for the instruction have been selected because they sound like an abbreviation for the word "timer." The mnemonic is understood by the microprocessor, and it is easily recognized by the technician when the program is reviewed. Remember, the program is used like a ladder diagram to troubleshoot the system when it malfunctions, so it is important that all the instructions in the program are easily recognized.

Each timer in the P/C program will be numbered, such as TMR1, TMR2, and so on. The timer

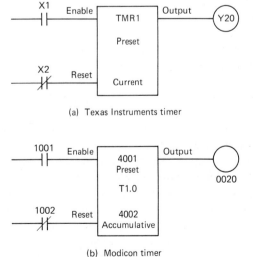

FIGURE 16-4 Texas Instruments and Modicon timers as a box instruction.

instruction box also provides other important information, such as preset value, time base, and current value. The preset value is the amount of time that you want the timer to delay. When you place a value into the preset location, it is like adjusting the setting on an Eagle timer. The time base value allows the timer to keep track of the time delay in 0.1 ($\frac{1}{10}$) second or in milliseconds. The millisecond time base is used when more accuracy is required. The current value is used to display the actual amount of delay the timer has timed at the moment you are watching it. This means that the current time is continually changing (timing) while the timer is activated. In the case of the TI 530 timer, the current time starts out equal to the preset time and counts down to zero. When the current time reaches zero, the timer has timed out.

The timer in the example is using 0.1 second as a time base. The preset time is set at 100, so this means that the timer will delay 100 tenths of a second, which is equivalent to 10 seconds. You can watch the current time on the CRT as it times down toward zero. This is similar to watching the dial on the motor-driven timer. When the current time reaches zero, the output is energized in a manner similar to the motor-driven timer timing out.

The other important parts of the programmed timer are the inputs, which control the operation of the timer, and the outputs, which signal the program that the timer has timed out. The functions of these inputs and outputs are similar for TI and Modicon timers. The contacts on the top left side of the instruction box are called *enable contacts*. These are similar to a set of contacts in series with the timer motor in an Eagle timer. The top terminal on the left side of the timer box is used to turn the timer on and off.

The lower set of contacts on the left side of the timer instruction box are called the *reset contacts*. They control the reset line on the timer instruction box. The reset line must be "high" or energized for the timer to operate. When this line is made "low" by opening the contacts, the current time is reset to equal the preset time. This means that if you want the timer to operate, the reset line must be "high" and the enable line must be "high." When both sets of contacts on the left side of the timer instruction are closed, this condition will be met and the timer will begin to time the amount of time delay.

The terminal on the top right side of the timer instruction box is called the *timer output*. Any time the timer current value has timed down to zero, this output will be energized. The output coil that is connected to this timer terminal can be memory output, or it can be a real output that sends a signal to an output module address. The contacts of this output can be used all over the program as timer contacts, even though they will have the address of the output. Since the contacts can be normally open or normally closed, these contacts are identical to the timed contacts in an Eagle timer. The actual operation of a delay-on timer is presented later in this section after we discuss the Modicon timer.

Modicon Timers. The Modicon timer is shown in Figure 16-4b. You should notice several differences between this timer and the TI timer. The Modicon timers are located in addresses starting with 4001. Since other functions, such as counters and sequencers, also use these addresses, you must look for the mnemonic T1.0, T.1, or T.01, which indicates that this instruction is a timer. T1.0, T.1, and T.01 also indicate the time base in tenths, hundreds, or whole seconds that the timer will use. The timer function instruction is similar to the TI instruction, in that the preset value is listed at the top of the timer, and the current value, which Modicon calls the *accumulated value,* is listed at the bottom of the instruction box. From the figure you will notice that 4001 is listed at the top of the timer instruction and 4002 is shown as the bottom value. 4001 is not the amount of preset time; rather, it is the memory address location where the amount of time is stored. When you enter the timer instruction into the program, you will list the address location first, then the value that you want the preset time to be, such as 100. The value 100 will then be stored at address 4001 and be used by the timer as the preset value. When the timer begins to time, the accumulated value will be stored in address 4002. Modicon refers to these addresses as *registers*.

The accumulated value in the Modicon timer starts at zero and times up until the accumulated time equals the preset time. At that point, the timer is considered to be timed out. The terminals on the side of the timer operate exactly like the TI 530 timer. The top terminal is used to enable the timer, and the bottom terminal is used to reset the timer. Both the enable and the reset terminal must be "high" for the timer to begin operation. Whenever the reset line is made "low," the accumulated value will be reset to zero. Any time the enable line is low, the time accumulated value will freeze and the timer will stop timing. When the enable line is set high again, the timer will begin timing again and the accumulated value will pick up where it left off and continue timing until it equals the preset time, which indicates that the timer has timed out.

The output terminal of the timer will be energized any time the accumulated time equals the preset time. The timer will stop timing at this point

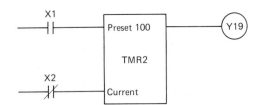

FIGURE 16-5 Examples of an on delay type timer programmed for a Texas Instruments controller.

and the output will be energized. The reset line must be made low to reset the timer and allow the time delay to start again.

On-Delay Operation. Figure 16-5 shows an on-delay timer for a TI 530. You can see that contacts X1 control the timer enable line, and contacts X2 control the reset line. When X1 is closed the timer will begin to operate since contacts X2 are programmed normally closed. The time base for this timer is set for tenths of a second, so the preset value of 100 will cause a 10-second time delay. This means that the output connected to Y19 will be energized 10 seconds after contacts X1 are closed. Whenever time contacts X2 are opened and closed, the timer will reset, and whenever contacts X1 are opened while the timer is still timing, the timer will be stopped and the current time will be frozen. When X1 is closed again, the time will time out the remainder of the time delay.

Off-Delay Operation. The timers in the programmable controller can also be programmed to provide off-delay operation. Figure 16-6 shows the off-delay timer function, programmed into a TI 530 P/C. In this circuit you can see that X3 has been programmed normally open and wired normally closed. This means that the timer will not begin to time until the real contacts connected to input module address X1 are deenergized. When this occurs, the programmed contact will pass power to the timer, and the output of the timer will energize C11 10 seconds later. Since the real output at Y20 is controlled by the normally closed C11 contacts, it will be deenergized 10 seconds after the switch at X3 has been deenergized.

The Modicon timer could have been programmed exactly like the TI timer and the operation would have been identical. There are several other variations to the off-delay and on-delay timer that can easily be made by adding other normally open or normally closed contacts.

Allen-Bradley Timers. The Allen-Bradley PLC 2 family does not use boxes in their program to represent timers; instead, their timer instruction looks similar to the motor-driven timer in a hard-wired circuit. The larger Allen-Bradley programmable controllers, including the PLC-3 family of processors, use the box instruction as in the Modicon and Texas Instruments P/Cs.

Figure 16-7a and b show an on-delay timer and off-delay timer programmed in the Allen-Bradley

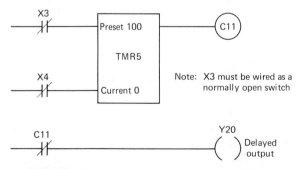

FIGURE 16-6 Diagrams the provide the off-delay timer function.

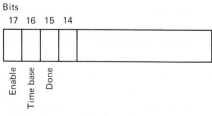

FIGURE 16-7 On delay and off delay timer instructions for an Allen-Bradley P/C.

Programmed Functions 445

PLC-2 format. From the diagram you can see that timer instruction looks similar to a normal output instruction. Notice the TON mnemonic represents the instruction "timer on delay." The timer has a preset value and an accumulative value just like the other types of timers. The timers are identified by addresses instead of using TMR1, TMR2, and so on. These addresses start at the address after the last rack. In this case the P/C is using two racks, so the first timer is located in address 030.

The timer's time base is displayed directly below the TON instruction as 1.0, which means that the time base is 1 second. The accumulative time will be displayed directly below the time base when the timer begins to operate. The timer will begin to time when the contacts (111/01) are closed. The accumulative time will increase from zero until the accumulative time equals the preset time. When this occurs, the timer times out and activates an output which energizes contacts. Whenever the contacts are opened, the timer will reset to the accumulative value to zero.

To this point, the operation of the Allen-Bradley timer is very similar to timers found in Modicon and Texas Instruments systems. The method that Allen-Bradley uses to activate its contacts when the timer is timing, times out, and is reset is slightly different. Instead of using output instructions in the ladder logic program, the Allen-Bradley system uses a series of four bits to control these operations. A diagram of these bits is shown in Figure 17-7c. Bit 17 is called the enable bit, bit 16 is called the time base bit, bit 15 is called the done bit, and bit 14 is called the timer timing bit.

Bit 17, the enable bit, will be high any time the timer is enabled (receiving power) by the contacts in series with the timer instruction. Bit 16, the time base bit, will pulse on and off (from a 1 to a 0) at the rate of the time base that has been selected. Bit 15, the timer done bit, will be set to a 1 when the timer has timed out. This will occur any time the accumulated value equals the preset value, and any time the timer is not timed out, this bit will be a 0. Bit 14, the timer timing bit, will be set to a 1 whenever the timer is timing. This value will be set to a 0 when the timer times out or when the timer has been disabled.

These bits function similarly to an output coil. Any contact that has the timer number and bit number assigned to it will be activated by the status of that bit. For example, in the diagram for the TON timer you should notice that the normally open set of contacts in the line below the timer is identified as 030/15. This means that these contacts are controlled by bit 15 of timer 030. These contacts will be closed whenever timer 030 has timed out (done).

This means that output 011/02 will be energized when the timer times out. If the timer was numbered 031 or 032, the contacts would be addressed 031/15 or 032/15.

If the timer application needs instantaneous contacts, they would be controlled by bit 14, the timer timing bit, or bit 17, the timer enable bit. You should also understand that the program can have multiple contacts controlled by timer 030 bit 15, just like applications in other P/Cs.

Off-Delay Timers. A different timer instruction is used for an off-delay timer. This instruction is called a TOF instruction, and Figure 16-7 shows it used in a program. From this figure you can see that the TOF instruction is similar to the TON instruction in that it has a preset value and an accumulated value. The TOF instructions use the same addresses as the TON. Operation of the off-delay timer is similar to a motor drive timer. When contact 111/05 is energized, the TOF timer will not be timing. The TOF contacts at 032/15 will be energized and the output 011/03 will also be energized. Since the timer is an off-delay timer, the opening of contacts 032/15 will be deenergized 10 seconds after the TOF instruction has been deenergized. As you have noticed, bit 15, the done bit, operates slightly different for the TOF than it did for the TON instruction. In the TOF instruction, the done bit is energized whenever the timer is enabled, and it will remain set to a 1 until the time delay has expired. The time delay will not start until the timer instruction is deenergized, and it will continue until the 10 seconds has been timed. At that time, the done bit will be returned to a 0, which will deenergize contacts 032/15.

Reset Timers. Another type of timer application is used frequently in programmable controller applications. There are no timer instructions for a reset timer; instead, the reset function can be programmed into the P/C by connecting contacts in series and parallel with the normal timer functions. Figure 16-8 shows an example of a reset timer for an Allen-Bradley and a Texas Instruments system. The reset timer in this application is used in a lubrication circuit, where a lubrication solenoid is activated once every 10 seconds.

In Figure 16-8a an Allen-Bradley TON timer is used as a reset timer. In this application the normally closed 032/15 contacts will energize the TON timer and it will begin to time. When 10 minutes has elapsed, the timer's done bit (bit 15) will become energized and the output at 011/01 will be energized and activate the lubrication solenoid. The normally closed 032/15 contacts will pulse open when the

446   Chap. 16 / Programmable Controllers

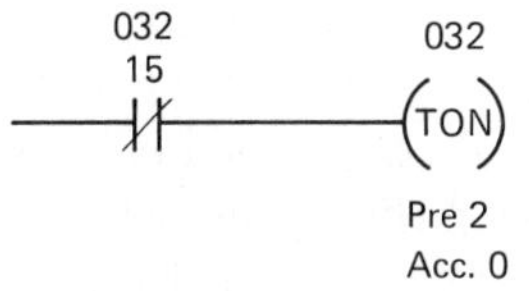

```
       032
        15            032
    ────┤/├──────────(TON)
                      Pre 2
                      Acc. 0
```

(a) Allen-Bradley auto resetting timer

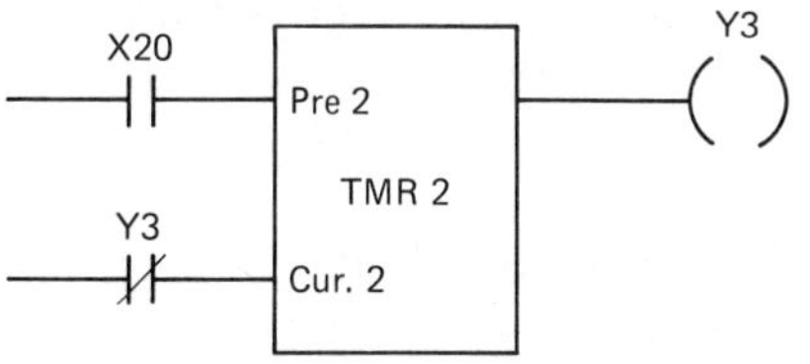

```
     X20    ┌─────────┐       Y3
    ──┤├────┤ Pre 2   ├──────( )
            │         │
            │  TMR 2  │
     Y3     │         │
    ──┤/├───┤ Cur. 2  │
            └─────────┘
```

(b) Texas Instruments auto resetting timer

FIGURE 16-8 Programs of reset timers for Allen-Bradley and Texas Instruments programmable controller.

timer times out, which will reset the TON timer. When the TON timer is reset, the normally closed 032/15 contacts return to their normally closed position, and the timer will begin to time again.

Figure 16-8b shows the same resetting timer application programmed for a Texas Instruments programmable controller. In this figure the circuit is activated when contacts X20 are closed. When these contacts close, the timer begins timing the 10-second time delay. At the end of the time delay, output Y16 is energized, which activates the lubrication solenoid. This output also opens the normally closed Y3 reset contacts, which resets the timer. When the timer is reset, the output is deenergized and the normally closed Y16 contacts are closed again, which starts the timer's time delay again. The timer in this application times out and resets every 10 seconds. This also activates the lubrication solenoid once every 10 seconds.

Retentive Timers. A retentive timer is a timer that has the ability to retain its accumulative value even when the timer enable line has been disabled. When the enable line is energized again, it will continue timing from the value where it stopped timing, rather than reset to zero.

This type of timer would be used to keep track of the total running time of a machine. For instance, the machine may be energized and operate for 3 hours today and 4 hours tomorrow. If you used a reset timer to keep track of this time, the total time would be reset to zero each time the machine was deenergized. If a retentive timer was used, it would accumulate 3 hours on the first day and stop timing when the machine was deenergized. When the machine was energized again the second day, the timer would begin adding time to the value of 3, and at the end of the second day the timer would show 7 total hours of running time. If you wanted to reset the timer, the reset line would have to be toggled.

This function is built into the box instruction timer used in Texas Instruments and Modicon P/Cs. This means that the timer will remember the current time or accumulative time whenever the enable line is deenergized, and no other special conditions must be made. If you want to reset the timer, the reset line must be deenergized and then set to a "1" again.

If you want to program a retentive timer in an Allen-Bradley P/C, you would need to use a RTO timer instruction instead of the TON or the TOF, since these are reset timers. The RTO is a special retentive timer that will retain the accumulative value whenever the enable line is opened. If you want to reset a RTO timer, you will need to add a RST instruction with the same address as the RTO timer. The RST instruction is a mnemonic for "reset," and whenever the reset instruction is energized, the RTO timer's accumulative value will be reset to zero.

Viewing Timer Preset and Accumulative Values. At different times you will need to view the preset and current values in a timer. You may also be asked to adjust the preset time for these timers. To perform these operations, you will need to understand how the preset and accumulative values are stored in the P/C memory. In the Texas Instruments P/C, the preset value for TMR1 will be stored in address TCP1. TCP in this address stands for "timer counter preset." The number in the address will identify which timer or counter the preset value is for. The current value will be stored in address TCC1, which stands for "timer counter current address." You can change the preset value by changing the value in the TCP address, and you can see the current time in the TCC address when the P/C is running and in the debug mode.

The Modicon system stores the preset value in the address that is shown at the top of the timer instruction box, and the accumulated value will be displayed in the address shown at the bottom of the box. This system displays the accumulative value and the preset value in the reference areas of the CRT when the program is running. The preset value will also be displayed at the top of the box instruction if a fixed value is used instead of a variable. If a variable is being used for the preset value, the address of the register will be displayed in the timer preset value instead of the actual value. When an address is used for the preset value, it allows the timer to accept any value in this address as its preset time. This value can be sent to the present

Programmed Functions **447**

address by a set of thumbwheels. This type of program is called *indirect addressing*.

The Allen-Bradley P/C system also uses two separate addresses to store the preset timer value and the accumulative timer value. The accumulative value will be stored in the address that is shown above the timer. In the previous examples, the timer address has been 030 or 031, which means that the accumulative value is stored at this address. The preset value is always stored at the 1XX address that matches the preset address. In the example where the accumulative value is stored at 030, the preset value is stored at address 130, and when the accumulative value is stored at address 031, the preset value is stored at 131. The preset and current value will be displayed directly under the timer instruction in the program when the P/C is in the run mode.

You can see that the preset value and the accumulative value can be viewed by looking at the value in their respective addresses. It is important to remember that new values can be sent to the preset address by connecting a set of thumbwheels to this address, and the accumulative value can be viewed by sending the value in the address to a seven-segment display.

Maximum Preset and Accumulative Timer Values. The maximum preset time that a time can have will depend on the type and model of the P/C. The Allen-Bradley PLC 2 and the Modicon 484 can have preset values up to 999 because their P/C uses a binary-coded decimal format to store the values. The Texas Instruments 530 can have values up to 32767 because the preset value and current values are stored in binary format. Other brands of P/Cs may allow larger or smaller values, depending on the format they use to store the values. Use the documentation provided with the P/C to determine these maximum values.

You can see from all of the timer examples provided in this section that each of these P/Cs can provide the same timer applications. Other timer applications can be programmed with additional contacts to provide a wide variety of timer applications. The important point to remember is that if you understand the operation of the enable, reset, and output lines, you will be able to design or troubleshoot most timer logic circuits.

Programmable Counters

Another instruction that P/Cs can execute very easily is the count instruction. The counter can be programmed as an up-counter, down-counter, reset counter, and totalizing counter. In some P/Cs, only one counter instruction is available, but like the timers, the different types of counter functions can be programmed with the aid of coils and contacts.

Figure 16-9 shows the counter instruction for the Texas Instruments TI 530 and the Allen-Bradley PLC 2 family. The Texas Instruments counter instruction uses a box format just like the timer. The counter is identified by the CTR mnemonic and the number will identify which counter is being programmed. It should be noted that timers and counters use the same address area in memory, so that once a number is used with a timer or counter, it cannot be used again. This ensures that the program will not have a TMR1 and a CTR1 instruction.

The top input to the box is called the *count signal*, and the bottom input is called the *reset signal*. The counter has a preset value and a current value that will be displayed in the box when the program is executed. The counter current value will start at the preset value and be decremented (decreased by 1) each time the count line is transitioned from low to high. This means that the contacts connected to the count line of the box instruction must start in the deenergized state and become energized for the counter to add on to its current count. Each time the contacts toggle, the count will be increased by one, until the current

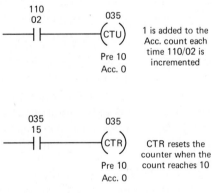

(a) Allen-Bradley up-counter instruction

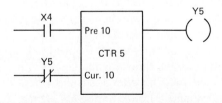

Counter is automatically reset when current count is equal to zero; count is incremented each time X4 is energized and deenergized

(b) Texas Instruments counter instruction

FIGURE 16-9 Counter instructions used for Texas Instruments and Allen-Bradley programmable controllers.

count equals the preset count. When this occurs, the counter output will become energized and the counter will not increment the current count any longer. The counter must then be reset by toggling the reset line before it can be counting again.

The counter in the Allen-Bradley P/C is very similar to the timer instruction. From Figure 16-9 you can see that the CTU mnemonic is used to indicate that this is an up-counter. The counter is identified by address 035, which means that its accumulative value is stored at that address, and the preset value will be stored in address 135. In this example, the preset value is 10.

The operation of this counter is similar to the other counters in that the accumulated value will be incremented each time the contacts prior to the counter instruction are opened and closed. The counter is reset by the RST instruction, which has the same address as the counter. Whenever the RST instruction is toggled, the accumulated count will be reset to zero.

Each counter controls a set of bits that are similar to the timer bits. These bits include bit 14, the overflow/underflow bit; bit 15, the done bit; bit 16, the down counter enable bit; and bit 17, the upcounter enable bit. As you can see in this example, the set of contacts that are labeled 035/15 will be energized when the accumulated counter value is equal to the preset value. These contacts can be used to provide a variety of logic functions.

Reset Counters. Counters can be used in P/C programs as reset counters and totalizing counters. The reset counter is used to keep track of a number of operations, and then resets itself automatically. This type of counter is especially useful in applications such as palletizing, where a new level should be started each time four boxes are placed on the pallet. The preset value in the counter is set at 4. Each time a box is placed on the pallet, the counter is incremented. When four boxes have been placed on the pallet, the counter will energize its output contacts, which will energize a solenoid that places a piece of cardboard between the layers. The contacts are also used to reset the counter accumulated value to zero.

Another reset counter can be used to energize a wrapping machine each time six layers of boxes have been stacked. This counter can have a preset value of 24 and count each box that is placed on the pallet, or it can have a preset of 6 and count the layers. In both cases the counter will energize the motor starter for the wrapping machine, which will wrap the entire pallet with plastic. The resetting type of counter can be used to keep track of any number of steps before the counter is automatically reset.

Totalizing Counters. Totalizing counters are used to keep count of the number of parts being made by a machine. This type of counter is programmed by making a very large preset value. If the accumulative value reaches the preset value, the counter will stop counting. The Texas Instruments counter can have a preset value up to 32767, which means that it can count up to that amount without being reset. The Allen-Bradley and Modicon counters can count values up to 999. If larger values must be counted, several counters must be daisy chained. Figure 16-10 shows an example of several counters used together so that they can totalize values up to 1 million. You can also use this format

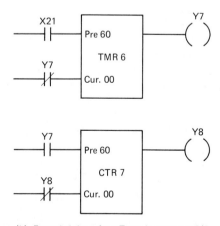

(a) Allen-Bradley extended timer using counters to extend length of time

(b) Extended timer for a Texas Instruments P/C

FIGURE 16-10 Totalizing counters with extended preset capabilities. A totalizing timers which uses a counter to extend the amount of time that can be stored.

Programmed Functions **449**

to extend the amount of time that a totalizing timer can keep track of. This extended timer can utilize a counter or several counters to increase the amount of accumulated time the timer can store.

Sequencer Instructions

Most P/Cs have some type of sequencer instruction which operates like the electromechanical drum sequencer. These sequencer instructions allow a set of steps to be programmed that will be executed in order. Each step of the sequencer will energize the outputs in the pattern indicated by the program instruction. This instruction is very useful for operating and troubleshooting complex sequential machines such as indexing tables and presses where the program must go through repetitive steps. Another application for a sequencer is a robot that is driven by air cylinders. The outputs of the sequencer instruction can be used to turn the air solenoids off and on to make the robot step through its routine. If the robot fails for any reason, the sequencer instruction will stop on that step and allow the technician to determine what inputs have not been enabled and are preventing the robot from continuing to the next step.

Texas Instruments provides an instruction called an *event drum* or *timed drum,* a sequencer that is incremented to its next step by an input bit. This bit is called *permissive* because it can be the output of a logic rung that has several switches that must be on or off for any given step of the sequence. The timed drum instruction allows time delay to be used with the permissive switches. The instruction can be programmed to use either a permissive switch or a time delay to advance the sequence.

The Modicon system uses a special set of counters that are stored at addresses 4051 through 4058. The CTR instruction is used because it is sequential in nature. You should remember that the accumulative counter value will increment by one each time a count signal is received at the top left side of the instruction box. When these counters are used as sequencers, the count line is used to step the sequencer ahead incrementally.

The eight counter instructions provide eight separate sequencers that can control outputs at addresses 2101 through 2832. The 21XX contact addresses are controlled by sequencer 1, the 22XX contacts are controlled by sequencer 2, and the system continues through contacts 28xx being controlled by sequencer 8. The XX indicates individual step numbers (1 to 32) within each sequencer. For instance, contacts 2101 will be turned on by sequencer 1 during step 1, contacts 2106 will be turned on by sequencer 1 during step 6, and contacts 2803 will be controlled by sequencer 8 and will be energized during step 3 of that sequence. Additional ladder logic must be used to add time-delay or permissive inputs for machine operation.

The Allen-Bradley sequencer instruction is very similar to the Texas Instruments in that a matrix is used to indicate which outputs will be energized during each step of the sequence. The sequencer instruction allows the programmer to indicate which outputs should be controlled and the sequence of their operation. It also provides an easy method of troubleshooting when the machine stops during the sequence.

Other Programmable Controller Functions

The P/C is capable of executing complex logic instructions that will provide complex machine operation. When this type of machine control was provided by control relays and hard-wired systems, the control panels and relay panels were very complex and difficult to troubleshoot. It was also very difficult to make changes to the machine's operation since these changes required the relays to be rewired. The P/C provided an easy way to reprogram this type of complex machine operation, since the only hard wiring involved each input switch to an input module and each output connection to an output module. The logic changes could be made in the P/C memory and stored on a storage device such as a floppy disk or magnetic tape.

The P/C can also perform routine and complex math functions such as addition, subtraction, multiplication, division, square, and square root. Larger P/Cs can also execute chained math functions which involve the use of exponents, logarithms, data tables, and other math functions.

BINARY AND BINARY-CODED DECIMAL NUMBERS

In previous examples in this chapter, the input and output signals have been digital (on/off). This means that the processor activated their address by energizing only one bit. In many other applications a value must be considered instead of just an individual address. An example of these values would include preset and accumulated values for timers and counters. The P/C stores these values as binary or binary-coded decimal numbers.

The typical binary value is stored as 16 consecutive bits. These 16 bits can be used to display numbers up to 65536. Since each bit of the binary system can be either on or off, the system is called the *base 2 numbering system.* This means that the binary numbering system has only two values. These values are 1 and 0. Since they are the only two values in the system, columns must be used to show position for weighting numbers above 1.

An example of the binary numbering system is shown in Figure 16-11. In Figure 16-11a you can see the binary system used to count up to 9 using four binary digits. From this example you can see that each bit represents a different column. These columns are called 1's, 2's, 4's, 8's, and so on. Figure 16-11b identifies each of the columns up to 32768, which allows 65535 to be the largest value to be counted with 16 bits.

The *binary-coded decimal* (BCD) *numbering system* is a combination of the binary and decimal numbering systems. An example of the BCD system is shown in Figure 16-11c. Only binary values are used in this system. Since the binary values are used to represent decimal numbers 0 to 9, only four bits are needed to represent each decimal number. By using a second set of four binary digits, a second decimal value can be displayed. This second decimal value represents the tens column in the decimal system. By adding four more binary digits another decimal value can be used to represent hundreds column for the decimal system.

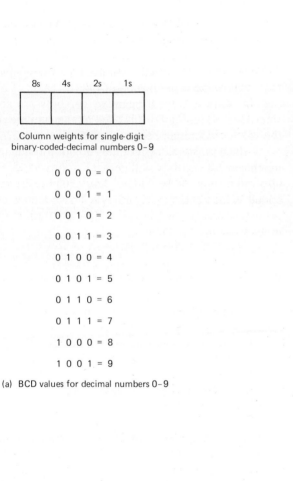

FIGURE 16-11 Examples of binary and binary coded decimal numbering systems.

Binary and Binary-Coded Decimal Numbers **451**

This combination of binary digits that are arranged in groups of four to represent the decimal system allows 12 bits to represent decimal values up to 999. If 16 bits were used, values up to 9999 could be represented.

The BCD numbering system is used extensively by the Allen-Bradley and Modicon system to store counter and timer values. The Texas Instruments system uses a straight binary system to store its timer and counter values, so that larger values can be stored.

Several components can be interfaced with the P/C using the binary or BCD numbering systems. Thumbwheels and LED displays use three- or four-place BCD values to send and receive values. The thumbwheel sends values to the P/C in BCD format and the LED receives values from the P/C to be displayed as decimal values. When 16 bits are used to represent the BCD values, 16 input or output addresses must be used. In some system all 16 bits are sent to a "word" module that sends all 16 bits with a strobe signal to indicate that these bits should be examined as a group (word) which represents a numeric value.

Some P/Cs, like the Texas Instruments TI 530, will display a value in either binary, BCD, or as a decimal value upon request. This allows the technician to select the numbering format for any value that is found in the system, and it can also easily be converted by asking for the conversion from the keyboard of the programming panel.

ANALOG INPUTS AND OUTPUTS

In all the examples so far in this chapter the input and output signals have been digital. This means that they turn one bit on or off. For instance, when a float switch is connected to a P/C input module, the signal that it sends to the processor is either on or off. This means that the address bit can be either a 1 or a 0. When a solenoid valve is connected to an output module, it can only be turned on or off.

Since these input and output signals can only be turned on or off, they are called *digital signals*. In some applications, to measure the water in a tank it can only indicate that the tank is full or empty. It may be important to know when an input is between full on and off. For instance, when the level of water in a tank is being measured, it is important to know the exact amount of water in the tank rather than to know when the tank is full or empty. It is possible to measure the water level from minimum to maximum using an analog signal to indicate and convert this value into a variable voltage or a variable current.

The typical voltage level of this signal is 0 to 5 V or 0 to 10 V, and the typical current signal is 0 to 20 mA. These signals can also use a 20 percent offset so that the voltage signals will be 1 to 5 V or 2 to 10 V and the current signal will be 4 to 20 mA. Figure 16-12b shows a scale for these signals. From this figure you can see that when the 0 to 10 V span is used to indicate the level of water in the tank, a 5.0-V signal would mean that the tank is half full. If the voltage is 2.50 V, the tank would be one-quarter full, and 7.5 V would signify that the tank is three-quarters full.

The analog input signal must be converted to a digital value before the microprocessor in the P/C can use it as a value. The analog input module converts the analog voltage or milliampere signal to a digital value. The digital value is converted into 12 or 16 bits. When these bits are placed together they form a value called a *word*. Values between 0 and 65535 can be measured (counted) with 16 binary bits. This value is determined from the value of 2 to the sixteenth power. If negative numbers are considered, the sixteenth bit is used as the sign bit and the values will range between −32768 and +32767.

Since most analog modules do not provide this much resolution, they will generally convert 12 bits into binary values between 0 and 32000 or binary-coded values between 0 and 999. These values can be scaled directly to the 0- to 10-V or 0- to 20-mA input signal.

Analog Output Signals

P/Cs can also convert 12- or 16-bit digital signals to analog voltage and current values. These analog output signals can be used to control variable outputs such as variable-speed motor drives, analog motorized valves, and proportional hydraulic valve.

These outputs can be used in conjunction with analog inputs and variables in the program. Figure 16-12a shows a scale of the values in the program and the amount of voltage and current they would produce at the output module. Figure 16-12a shows binary-coded-decimal values in the program that range from 0 to 999 and voltage and current values that range from 0 to 10 V and 0 to 20 mA. You can see that the one-quarter, one-half, three-quarter, and full scale are indicated as 250, 500, 750, and 999, respectively. This means that if you sent a value of 500 to an analog output module, the module would produce 5 V or 10 mA. This voltage or current could then be sent to a variable-frequency motor drive and the motor would operate at 50 percent of full speed. If the P/C sent a value of 250, the output module would produce 2.5 V or 5 mA

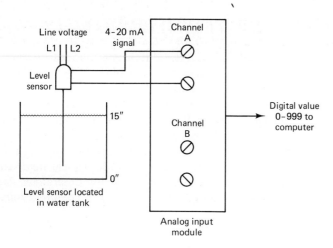

(a) Analog-to-digital converter module

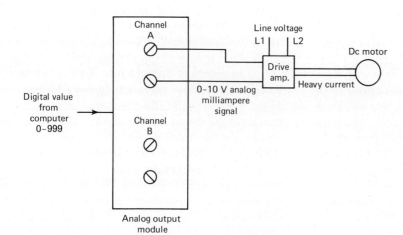

(b) Digital-to-analog converter module

FIGURE 16-12 Examples of A/D and D/A converter modules.

and the motor would operate at 25 percent of full speed.

This method can be used to send values to the analog output module in Texas Instruments processors, too. Their P/C will scale the value in the program from 0 to 32000. Values of 8000, 16000, 24000, and 32000 will produce 2.5, 5, 7.5, and 10 V, respectively, or 5, 10, 15, and 20 mA. This means that if the P/C sent the output module a value of 24000, the module would produce 7.5 V or 15 mA and a motor drive that is connected to this module would operate at 75 percent full speed. You should also understand that these module can produce both a milliampere and a voltage signal at the same output module address. This will provide the user with a choice of the type of signal they would like to use. In most cases, this choice is determined by the type of control that is being used and the type of signal it expects to see.

Typical Analog Module

Each P/C manufacturer has an analog module that can produce a specific amount of voltage or current in response to a value that is sent from the P/C program. The value can be varied within the program with math functions or other instructions and then sent to the analog module to produce the amount of current. Since the value coming from the microprocessor of the P/C is a digital value, it is sent through a data bus on terminals D0 through D15 as a 16-bit word. Allen-Bradley and Modicon use three-place binary-coded-decimal values (0 to 999) on their smaller systems and four-place values (0 to 9999) on their larger systems, to send values to their analog modules. Texas Instruments uses a binary value (0 to 32000) to control their analog output module. Each of these modules then produce the analog value with a digital-to-analog con-

Analog Inputs and Outputs   453

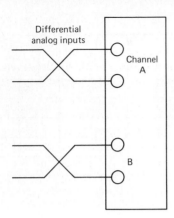 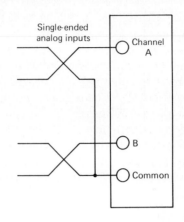

FIGURE 16-13 Typical field wiring diagram for differential and single point analog input modules.

verter. Some of the modules use jumpers or dip switches to select the type of voltage or current output that is desired. Other modules use a common terminal with voltage and current terminals to provide the voltage and current choice.

These modules can produce four or eight separate analog signals at the module. Differential signals use module signals to provide four separate signals to four channels on the modules. Modules are also available with eight nonisolated channels. Figure 16-13 shows a typical analog output module field wiring diagram. From this diagram you can see that each channel on the module is isolated from all the others. This allows the user to connect these as differential signals on isolated channels.

Applications for Analog Signals in Modern Industrial Controls

Analog input and output modules provide the P/C with the ability to interface in control systems where process control and motion control are needed with basic motor control. In industrial systems today it is not uncommon to find robots that use motion control and processes that require process control in the same P/C program. The complex operation can be controlled by sending and receiving analog signals along with the typical on/off motor control signals.

In some systems, analog sensors are used to indicate levels, pressures, temperature, flows, weights, and speeds. These signals are compared to set points in the system where the P/C makes decisions to turn motor starters on and off. Since a motor starter is used as the output, its signal must be digital since it can either be turned on or turned off.

In one application, a level indicator is used to control the level of water in a tank. The level indicator is connected to channel 1 on the input analog module. Its signal will vary from 0 to 10 V in response to the level of water in the tank. The transmitter will send a value of 10 V when the tank is full (1000 gallons), and will send a signal of 5 V when the tank is half-full (500 gallons). The transmitter will send a 0-V signal when the tank is empty.

The signal is brought into the P/C at address WX100. Remember that each analog address uses a 16-bit word to send the value that corresponds to the level of water in the tank. The value that represents the level is used in several "compare" statements. The first compare statement checks to see if the value is greater than the set point, and the second compare statement checks to see if the value is less than the set point. Since the desired water level can be set at any point, this application will use a dead band. This means that the level of the tank is going to be controlled between 400 and 600 gallons. If the level drops below 400 gallons, the motor starter will be energized and the water pump will begin to add water to the tank. When the tank level exceeds 600 gallons, the motor starter will be deenergized and the pump will stop pumping water. The pump will not be energized again until the water level falls below the 400-gallon level. You can see that this creates a dead band of 200 gallons.

Ladder logic is used to create the dead band. Contacts controlled by the outputs from the compare statements are used in this ladder logic. One set of contacts of will close by the compare instruction when the level is below 400, and open whenever the level is above 400. A second set of contacts will be closed whenever the level is below 600 and open any time the level is above 600.

When the level is below 400 both sets of contacts will close, the output will be energized, and the pump will begin to run. When the level increases above 400, the first contact will open, but the output contacts have sealed around the first contacts, like the hold-in contacts on a standard start/stop circuit. As the level continues to rise, the second contact will be opened when it exceeds 600 gallons. Since these contacts are in series with the output coil, the coil will be deenergized, and its hold-in contacts will also open. This action will deenergize the pump and the water level will begin to drop again as water is used in the manufacturing process. As the level drops below 600 gallons, the second contacts will be closed again by the compare statement, but since the first contacts are still open, the output will not be energized until the level drops below 400. At this time the first contacts will be energized again and the pump will again begin to increase the water level.

If the pump is turning on and off too often, the dead band can be enlarged by increasing or decreasing the high and low set points in the compare statements. These values can also be adjusted externally by connecting thumbwheels to the set-point addresses. In this way, an operator can adjust the level of the tank as conditions change in the system. This type of control system is also used in heating applications where electric heating elements will be controlled by the motor starter and the analog input sensors will be thermocouples or RTDs.

QUESTIONS

16-1. Identify the main parts of any programmable controller system.

16-2. Draw a diagram of a simple ladder diagram of a start-stop circuit that a programmable controller would use.

16-3. Draw the input and output module diagram showing the components of the start-stop circuit connected to them.

16-4. Explain why a programmable controller uses only symbols of contacts and coils instead of the JIC symbol for all switches and outputs.

16-5. What does a "1" indicate when it is sent to the image register of a programmable controller?

16-6. Explain the operation of the read and write cycle in a programmable controller.

16-7. Explain why the programmable controller uses addresses.

16-8. Explain the method Allen-Bradley uses to identify an input switch with the following address;

112
–] [–
03

16-9. Explain how Texas Instruments identifies its inputs and outputs.

16-10. Explain how Modicon identifies its inputs and outputs.

16-11. Explain the difference between a memory coil and a regular coil.

16-12. Explain the difference between a set of real contacts and memory contacts.

16-13. Explain the function of the preset value, the accumulative value for a timer.

16-14. Explain the function of the time base for a timer.

16-15. Explain the function of the enable contracts and the reset contacts for a Texas Instruments timer.

16-16. Explain how an Allen-Bradley type (TON) timer is reset.

16-17. Explain the operational difference between an on delay type timer and an off delay type timer.

16-18. Show a programming example of an on delay and an off delay type timer for an Allen-Bradley programmable controller.

16-19. Show a programming example of an on delay and an off delay type timer for a Texas Instruments programmable controller.

16-20. Explain what the word mnemonic means.

16-21. Explain the difference between a digital type on-off signal and an analog signal.

16-22. Explain the operation of a counter in a programmable controller and explain the terms preset and current values.

16-23. Show a program example of an Allen-Bradley counter that counts values above 999.

16-24. Explain the difference between a binary number, a BCD number, and a decimal number. Show an example of each by showing the value of the decimal number 36 in binary and in BCD.

Troubleshooting techniques and test equipment

17

INTRODUCTION

It is very important to understand all the methods and specialized test equipment that are available for you to troubleshoot and test modern motor control systems. In this chapter we provide information about these methods of testing and the equipment available to make them. You should understand that there is not just one way to test a circuit or component. The methods that you choose will depend on whether the machine is still under power or if it has been deenergized. The type of test you choose will also depend on whether you can get access to certain terminals, or how long it takes to gain access to the terminal that you need. Sometimes it is easier to try an alternative test because it is easier to perform.

Another consideration you must be aware of is your safety and the safety of the equipment you are working with. This means that some tests will not be the best to use when a machine is under power because of the safety risks. In other cases, the test you need to make must be made when full power is applied to the machine and you must have a helper to watch for unsafe conditions. It is vitally important that you remain aware of safety conditions at all times.

TYPES OF TESTS

In the previous chapters we have introduced many types of tests and explained when they should be used. In this section we review these tests and when you should use them. One of the tests that you should use to determine the condition of fuses and wires, and whether contacts are open or closed in switches that have been removed from a circuit, is the continuity test. This test involves the use of an ohmmeter or a test lamp. The meter should be set to zero prior to this test. You should also use the lowest ohms scale that is available on the meter. You should be aware that you should keep your fingers away from the tips of the meter leads since your body will introduce a certain amount of resistance in the test. Figure 17-1 shows two diagrams that indicate the proper method of using the meter to make this test. Be sure that all power is removed from the device or conductor that you are testing and that they are isolated from other devices or circuits so that the meter does not read current passing through an alternative path. If the contacts of a switch are closed or if a wire or fuse is in good condition, the amount of resistance measured by the meter should be near zero. If the meter indicates infinity (∞), the wire or fuse is open and should be replaced.

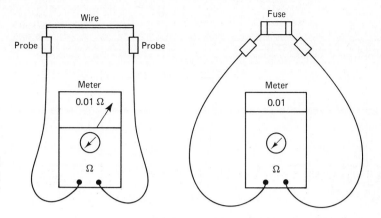

FIGURE 17-1 Diagrams showing proper methods of using analog and digital ohmmeters to test for continuity.

The ohmmeter can also be used to test solid-state devices, since the meter has a 30-V battery or a 1.5-V battery in its circuit when the meter is switched to ohms. This battery is useful in that it provides a polarized power supply with sufficient voltage to provide breakover voltage for the PN junctions. You should remember that the meter battery should be in good condition to have sufficient current available when you test the SCR and triacs since they need gate current to go into conduction. This means that you probably will need to use an analog meter (one with a needle) rather than a solid-state meter or digital ohmmeter. It is also important that you not try to test solid-state devices or circuits with an ohmmeter while they are still in a circuit or connected to a printed circuit board, since you may damage other components, such as MOS devices.

VOLTMETERS

You can test most circuits and devices with a voltmeter while they remain connected to other components. In most cases this is the preferred type of test, since you do not need to remove any conductors to complete the test. The main drawback about using this test is that you must constantly be aware of the voltage that is around you on the terminals of all the components while you are making the tests. If you come into contact with any of these terminals while you are making the test, you may be severely shocked. It is also possible to allow the meter terminals to touch several contacts of different potential, which will damage the components and may cause sparks that can harm you or personnel working with you.

Figure 17-2 shows three diagrams of methods of testing for voltage. When you make voltage tests, you should be sure that the meter is set for a voltage range that is higher than the voltage you are working with. In some cases you will not know what the voltage level is, so you should set the meter on the highest range available on the meter. You should also be aware that the meter has both ac and dc voltage scales available. If you are not sure if the circuit you are working with is an ac or dc circuit, you should place the meter to the highest dc range and reverse the meter leads for an instant. If the meter tries to read down scale (negative value on a digital meter), the voltage in the circuit is dc. If the needle on the meter remains the same when the leads are reversed, the voltage is ac. You should also make an attempt to determine the type of voltage from the wiring diagrams.

Figure 17-2 shows several types of tests that you can perform with the voltmeter. The most frequent test that is made with a voltmeter is the test for supply voltage. If the voltage is three-phase, you must test between phases A–B, B–C, and A–C (Figure 17-2a). In some cases you should also test between each of the phases and the ground or neutral potential. Remember: it is possible that voltage is back feeding through motors and transformers that are connected in the circuit. To protect against this, you should ensure that the disconnects are open in the circuit to isolate the system you are testing. It is also important to remember to replace all safety shields and covers that you remove during these tests.

The second type of test that is made with the voltmeter is the voltage loss or voltage drop test, which will indicate the exact location of an open circuit which may occur in a conductor or through a set of open contacts. This test is the most useful test that you can make to locate these type of faults. Detailed procedures are listed in Chapter 8. As you can see in Figure 17-2b, the most important point in using this test is to leave one meter terminal in place as a reference on one side of the circuit, and move the other terminal from point to point throughout the circuit until the voltage is lost. The point

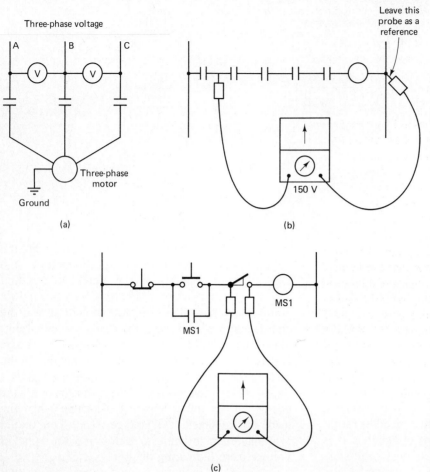

FIGURE 17-2 Diagrams of using a voltmeter to locate open circuits, measure the amount of supply voltage and to determine if contacts are open or closed.

where the voltage is lost indicates an open circuit between that terminal and the last terminal where voltage was recorded. You should also understand that you do not need to test every terminal in this procedure, since you can skip to the middle of a circuit. If voltage is present at this terminal, you have confirmed that all the components up to that point are operating correctly. If voltage is not present, you have narrowed the search for the open in the circuit to the half of the circuit that does not have power. By moving the terminal to the middle of the remaining portion of the circuit each time, you will eliminate half of the remaining components with each test.

Figure 17-2c shows a method of determining if a set of contacts are open or closed. If you place the terminals directly across a set of contacts, they will show that voltage is present if the contacts are open, and the voltage will indicate zero if the contacts are closed. This method can also be used to test for an open fuse. You must understand that this test will be successful only if there is voltage present in the circuit and if there is only one open set of contacts in the circuit. If more than one open is present in the circuit, this test will not provide accurate results.

When you are testing for voltage in a solid-state circuit or an analog circuit, it is vitally important to record the actual amount of voltage as accurately as possible. In some cases you should use a digital meter to determine that the proper amount of voltage is present. Some technicians get into the bad habit of assuming that the circuit is operating correctly as long as any amount of voltage is present. In these cases, if the needle moves, they assume that there is sufficient voltage present. Be sure that you read and record the exact amount of voltage that is being measured so that you can determine if the proper amount is present. In some electronic circuits it is also important to determine the polarity of the voltage that is present as well as the amount. This is especially critical in circuits that have bipolar sensors.

Testing for Voltage with a Test Lamp. The test lamp can also be used to determine if voltage is present in some circuits. In most high-voltage and control-voltage circuits the test lamp can determine if a voltage is present. You must be aware that the test lamp cannot determine if the amount of voltage is within specifications. Another important point to remember is that a pilot lamp will draw current when it is used to test for voltage. The amount of current that it will draw is sufficient to energize the input of a programmable controller when you are testing with the test lamp a set of switch contacts or the contacts of a pilot device that is connected to a P/C input module. The problem arises because the P/C input module is triggered by current flow. Since the test lamp allows several hundred milliamperes to flow, it is sufficient current to trigger the module circuit when the test lamp leads are connected across a set of open contacts on a pilot switch. This can cause serious damage to equipment or harm personnel if the input is energized inadvertently. This means that you should never use a test lamp to make any test on a programmable controller input module. You should also be aware that this same condition can occur if you are using an inexpensive voltmeter that does not provide at least 20,000 ohms per volt impedance to the circuit. The meter with low impedance will allow too much current to flow even though the meter is reading voltage.

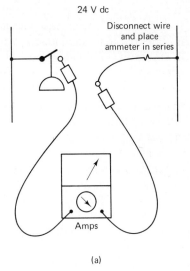

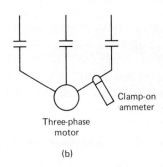

FIGURE 17-3 Applications for using AC and DC ammeters for troubleshooting.

AMMETERS

Ammeters are available for both ac and dc currents. They are also available in ranges for milli- and microampere measurements, as well as ranges for up to hundreds of amperes. The milliampere meter is very useful in determining the amount of current that is available for analog devices such as proportional hydraulic valves or variable-frequency motor drives. The meters with higher scales are needed to determine the amount of current an ac or dc motor is consuming.

The milliammeter is normally available as a dc meter. The ac ammeter is available as a special option on some meters, while the dc meter is generally provided as standard. Figure 17-3a shows how to use the milliammeter to measure the amount of control current that a circuit has. The main point to remember while using this type of meter is that it must be placed in series with the load to be able to measure the current. As you can see in the diagram, this means that one of the wires going to the load must be disconnected to create an open in the circuit. The milliammeter leads are then connected at the two ends of the open circuit. When current begins to flow, the meter will indicate the amount of current that is in the circuit. You should understand that it is important to ensure that the equipment is not operating when you remove the wire from the load terminal to create the open for the meter. If the machine is operating, the open circuit will cause it to malfunction during the time the wire is disconnected. This may cause unwanted machine operation and cause damage to the machine or to the personnel around the machine.

Since the milliammeter is generally a dc meter, you must also be aware of the polarity of the circuit you are measuring. If you have a digital meter, it will not matter which way you place the meter in the circuit, but if the meter has a needle (an analog type), you will cause damage to the meter if you connect it into the circuit in reverse polarity.

Figure 18-3b shows the correct way to make ac and dc high-current measurements. This can be accomplished by using a clamp-on meter for ac circuits. If the circuit is a dc circuit, you may need to open the circuit to make the measurement if you have a regular analog dc meter. Some new digital types of ammeters provide a clamp-on device for dc circuits also, but you will probably need to open the circuit to read the dc current.

Ammeters 459

The ac clamp-on ammeter operates like a transformer in that the meter has laminated segments in its claw that make an inductive circuit in a reduced or step-down configuration. This means that the current induced in the meter is a fraction of the actual current in the circuit. The meter has internal shunts that are switched into the metering circuit when you change the meter's range. This allows the meter to provide an accurate reading of the current in any ac line. Since the meter is a clamp-on meter, you can make measurements without disconnecting any wires in the circuit, merely by clamping the meter jaws around any wire in the system that has ac current flowing in it.

You can also read ac milliamperage with a standard high-current clamp-on ammeter. The simplest way of making a milliampere reading on a meter that has scales of 3, 10, 30, 100, and 300 A is to use the lowest range and design a small multiplier for the meter. As you can see in this figure, a length of 70 inches of 18 to 22 gage wire is used to make a multiplier coil. This wire should be coiled to make exactly 10 coils of wire. The 10 coils will provide a multiplier factor of 10. An alligator clip can be connected to each end of the wire in the coil to provide a simple means of connecting the coil into the circuit. In this case, the circuit where the measurement is being made must be opened just like the dc circuit. After you have disconnected one wire from the load, connect the two end of the multiplier coil to the two points where the open exists. Next, you should place the coil around one of the claws of the clamp-on ammeter and energize the circuit. The reading on the meter must be divided by 10 to derive an accurate measurement. This means that if the clamp-on ammeter is reading 1.5 A, the actual reading is 150 mA. If the amount of current is exceptionally small, it is possible to add any number of turns to the wire. Be sure to divide the reading on the scale of the clamp-on ammeter by the number of turns that are used in the multiplier coil.

Another application for the clamp-on ammeter is to determine if current is flowing through a conductor, fuse, or set of contacts. In an application of a three phase motor, motor starter, disconnect and fuse, if the motor single-phases (loses voltage in one phase), it will continue to operate but will tend to overheat. This test can also be used if the motor stops. In this test you are trying to determine which line of the motor the blown fuse or open conductor exists.

You can determine if a line is carrying current by placing an ammeter around each of the three lines (one at a time). If a line is continuous and all the fuses overload and contacts are carrying current, the meter will show the amount of current drawn. If a meter indicates that a line has no current, you can assume that the open is in that line, including an open circuit in a motor winding. This test will provide correct results only if only one of the phases is open. If two of the phases are open, the meter will indicate this by showing no current in all three phases.

Another sign that two or more of the phases are open is that the motor will not make any sound (hum) when current is applied. If the motor with the single-phase condition has stopped, it will hum loudly when power is applied to a circuit again. The reason for this is that the motor will not have sufficient torque to start and move the load. Since the rotor is in a locked-rotor state, it will draw excessive current and hum loudly.

As you can see, it is easy to locate the phase that has the open circuit in it, but you should understand that the ammeter will not indicate which component in the circuit has failed. This means that you have narrowed down the search for an open component to one of the three circuits, and you will need to continue testing with a voltmeter or an ohmmeter to locate the actual faulty component. Be sure to disconnect all voltage from the circuit if you choose to use an ohmmeter to make the additional tests.

Balancing Circuits With an Ammeter. In some plants it is necessary to ensure that the loads in all distribution panels and busways are balanced so that nearly the same amount of current is being drawn on each phase. If one of the phases is drawing considerably more or less current than the other two, the single-phase loads can be reconnected across the phases to balance the load. The ammeter is also useful in determining the maximum amount of current being drawn by a motor so that the proper-size fuse and overload can be determined. For this type of test, it is important to place the clamp-on ammeter on one of the phases and energize the motor. When the motor has come up to speed, you should record the amount of LRA current when the motor starts, and you should also record the amount of FLA current the motor is drawing when it is under maximum load. These values can be used to compare the amount of current that is indicated on its nameplate. If the current is excessive, the load must be adjusted to lower the current. If the current is lower than the nameplate data, you may want to reduce the size of heaters and fuses in the circuit to protect the motor against overloads.

PANEL METERS

The voltage and current meters discussed so far may be connected directly to the circuit at all times as panel meters. In these applications it is necessary to monitor the amount of voltage and current at all times with the panel meters. In some cases, these meters will also send a sample of their value to a programmable controller modules, so that the values can be archived for later reference.

SPECIALIZED TEST EQUIPMENT

A number of specialized meters and test sets have been developed for the specific needs of the factory technician who must test motors and motor control circuits. These meters are introduced individually below and a diagram presented to show how the tester should be used.

Phase Rotation Indicator

The phase rotation indicator is used to identify the three phases of a three-phase voltage system. Figure 17-4 shows the connections that should be used to utilize this type of meter properly. From the diagram you can see that the meter has three leads. These leads should be connected to the three voltage phases. The tester has three different-colored LEDs that identify when each of the leads is connected to the correct phase. This means that the LED for L1 will light up when the correct lead is connected to L1. If the L1 lead is connected to one of the other lines, the LED will not be illuminated. When the leads are all connected to the proper lines, all three LEDs will be illuminated. The meter also has two additional LEDs that indicate the rotation of the phase sequence. One of these LEDs will be illuminated at all times when the leads are connected to line voltage to help you determine which terminals should be switched to provide the proper phase rotation. If the lines are connected to the disconnect panel out of phase sequence, the power must be deenergized while those cables are reversed. You must understand that if a transformer or power distribution equipment is replaced or added to existing equipment, its phase sequence must be determined before it is connected to existing equipment and energized. If the new or replacement equipment does not have the exact phase sequence as the existing equipment, a tremendous short circuit will be caused when the new gear is switched into the circuit. This is also a concern when large motors that must rotate in the proper direction for machine tools, pumps, exhaust fans, and air supply fans are connected to the power distribution system. You can imagine the severe damage that could result if the replacement transformer on the power distribution system is connected with one phase reversed. This means that all three-phase motors on that system will rotate in the reverse direction. If the motor is providing combustion air for a blast furnace, it will draw the fire back toward the motor and fan instead of providing combustion air, which would result in a serious fire. If the motors are operating machine tools, the wrong rotation would damage the machines severely. The same condition would occur if the motor is driving a pump. From these examples you can see that it is vitally important to determine the phase rotation of any connections in the entire power distribution system before power is applied after making any changes.

Motor Rotation Indicator

The motor phase rotation indicator is very similar to the phase rotation indicator. It provides a phase sequence meter for three-phase applications that is similar to the one indicated previously. The motor rotation meter is added to this meter to provide a complete tester in one case that can be used to determine the rotation of the voltage and the motor. As you know, the motor's rotation can be reversed

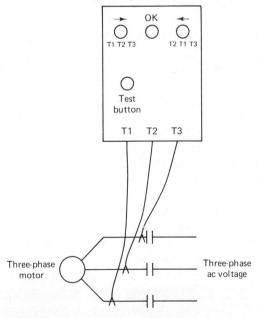

FIGURE 17-4 Phase rotation meter shown connected to a circuit to determine the proper phase sequence.

Specialized Test Equipment **461**

by reversing any two of the three-phase motor leads. Even though this method allows the motor's rotation to be reversed, it does not ensure that the voltage is applied to the motor in the proper phase sequence. This means that you should use the meter to determine that the voltage phase sequence is correct before you begin to reverse the motor leads. This will ensure that all phase-sensitive equipment connected to the power distribution system will be installed correctly.

The motor rotation indicator is also available as an individual tester to ensure that the motor is connected to the proper phases before it is started. These meters contain solid-state circuits that provide complete isolation between phases, and the clips on the end of the terminals are heavily insulated so that they can be connected directly to the leads of an energized system if necessary.

Phase Monitor Indicator and Tester

The phase monitor indicator is available as a portable test unit or as a permanently connected indicator. The portable tester has the advantage of being used on any motor or power distribution system that is suspected of having problems. This type of meter can detect loss of phase, undervoltage conditions, voltage imbalances, the reversal of any phases, and the shift of any phases. When these sensors are used in permanent applications, built-in relays are connected in the circuit as a safety

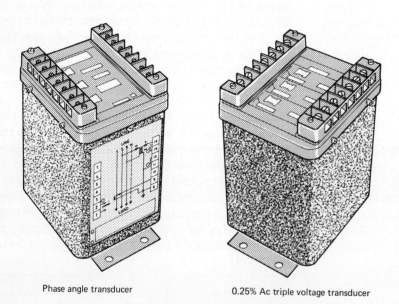

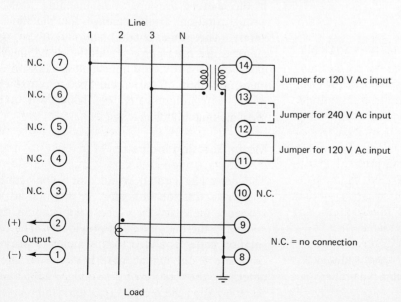

FIGURE 17-5 Phase protection devices used to detect loss of phase, low voltage, reversed phases, or excessive phase shift. (Courtesy of Square D)

control circuit which will deenergize control voltage to the motor starter coil. Figure 17-5 shows several examples of these types of sensors and a typical circuit for this type of control. The relay contacts can also be connected to a programmable controller, where the input signal can be used to warn technicians that an unsafe condition has occurred, or they can be connected to deenergize the motor when any of the previously stated conditions exists.

Since this type of sensor is also available as a permanent control, it can be connected in the system to protect one of the motors or the entire power distribution system for a set of motors.

Cable-Length Meter

A cable-length meter is a specialized solid-state tester that can determine the length of a single conductor by sending a signal down the length of the conductor and measuring the equivalent resistance. The meter has a selector switch on its face to select the proper cable type (copper or aluminum) and the cable gage (1, 2, 4, 6, 8, 10, 12, or 14).

This type of meter allows you to measure the resistance of a single conductor that must be replaced to determine its length, and compare the length to the amount of wire that remains on a spool. When you install extremely long lengths of cable, you will understand the inconvenience caused when you must physically unwind the wire from a spool to measure it. If the wire is too short, it cannot be used and must be returned to the spool, which is time consuming.

Wire Sorting Meter

Another type of test unit is available to help you sort wires and keep track of them during installation. This type of tester has two units (one for each end of the cable). One of the units has multiple leads (up to 10) which are connected individually to each conductor. The other end of the test set has one ground lead and a digital display which will indicate which wire the ground probe is touching. The two units can be separated by up to 3 miles, which allows one person to sort out the cables. As the ground probe is touched to each cable, the digital readout will indicate the cable number that corresponds to the connector at the other end. The technician can attach a permanent wire maker to each conductor as it is identified.

Megohmmeter

Another type of specialized test meter for cables is a megohmmeter. This meter, also called a *meggar,* can measure the amount of resistance between the insulation of cables that are in close proximity to each other. This is an important test when you suspect that the insulation between cables in the same conduit is breaking down. If the insulation begins to break down, a short circuit can develop inside the conduit that may damage all the conductors in it.

The meggar is also useful in measuring the insulation value of all wires in a motor winding. This test is made if a motor begins to show signs of insulation breakdown, such as overheating. If you can determine that one winding of a motor is beginning to show a decrease in the amount of insulation in its windings, the motor can be removed from service and have the bad winding replaced before the entire motor is damaged. In some cases the decrease in insulation may be caused by moisture accumulating in the winding. This occurs in motors and transformers and can destroy the device and cause damage to the entire power distribution system and all switchgear in the circuit if a short circuit occurs. If the cause of the decreased resistance between windings is caused by moisture, the motor or transformer can be dried out by heating the winding in a furnace or by adding dry nitrogen to the winding to absorb the excess moisture. The meggar is normally used during preventive maintenance tests as well as for troubleshooting.

CONCLUSIONS

You should begin to understand that a wide variety of test equipment is available for troubleshooting and diagnostic tests that must be made to motor control circuits. You should become familiar with the more widely used testers, which will save you enormous amounts of time during the test. These meters will also provide safe and accurate results that will cut down on the amount of time you will use to locate faults in a system. Many other specialized testers are available through your electrical supply house, and you will need to discuss your needs with the representative that will call on your shop. Any time you find that you are spending large amounts of time performing tests without the proper test equipment, you should inquire into the possibility of acquiring the tester specially made for that situation.

QUESTIONS

17-1. Explain when you would choose to make a test on a circuit with power applied. Why is this necessary?

17-2. When would you test something with no power applied?

17-3. Explain why it is important to keep one probe at a reference point when making voltage tests on a large circuit.

17-4. Explain when you could use a test lamp to locate faults in a circuit.

17-5. When is it necessary to use an ammeter to make tests?

17-6. Explain what you must do to a dc circuit to read amperage.

17-7. Explain why a clamp-on ammeter can be used in an ac circuit.

17-8. Explain how you would balance circuits with an ammeter.

17-9. Explain what panel meters are and why they are useful.

17-10. Explain what a phase rotation meter is and what it is used for.

17-11. Explain what a motor rotation indicator is and what it is used for.

17-12. Explain what a phase-loss monitor is and what it is used for.

17-13. What is a cable-length meter used for?

17-14. What is a megohm (meggar) meter, and what is it used for?

Glossary

Absolute encoder A device that is used to measure the position of a motor shaft. In typical encoders light is passed through a optical disk and sensed by photoelectric component which translates the location of the motor shaft into a digital signal. The digital signal provides information in regards to specific location and direction of rotation.

Accuracy The ability of a sensor to determine the actual value that is being measured.

Closed loop control This is a control system that uses a feedback sensor to determine the condition or position of the output device. This information is sent to the controller as feedback so that it can determine the next move. A simple example of a closed loop control system is the cruise control that is found on an automobile. In the cruise control system the operator of the vehicle sets the desired speed and a sensor on one of the tires sends information about wheel speed back to the control module. If the car is running below the desired speed, the control module will send a signal that causes the engine to increase the vehicle's speed. When the vehicle reaches the desired speed, the control module will sustain the output signal to maintain that speed.

CMOS A family of semiconductor devices called complimentary metal oxide semiconductors. The main characteristic of these semiconductors is their potential to control and switch voltage at extremely low current levels. The CMOS family of devices are used extensively in printed circuit board components for a variety of control and memory devices. When CMOS devices are used in memory boards of programmable controllers they provide the advantage of remaining in a charged state after main system power has been turned off because they use a very small amount of current from the battery back up voltage. Common abbreviations for organization, standards and governing bodies associated with the motor controls industry:

AWG American Wire Gage

IEC The International Electro-technical Commission

JIC The Joint Industrial Council

NEC The National Electric Code (also known as the National Fire Protection Association).

NEMA The National Electrical Manufacturer's Association

UL Underwriters Laboratory

Counter EMF Is the voltage (electro motive force) that is created by the rotation of an armature when it passes a magnetic field. This voltage is very useful to motors since it is generally out of phase with the applied voltage and has the effect of countering the applied voltage. When the motor armature is rotating at full speed the counter EMF is at its maximum which will tend to cause the amount of current the motor uses to be at its minimum. When a load is applied to the motor shaft it will slow down and the amount of CEMF will drop. This allows the motor to draw additional current to create the extra torque needed to drive the additional load.

Detecting distance The distance at which a proximity or photoelectric switch will operate. This distance is measured from the tip of the sensor to the target (the target is the part being detected: for example, bottles on a bottling line). The detecting distance is calculated in the specifications using a standard-sized target or reflector.

Fail safe condition When a motor control device, control circuit, or energy converter is designed, it is important to study the conditions that exist when a power failure occurs. The fail safe condition is determined to be a condition where the system can exist without causing damage to personnel or equipment if power was abruptly discontinued. Conditions such as gravity, springs and pressure must be considered to attain a safe state.

FLA (full load amperage) FLA is also called full load current and it is the amount of current a motor or other device will use when it is operating at full load condition. The full load current specification is generally stamped on the motor data plate, or it can be determined from tables if the motor's horse power is known.

Led A light emitting diode is a solid state device that creates a small amount of light when voltage is applied to it in the forward bias direction. It is used frequently in solid state controls as an indicator lamp.

LRA (locked rotor amperage) LRA is also called locked rotor current. This is typically the amount of current that a motor will use when power is first applied and the shaft is starting to turn. As the motor's shaft increases speed, the counter EMF produced by the armature will oppose the applied voltage and cause the amount of current the motor is using to drop substantially. Locked rotor current is typically 5 to 10 times larger than the rated full load current for the motor, but it will rarely damage the motor because it normally only occurs for 1–2 seconds while the motor is started. If the motor shaft becomes jammed so that it cannot turn, locked rotor current will occur and severely damage the motor unless it is protected against over current conditions.

Magnetic motor starter A control device that is used to start and protect a motor. Its main parts consist of a coil, large current carrying contacts, small auxiliary contacts, and an overload device that includes a heater. When the coil of the motor starter is energized with the correct amount of applied voltage, it causes the coil to produce a strong magnetic field that makes the main contacts move to the close position. Any auxiliary contacts on the motor starter will also change their status when the coil produces its magnetic field. As the name of the control implies, a motor is the main load that is connected to the contacts. The heater and overload assembly will sense any excess current that the motor may draw, and cause the coil circuit to deenergize which protects the motor.

Motor control system An electrical system that is used to provide safety for the operator of the system, safety of the equipment in the system, and operational control of a system. The typical motor control circuit will have one or more safety or control switches and one or more output elements which are typically motors or solenoids. The elements of a motor control system can also be classified as sensing elements, decision elements, and action elements.

Open loop control This is a control system that does not use a sensor to provide a feedback signal. The control of the system is provided by the operator. An example of this type of control system is an automobile without cruise control. The operator must determine the maximum speed from looking at speed limit signs. If the vehicle is operating at speeds below the speed limit, the operator will depress the gas pedal to increase the speed. The operator must continually view the speedometer to determine the vehicle's speed and make adjustments to reach the desired speed. The amount of adjustments made by the operator will depend on how closely the speed is being controlled.

Pilot device Is a classification of control devices used in motor control circuits that typically have control small currents. The contacts in a pilot device can be normally open or normally closed and can be changed to the opposite condition with a minimal amount of change to its actuator. The contacts can be mounted in the pilot device so that any changing parameter found in the control environment can be sensed and controlled. For example, typical pilot devices can sense machine movement (limit switch), the change in temperature (thermal switch), the change in pressure (pressure or vacuum switch), or the change in level (float switch). Pilot devices are typically used in the control portion of the circuit, but in some systems they can have larger size contacts which allows the device to connected directly to a small motor instead of a relay or motor starter.

Programmable controller A simple programmable controller is a reprogrammable device that has one or more input signals that are conditioned by an input module, one or more output signals that are conditioned by an output module, and a central processing unit (CPU). The program in the controller is unique in that it is generally written in the form of contacts that logically control the condition of output coils. This program tends to look exactly like electrical ladder diagrams that are used for traditional hard wired systems. The controller can have a variety of memory media and can also provide more complex functions such as time delay, counts and sequences. Addition types of controls are generally available to provide motion control, process control and complex mathematical calculation. A display device such as a cathode ray tube (CRT) or computer is generally used to program the programmable controller and view the status of inputs and outputs in a real time display.

Resolver A device that is used to provide positional information. A typical resolver has two stator windings that are mechanically displaced at 90 degrees to each other and to the rotor winding. The resolver shaft is coupled directly or connected through gears to the shaft of the motor that is being controlled. The resolver's rotor winding and one of its stator windings are excited by an applied voltage and the remaining stator winding is shorted. When the motor shaft turns, the resolver's rotor turns and the relationship between the resulting sinusoidal wave forms can be measured to determine accurate positional data. The resolver can be used to measure rotational position or linear position. If the motor uses more than one rotation to complete the full distance of travel, a course and fine resolver may be used to accomplish the measurement.

Response frequency The time it takes a proximity or photoelectric switch to cycle from an off condition to an on condition. This is the time a switch is activated and then deactivated and ready to be activated again. This particular term is critical in high speed sensing application.

Response time The time it takes a proximity switch to locate a target and activate its output.

Safety controls These are controls whose main prupose in a motor control circuit is to sense unsafe conditions for personnel who operate or come in close proximity to the system, or unsafe conditions for the machinery and circuitry in the system. The safety of personnel is always the top priority of any safety controls.

Servo system Provides one or more feedback signal that are used to indicate velocity or position. Energy conversion for a servo system can be provided by rotary devices such as electrical motors and hydraulic motors, or by linear devices such as rack and pinion systems or hydraulic cylinders. The typical servo system will usually have the following major components; input signal which is called set point or command, a feedback signal, a digital or analog controller, an output signal (which usually requires an amplifier), and an output device. In an analog system a summing junction is used to determine the difference (error) between the command signal and the feedback signal.

Short circuit This is a condition that occurs when the amount of resistance in a load or circuit is allowed to drop to a near zero condition. This condition may occur at the load or at a point between the applied voltage and ground potential. When a short circuit occurs, current is allowed to increase to a dangerously high level (up to 100,000 amps if the circuit is not protected by fuses or circuit breakers. If the control circuit and motors are not properly grounded, a condition will occur which could cause a severe or lethal electrical shock.

Index

A

Absolute encoder, 428
Acceleration circuits, 372
Accumulated value, 444
Accuracy, 370, 423
AC motor, 300, 284
Across the line starter, 181
AC single phase motors, 298
AC variable speed drive, 411
AC voltage, 123
Adjustable ramp, 386
Alarm contacts, 154
Allen-Bradley, 14, 16, 35, 76, 77, 86, 87, 92, 93, 95–97, 99–101, 105, 122, 132, 134, 140, 150, 151–156, 159, 168, 169, 171, 173–175, 178, 179, 184, 186, 187, 192, 196, 208, 209, 237, 240, 317–330, 337, 339, 341, 344, 345, 350, 352, 357, 365, 391, 394, 395, 408, 409, 422, 445–448,
Alnico, 110
Ambient conditions, 363
American Wire Gage (AWG), 80
Analog inputs, 452
Analog module, 453
Analog output, 452
Analog signal, 409
AND gate, 361

Anti plugging, 393
Anti plugging switch, 393
Arc hood, 87, 149
Arc quencher, 149
Arc supression, 151
Armature, 49, 122, 124, 268
Armature assemblies, 124
Armor cable, 78
Auto transformers, 58, 98
Auxillarycontacts, 154, 170, 177, 322, 329, 393

B

Back EMF, 289
Ball bearing, 272, 434
Ball screw, 421
Ball screw mechanism, 433
Base assembly, 85
BCD binary coded decimal, 450
Bell crank assembly (relay type), 125
Bellows pressure switch, 218
Bimetal overload, 175
Bimetal strip, 223, 224
Binary numbers, 450
Bourdon tube, 218
Box lug, 89
Brakes and clutches, 395

Brakeshoes, 395
Brake solenoid, 396
Braking methods, 389
Break, (contacts), 129, 194
Break away torque, 386, 414
Brushes, 272
Brush rigging, 272
Buck boost transformer, 58
Bus duct systems, 74

C

Cable length meter, 463
Cable limiting fuses, 65
Cable tray, 75, 80
Cadmium selenide (CdSe), 350
Cadmium sulfide cells, 349
Capacitive proximity switch, 366
Capillary tube, 225
Capillary tube control, 224
Cathode, 348
Cathode ray tube CRT, 335, 346, 437
Cavity enclosures, 105
Celsius rise, 297
Changing torque (wiring motor for), 274
Circuit breaker, 51, 61
Clamp on ammeter, 121, 317
Clapper (relay type), 124
Clutch drives, 400
Clutch solenoid, 239
CMOS, 360
CMOS counters, 257
CMOS devices, 134
Coarse resolver, 427
Coil, 154
Coil, set, 137
Coil, shading, 124
Coil current (amperage) data, 133
Coil currents, 112, 154,
Coil ratings, 177
Coil shunt plate, 154
Coil voltages, 133
Color marks, 351
Combination starters, 181
Command signal, 428
Common point, 299
Compound motor, 269
Conduction angle, 385
Conductors, 80
Contact, block, 199
Contact assembly, 85
Contact carrier, 85, 149, 170

Contact coating, 149
Contactor, 148
Contactor, heating, 161
Contactor, lighting, 161
Contact ratings, 131, 133
Contacts, reversible, 132
Contacts, reversing, 184
Contacts, seal in, 318
Contacts, single break, 129
Contacts, single make, 129
Contacts, stationary, 149
Contact size, 151
Contacts replacement, 192
Continuity, 65
Continuous plugging circuit, 161, 389
Control base, 361
Control circuit, 4, 128, 319, 382
Control device, 323
Control relay, 123
Control transformer, 60, 182, 319
Core losses, 270
Counter, 234, 256, 260
Counter, solid state, 257
Counter, totalizing, 448
Counter (down count), 257
Counter EMF, 289, 303
Counter preset value, 447
Counter reset, 448
Count signal, 448
Cross roller actuator, 203
CRT, 335, 346, 437
CSI drive, 417
Cube timers, 247
Current relay operation, 306

D

D/A converter, 424
Dark activated (photoelectric), 351
DC braking, 389
DC compound motors, 281
DC motor, 266
DC motor speed control, 406
DC motor theory, 268
DC series motors, 275
DC shunt motors, 277
Deceleration, 389, 395
Deceleration circuits, 372
Delta connected motors, 290
Detecting distance, 370
Diagrams, ladder, 23, 36, 130, 171, 329
Diagrams, machine, 45

Diagrams, plant layout, 45
Diagrams, sequence of operation, 40
Diagrams, timing, 242, 243
Differential amplifier, 410
Differential compound motor, 281
Diffuse scan, 347
Digital signals, 452
Digital timers, 244, 246
Digital to analog (D/A) conversion, 422
Diode, 144
Disconnect switches, 84
Domains, magnetic, 109
Double break contacts, 88, 129
Double make contacts, 129
Doughnut transformers, 73
Down counter, 257
Drives, motor, 403
Drop out voltage (coil), 126
Drum switch, 105
Dual element fuse, 63, 64
Dynamic braking, 393
Dynamic torque, 432

E

Eagle Signal, 241–243, 250, 252, 253, 258, 259, 260, 261, 262, 264
Eaton\Cutler-Hammer, 25–27, 59, 79, 103, 164–166, 181, 183, 198, 207, 212, 236, 239, 265, 355, 364, 366, 368, 370, 374, 379, 381, 383, 385, 387, 388, 397, 404, 412–419
Eddy current drives, 411
Eddy currents, 270
Electrical diagram, 33
Electrical generator, 49
Electrical safety, 10
Electric braking, 393
Electromagnets, 110
Electronic motor drives, 403
Elementary diagram, 34
Emitter, 358
Enable contacts, 444
Enclosure, 76, 78
Encoder, 424
Encoder, optical, 429
Encoder, pulse, 428
End plates, 272
End switch, 301
Equipment safety ground, 12
Error, 407, 421, 424
Eutectic alloy heater, 174

Eutectic alloy overload, 174
Event drum, 450

F

Fail safe brakes, 397
Feedback loop, 407
Fiber optic controls, 353
Fiber optics, 358
Field, 49, 268
Field poles, 271
Field wiring terminals, 149
Fine resolver, 427
Flexible conduit, 78
Float control, 317
Float switch, 214
Float switch closed tank, 216
Float switch open level, 216
Flow switches, 220
Flush mounted panels, 101
Flux lines, 49, 109
Flux losses, 270
Fly weights, 301
Foot switch, 213
Fork lever actuator, 203
Forward contacts, 184
Four way solenoid valves, 116
Friction brakes, 395, 397
Friction pads, 399
Full load amperage (FLA), 132
Full load current, 61, 63, 303
Full voltage starters, 177
Fused disconnect, 85
Fuses, 60
Fuses, dual element, 63, 64
Fuses, interruption capacity, 51, 64

G

Gate turn off thyristers GTOs
General purpose relay, 137
Grooved head sensor, 370
Grounded winding, 11
Ground fault interruptor GFI, 12
Ground wire, 54

H

Hall effect sensor, 370
Hard wired circuit, 336
Heating contactors, 161
Helix coils, 111
Hold down clips, 142

Holding current, 112
Holding torque, 432
Honeywell Micro Switch, 204, 205, 210, 227, 348, 349, 356, 358, 359, 366, 369
Horse power ratings, 89
Hybrid stepper motor, 432

I

IEC (International Electro-technical Commission), 162
Image register, 439
Incandescent controls, 347
Incremental encoder, 427
Indicator lamp, 225, 319
Indirect addressing, 448
Induction motors, 286
In line solenoid valve, 113
Input and output modules, 229
Inrush current, 61, 126, 194, 384
Installing relays, 139
Instantaneous contacts, 239
Interlock contacts, 154
Interlock, electrical, 160
Interlock, mechanical, 160
Interlocks, 322
Interruption capacity (IC), 51, 64
Isolation transformer, 53

J

Jacks, 40
JIC (Joint Industrial Council), 437
JIC symbols, 26–30, 437
Jogging, 84, 342
Jogging circuit, 158, 341
Jogging motors, 326
Jog/run switch, 327
Joint Industrial Council (JIC), 437
Joy stick, 211

K

Knife switch, 83

L

Ladder diagram, 23, 36, 130, 171, 329,
Ladder logic, 43, 131, 332, 340, 345 (*See also* relay ladder logic)
Laminated steel core, 123
Latch contacts, 154
Latch current, 154

Latching relays, 138
Lead screw, 433
LED (light emitting diode), 142, 260, 226, 350, 368, 433
LED indicator, 226
Left hand rule, 111
Level switch, 214
Light emitting diode (LED). *See* LED
Lighting contactors, 161
Light rejection, 350
Limit switch, 201, 203, 323
Limit switch, cat whisker, 205
Limit switch, cross lever, 203
Limit switch, forklever, 203
Limit switch, roller arm, 203
Limit switch, wobble head, 205
Linkage effect, 411
Load circuit, 4, 128
Locked rotor amperage, (see LRA)
Lockout device, 20
Logic, 332
Logic card, 361
Logic diagram, 130
Loss of phase, 176
Low pressure switch, 219
LRA (locked rotor amperage), 132, 151, 181, 323, 331 (*see also* Locked rotor current)

M

Machine diagrams, 45
Magnet, 108
Magnet, alnico, 110
Magnet assembly, 124
Magnetic field, 122
Magnetic motor starter, 167
Magnetic theory, 267
Magnetite, 108
Magnets, 108
Make, (contact closure) 194
Manual control, 5
Manual controllers, 83
Manual drum switches, 103
Manual motor starters, 84, 89
Manual starting switches, 95
Mechanical interlock, 97
Meggar, 11, 463
Megohm meter (meggar), 11, 463
Memory coil, 341, 342
Memory contacts, 442
Meter, clamp on ammeter, 121, 317

Meter, megohm, 11, 463
Meter, milliamp, 459
Meter, phase rotation, 461
Meter, volt, 106, 200, 457
Meter, wire sorting, 463
Metering panels, 73
Meters, 461, 463
Microprocessor, 7, 229
Milli ammeter, 459
Miniature switch, 197
Mnemonic, 443
Modicon P/C addressing, 441
Momentary pushbutton switch, 88, 197
Motion control, 5
Motors:
Motor, compound, 281
Motor, DC, 266
Motor, PM, (permanet magnet), 425
Motor, series, 275
Motor, servo, 425
Motor, shaded pole, 309
Motor, shunt, 277
Motor, split phase, 302
Motor, star connected, 290
Motor, start winding, 299
Motor, stator, 284
Motor, stepper, 430, 432
Motor, synchronous, 298
Motor, two capacitor, 302
Motor, wound rotor, 302
Motor, wye connected, 289
Motor code, 297
Motor control centers, 182
Motor data plates, 296
Motor drives, 403
Motor end plates, 288
Motor rating, 297
Motor starter, magnetic, 167
Motor starter, manual, 84, 89
Motor starter, part winding, 380
Motor starter, reversing, 98, 185, 321
Motor starter, star-delta (wye-delta), 378
Motor starter, two speed, 185, 342
Motor starter, wye-delta, 378
Motor starters, 84, 89, 167, 188
Movable core, 122
Mushroom head switch, 197

N

NPN transistor, 351
Name plate, 296

National Controls Company, 248, 249
National Electrical Code, (NEC) 13, 64
National Electrical Manufacturer's
    Association, (NEMA), 162
NEC, (National Electric Code), 64
NEMA, (National Electric Manufacturer's
    Association, 162
NEMA rating for contacts, 151
NEMA starters, 154
Neutral, 54
Normally open contacts, 128, 154

O

Octal base socket, 139
Octal base timers, 247
OEM, (original equipment manufacture), 164
Ohmmeter, 106, 457
Omron, 135–138, 141, 142, 145, 163, 235, 353,
    354, 357, 360, 362, 363, 367
Open loop, 409
Open transition starter, 373
Operational amplifier (op-amp)
Operational controls, 3
Operator assembly, 85
Optical encoder, 429
OR gate, 361
Over current circuit, 432
Overload assembly, 172
Overload contacts, 171
Overload device, 90

P

Palm button, 197
Panel meters, 461
Parts list, 41
Part winding starter, 380
Pawl, 174
P/C, (programmable controller), 436
P/C address, 437
P/C counters, 448
P/C off delay timer, 445
P/C on delay timer, 445
P/C output module, 230
P/C register, 444
P/C reset timers, 446
P/C retentive timers, 447
P/C scan, 440
P/C sequencers, 450
P/C write cycle, 439
Peak demands, 50

Peaking generators, 50
Permalloy, 110
Permanent magnet field, 397
Permanent magnet motor, 425
Permissive contacts, 450
Phase monitor indicator, 462
Phase rotation meter, 461
Phases, 50
Photoelectric, specular scan, 358
Photoelectric, through beam scan, 347
Photoelectric applications, 263
Photoelectric controls, 263
Photoelectric head, 364
Photoelectric switches, 347
Photoelectric switches, retroreflective, 347
Phototransistor, 351
Photovoltaic cells, 349
Pick up voltage, 126
Pilot devices, 194
Pilot duty contacts, 177
Pilot lamp, 96, 337
Pins, 40
Pivot, 174
Plant layout diagrams, 45
Plugging, 391, 392
Plugging circuit, 161, 389
Plugging switch, 390
Plunger, 203
PM motor, (permanent magnet motor), 425
PM motor (*see* permanet magnet motor)
Pneumatic time delay, 140
Pneumatic time delay element, 140
Pneumatic timers, 234, 236
PNP transistor, 351
Pole, 309
Potential relay, 307
Power flow, 335
Power rail, 39
Preset time, 241
Press to test lamp, 320
Pressure switch, 217, 317
Pressure switch, low, 219
Pressure switch bellows, 218
Process control, 6
Producing back emf, 275
Programmable controller (*see also* P/C), 6, 7
Programmable timers, 443
Program scan, 440
Proximity, 358
Proximity switch, short barrel, 369
Proximity switch standard target, 370

Proximity switches, 366
Pulse encoder, 428
Pushbutton switch, 88, 197

R

Ramping, 385
Ratchet, 174
Read cycle, 439
Receiver, 358
Reduced voltage starter, 373
Reed switch, 195
Regenerative braking, 389
Registers, 444
Relay, 121, 122
Relay, socket mounted, 142
Relay contacts, 128, 144
Relay ladder logic (RLL), 43, 332 (*See also* ladder logic)
Remote control, 5
Remote sensor, 369
Repeat cycle timers, 234
Replacing contacts and coils, 192
Repulsion start motor, 302, 310
Reset coil, 137
Reset contacts, 444
Reset counter, 448
Reset signal, 448
Reset timers, 234
Resolver, 423, 425
Resolver resolution, 423
Response frequency, 370
Response time, 370
Retroflective scan, 347
Reverse contacts, 184
Reversible contacts, 132
Reversing, 84
Reversing a motor, 160
Reversing circuit, 338
Reversing motors, 322
Reversing motor starter, 321
Reversing starter, 98, 185
Reversing the motor, 274
RF field, 366
Rigid conduit, 78
Robot, 46, 143, 144, 190, 191, 228, 231, 232
Robot applications, 191
Roller arm acturator, 203
Rotor, 284
RTD's, 224
Run winding, 299

S

Saddle clamp, 89
Safety circuit, 2
Safety devices, 4
Safety inspections, 20
Schmitt trigger, 366
SCR, (silicon controlled rectifier), 283, 351, 60, 368, 384
SCR drive, 425
SCR motor speed control, 405
SCR wave forms, 410
Sealed current, 126
Seal in contacts, 318
Secondary resistor, 383
Selector switch, 88, 210
Selector switch, two position, 210
Self contained sensor, 369
Sensing bulb control, 224
Sensor, under voltage, 94
Sensors, short barrel, 369
Sensors, standard target, 370
Sequence control, 328, 329
Sequence of operation, 40
Sequencer, 234, 263
Series motor, 269
Service factor, 298
Servo loop, 420
Servo motors, 425
Servo systems, 420
Set coil, 137
Shaded pole motors, 309
Shading coil, 124
Short barrel sensor, 369
Short circuit, 11
Short circuit current, 60
Shunt motor, 269, 404
Shunt motor's torque, 279
Single break contacts, 129
Single make contacts, 129
Single pole switches, 197
Sizing heaters, 176
Sleeve, 272
Sleeve type bearings, 288
Slow over currents, 60
Socket mounted relays, 142
Soft start, 414
Solar cell, 349
Solder pot overload, 90
Solenoid, 112
Solenoid action, 112

Solid state counters, 257
Solid state timer, 140, 244
Specular scan, 358
Speed feedback signal, 409
Speed reducer, 399, 404
Speed reference signal, 417
Speed switch, 390
Speed of the motor, 274
Split phase motor, 302
Square D Company, 13, 30, 34, 73, 74, 75, 85, 124, 127, 129, 180, 195, 201, 214, 217, 294, 295, 375, 377
Squirrel cage rotor, 301
Standard target, 370
Star connected motor, 290
Star delta starter, 378
Start winding, 299
Stationary contacts, 149
Stator, 284, 299
Step angles, 431
Step down transformer, 319
Stepper motors, 430
Stop buttons, 182
Summing junction, 407, 421, 424
Switch contact block, 199
Switch gear, 51
Switch legend plates, 199
Switch mounting ring, 199
Switch symbols, 198
Switch types:
Switch, float, 214
Switch, flow, 220
Switch, foot, 213
Switch, level, 214
Switch, limit, 201
Switch, low pressure, 219
Switch, miniature, 197
Switch, mushroom, 197
Switch, overload, 90
Switch, palm button, 197
Switch, photoelectric, 347
Switch, plugging, 390
Switch, pressure, 217
Switch, proximity, 369
Switch, reed, 195
Switch, selector, 210
Switch, single pole, 197
Switch, speed, 390
Switch, stop button, 182
Switch, temperature, 222
Switch, toggle, 88

Switch, vacuum, 217
Synchronous motor, 298

T

Tachometer, 407
Temperature rise, 297
Temperature switches, 222
Test circuit, 337
Testing fuses, 65
Test lamp, 458
Thermal cut out, 154
Thermal overload protector 154
Thermistors, 224
Thermocouple, 224
Three phase voltage, 50, 285
Three wire circuit, 157, 334, 361
Three wire control, 361
Through beam scan, 347
TI (Texas Instruments) addressing, 441
Through beam switches, 351
Time delay applications, 254
Time delay relay, 390
Time delay unit, 140
Timed out, 244
Timer motor, 331
Timer operations, 443
Timer output, 444
Timers, 234
Timer, reset cycle, 324
Timer, reset, 234
Timer, solid state, 140, 244
Timer, totalizing, 234
Timing, 244
Timing diagrams, 242, 243
TI (Texas Instruments) programmable controllers, 42, 43, 228, 334, 336, 338, 340, 342, 343, 437, 438, 439, 440, 441, 443, 447, 449, 450, 454
Toggle switch, 88
TON timers, 446
Totalizing counters, 448
Totalizing timers, 234
Transformer, auto, 59, 98
Transformer, buck-boost, 59
Transformer connections, 56
Transformer losses, 54

Transformers, 52
Transformer substation, 51
Transformer turns ratio, 53
Troubleshooting timers, 255
TTL, (transistor-transistor logic), 360
TTL circuits, 134
TTL voltage, 189
Tube base socket, 139
Turns ratio, 53
Two capacitor motor, 302
Two speed motor starter, 185, 342
Two wire circuit, 157
Two wire control, 361
Two wire control circuit, 316
Types of fires, 13

U

UL (Underwriters Laboratory), 162
Under voltage protection, 91
Under voltage sensor, 94
Unlatch contacts, 154

V

Vacuum switch, 217
Variable current drives, 403
Variable voltage drives, 403
Velocity feedback, 408
Velocity loop, 424
Voltmeter, 106, 200, 457

W

Warner Electric Company, 398, 399, 400, 401, 405, 406, 421
Wear bar, 273
Wear mark, 273
Wild leg voltage, 54
Wire sorting meter, 463
Wireways, 75
Wiring diagram, 170, 322
Wobble head actuator, 205
Word, 452
Wound rotor motor, 302
Write cycle, 439
Wye connected motors, 289
Wye delta starter, 378